Lecture Notes in Computer S

Commenced Publication in 1973
Founding and Former Series Editors:
Gerhard Goos, Juris Hartmanis, and Jan van Leeu

Advanced Research in Computing and Software Science

Subline of Lectures Notes in Computer Science

Yiling Chen   Nicole Immorlica (Eds.)

Web and Internet Economics

9th International Conference, WINE 2013
Cambridge, MA, USA, December 11-14, 2013
Proceedings

 Springer

Volume Editors

Yiling Chen
Cambridge, MA, USA
E-mail: yiling@eecs.harvard.edu

Nicole Immorlica
Cambridge, MA, USA
E-mail: nicimm@microsoft.com

ISSN 0302-9743   e-ISSN 1611-3349
ISBN 978-3-642-45045-7   e-ISBN 978-3-642-45046-4
DOI 10.1007/978-3-642-45046-4
Springer Heidelberg New York Dordrecht London

Library of Congress Control Number: 2013952968

CR Subject Classification (1998): H.3, F.2, G.1, C.2, K.4.4, F.1, J.4

LNCS Sublibrary: SL 1 – Theoretical Computer Science and General Issues

© Springer-Verlag Berlin Heidelberg 2013

Typesetting: Camera-ready by author, data conversion by Scientific Publishing Services, Chennai, India

Printed on acid-free paper

Springer is part of Springer Science+Business Media (www.springer.com)

Preface

This volume contains the papers and extended abstracts for work presented at WINE 2013: The 9th Conference on Web and Internet Economics held during December 11–14, 2013, at Harvard University, Cambridge, Massachusetts, USA.

Over the past decade, researchers in theoretical computer science, artificial intelligence, and microeconomics have joined forces to tackle problems involving incentives and computation. These problems are of particular importance in application areas like the Web and the Internet that involve large and diverse populations. The Conference on Web and Internet Economics (WINE) is an interdisciplinary forum for the exchange of ideas and results on incentives and computation arising from these various fields.

WINE 2013 built on the success of the Workshop on Internet and Network Economics, which had the same acronym, WINE. The workshop was held annually from 2005 to 2012 and published archival proceedings. To accommodate the growing research interests and emphasize its archival nature, WINE was renamed a conference with the same acronym in 2013.

WINE 2013 received 150 submissions. All submissions were rigorously peer-reviewed and evaluated on the basis of originality, soundness, significance, and exposition. The committee decided to accept 36 papers. The program also included four invited talks by Dirk Bergemann (Yale University), Joe Halpern (Cornell University), Ehud Kalai (Microsoft Research and Northwestern University), and Eva Tardos (Cornell University). In addition, WINE 2013 featured four tutorials on December 11: Price of Anarchy in Auctions, by Jason Hartline (Northwestern University), Online Behavioral Experiments, by Andrew Mao (Harvard University) and Siddharth Suri (Microsoft Research), Budget Feasible Mechanisms, by Nick Gravin (Microsoft Research) and Yaron Singer (Harvard University), and Computational Social Choice, by Lirong Xia (Rensselaer Polytechnic Institute).

We would like to thank Microsoft Research, Facebook, and Google Research for their generous financial support to WINE 2013 and Harvard University for hosting the event. We thank David Parkes, the general chair of the conference, and Ann Marie King for their excellent local arrangements work and Andrew Mao for his help with the conference website.

We also acknowledge the work of the Program Committee, Anna Kramer at Springer for helping with the proceedings, and the EasyChair paper management system.

October 2013

Yiling Chen
Nicole Immorlica

Conference Organization

General Chair

David C. Parkes   Harvard University, USA

Program Committee Chairs

Yiling Chen   Harvard University, USA
Nicole Immorlica   Microsoft Research

Program Committee

Saeed Alaei	Cornell University, USA
Itai Ashlagi	Massachusetts Institute of Technology, USA
Moshe Babaioff	Microsoft Research
Richard Cole	New York University, USA
Nikhil Devanur	Microsoft Research
Shaddin Dughmi	University of Southern California, USA
Michal Feldman	Tel Aviv University, Israel
Amos Fiat	Tel Aviv University, Israel
Felix Fischer	University of Cambridge, UK
Hu Fu	Microsoft Research
Gagan Goel	Google Research
Mingyu Guo	University of Liverpool, UK
Ian Kash	Microsoft Research
David Kempe	University of Southern California, USA
Robert Kleinberg	Cornell University, USA
Jochen Koenemann	University of Waterloo, Canada
Scott Duke Kominers	Harvard University, USA
Sébastien Lahaie	Microsoft Research
Stefano Leonardi	Sapienza University of Rome, Italy
Katrina Ligett	California Institute of Technology, USA
Brendan Lucier	Microsoft Research
Mohammad Mahdian	Google Research
David Malec	University of Wisconsin - Madison, USA
Yishay Mansour	Tel Aviv University, Israel
Evangelos Markakis	Athens University of Economics and Business, Greece
Reshef Meir	The Hebrew University of Jerusalem, Israel
Vahab Mirrokni	Google Research

Thanh Nguyen	Northwestern University, USA
Renato Paes Leme	Microsoft Research
Mallesh Pai	University of Pennsylvania, USA
Aaron Roth	University of Pennsylvania, USA
Tim Roughgarden	Stanford University, USA
Michael Schapira	The Hebrew University of Jerusalem, Israel
Yaron Singer	Harvard University, USA
Rakesh Vohra	University of Pennsylvania, USA
Michael Wellman	University of Michigan, USA
Aviv Zohar	The Hebrew University of Jerusalem, Israel

External Reviewers

Bruno Abrahao
Shipra Agrawal
Kareem Amin
Haris Aziz
Yoram Bachrach
Ashwinkumar Badanidiyuru
Eytan Bakshy
Anand Bhalgat
Umang Bhaskar
Kshipra Bhawalkar
Davide Bilò
Michael Brautbar
Simina Brânzei
Ozan Candogan
Jing Chen
Giorgos Christo
Riccardo Colini-Baldeschi
Rachel Cummings
Bart De Keijzer
Alan Deckelbaum
Argyrios Deligkas
Miroslav Dudík
Paul Dütting
Lili Dworkin
Yuval Emek
Piotr Faliszewski
Fei Fang
Linda Farczadi
Andreas Emil Feldmann
Jugal Garg
Konstantinos Georgiou
Vasilis Gkatzelis

David Gleich
Renato Gomes
Ragavendran Gopalakrishnan
Nick Gravin
Nima Haghpanah
Paul Harrenstein
Xinran He
Hoda Heidari
Chien-Ju Ho
Martin Hoefer
Justin Hsu
Zhiyi Huang
Patrick Hummel
Atsushi Iwasaki
Shaili Jain
Max Klimm
Spyros Kontogiannis
Nitish Korula
Janardhan Kulkarni
Silvio Lattanzi
Omer Lev
Vahid Liaghat
Hamid Mahini
Ruta Mehta
Debasis Mishra
Jamie Morgenstern
Hamid Nazerzadeh
Ilan Nehama
Rad Niazadeh
Afshin Nikzad
Lev Omer
Joel Oren

Sigal Oren
Michael Ostrovsky
Georgios Piliouras
Emmanouil Pountourakis
Ariel Procaccia
Heiko Röglin
Daniela Saban
Anshul Sawant
Guido Schaefer
Lior Seeman
Nisarg Shah
Ankit Sharma
Peng Shi
Adish Singla
Balasubramanian Sivan
Alexander Skopalik
Greg Stoddard

Vasilis Syrgkanis
Inbal Talgam-Cohen
Omer Tamuz
Orestis Telelis
Nithum Thain
Dave Thompson
Pushkar Tripathi
Christos Tzamos
Laci Vegh
Carmine Ventre
S. Matthew Weinberg
Omri Weinstein
Zhiwei Wu
Qiqi Yan
Jinshan Zhang
Yair Zick

Table of Contents

The Asymmetric Matrix Partition Problem

Noga Alon[1], Michal Feldman[1], Iftah Gamzu[2], and Moshe Tennenholtz[3]

[1] Tel Aviv University and Microsoft Research
{nogaa,mfeldman}@tau.ac.il
[2] Yahoo! Research
iftah.gamzu@yahoo.com
[3] Microsoft Research and Technion-Israel Institute of Technology
moshet@microsoft.com

Abstract. An instance of the asymmetric matrix partition problem consists of a matrix $A \in \mathbb{R}_+^{n \times m}$ and a probability distribution p over its columns. The goal is to find a partition scheme that maximizes the resulting partition value. A partition scheme $\mathcal{S} = \{\mathcal{S}_1, \ldots, \mathcal{S}_n\}$ consists of a partition $\mathcal{S}_i$ of $[m]$ for each row i of the matrix. The partition $\mathcal{S}_i$ can be interpreted as a smoothing operator on row i, which replaces the value of each entry in that row with the expected value in the partition subset that contains it. Given a scheme $\mathcal{S}$ that induces a smoothed matrix A', the partition value is the expected maximum column entry of A'.

We establish that this problem is already APX-hard for the seemingly simple setting in which A is binary and p is uniform. We then demonstrate that a constant factor approximation can be achieved in most cases of interest. Later on, we discuss the symmetric version of the problem, in which one must employ an identical partition for all rows, and prove that it is essentially trivial. Our matrix partition problem draws its interest from several applications like broad matching in sponsored search advertising and information revelation in market settings. We conclude by discussing the latter application in depth.

1 Introduction

An instance of the *asymmetric matrix partition* problem consists of a matrix $A \in \mathbb{R}_+^{n \times m}$ of non-negative values and a probability distribution p over its columns, namely, $p \in [0,1]^m$ such that $\sum_{j=1}^m p_j = 1$. The objective is to find a partition scheme $\mathcal{S}$ that maximizes the resulting partition value $v_{\mathcal{S}}$. A *partition scheme* $\mathcal{S} = \{\mathcal{S}_1, \ldots, \mathcal{S}_n\}$ consists of a partition $\mathcal{S}_i$ of $[m] = \{1, \ldots, m\}$ for each row i of the matrix, namely, $\mathcal{S}_i$ is a collection of pairwise disjoint subsets $S_{i1}, \ldots, S_{ik_i} \subseteq [m]$ such that $S_{i1} \dot\cup \cdots \dot\cup S_{ik_i} = [m]$. Note that the partitions within a scheme may be different, and hence, it is referred to as an *asymmetric* scheme. The partition $\mathcal{S}_i$ can be interpreted as a smoothing operator on row i, which replaces the value of each entry in that row with the expected value in the partition subset that contains it. Formally, the smoothed value for each $j \in S_{ik}$ is $A'_{ij} = \sum_{\ell \in S_{ik}} p_\ell A_{i\ell} / \sum_{\ell \in S_{ik}} p_\ell$. Given a partition scheme $\mathcal{S}$ that induces a smoothed matrix A', the resulting *partition value* is the expected maximum column entry, that is, $v_{\mathcal{S}} = \sum_{j \in [m]} p_j \cdot \max_{i \in [n]} A'_{ij}$. The *contribution* of a column j to the partition value is $p_j \cdot \max_{i \in [n]} A'_{ij}$, and similarly, $\operatorname{argmax}_{i \in [n]} A'_{ij}$ is referred to as the entry of column j that contributes to the partition value.

Y. Chen and N. Immorlica (Eds.): WINE 2013, LNCS 8289, pp. 1–14, 2013.

For the purpose of illustrating the above setting, let us focus on the simple scenario in which the input instance consists of an $n \times n$ matrix such that all the entries in the first column have a value of 1 and all remaining entries have a value of 0. Furthermore, the probability distribution over the columns of this matrix is uniform. One partition scheme that naturally comes to mind is the identity scheme, which results in a smoothed matrix that is identical to the original matrix. This identity scheme sets all the partitions to consist of singletons, namely, each $S_i = \{\{1\}, \ldots, \{n\}\}$. One can easily validate that the resulting partition value in this case is $1/n$. Another extreme partition scheme is the one in which all partitions consist of one subset, that is, each $S_i = \{[n]\}$. This scheme gives rise to a smoothed matrix in which all the entries of each row have the same value. In our case, all the entries of the resulting matrix are $1/n$, and accordingly, it is easy to validate that the partition value is again $1/n$. Finally, one can demonstrate that there is a partition scheme that exhibits a significant improvement over the above-mentioned schemes. This scheme consists of the partitions $S_i = \{\{1, i\}, [n] \setminus \{1, i\}\}$, namely, it joins together the 1-value of each row $i \neq 1$ with the 0-value of column i in that row, resulting in a smoothed value of $1/2$ for both entries. One can verify that the resulting partition value in this case is roughly $1/2$. The above scenario is presented in the figure below.

$$
\begin{pmatrix} 1 & 0 & \cdots & 0 \\ 1 & 0 & \cdots & 0 \\ \vdots & \vdots & \ddots & \vdots \\ 1 & 0 & \cdots & 0 \end{pmatrix}
\qquad
\begin{pmatrix} \boxed{1} & 0 & \cdots & 0 \\ \boxed{1} & \boxed{0} & \cdots & 0 \\ \vdots & \vdots & \ddots & \vdots \\ \boxed{1} & 0 & \cdots & \boxed{0} \end{pmatrix}
\qquad
\begin{pmatrix} 1 & 0 & \cdots & 0 \\ \frac{1}{2} & \frac{1}{2} & \cdots & 0 \\ \vdots & \vdots & \ddots & \vdots \\ \frac{1}{2} & 0 & \cdots & \frac{1}{2} \end{pmatrix}
$$

Fig. 1. Given the input matrix on the left with a uniform distribution over its columns, one can utilize the partition scheme illustrated on the middle, and obtain the smoothed matrix on the right. Note that the boxes in each row of the middle matrix represent entries that are joined together in the same subset; the remaining entries of each row are clustered together in a different subset.

Application I: Personalized Broad Matching in Sponsored Search Advertising. The asymmetric matrix partition problem draws its interest from several applications. One such application relates to sponsored search advertising, namely, advertising on a web search result page, where the ads are driven by the originating query. In the basic model, there are advertisers, each of which has keywords relevant to her ad. Each advertiser also associates some valuation with each of her keywords, indicating the gain she derives when a user clicks on her ad. This valuation underlies a bid that the advertiser reports to the search engine, expressing the maximum amount that she is willing to pay for a click. When a user queries the search engine for some keyword, the engine runs an auction among all the advertisers interested in that keyword. The advertiser that wins this auction is allocated the ad slot, and she is required to pay some amount if the user clicks on her ad. This amount is determined by her bid and the payment rule of the engine.

Advertisers can realistically only identify a small set of keywords due to the effort involved, and therefore, search engines recently introduced broad matching. This feature enables an advertiser to automatically target a broader range of queries that the search engine deems relevant to match her ad, and not only the keywords specified by her. Such relevant queries can be modifications of the specified keywords (like synonyms, singular and plural forms, misspellings, reordering, etc.), or can even be a completely different set of keywords, which are conceptually related to the specified keywords. This feature clearly has potential to help advertisers reach wider audience, while spending less time on building their keyword lists. On the other hand, a search engine can utilize the flexibility in expanding the set of keywords specified by an advertiser to optimize its revenue. Understanding the power of flexibility in broad matching seems an interesting research goal.

We consider a stylized non-strategic version of broad matching. In the underlying setting, there is a single ad slot, and a set of advertisers, each of which interested in one keyword from a set of possible keywords $\{k_1, \ldots, k_m\}$, where keyword k_j is queried by users with probability p_j. The search engine keeps a relevance distance measure between keywords, $\alpha(i, j)$, that has the following semantics: if an advertiser has valuation v for her specified keyword k_i, then her valuation for each keyword k_j is $v \cdot \alpha(i, j)$. The goal is to develop a personalized broad matching scheme that maximizes the expected revenue of the search engine. Specifically, we are interested in a scheme that assigns each advertiser a partition of keywords to disjoint subsets, such that all keywords in each subset are automatically bid with the expected valuation in that subset whenever a user queries a keyword from that subset. We assume that the search engine knows the valuation that each advertiser has for her specified keyword, i.e., a non-strategic setting in which there is no need to incentivize the advertisers, and a winning advertiser pays her expected valuation. Consequently, given a query, the search engine selects the advertiser that has the highest bid. One can validate that our asymmetric matrix partition problem captures the problem of designing a personalized broad matching that maximizes the expected revenue.

Application II: Signaling in Take-It-or-Leave-It Sales. Another application of the asymmetric matrix partition problem relates to a question of information revelation in market settings. In many sale scenarios, a seller has much more accurate information about an item for sale than the buyers. As an example, consider a used-car dealer or an Internet liquidation site, both of which receive or purchase items for sale. The seller in these scenarios may have quite adequate information about the particular item for sale (e.g., by checking it in detail), while the potential buyers may only have probabilistic information about the item, relying, for example, on some publicly-available statistical information. It seems of the essence to study how a seller can utilize her informational superiority to optimize her revenue.

The above-mentioned scenario can be modeled by considering a take-it-or-leave-it sale of a probabilistic item among multiple buyers. More precisely, a single item is chosen randomly from a set of m possible items according to some known probability distribution p, and the seller approaches a buyer with a monetary offer of delivering the item for a specified payment. There are n buyers, each of which has her own valuation for every item in the set. While the buyers only know the probability distribution over the items, the seller knows the actual realization of the probabilistic item. In an attempt to increase her revenue, the seller may partially reveal some information about the item to the buyers. The question that concerns us is how much information should the seller reveal to every buyer in order to maximize her expected revenue.

The information revelation is materialized by means of a buyer-specific signaling scheme. For each buyer, the seller partitions the set of items into pairwise disjoint subsets, and reports this partition to the buyer. After the signaling scheme has been declared, an item j is randomly chosen by nature, and the seller reveals to each buyer i, the subset that contains j according to i's partition. Upon being signaled, a buyer can update her belief regarding the probability distribution p conditioned on the choice of some item in her signaled subset. A key assumption in our model is that the buyers are unaware of the environment, namely, each buyer knows her own valuation and partition, but is unaware of the existence of the other buyers and their associated valuations and partitions. Hence, the conditional probability of every item j that is contained in a signaled subset is the ratio between p_j and the overall probability of items in that subset, and 0 in case j is not in the signaled subset. It is clear that the maximal take-it-or-leave-it offer that a buyer will accept is her expected valuation for the item under the new probability distribution induced by the received signal. Consequently, upon the realization of an item, the seller will choose to make such offer to a buyer that has the highest expected valuation. One can validate that our asymmetric matrix partition problem captures the task of designing a buyer-specific signaling scheme that maximizes the expected revenue.

Our Contribution. We begin by studying the asymmetric matrix partition problem when the input matrix is binary, namely, $A \in \{0,1\}^{n \times m}$. We prove that this seemingly simple setting is already APX-hard when the probability distribution p is uniform. Specifically, we show a gap-preserving reduction that proves that the problem is NP-hard to approximate to within a factor of 1.0001. We also establish that the binary setting admits a constant factor approximation; thus, settling the complexity of this setting to within constant factors. In particular, we demonstrate that there is a 1.775-approximation algorithm when p is uniform, and there is a 13-approximation algorithm when p is arbitrary. We further study several interesting special scenarios. For example, we prove that when the number of rows n is fixed then the uniform distribution case can be solved to optimality in polynomial-time, whereas the general distribution case remains NP-hard even when $n = 4$. This result separates the uniform distribution setting from the general distribution setting. The specifics of these results are presented in Section 2. We then consider the problem in its utmost generality, that is, when the input matrix $A \in \mathbb{R}_+^{n \times m}$ consists of arbitrary non-negative values. We present a 2-approximation algorithm for the case that p is uniform, and a logarithmic approximation when p is arbitrary under some practical assumptions. These results appear in Section 3. Later on, in Section 4, we discuss the symmetric version of our problem in which one must employ an identical partition for all rows. We demonstrate that this problem is essentially trivial, and establish a tight bound on the advantage that asymmetric schemes have over symmetric ones. Finally, we formally model the application of signaling in take-it-or-leave-it sales with its connection to our problem, and discuss some of our modeling decisions. These application details are provided in Section 5. Due to space constraints, all proofs are omitted from this extended abstract and may be found in the full version of the paper.

2 The Binary Matrix Case

In this section, we study the problem when the input matrix is binary, namely, $A \in \{0, 1\}^{n \times m}$. We prove that this setting is APX-hard even when the probability distribution p is uniform. We also establish that this setting admits a constant factor approximation; thus, settling the complexity of this setting to within constant factors. We further study several interesting special scenarios.

We begin by introducing a notation and terminology that will be used in the remainder of this section. Let $C^+ = \{j \in [m] : \exists i \text{ such that } A_{ij} = 1\}$ be the set of columns that consist of at least one 1-value entry, and $C^0 = [m] \setminus C^+$ be the set of remaining all-zero columns. Moreover, let $r = \sum_{j \in C^+} p_j$ be the total probability of the columns in C^+. Similarly, we denote the set of columns that have a 1-value entry in row i by $C_i^+ = \{j \in [m] : A_{ij} = 1\}$, and use $r_i = \sum_{j \in C_i^+} p_j$ to denote their total probability. We say that a partition scheme $\mathcal{S}$ covers C^+ if it covers each of the columns in C^+. A column $j \in C^+$ is said to be *covered* by $\mathcal{S}$ if there is some row i such that $A_{ij} = 1$ and the partition scheme consists of a singleton subset of column j in row i, namely, $\{j\} \in \mathcal{S}_i$. Note that a partition scheme that covers C^+ can be easily computed in polynomial-time. Finally, we say that a subset is *mixed* if it consists of both 1-value and 0-value entries.

We now turn to identify several interesting structural properties of partition schemes for the binary case. These properties will be utilized later when establishing our primary technical results.

Lemma 1. *Let S, T be two disjoint subsets of columns, and let S^+ (resp., T^+) and S^0 (resp., T^0) be the respective 1-value entries and 0-value entries of S (resp., T) in some row i. Suppose that all the entries of S^0 and T^0 contribute to the partition value when the partition of row i consists of S and T. Then, the overall contribution of those entries when the partition consists of a unified subset $S \cup T$ is at least as large.*

Note that a useful corollary of the above lemma is that given some fixed covering of C^+, the optimal way to complete the partition scheme is to join together all the remaining 1-value entries of each row in a single subset with some additional 0-value entries. We can also utilize the above lemma and prove the following.

Lemma 2. *Given an instance of the asymmetric matrix partition problem in which A is binary and p is uniform, there is an optimal solution that covers C^+.*

2.1 A Uniform Distribution

We study the binary matrix setting when the distribution over the columns is uniform, namely, each $p_j = 1/m$. We establish that this seemingly simple setting is already APX-hard, that is, it is NP-hard to approximate to within some constant. On the algorithmic side, we identify a simple algorithmic procedure that guarantees 2-approximation, and then develop an algorithm that achieves a better approximation ratio. We also prove that the case that the number of rows n is constant can be solved to optimality in polynomial-time. We emphasize that for simplicity of presentation, we neglect the uniform probability term $1/m$ from the partition value contribution terms in the rest of this subsection.

Theorem 1. *Given an instance of the asymmetric matrix partition problem in which A is binary and p is uniform, it is NP-hard to attain an approximation ratio better than* 1.0001.

Approximation Algorithms. We begin by presenting a simple 2-approximation algorithm for the problem under consideration. Later on, we develop a different algorithm that attains an improved approximation ratio. Our 2-approximation algorithm begins by covering C^+. This ensures that the contribution of the columns of C^+ to the resulting partition value is exactly r. Subsequently, the algorithm goes over the rows, one after the other, and for each row that has ℓ remaining 1-value entries (after the covering), it creates a subset in that row that consists of these entries and ℓ entries of distinct all-zero columns. In case there are no more all-zero columns left to match to some row then this step ends. Finally, all remaining entries of each row are clustered together.

For the purpose of analyzing this algorithm, notice that a straight-forward upper bound on the partition value of the optimal scheme is $\text{OPT} \leq \min\{1, \sum_{i=1}^{n} r_i\}$. Now, consider the following two complementary cases: (case 1) if we matched every all-zero column to some row, then the contribution of each column j is 1 if $j \in C^+$, or at least $1/2$ if $j \in C^0$. Hence, the partition value is at least $r + (1 - r)/2 \geq 1/2 \geq \text{OPT}/2$; (case 2) if we did not match all all-zero columns, then the partition value is at least $r + (\sum_{i=1}^{n} r_i - r)/2 \geq \sum_{i=1}^{n} r_i/2 \geq \text{OPT}/2$. This implies that the partition scheme achieves 2-approximation.

A Greedy Completion Procedure. Before we turn to improve the above algorithm, we study the following greedy procedure that given a fixed covering of C^+ completes the partition scheme by matching all-zero columns to partition subsets. The *greedy procedure* begins by associating a subset S_i to each row i. This subset is initialized with all the columns corresponding to 1-value entries in row i that were not used in the covering of C^+. Then, it proceeds by going over the all-zero columns, one after the other, and adding a column to the subset S_i that maximizes the marginal contribution from the all-zero columns. Specifically, the *marginal contribution* of some all-zero column that is added to a subset that already consists of x and y columns corresponding to 0-value entries and 1-value entries, respectively, is

$$\Delta(x,y) = (x+1)\frac{y}{x+y+1} - x\frac{y}{x+y} = y^2 \left(\frac{1}{x+y} - \frac{1}{x+y+1} \right).$$

Note that $\Delta(x,y) \geq 0$ for any non-negative x, y, and that Δ is monotonically non-increasing in x for any fixed y, that is, $\Delta(x,y) \geq \Delta(x+1,y)$. The following lemma establishes that once C^+ is covered in some way, the greedy procedure yields the optimal contribution from the all-zero columns. Notice that this result implies, in conjunction with Lemma 2, that the computational hardness of the underlying setting of the problem resides in finding the right way to cover C^+.

Lemma 3. *Given some fixed covering of C^+, the greedy procedure yields the optimal contribution from the all-zero columns.*

The greedy procedure can be leveraged to construct a 1.775-approximation algorithm for our problem. We emphasize that our main effort is to improve upon the previous 2-approximation algorithm, and we have not tried to optimize the constants in our analysis. Let $r^* = 0.127$ and $\sigma^* = 2(1 - r^*)/3 = 0.582$. The algorithm computes a partition scheme according to the following cases, which depend on the values of r and $\sum_{i=1}^{n} r_i$ in a given instance:

Case I: When $r \geq r^*$. The algorithm first covers C^+ in some arbitrary way. Then, the algorithm goes over the rows, one after the other, and for each row that has ℓ remaining 1-value entries after the covering, it creates a subset in that row that consists of these entries and ℓ entries of distinct all-zero columns. Note that in case there are no more all-zero columns left to match to some row then this step ends. Finally, all the remaining entries of each row are clustered together.

Case II: When $r < r^*$ and $\sum_{i=1}^{n} r_i \leq \sigma^*$. The algorithm forms a subset on top of every 1-value entry. Specifically, given a 1-value entry (i, j), the algorithm forms a subset in row i that consists of column j and some additional distinct all-zero columns. Half of the 1-value entries are clustered together with two distinct two all-zero columns, and the other half of the 1-value entries are clustered together with a single all-zero column. Then, all the remaining entries of each row are clustered together.

Case III: When $r < r^*$ and $\sum_{i=1}^{n} r_i > \sigma^*$. The algorithm executes the previously-mentioned greedy procedure over the given instance (without covering C^+ first). After this procedure ends, all the remaining entries of each row are clustered together.

Theorem 2. *Given an instance of the asymmetric matrix partition problem in which A is binary and p is uniform, our algorithm computes a partition scheme whose resulting partition value is a 1.775-approximation for the optimal one.*

An Optimal Solution for a Fixed Number of Rows. We prove that an optimal partition scheme can be computed in polynomial-time when the number of rows n is fixed. We emphasize that this result separates the uniform distribution setting from the general distribution setting since we establish that the latter setting is NP-hard in Theorem 4.

Theorem 3. *Given an instance of the asymmetric matrix partition problem in which A is binary and p is uniform, an optimal partition scheme can be constructed in polynomial-time when n is fixed.*

2.2 A General Distribution

We next study the binary matrix setting when the distribution over the columns is arbitrary. We first demonstrate that this setting is NP-hard even when the number of rows is fixed. This result separates this setting from the uniform distribution setting, which admits a polynomial-time optimal solution when the number of rows is fixed. Later on, we present a constant factor approximation algorithm for this setting.

Theorem 4. *Given an instance of the asymmetric matrix partition problem in which A is binary and p is general, and a positive number α, it is NP-hard to determine if there is a partition scheme whose resulting partition value is at least α, even when $n = 4$.*

An Approximation Algorithm. We develop a polynomial-time constant approximation algorithm for the problem under consideration. Specifically, we present three algorithms whose performance depends on different parameters of the input instance and the optimal solution. We then demonstrate that given any input instance, one of these algorithms is guaranteed to compute a 13-approximation partition scheme. Hence, by executing all three algorithms and selecting the scheme that attains the maximal resulting partition value, we achieve a 13-approximation solution.

Recall that C^+ denotes the set of columns that consist of at least one 1-value entry, C^0 is the set of remaining all-zero columns, and C_i^+ marks the set of columns having 1 in row i. Furthermore, recall that r is the total probability of columns in C^+, and r_i is the total probability of columns in C_i^+. Note that the optimal partition value can be trivially bounded by $\mathrm{OPT} \leq r + \mathrm{OPT}_0$, where OPT_0 is the overall contribution of the all-zero columns of C^0 to the optimal partition value.

Algorithm 1. The first algorithm attains to the case in which the input instance has a relatively large r, namely, $r \geq \mathrm{OPT}_0/12$. In this case, a constant approximation can be obtained by simply covering C^+. That is, for every column $j \in C^+$, the algorithm arbitrarily selects some row i such that $A_{ij} = 1$ and forms a singleton subset of entry j in row i. Then, all the remaining entries of each row are clustered together. One can easily validate that the resulting scheme has a partition value which is a 13-approximation to the optimal one. This follows since the resulting partition value is at least r, while $\mathrm{OPT} \leq r + \mathrm{OPT}_0 \leq 13r$.

In the remainder of the subsection, we focus on the case that r is relatively small, namely, $r < \mathrm{OPT}_0/12$. Since r is small, we concentrate on designing partition schemes that yield high contribution from the all-zero columns. We say that an all-zero column j is *large* for a row i in case $p_j \geq r_i$; otherwise, j is said to be *small* for i. We consider two complementary cases and develop constant factor approximation algorithms for both. The first case is when a large fraction of the contribution of all-zero columns to the optimal partition value comes from such columns that are large for the rows that realize their contribution.

Algorithm 2. The algorithm begins by constructing an undirected bipartite graph $G = (V_R, V_L, E)$ with a weight function $w : E \to \mathbb{R}_+$ on its edges. Specifically, V_R is a set of n vertices that correspond to the rows, V_L is a set of $|C^0|$ vertices that correspond to the all-zero columns, and $E = \{(i,j) \in V_R \times V_L : r_i \leq p_j\}$ is the edge set. Moreover, the weight function sets $w(i,j) = r_i$, for every $(i,j) \in E$. With these definitions in mind, the algorithm finds a maximal weighted matching M with respect to the constructed bipartite graph. Then, for each edge $(i,j) \in M$, the algorithm forms the subset $C_i^+ \cup \{j\}$ in row i. Subsequently, all the remaining entries of each row are clustered together.

Lemma 4. *Algorithm 2 computes a partition scheme that yields at least $1/2$ of the optimal contribution of all-zero columns that are large for the rows that realize their contribution.*

Lemma 4 implies that in case that at least $1/6$ of the optimal contribution of all-zero columns comes from such columns that are large for the rows that realize their contribution then we obtain a 13-approximation solution. Formally, one can utilize Lemma 4 to claim that the partition value of the computed scheme is at least $OPT_0/12$. On the other hand, $OPT \leq r + OPT_0 \leq 13/12 \cdot OPT_0$, where the last inequality follows from the assumption that $r < OPT_0/12$. We now turn to consider the remaining case in which at least $5/6$ of the optimal contribution of all-zero columns comes from such columns that are small for the rows that realize their contribution.

Algorithm 3. Similarly to the previous algorithm, this algorithm forms two subsets for each row; one mixed subset and an additional subset that consists of the remaining row entries. The mixed subset of row i consists of the columns in C_i^+ and some additional all-zeros columns. To decide which all-zeros columns are added to each mixed subset, the algorithm goes over the all-zero columns in an arbitrary order, and adds the column j to the mixed subset of row i if (1) j is small for i and (2) the total probability of the all-zero columns already added to this subset is no more than r_i. We emphasize that each column is added to at most one mixed subset, and a column is neglected only if the algorithm cannot add it to any of the mixed subsets.

Lemma 5. *Algorithm 3 computes a partition scheme that yields at least $1/10$ of the optimal contribution of all-zero columns that are small for the rows that realize their contribution.*

Lemma 5 implies that in case that at least $5/6$ of the optimal contribution of all-zero columns comes from such columns that are small for the rows that realize their contribution then we also attain a 13-approximation solution. More precisely, one can utilize Lemma 5 to claim that the partition value of the computed scheme is at least $OPT_0/12$, while $OPT \leq r + OPT_0 \leq 13/12 \cdot OPT_0$. Reviewing the algorithms and the case analysis, we can conclude with the following theorem.

Theorem 5. *Given an instance of the asymmetric matrix partition problem in which A is binary and p is general, there is an algorithm that computes a partition scheme whose resulting partition value is a 13-approximation for the optimal one.*

3 The General Matrix Case

In this section, we study the problem in its utmost generality, i.e., when the input matrix $A \in \mathbb{R}_+^{n \times m}$ consists of arbitrary non-negative values. We develop a constant factor approximation algorithm for the case that the probability distribution p over the columns is uniform, and a logarithmic approximation for the general case under some practical assumptions.

3.1 A Uniform Distribution

We present an algorithm that computes a partition scheme whose resulting partition value is a 2-approximation for the optimal one. Let M be the set of m largest entries in the matrix A. Our algorithm first forms a singleton subset of every entry $(i, j) \in M$ that is maximal for the corresponding column. In case there are several maximal entries for some column then one of them is selected arbitrarily. We say that column j was *covered* if there was an entry (i, j) that was clustered as a singleton. Subsequently, for every entry $(i, j) \in M$ that was not clustered in the first step, the algorithm forms a subset in row i consisting of column j and a distinct column that was not covered. Then, all the remaining entries of each row are clustered together.

Theorem 6. *Given an instance of the asymmetric matrix partition problem in which A is general and p is uniform, our algorithm computes a partition scheme whose resulting partition value is a 2-approximation for the optimal one.*

3.2 A General Distribution

We present an algorithm that achieves a logarithmic approximation under some practical assumptions. Specifically, the algorithm achieves $O(\log m)$-approximation if the column probabilities are at most polynomially small, namely, when each $p_j \geq 1/m^c$ for some constant c.

Let $A_{\max}$ be the value of the largest entry of an input matrix A. Our algorithm begins by manipulating the matrix A to construct the matrix B as follows: All the entries whose value is smaller than $A_{\max}/m^{c+2}$ are replaced by 0, and all the values of the remaining entries are rounded down to the closest power of 2. For example, if $2^{-k} \leq A_{ij} < 2^{1-k}$ then $B_{ij} = 2^{-k}$. Notice that after this manipulation, the matrix B is populated with at most $K = O(\log m^c) = O(\log m)$ types of positive values $\{v_1, \ldots, v_K\}$ in addition to a 0-value. As a result, we can express the matrix B as a sum of K matrices where the kth matrix consists of the values $\{0, v_k\}$. That is, the kth matrix has a value of v_k in each entry that B has a value of v_k, and 0 in all remaining entries. Each of the K matrices, together with the probability distribution p, can be considered to be an instance of our problem with a binary matrix and a general distribution. Hence, we can apply the algorithm from Theorem 5 on each of these K instances to obtain K partition schemes. Finally, the algorithm selects the partition scheme that obtains the maximal partition value from the original instance.

Theorem 7. *Given an instance of the asymmetric matrix partition problem in which A and p are general such that each $p_j \geq 1/m^c$ for some constant c, our algorithm computes a partition scheme whose resulting partition value is a $O(\log m)$-approximation for the optimal one.*

4   Symmetric Partition Schemes

In this section, we discuss the symmetric version of our matrix partition problem, and most notably, compare between the performance guarantees of symmetric and asymmetric partition schemes. The *symmetric matrix partition* problem is identical to the asymmetric matrix partition problem with the exception that the underlying partition scheme must be symmetric. A *symmetric* partition scheme consists of a single partition S of $[m]$ that is used as the smoothing operator of all the rows. A variant of the symmetric matrix partition problem has been studied in a series of works [7,3,4,14]; A more detailed discussion is given in Section 5.

An easy argument shows that the symmetric matrix partition problem is essentially trivial as the partition scheme that consists only of singletons always achieves the optimal partition value. To establish this argument, suppose by way of contradiction that the partition scheme of singletons does not attain an optimal outcome. Consider the optimal partition scheme $\mathcal{S}$. This scheme must consist of a subset $S \in \mathcal{S}$ whose cardinality is greater than 1. Notice that the contribution of all the columns in S to the resulting partition value is realized in the same row i. The overall contribution of those columns is exactly $\sum_{j \in S} p_j A_{ij}$. Now, observe that if one replaces the instance of S in the optimal partition scheme with the collection of singleton subsets of the columns in S, the overall contribution of the columns in S may only improve to $\sum_{j \in S} p_j \cdot \max_{i \in [n]} A_{ij}$, and the contribution of any other column in $[m] \setminus S$ does not change. Applying this argument repeatedly as long as $\mathcal{S}$ has subsets whose cardinality is greater than 1 results in an optimal partition scheme that consists only of singletons; a contradiction.

In light of this state of affairs, we next focus on quantifying the advantage that asymmetric partition schemes have over symmetric schemes. Given an instance of our matrix partition problem, let OPT_{sym} and OPT_{asym} denote the optimal partition values that can be achieved by symmetric and asymmetric partition schemes, respectively. Clearly, $\text{OPT}_{\text{asym}}/\text{OPT}_{\text{sym}} \geq 1$. However, we are also interested to establish a tight upper bound on this ratio.

Lemma 6. *Given a matrix partition instance in which A and p are general, the ratio* $\text{OPT}_{\text{asym}}/\text{OPT}_{\text{sym}} \leq m$. *Furthermore, there are instances for which the ratio can be arbitrarily close to m.*

5   An Application: Signaling in Take-It-or-Leave-It Sales

In this section, we formally model the application of signaling in take-it-or-leave-it sales, and explain its connection to our asymmetric matrix partition problem. Later on, we discuss the previous literature on signaling and some of our modeling decisions.

5.1   The Model

A *probabilistic single-item sale* is formally depicted by a valuations matrix $A \in \mathbb{R}_+^{n \times m}$, and a probability distribution p over its columns. More precisely, there are n *agents* and m distinct indivisible *items*. Each entry A_{ij} of the matrix captures the *valuation* of the row-agent i for the column-item j. We assume that each agent knows her valuation vector but is *unaware* of the rest of the valuation matrix. A single item j is chosen by nature according to the distribution p and then offered for sale. This one-time sale is conducted via a personalized *take-it-or-leave-it* rule: The *seller* gives a take-it-or-leave-it offer to some agent i. If the selected agent is interested, the chosen item is sold to her for the suggested price.

A Signaling Scheme. Although the agents know the distribution p, they do not know its actual realization, which is only observed by the seller. In an attempt to increase her expected *revenue*, the seller may partially reveal the realization to the agents. This is performed via the following *asymmetric signaling scheme*: For every agent i, the seller partitions the items into a collection of pairwise disjoint subsets $S_{i1} \dot\cup \cdots \dot\cup S_{ik_i} = [m]$, and reports this partition to agent i; we denote the partition of agent i by $\mathcal{S}_i$. Crucially, the seller can use different partitions for different agents, i.e., a buyer-specific signaling scheme. We emphasize that the seller decides on a signaling scheme prior to nature's random choice of an item. When item j is randomly chosen, every agent i is *signaled* with the subset S_{ik} that contains j. The agent can then update her belief to the probability distribution p conditioned on the choice of some item in S_{ik}. In other words, each agent i knows that none of the items in $[m] \setminus S_{ik}$ was chosen, and can calculate the conditional probability $\mathbb{P}(j : S_{ik}) = p_j/\mathbb{P}(S_{ik})$, for every $j \in S_{ik}$.

The Optimization Problem. Consider some probabilistic single-item sale $\langle A, p \rangle$ and a signaling scheme $\mathcal{S} = (\mathcal{S}_1, \dots, \mathcal{S}_n)$. Clearly, the maximal take-it-or-leave-it offer that agent i will accept under the signal S_{ik} is given by $\mathbb{E}_p[A_{ij} : S_{ik}]$. Therefore, given an asymmetric signaling scheme $\mathcal{S}$, when item j is randomly chosen, the seller will choose to make a take-it-or-leave-it offer to agent i that maximizes $\mathbb{E}_p[A_{ij} : S_{ik}]$, where S_{ik} is the subset that contains j for agent i. In what follows, we denote by $S^i(j)$ the subset S_{ik} of agent i that contains j. Hence, the expected *revenue* of the seller is given by

$$\sum_{j \in [m]} p_j \cdot \max_{i \in [n]} \left\{ \sum_{\ell \in S^i(j)} \mathbb{P}(\ell : S^i(j)) \cdot A_{i\ell} \right\} = \sum_{j \in [m]} p_j \cdot \max_{i \in [n]} \left\{ \frac{\sum_{\ell \in S^i(j)} p_\ell A_{i\ell}}{\sum_{\ell \in S^i(j)} p_\ell} \right\}. \quad (1)$$

This raises the following combinatorial optimization problem: given a probabilistic single-item sale $\langle A, p \rangle$, construct the asymmetric signaling scheme $\mathcal{S}$ that maximizes the expected revenue.

We believe that the mapping of the above problem to the asymmetric matrix partition problem is straightforward. Yet, we wish to emphasize that the expression that is maximized in Equation 1 is essentially the smoothed value A'_{ij}, which was defined when we formalized the problem.

5.2   Related Work

The literature on signaling in economics is very broad. Our approach can be viewed as related to the study of strategic information transmission, originated in the seminal work of Crawford and Sobel [2]. More specifically, our approach deals with the idea that a seller knows some information about the valuations of the buyers, via information about the item, and may use strategic information transmission to exploit this knowledge [11]. As in that work, our model depart from the classic literature of Milgrom and Weber [12,13], who showed the superiority of full revelation of information. Information revelation in online markets has been recently studied also in [9], where it is shown that in an environment with multiple publishers, a publisher may prefer not to share user information with the advertiser, due to information leakage, where the advertiser may target the same user through a cheaper publisher. Our approach has some of the flavor of the work on the value of information in conflicts [10]. One special distinction of the current work is its focus on algorithmic issues.

This work is also closely related to the study of revenue maximization via signaling in second-price auctions [7,3,4,14]. The are few fundamental differences between the model considered by those papers and ours. First, rather than a take-it-or-leave-it sale, the sale is conducted by means of a *second-price* auction; i.e., each agent places her bid and the chosen item is sold to the bidder that placed the highest bid for the price of the second highest bid. Second, rather than a *buyer-specific asymmetric* signaling scheme, the signaling is performed via a *symmetric* partition, where the auctioneer partitions the items into pairwise disjoint clusters and reports this partition to all the bidders.

A point of interest in our approach is the assumption of unawareness. Classical economic and game-theoretic approaches assume that buyers are aware of other buyers and the take-it-or-leave-it offers they may be given. As a result, the (deduced) probabilistic information that each buyer holds about the item may be affected by her awareness to the cases when she is given an offer versus the cases other buyers are given an offer. While this approach is natural, we believe it is interesting to consider the complementary attitude of competition-unaware buyers, who disregard the existence of other buyers. This approach clearly gives much power to the seller. Indeed, the fact that decision-makers may be unaware of aspects of a strategic situation, and in particular, of actions and even existence of other players is a puzzle game theorists were concerned with. Most efforts so far have been concentrated on trying to find general models that incorporate such reasoning. For an example of the modeling challenges encountered when considering unaware agents, one may consult the work of Halpren and Rego [8] on extensive games with possibly unaware players, or the work by Feinberg [5] on games with unawareness. Our approach is complementary, as it emphasizes the combinatorial and algorithmic issues that arise in such settings.

5.3 Awareness vs. Unawareness

A natural question one may ask is what would be the ramifications when considering a setting in which the agents are aware of one another, and more generally, when the whole setting is common-knowledge. Interestingly, in what follows, we observe that in the latter case, the seller maximizes her revenue by fully revealing all information, essentially revealing the realization of the probabilistic item. This result is in the spirit of the famous 'Linkage Principle' of Milgrom and Weber [12,13].

In a *competition-aware* model, each agent is aware of the valuations of other agents and the signaling scheme that the seller runs. As a result, any agent can calculate, for each item, which agent will be given the take-it-or-leave-it offer and in which price. Suppose some agent i is signaled a subset S. How would she evaluate her expected value? Clearly, agent i should compute the expectation only over the items $j \in S$ such that she would be given the take-it-or-leave-it offer. For that reason, when analyzing an asymmetric partition scheme in the competition-aware model, it can be assumed without loss of generality that if some subset is a winning subset for some item then it is a winning subset for all its items. One can also verify that this implies that there is only one winning subset for each agent. Using these observations, we next show that in the competition-aware model, an optimal signaling scheme obtains the same expected revenue as a signaling scheme that is symmetric and partitions the items into singleton subsets. Conceptually, this implies that the best interest of the seller is to fully reveal which item arrived when the buyers are competition-aware.

Lemma 7. *In the competition-aware model, the optimal asymmetric signaling scheme obtains the same expected revenue as a signaling scheme that is symmetric and partitions the items into singleton subsets.*

References

1. Chlebík, M., Chlebíková, J.: Approximation Hardness for Small Occurrence Instances of NP-Hard Problems. In: Petreschi, R., Persiano, G., Silvestri, R. (eds.) CIAC 2003. LNCS, vol. 2653, pp. 152–164. Springer, Heidelberg (2003)
2. Crawford, V., Sobel, J.: Strategic Information Transmission. Econometrica 50(6), 1431–1451 (1982)
3. Emek, Y., Feldman, M., Gamzu, I., Paes Leme, R., Tennenholtz, M.: Signaling schemes for revenue maximization. In: ACM Conference on Electronic Commerce, pp. 514–531 (2012)
4. Emek, Y., Feldman, M., Gamzu, I., Tennenholtz, M.: Signaling Schemes for Revenue Maximization. In: AdAuctions Workshop (2011)
5. Feinberg, Y.: Games with Unawareness (2009) (manuscript)
6. Garey, M.R., Johnson, D.S.: Computers and Intractability: A Guide to the Theory of NP-Completeness. W.H. Freeman (1979)
7. Ghosh, A., Nazerzadeh, H., Sundararajan, M.: Computing optimal bundles for sponsored search. In: Deng, X., Graham, F.C. (eds.) WINE 2007. LNCS, vol. 4858, pp. 576–583. Springer, Heidelberg (2007)
8. Halpern, J.Y., Rego, L.C.: Generalized Solution Concepts in Games with Possibly Unaware Players. International Journal of Game Theory 41(1), 131–155 (2012)
9. Mahdian, M., Ghosh, A., McAfee, R.P., Vassilvitskii, S.: To match or not to match: economics of cookie matching in online advertising. In: ACM Conference on Electronic Commerce, pp. 741–753 (2012)
10. Kamien, M., Tauman, Y., Zamir, S.: Information Transmission. In: Game Theory and Applications, pp. 273–281. Academic Press
11. Kaplan, T.R., Zamir, S.: The Strategic Use of Seller Information. Private-Value Auctions, Hebrew University Center For Rationality Working Paper No. 221 (2000)
12. Milgrom, P.R., Weber, R.J.: A Theory of Auctions and Competitive Bidding. Econometrica 50(5), 1089–1122 (1982)
13. Milgrom, P.R., Weber, R.J.: The Value of Information in a Sealed-Bid Auction. Journal of Mathematical Economics 10(1), 105–114 (1982)
14. Miltersen, P.B., Sheffet, O.: Send mixed signals: earn more, work less. In: ACM Conference on Electronic Commerce, pp. 234–247 (2012)

Polylogarithmic Supports Are Required for Approximate Well-Supported Nash Equilibria below 2/3

Yogesh Anbalagan[1], Sergey Norin[2], Rahul Savani[3], and Adrian Vetta[4]

[1] School of Computer Science, McGill University
yogesh.anbalagan@mail.mcgill.ca
[2] Department of Mathematics and Statistics, McGill University
snorin@math.mcgill.ca
[3] Department of Computer Science, University of Liverpool
rahul.savani@liverpool.ac.uk
[4] Department of Mathematics and Statistics, and School of Computer Science,
McGill University
vetta@math.mcgill.ca

Abstract. In an ϵ-approximate Nash equilibrium, a player can gain at most ϵ *in expectation* by unilateral deviation. An ϵ-*well-supported* approximate Nash equilibrium has the stronger requirement that every pure strategy used with positive probability must have payoff within ϵ of the best response payoff. Daskalakis, Mehta and Papadimitriou [8] conjectured that every win-lose bimatrix game has a $\frac{2}{3}$-well-supported Nash equilibrium that uses supports of cardinality at most three. Indeed, they showed that such an equilibrium will exist *subject to* the correctness of a graph-theoretic conjecture. Regardless of the correctness of this conjecture, we show that the barrier of a $\frac{2}{3}$ payoff guarantee cannot be broken with constant size supports; we construct win-lose games that require supports of cardinality at least $\Omega(\sqrt[3]{\log n})$ in any ϵ-well supported equilibrium with $\epsilon < \frac{2}{3}$. The key tool in showing the validity of the construction is a proof of a bipartite digraph variant of the well-known Caccetta-Häggkvist conjecture [4]. A probabilistic argument [13] shows that there exist ϵ-well-supported equilibria with supports of cardinality $O(\frac{1}{\epsilon^2} \cdot \log n)$, for any $\epsilon > 0$; thus, the polylogarithmic cardinality bound presented cannot be greatly improved. We also show that for any $\delta > 0$, there exist win-lose games for which no pair of strategies with support sizes at most two is a $(1 - \delta)$-well-supported Nash equilibrium. In contrast, every bimatrix game with payoffs in $[0, 1]$ has a $\frac{1}{2}$-approximate Nash equilibrium where the supports of the players have cardinality at most two [8].

1 Introduction

A Nash equilibrium of a bimatrix game is a pair of strategies in which the supports of both players consist only of best responses. The apparent hardness of computing an exact Nash equilibrium [6,5] even in a bimatrix game has led to

Y. Chen and N. Immorlica (Eds.): WINE 2013, LNCS 8289, pp. 15–23, 2013.
© Springer-Verlag Berlin Heidelberg 2013

work on computing approximate Nash equilibria, and two notions of approximate Nash equilibria have been developed. The first and more widely studied notion is of an ϵ-*approximate Nash equilibrium* (ϵ-Nash). Here, no restriction is placed upon the supports; any strategy can be in the supports provided each player achieves an expected payoff that is within ϵ of a best response. Therefore, ϵ-Nash equilibria have a practical drawback: a player might place probability on a strategy that is arbitrarily far from being a best response. The second notion, defined to rectify this problem, is called an ϵ-*well supported approximate Nash equilibrium* (ϵ-WSNE). Here, the content of the supports are restricted, but less stringently than in an exact Nash equilibrium. Specifically, both players can only place positive probability on strategies that have payoff within ϵ of a pure best response. Observe that the latter notion is a stronger equilibrium concept: every ϵ-WSNE is an ϵ-Nash, but the converse is not true.

Approximate well-supported equilibria recently played an important role in understanding the hardness of computing Nash equilibria. They are more useful in these contexts than ϵ-Nash equilibria because their definition is more combinatorial and more closely resembles the *best response condition* that characterizes exact Nash equilibria. Indeed, approximate well-supported equilibria were introduced in [12,6] in the context of PPAD reductions that show the hardness of computing (approximate) Nash equilibria. They were subsequently used as the notion of approximate equilibrium by Chen et al. [5] that showed the PPAD-hardness of computing an exact Nash equilibrium even for bimatrix games.

Another active area of research is to investigate the best (smallest) ϵ that can be guaranteed in polynomial time. For ϵ-Nash, the current best algorithm, due to Tsaknakis and Spirakis [17], achieves a 0.3393-Nash equilibrium; see [8,7,3] for other algorithms. For the important class of win-lose games – games with payoffs in $\{0, 1\}$ – [17] gives a $\frac{1}{4}$-Nash equilibrium. For ϵ-WSNE, the current best result was given by Fearnley et al. [10] and finds a $(\frac{2}{3} - \zeta)$-WSNE, where $\zeta = 0.00473$. It builds on an approach of Kontogiannis and Spirakis [13], which finds a $\frac{2}{3}$-WSNE in polynomial time using linear programming. The algorithm of Kontogiannis and Spirakis produces a $\frac{1}{2}$-WSNE of win-lose games in polynomial time, which is best-known (the modifications of Fearnley et al. do not lead to an improved approximation guarantee for win-lose games).

It is known that this line of work cannot extend to a fully-polynomial-time approximation scheme (FPTAS). More precisely, there does not exist an FPTAS for computing approximate Nash equilibria unless PPAD is in P [5]. Recall, an FPTAS requires a running time that is polynomial in both the size of the game input and in $\frac{1}{\epsilon}$. A polynomial-time approximation scheme (PTAS), however, need not run in time polynomial in $\frac{1}{\epsilon}$. It is not known whether there exists a PTAS for computing an approximate Nash equilibrium and, arguably, this is the biggest open question in equilibrium computation today. While the best-known approximation guarantee for ϵ-Nash that is achievable in polynomial time is much better than that for ϵ-WSNE, the two notions are polynomially related: there is a PTAS for ϵ-WSNE if and only if there is a PTAS for ϵ-Nash [5,6].

1.1 Our Results

The focus of this paper is on the combinatorial structure of equilibrium. Our first result shows that well-supported Nash equilibria differ structurally from approximate Nash equilibria in a significant way. It is known that there are $\frac{1}{2}$-Nash equilibria with supports of cardinality at most two [12], and that this result is tight [11]. In contrast, we show in Theorem 2 that for any $\delta > 0$, there exist win-lose games for which no pair of strategies with support sizes at most two is a $(1 - \delta)$-well-supported Nash equilibrium.[1]

With supports of cardinality three, Daskalakis et al. conjectured, in the first paper that studied algorithms for finding ϵ-WSNE [8], that $\frac{2}{3}$-WSNE are obtainable in every win-lose bimatrix game. Specifically, this would be a consequence of the following graph-theoretic conjecture.

Conjecture 1 ([8]). Every digraph either has a cycle of length at most 3 or an undominated set[2] of 3 vertices.

The main result in this paper, Theorem 1, shows that one cannot do better with constant size supports. We prove that there exist win-lose games that require supports of cardinality at least $\Omega(\sqrt[3]{\log n})$ in any ϵ-WSNE with $\epsilon < \frac{2}{3}$. We prove this existence result probabilistically. The key tool in showing correctness is a proof of a bipartite digraph variant of the well-known Caccetta-Häggkvist conjecture [4]. A polylogarithmic cardinality bound, as presented here, is the best we can hope for – a probabilistic argument [13] shows that there exist ϵ-WSNE with supports of cardinality $O(\frac{1}{\epsilon^2} \cdot \log n)$, for any $\epsilon > 0$.[3]

2 A Lower Bound on the Support Size of Well Supported Nash Equilibria

We begin by formally defining bimatrix win-lose games and well-supported Nash equilibria. A *bimatrix game* is a 2-player game with $m \times n$ payoff matrices A and B; we may assume that $m \leq n$. The game is called *win-lose* if each matrix entry is in $\{0, 1\}$.

A pair of mixed strategies $\{\mathbf{p}, \mathbf{q}\}$ is a *Nash equilibrium* if every pure row strategy in the support of $\mathbf{p}$ is a best response to $\mathbf{q}$ and every pure column strategy in the support of $\mathbf{q}$ is a best response to $\mathbf{p}$. A relaxation of this concept is the following. A pair of mixed strategies $\{\mathbf{p}, \mathbf{q}\}$ is an *ϵ-well supported Nash equilibrium* if every pure strategy in the support of $\mathbf{p}$ (resp. $\mathbf{q}$) is an ϵ-approximate best response to $\mathbf{q}$ (resp. $\mathbf{p}$). That is, for any row r_i in the support of $\mathbf{p}$ we have

$$\mathbf{e_i}^T A\mathbf{q} \geq \max_{\ell} \mathbf{e_\ell}^T A\mathbf{q} - \epsilon$$

---

[1] Random games have been shown to have exact equilibria with support size 2 with high probability; see Bárány et al. [2].

[2] A set S is undominated if there is no vertex v that has an arc to every vertex in S.

[3] Althöfer [1] and Lipton at al. [15] proved similar results for ϵ-Nash equilibria.

and, for any column c_j in the support of $\mathbf{q}$ we have

$$\mathbf{p}^T B \mathbf{e_j} \geq \max_{\ell} \mathbf{p}^T B \mathbf{e_\ell} - \epsilon .$$

In this section we prove our main result.

Theorem 1. *For any $\epsilon < \frac{2}{3}$, there exist win-lose games for which every ϵ-well-supported Nash equilibrium has supports of cardinality $\Omega(\sqrt[3]{\log n})$.*

To prove this result, we first formulate our win-lose games graphically. This can be done in a straight-forward manner. Simply observe that we may represent a 2-player win-lose game by a directed bipartite graph $G = (R \cup C, E)$. There is a vertex for each row and a vertex for each column. There is an arc $(r_i, c_j) \in E$ if and only if $(B)_{ij} = b_{ij} = 1$; similarly there is an arc $(c_j, r_i) \in E$ if and only if $(A)_{ij} = a_{ij} = 1$.

Consequently, we are searching for a graph whose corresponding game has no high quality well-supported Nash equilibrium with small supports. We show the existence of such a graph probabilistically.

The Construction.
Let $T = (V, E)$ be a random tournament on N nodes. Now create from T an auxiliary bipartite graph $G(T) = (R \cup C, A)$ corresponding to a 2-player win-lose game as follows. The auxiliary graph has a vertex-bipartition $R \cup C$ where there is a vertex of R for each node of T and there is a vertex of C for each set of k distinct nodes of T. (Observe that, for clarity we will refer to nodes in the tournament T and vertices in the bipartite graph G.) There are two types of arc in $G(T)$: those oriented from R to C and those oriented from C to R. For arcs of the former type, each vertex $X \in C$ will have in-degree exactly k. Specifically, let X correspond to the k-tuple $\{v_1, \dots, v_k\}$ where $v_i \in V(T)$, for all $1 \leq i \leq k$. Then there are arcs (v_i, X) in G for all $1 \leq i \leq k$. Next consider the latter type of arc in G. For each node $u \in R$ there is an arc (X, u) in G if and only if u dominates $X = \{v_1, \dots, v_k\}$ in the tournament T, that is if (u, v_i) are arcs in T for all $1 \leq i \leq k$. This completes the construction of the auxiliary graph (game) G.

We say that a set of vertices $W = \{w_1, \dots, w_t\}$ is *covered* if there exists a vertex y such that $(w_j, y) \in A$, for all $1 \leq j \leq t$. Furthermore, a bipartite graph is *k-covered* if every collection of k vertices that lie on the same side of the bipartition is covered. Now with positive probability the auxiliary graph $G(T)$ is k-covered.

Lemma 1. *For all sufficiently large n and $k \leq \sqrt[3]{\log n}$, there exists a tournament T whose auxiliary bipartite graph $G(T)$ is k-covered.*

Proof. Observe that the payoff matrices that correspond to $G(T)$ have $m = N$ rows and $n = \binom{N}{k}$ columns. Furthermore, by construction, any set of k vertices in R is covered. Thus, first we must verify that any set of k vertices in C is also covered.

So consider a collection $\mathcal{X} = \{X_1, \ldots, X_k\}$ of k vertices in C. Since each $X_i \in C$ corresponds to a k-tuple of nodes of T, we see that $\mathcal{X}$ corresponds to a collection of at most k^2 nodes in T. Thus, for any node $u \notin \cup_i X_i$, we have that u has an arc in T to every node in $\cup_i X_i$ with probability at least 2^{-k^2}. Thus with probability at most $(1 - \frac{1}{2^{k^2}})^{N-k^2}$ the subset $\mathcal{X}$ of C not covered in $G(T)$. Applying the union bound we have that there exists the desired tournament if

$$\binom{n}{k} \cdot \left(1 - \frac{1}{2^{k^2}}\right)^{N-k^2} < 1 \tag{1}$$

Now set $k = \log^{\frac{1}{3}} n$. Therefore $\log n^{\frac{1}{k}} = \log^{\frac{2}{3}} n = k^2$.

In addition, because $n = \binom{N}{k}$, we have that $N \geq \frac{k}{e} \cdot n^{\frac{1}{k}}$. Hence, $N - k^2 > n^{\frac{1}{k}}$. (Note that, since $N \geq k$ this implies that $G(T)$ is defined.) Consequently,

$$\binom{n}{k} \cdot \left(1 - \frac{1}{2^{k^2}}\right)^{N-k^2} \leq n^k \cdot \left(1 - \frac{1}{2^{k^2}}\right)^{n^{\frac{1}{k}}}$$

$$\leq n^k \cdot e^{-\frac{1}{2^{k^2}} \cdot n^{\frac{1}{k}}}$$

$$\leq n^k \cdot e^{-\frac{1}{e^{k^2 \cdot \log 2}} \cdot n^{\frac{1}{k}}}$$

Thus, taking logarithms, we see that Inequality (1) holds if

$$e^{k^2 \cdot \log 2} \cdot k \cdot \log n < n^{\frac{1}{k}} \tag{2}$$

But $n^{\frac{1}{k}} = e^{k^2}$, so Inequality (2) clearly holds for large n. The result follows.

A property of the auxiliary graph $G(T)$ that will be very useful to us is that it contains no cycles with less than six vertices.

Lemma 2. *The auxiliary graph $G(T)$ contains no digons and no 4-cycles.*

Proof. Suppose $G(T)$ contains a digon $\{w, X\}$. The arc (w, X) implies that $X = \{x_1, \ldots, x_{k-1}, w\}$. On the other-hand, the arc (X, w) implies that w dominates X in T and, thus, $w \notin X$.

Suppose $G(T)$ contains a 4-cycle $\{w, X, z, Y\}$ where w and z are in R and where $X = \{x_1, \ldots, x_{k-1}, w\}$ and $Y = \{y_1, \ldots, y_{k-1}, z\}$ are in C. Then z must dominate X in T and w must dominate Y in T. But then we have a digon in T as (w, z) and (z, w) must be arcs in T. This contradicts the fact that T is a tournament.

Lemmas 1 and 2 are already sufficient to prove a major distinction between approximate-Nash equilibria and well-supported Nash equilibria. Recall that there always exist $\frac{1}{2}$-Nash equilibria with supports of cardinality at most two [8]. In sharp contrast, for supports of cardinality at most two, no constant approximation guarantee can be obtained for ϵ-well-supported Nash equilibria.

Theorem 2. *For any $\delta > 0$, there exist win-lose games for which no pair of strategy vectors with support sizes at most two is a $(1 - \delta)$-well-supported Nash equilibrium.*

Proof. Take the auxiliary win-lose game $G(T)$ from Lemma 1 for the case $k = 2$. Now consider any pair of strategy vectors $\mathbf{p_1}$ and $\mathbf{p_2}$ with supports of cardinality 2 or less. Since $G(T)$ is 2-covered, the best responses to $\mathbf{p_1}$ and $\mathbf{p_2}$ both generate payoffs of exactly 1. Thus $\{\mathbf{p_1}, \mathbf{p_2}\}$ can be a $(1 - \delta)$-well-supported Nash equilibrium only if each strategy in the support of $\mathbf{p_1}$ is a best response to at least one of the pure strategies in the support of $\mathbf{p_2}$ and vice versa. Therefore, in the subgraph H of $G(T)$ induced by the supports of $\mathbf{p_1}$ and $\mathbf{p_2}$, each vertex has in-degree at least one. Thus, H contains a directed cycle. But $G(T)$ has no digons or 4-cycles, by Lemma 2. Hence, we obtain a contradiction as H contains at most four vertices.

In light of Lemma 2, we will be interested in the minimum in-degree required to ensure that a bipartite graph contains a 4-cycle. The following theorem may be of interest on its own right, as it resolves a variant of the well-known Caccetta-Häggkvist conjecture [4] for bipartite digraphs. For Eulerian graphs, a related but different result is due to Shen and Yuster [16].

Theorem 3. *Let $H = (L \cup R, A)$ be a directed $k \times k$ bipartite graph. If H has minimum in-degree $\lambda \cdot k$ then it contains a 4-cycle, whenever $\lambda > \frac{1}{3}$.*

Proof. To begin, by removing arcs we may assume that every vertex has in-degree exactly $\lambda \cdot k$. Now take a vertex v with the maximum out-degree in H, where without loss of generality $v \in L$. Let A_1 be the set of out-neighbours of v, and set $\alpha_1 \cdot k = |A_1|$. Similarly, let B_t be the set of vertices with paths to v that contain exactly t arcs, for $t \in \{1, 2\}$, and set set $\beta_t \cdot k = |B_t|$. Finally, let C_1 be the vertices in R that are not adjacent to v, namely $C_1 = L - (A_1 \cup B_1)$. Set $\gamma_1 \cdot k = |C_1|$.

These definitions are illustrated in Figure 1.

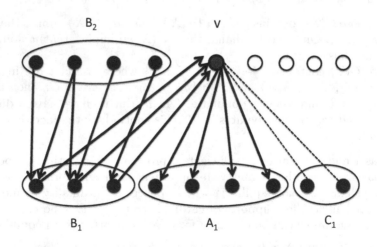

Fig. 1.

Observe that we have the following constraints on α_1, β_1 and γ_1. By assumption, $\beta_1 = \lambda$. Thus, we have $\gamma_1 = 1 - \alpha_1 - \lambda$. Moreover, by the choice of v, we have $\alpha_1 \geq \lambda$, since the maximum out-degree must be at least the average in-degree.

Note that if there is an arc from A_1 to B_2 then H contains a 4-cycle. So, let's examine the in-neighbours of B_2. We know B_2 has exactly $\lambda \cdot k \cdot |B_2|$ incoming arcs. We may assume all these arcs emanate from $B_1 \cup C_1$. On the other-hand, there are exactly $\lambda \cdot k \cdot |B_1|$ arcs from B_2 to B_1. Thus, there are at most $|B_1| \cdot (|B_2| - \lambda \cdot k)$ arcs from B_1 to B_2. So the number of arcs from C_1 to B_2 is at least

$$\lambda \cdot k \cdot |B_2| - |B_1| \cdot (|B_2| - \lambda \cdot k) = \lambda \cdot k \cdot \beta_2 \cdot k - \beta_1 \cdot k \cdot (\beta_2 \cdot k + \lambda \cdot k)$$
$$= \lambda \cdot k \cdot \beta_2 \cdot k - \lambda \cdot k \cdot (\beta_2 \cdot k + \lambda \cdot k)$$
$$= \lambda^2 \cdot k^2$$

Since the maximum degree is $\alpha_1 \cdot k$, the number of arcs emanating from C_1 is at most $\gamma_1 \cdot \alpha_1 \cdot k^2$. Thus $\gamma_1 \cdot \alpha_1 \cdot (1 - \alpha_1 - \lambda) \geq \lambda^2$. Rearranging we obtain the quadratic inequality

$$\alpha_1^2 - \alpha_1(1 - \lambda) + \lambda^2 \leq 0$$

The discriminant of this quadratic is $1 - 2\lambda - 3\lambda^2$. But $1 - 2\lambda - 3\lambda^2 = (1 - 3\lambda)(1 + \lambda)$ and this is non-negative if and only if $\lambda \leq \frac{1}{3}$. This completes the proof.

We may now prove our main result: no approximation guarantee better than $\frac{2}{3}$ can be achieved unless the well-supported equilibria has supports with cardinality $\Omega(\sqrt[3]{\log n})$.

Proof of Theorem 1. Take a tournament T whose auxiliary bipartite graph is k-covered. By Lemma 1, such a tournament exists. Consider the win-lose game corresponding to the auxiliary graph $G(T)$, and take strategy vectors $\mathbf{p_1}$ and $\mathbf{p_2}$ with supports of cardinality k or less. Without loss of generality, we may assume that $\mathbf{p_1}$ and $\mathbf{p_2}$ are rational. Denote these supports as $S_1 \subseteq R$ and $S_2 \subseteq C$, respectively. As $G(T)$ is k-covered, there is a pure strategy $c^* \in C$ that covers S_1 and a pure strategy $r^* \in R$ that covers S_2. Thus, in the win-lose game, $c^* \in C$ has an expected payoff of 1 against $\mathbf{p_1}$ and $r^* \in R$ has an expected payoff of 1 against $\mathbf{p_2}$.

Suppose $\mathbf{p_1}$ and $\mathbf{p_2}$ form a ϵ-well-supported equilibrium for some $\epsilon < \frac{2}{3}$. Then it must be the case that each $r_i \in S_1$ has expected payoff at least $1 - \epsilon > \frac{1}{3}$ against $\mathbf{p_2}$. Similarly, each $c_j \in S_2$ has expected payoff at least $1 - \epsilon > \frac{1}{3}$ against $\mathbf{p_1}$. But this cannot happen. Consider the subgraph of $G(T)$ induced by $S_1 \cup S_2$ where each $r_i \in S_1$ has weight $w_i = p_1(r_i)$ and each $c_j \in S_2$ has weight $w_j = p_2(c_j)$. We convert this into an unweighted graph H by making $L \cdot w_v$ copies of each vertex v, for some large integer L. Now H is an $L \times L$ bipartite graph with minimum in-degree $(1 - \epsilon) \cdot L > \frac{1}{3} \cdot L$. Thus, by Theorem 3, H contains a 4-cycle. This is a contradiction, by Lemma 2. $\qquad\square$

We remark that the $\frac{2}{3}$ in Theorem 1 cannot be improved using this proof technique. Specifically the minimum in-degree requirement of $\frac{1}{3} \cdot k$ in Theorem 3 is tight. To see this, take a directed 6-cycle C and replace each vertex in C by $\frac{1}{3} \cdot k$ copies. Thus each arc in C now corresponds to a complete $\frac{k}{3} \times \frac{k}{3}$ bipartite graph with all arc orientations in the same direction. The graph H created in this fashion is bipartite with all in-degrees (and all out-degrees) equal to $\frac{1}{3} \cdot k$. Clearly the minimum length of a directed cycle in H is six.

3 Conclusion

An outstanding open problem is whether any constant approximation guarantee better than 1 is achievable with constant cardinality supports. We have shown that supports of cardinality two cannot achieve this; a positive resolution of Conjecture 1 would suffice to show that supports of cardinality three can. However, Conjecture 1 seems a hard graph problem and it is certainly conceivable that it is false.[4] If so, that would lead to the intriguing possibility of a very major structural difference between ϵ-Nash and ϵ-WSNE; namely, that for any $\delta > 0$, there exist win-lose games for which no pair of strategies with constant cardinality supports is a $(1 - \delta)$-well-supported Nash equilibrium.

The existence of small support ϵ-WSNE clearly implies the existence of of polynomial time approximation algorithms to find such equilibria. Obtaining better approximation guarantees using more complex algorithms is also an interesting question. As discussed, the best known polynomial-time approximation algorithm for well-supported equilibria in win-lose games finds a $\frac{1}{2}$-well supported equilibrium [13] by solving a linear program (LP). For games with payoffs in $[0, 1]$ that algorithm finds a $\frac{2}{3}$-well-supported equilibrium. The algorithm has been modified in [10] to achieve a slightly better approximation of about $\frac{2}{3} - \zeta$ where $\zeta = 0.00473$. That modification solves an almost identical LP as [13] and then either transfers probability mass within the supports of a solution to the LP or returns a small support strategy profile that uses at most two pure strategies for each player. The results of this paper show that both parts of that approach are needed, and any improvement to the approximation guarantee must allow for super-constant support sizes.

Acknowledgements. We thank John Fearnley and Troels Sørensen for useful discussions. The first author is supported by a fellowship from MITACS and an NSERC grant. The second author is supported by an NSERC grant 418520. The third author is supported by an EPSRC grant EP/L011018/1. The fourth author is supported by NSERC grants 288334 and 429598.

References

1. Althöfer, I.: On sparse approximations to randomized strategies and convex combinations. Linear Algebra and its Applications 199, 339–355 (1994)
2. Bárány, I., Vempala, S., Vetta, A.: Nash equilibria in random games. Random Struct. Algorithms 31(4), 391–405 (2007)

---

[4] For example, the conjecture resembles a question about the existence of k-existentially complete triangle-free graphs for $k > 3$ referred to in [9], which the authors consider to be wide open.

3. Bosse, H., Byrka, J., Markakis, E.: New algorithms for approximate Nash equilibria in bimatrix games. Theoretical Computer Science 411(1), 164–173 (2010)
4. Caccetta, L., Häggkvist, R.: On minimal digraphs with given girth. Congressus Numerantium 21, 181–187 (1978)
5. Chen, X., Deng, X., Teng, S.: Settling the complexity of computing two-player Nash equilibria. Journal of the ACM 56(3), 1–57 (2009)
6. Daskalakis, C., Goldberg, P., Papadimitriou, C.: The complexity of computing a Nash equilibrium. SIAM Journal on Computing 39(1), 195–259 (2009)
7. Daskalakis, C., Mehta, A., Papadimitriou, C.: Progress in approximate Nash equilibria. In: ACM Conference on Electronic Commerce (EC), pp. 355–358 (2007)
8. Daskalakis, C., Mehta, A., Papadimitriou, C.: A note on approximate Nash equilibria. Theoretical Computer Science 410(17), 1581–1588 (2009)
9. Even-Zohar, C., Linial, N.: Triply existentially complete triangle-free graphs, arxiv.org/1306.5637 (2013)
10. Fearnley, J., Goldberg, P., Savani, R., Sørensen, T.: Approximate well-supported Nash equilibria below two-thirds. In: International Symposium on Algorithmic Game Theory (SAGT), pp. 108–119 (2012)
11. Feder, T., Nazerzadeh, H., Saberi, A.: Approximating Nash equilibria using small-support strategies. In: ACM Conference on Electronic Commerce, EC (2007)
12. Goldberg, P., Papadimitriou, C.: Reducability among equilibrium problems. In: STOC (2006)
13. Kontogiannis, S., Spirakis, P.: Well supported approximate equilibria in bimatrix games. Algorithmica 57, 653–667 (2010)
14. Kontogiannis, S.C., Spirakis, P.G.: Efficient algorithms for constant well supported approximate equilibria in bimatrix games. In: Arge, L., Cachin, C., Jurdziński, T., Tarlecki, A. (eds.) ICALP 2007. LNCS, vol. 4596, pp. 595–606. Springer, Heidelberg (2007)
15. Lipton, R., Markakis, E., Mehta, A.: Playing large games using simple startegies. In: ACM Conference on Electronic Commerce (EC), pp. 36–41 (2003)
16. Shen, J., Yuster, R.: A note on the number of edges guaranteeing a C_4 in Eulerian bipartite digraphs. Electronic Journal of Combinatorics 9(1), Note 6 (2002)
17. Tsaknakis, H., Spirakis, P.: An optimization approach for approximate Nash equilibria. Internet Mathematics 5(4), 365–382 (2008)

The Computational Complexity
of Random Serial Dictatorship

Haris Aziz[1], Felix Brandt[2], and Markus Brill[2]

[1] NICTA and UNSW, Kensington 2033, Sydney, Australia
haris.aziz@nicta.com.au
[2] Technische Universität München, 85748 München, Germany
{brandtf,brill}@in.tum.de

Abstract. In social choice settings with linear preferences, *random dictatorship* is known to be the only social decision scheme satisfying strategyproofness and *ex post* efficiency. When also allowing indifferences, *random serial dictatorship (RSD)* is a well-known generalization of random dictatorship that retains both properties. *RSD* has been particularly successful in the special domain of random assignment where indifferences are unavoidable. While *executing RSD* is obviously feasible, we show that *computing* the resulting probabilities is #P-complete and thus intractable, both in the context of voting and assignment.

1 Introduction

Social choice theory studies how a group of agents can make collective decisions based on the—possibly conflicting—preferences of its members. In the most general setting, there is a set of abstract *alternatives* over which each agent entertains *preferences*. A *social decision scheme* aggregates these preferences into a probability distribution (or *lottery*) over the alternatives.

Perhaps the most well-known social decision scheme is *random dictatorship*, in which one of the agents is uniformly chosen at random and then picks his most preferred alternative. Gibbard [3] has shown that random dictatorship is the only social decision scheme that is strategyproof and *ex post* efficient, i.e., it never puts positive probability on Pareto dominated alternatives. Note that random dictatorship is only well-defined when there are no ties in the agents' preferences. However, ties are unavoidable in many important domains of social choice such as assignment, matching, and coalition formation since agents are assumed to be indifferent among all outcomes in which their assignment, match, or coalition is the same (e.g., [4]).

In the presence of ties, random dictatorship is typically extended to *random serial dictatorship* (*RSD*), where dictators are invoked sequentially and ties between most-preferred alternatives are broken by subsequent dictators.[1] *RSD* retains the important properties of *ex post* efficiency and strategyproofness and is well-established in the context of random assignment (see e.g., [2]).

---

[1] *RSD* is referred to as *random priority* by Bogomolnaia and Moulin [2].

Y. Chen and N. Immorlica (Eds.): WINE 2013, LNCS 8289, pp. 24–25, 2013.
© Springer-Verlag Berlin Heidelberg 2013

In this paper, we focus on two important domains of social choice: (1) the *voting setting*, where alternatives are candidates and agents' preferences are given by rankings over candidates, and (2) the aforementioned *assignment setting*, where each alternative corresponds to an assignment of houses to agents and agents' preferences are given by rankings over houses. Whereas agents' preferences over alternatives are listed explicitly in the voting setting, this is not the case in the assignment setting. However, preferences over houses can be easily extended to preferences over assignments by assuming that each agent only cares about the house assigned to himself and is indifferent between all assignments in which he is assigned the same house. As a consequence, the assignment setting is a special case of the voting setting. However, due to the different representations, *computational* statements do not carry over from one setting to the other.

We examine the computational complexity of *RSD* and show that computing the *RSD* lottery is #P-complete *both* in the voting setting and in the assignment setting. Loosely speaking, #P is the counting equivalent of NP—the class of decision problems whose solutions can be verified in polynomial time. #P-completeness is commonly seen as strong evidence that a problem cannot be solved in polynomial time. As mentioned above, neither of the two results implies the other. We furthermore present a polynomial-time algorithm to compute the *support* of the *RSD* lottery in the voting setting. This is not possible in the assignment setting, because the support of the *RSD* lottery might be of exponential size. However, we can decide in polynomial time whether a given alternative (i.e., an assignment) is contained in the support or not.

A preprint of the complete paper is available from http://dss.in.tum.de/files/brandt-research/rsd.pdf. The paper has been accepted for publication in *Economics Letters* [1].

Acknowledgments. This material is based upon work supported by the Australian Government's Department of Broadband, Communications and the Digital Economy, the Australian Research Council, the Asian Office of Aerospace Research and Development through grant AOARD-124056, and the Deutsche Forschungsgemeinschaft under grants BR 2312/7-1 and BR 2312/10-1.

References

1. Aziz, H., Brandt, F., Brill, M.: The computational complexity of random serial dictatorship. Economics Letters (forthcoming, 2013)
2. Bogomolnaia, A., Moulin, H.: A new solution to the random assignment problem. Journal of Economic Theory 100(2), 295–328 (2001)
3. Gibbard, A.: Manipulation of schemes that mix voting with chance. Econometrica 45(3), 665–681 (1977)
4. Sönmez, T., Ünver, M.U.: Matching, allocation, and exchange of discrete resources. In: Benhabib, J., Jackson, M.O., Bisin, A. (eds.) Handbook of Social Economics, vol. 1, ch. 17, pp. 781–852. Elsevier (2011)

Incentives and Efficiency in Uncertain Collaborative Environments*

Yoram Bachrach[1], Vasilis Syrgkanis[2,**], and Milan Vojnović[1]

[1] Microsoft Research Cambridge, UK
{yobach,milanv}@microsoft.com
[2] Cornell University
vasilis@cs.cornell.edu

Abstract. We consider collaborative systems where users make contributions across multiple available projects and are rewarded for their contributions in individual projects according to a local sharing of the value produced. This serves as a model of online social computing systems such as online Q&A forums and of credit sharing in scientific co-authorship settings. We show that the maximum feasible produced value can be well approximated by simple local sharing rules where users are approximately rewarded in proportion to their marginal contributions and that this holds even under incomplete information about the player's abilities and effort constraints. For natural instances we show almost 95% optimality at equilibrium. When players incur a cost for their effort, we identify a threshold phenomenon: the efficiency is a constant fraction of the optimal when the cost is strictly convex and decreases with the number of players if the cost is linear.

1 Introduction

Many economic domains involve self-interested agents who participate in multiple joint ventures by investing time, effort, money or other personal resources, so as to produce some value that is then shared among the participants. Examples include traditional surplus sharing games [18, 6], co-authorship settings where the wealth produced is in the form of credit in scientific projects, that is implicitly split among the authors of a paper [14] and online services contexts where users collaborate on various projects and are rewarded by means of public reputation, achievement awards, badges or webpage attention (e.g. Q&A Forums such as Yahoo! Answers, Quora, and StackOverflow [7–9, 5, 2, 13], open source projects [19, 24, 27, 11, 28]).

We study the global efficiency of simple and prefixed rules for sharing the value locally at each project, even in the presence of incomplete information on

---

* For a full version see "Y. Bachrach, V. Syrgkanis, M. Vojnovic, Incentives and Efficiency in Uncertain Collaborative Environments, arXiv:1308.0990, 2013".

** Work performed in part while an intern with Microsoft Research. Supported in part by ONR grant N00014-98-1-0589 and NSF grants CCF-0729006 and a Simons Graduate Fellowship.

Y. Chen and N. Immorlica (Eds.): WINE 2013, LNCS 8289, pp. 26–39, 2013.

the player's abilities and private resource constraints and even if players employ learning strategies to decide how to play in the game.

The design of simple, local and predetermined mechanisms is suitable for applications such as sharing attention in online Q&A forums or scientific co-authorship scenarios, where cooperative solution concepts, that require ad-hoc negotiations and global redistribution of value, are less appropriate.

Robustness to incomplete information is essential in online applications, where players are unlikely to have full knowledge of the abilities of other players. Instead, participants have only distributional knowledge about their opponents. Additionally, public signals, such as reputation ranks, achievement boards, and history of accomplishments result in a significant asymmetry in the beliefs about a player's abilities. Therefore, efficiency guarantees should carry over, even if player abilities are arbitrary asymmetrically distributed.

In our main result we show that if locally at each project each player is awarded at least his marginal contribution to the value, then every equilibrium is a 2-approximation to the optimal outcome. This holds even when players' abilities and resource constraints are private information drawn from commonly known distributions and even when players use no-regret learning strategies to play the game. We portray several mechanisms that satisfy this property, such as sharing proportionally to the quality of the submission. We also give a generalization of our theorem, when players don't have hard constraints on their resources, but rather have soft constraints in the form of convex cost functions. Finally, we give classes of instances where near optimality is achieved in equilibrium.

Our Results. We consider a model of collaboration where the system consists of set of players and a set of projects. Each player has a budget of time which he allocates across his projects. If a player invests some effort in some project, this results in some submission of a certain quality, which is a player and project specific increasing concave function of the effort, that depends on the player's abilities. Each project produces some value which is a monotone submodular function of the qualities of the submissions of the different participants. This common value produced by each project is then shared among the participants of the project according to some pre-specified sharing rule, e.g. equal sharing, or sharing proportionally to quality.

1. Marginal Contribution and Simple Sharing Rules. We show that if each player is awarded at least his marginal contribution to the value of a project, locally, then every Nash equilibrium achieves at least half of the optimal social welfare. This holds at coarse correlated equilibria of the complete information game when player's abilities and budget are common knowledge and at Bayes-Nash equilibria when these parameters are drawn independently from commonly known arbitrary distributions. Our result is based on showing that the resulting game is universally $(1, 1)$-smooth game [21, 22, 25] and corresponds to a generalization of Vetta's [26] valid utility games to incomplete information settings. We give examples of simple sharing rules that satisfy the above condition, such as proportional to the marginal contribution or based on the Shapley value or proportional to the quality. We show that this bound is tight for very special cases of the class of games that we study and holds even for the best pure Nash equilibrium of the complete information setting and even when the equilibrium is unique. We also analyze ranking-based sharing rules and show that they can approximately satisfy the marginal contribution condition, leading only to logarithmic in the number of players loss.

2. Near Optimality for Constant Elasticity. We show that for the case when the value produced at each project is of the form $v(x) = w \cdot x^\alpha$ for $\alpha \in (0,1)$, where x is the sum of the submission qualities, then the simple proportional to quality sharing rule achieves almost 95% of the optimal welfare at every pure Nash equilibrium of the game, which always exists.

3. Soft Budget Constraints and a Threshold Phenomenon. When the players have soft budget constraints in the form of some convex cost function of their total effort, we characterize the inefficiency as a function of the convexity of the cost functions, as captured by the standard measure of elasticity. We show that if the elasticity is strictly greater than 1 (strictly convex), then the inefficiency both in terms of produced value and in terms of social welfare (including player costs) is a constant independent of the number of players, that converges to 2 as the elasticity goes to infinity (hard budget constraint case). This stands in a stark contrast with the case when the cost functions are linear, where we show that the worst-case efficiency can decrease linearly with the number of players.

Applications. In the context of *social computing* each project represents a specific topic on a user-generated website such as Yahoo! Answers, Quora, and StackOverflow. Each web user has a budget of time that he spends on such a web service, which he chooses how to split among different topics/questions that arise. The quality of the response of a player is dependent on his effort and on his abilities which are most probably private information. The attention produced is implicitly split among the responders of the topic in a non-uniform manner, since the higher the slot that the response is placed in the feed, the higher the attention it gets. Hence, the website designer has the power to implicitly choose the attention-sharing mechanism locally at each topic, by strategically ordering the responses according to their quality and potentially randomizing, with the goal of maximizing the global attention on his web-service.

Another application of our work is in the context of *sharing scientific credit* in paper co-authorship scenarios. One could think of players as researchers splitting their time among different scientific projects. Given the efforts of the authors at each project there is some scientific credit produced. Local sharing rules translate to scientific credit-sharing rules among the authors of a paper, which is implicitly accomplished through the order that authors appear in the paper. Different ordering conventions in different communities correspond to different sharing mechanisms, with the alphabetical ordering corresponding to equal sharing of the credit while the contribution ordering is an instance of a sharing mechanism where a larger credit is rewarded to those who contributed more.

Related Work. Our model has a natural application in the context of online crowdsourcing mechanisms which were recently investigated by Ghosh and Hummel [7, 8], Ghosh and McAfee [9], Chawla, Hartline and Sivan [4] and Jain,

Chen, and Parkes [13]. All this prior work focuses on a *single* project. In contrast, we consider multiple projects across which a contributor can strategically invest his effort. We also allow a more general class of project value functions. Having multiple projects creates endogenous outside options that significantly affect equilibrium outcomes. DiPalantino and Vojnović [5] studied a model of crowdsourcing where users can choose exactly one project out of a set of multiple projects, each offering a fixed prize and using a "winner-take-all" sharing rule. In contrast, we allow the value shared to be increasing in the invested efforts and allow individual contributors to invest their efforts across multiple projects.

Splitting scientific credit among collaborators was recently studied by Kleinberg and Oren [14], who again examined players choosing a single project. They show how to globally change the project value functions so that optimality is achieved at some equilibrium of the perturbed game.

There have been several works on the efficiency of equilibria of utility maximization games [26, 10, 16], also relating efficiency with the marginal contribution property. However, this body of related work focused only on the complete information setting. For general games, Roughgarden [21] gave a unified framework, called smoothness, for capturing most efficiency bounds in games and showed that bounds proven via the smoothness framework automatically extend to learning outcomes. Recently, Roughgarden [22] and Syrgkanis [25] gave a variation of the smoothness framework that also extends to incomplete information settings. Additionally, Roughgarden and Schoppmann gave a version of the framework that allows for tighter bounds when the strategy space of the players is some convex set. In this work we utilize these frameworks to prove our results.

Our collaboration model is related to the contribution games of [1]. However, in [1], it is assumed that all players get the same value from a project. This corresponds to the special case of equal sharing rule in our model. Moreover, they mainly focus on network games where each project is has only two participants.

Our model is also related to the bargaining literature [12, 15, 3]. The main question in that literature is similar to what we ask here: how should a commonly produced value be split among the participants. However, our approach is very different than the bargaining literature as we focus on simple mechanisms that use only local information of a project and not global properties of the game.

2   Collaboration Model

Our model of collaboration is defined with respect to a set N of n players and a set M of m available projects. Each player i participates in a set of projects M_i and has a budget of effort B_i, that he chooses how to distribute among his projects. Thus the strategy of player i is specified by the amount of effort $x_i^j \in \mathbb{R}_+$ that he invests in project $j \in M_i$.

Player Abilities. Each player i is characterized by his type t_i, which is drawn from some abstract type space T_i, and which determines his abilities on different projects as well as his budget. When player i invests an effort of x_i^j on project

j this results in a submission of quality $q_i^j(x_i^j; t_i)$, which depends on his type, and which we assume to be some continuously differentiable, increasing concave function of his effort that is zero at zero.

For instance, the quality may be linear with respect to effort $q_i^j(x_i^j) = a_i^j \cdot x_i^j$, where a_i^j is some project-specific ability factor for the player that is part of his type. In the context of Q&A forums, the effort x_i^j corresponds to the amount of time spent by a participant to produce some answer at question j, the budget corresponds to the amount of time that the user spends on the forum, the ability factor a_i^j corresponds to how knowledgeable he is on topic j and q_i^j corresponds to the quality of his response.

Project Value Functions. Each project $j \in M$ is associated with a value function $v_j(q^j)$, that maps the vector of submitted qualities $q^j = (q_i^j)_{i \in N_j}$ into a produced value (where N_j is the set of players that participate in the project). This function, represents the profit or revenue that can be generated by utilizing the submissions. In the context, of Q&A forums $v_j(q^j)$ could for instance correspond to the webpage attention produced by a set of responses to a question.

We assume that this value is *increasing in the quality* of each submission and satisfies the *diminishing marginal returns property*, i.e. the marginal contribution of an extra quality decreases as the existing submission qualities increase. More formally, we assume that the value is submodular with respect to the lattice defined on $\mathbb{R}^{|N_j|}$: for any $z \geq y \in \mathbb{R}^{|N_j|}$ (coordinate-wise) and any $w \in \mathbb{R}^{|N_j|}$:

$$v_j(w \vee z) - v_j(z) \leq v_j(w \vee y) - v_j(y), \tag{1}$$

where $\vee$ denotes the coordinate-wise maximum of two vectors. For instance, the value could be any concave function of the sum of the submitted qualities or it could be the maximum submitted quality $v_j(q^j) = \max_{i \in N_j} q_i^j$.

Local Value Sharing. We assume that the produced value $v_j(q^j)$ is shared locally among all the participants of the project, based on some predefined redistribution mechanism. The mechanism observes the submitted qualities q^j and decides a share $u_i^j(q^j)$ of the project value that is assigned to player i, such that $\sum_{i \in N_j} u_i^j(q^j) = v_j(q^j)$. The utility of a player i is the sum of his shares across his projects: $\sum_{j \in M_i} u_i^j(q^j)$.

In the context of Q&A forums, the latter mechanism corresponds to a local sharing rule of splitting the attention at each topic. Such a sharing rule can be achieved by ordering the submissions according to some function of their qualities and potentially randomizing to achieve the desired sharing portions.

From Effort to Quality Space. We start our analysis by observing that the utility of a player is essentially determined only by the submitted qualities and that there is a one-to-one correspondence between submitted quality and input effort. Hence, we can think of the players as choosing target submission qualities for each project rather than efforts. For a player to submit a quality of q_i^j he has to exert effort $x_i^j(q_i^j; t_i)$, which is the inverse of $q_i^j(\cdot; t_i)$ and hence is some increasing convex function, that depends on the player's type. Then the strategy space of a player is simply be the set:

$$Q_i(t_i) = \left\{ q_i = (q_i^j)_{j \in M_i} : \sum_{j \in M_i} x_i^j(q_i^j; t_i) \leq B_i(t_i) \right\} \tag{2}$$

From here on we work with the latter representation of the game and define everything in quality space rather than the effort space. Hence, the utility of a player under a submitted quality profile q, such that $q_i \in Q_i(t_i)$ is:

$$u_i(q; t_i) = \sum_{j \in M_i} u_i^j(q^j). \tag{3}$$

and minus infinity if $q_i \notin Q_i(t_i)$.

Social Welfare. We assume that the value produced is completely shared among the participants of a project, and thus, the social welfare is equal to the total value produced, assuming players choose feasible strategies for their type:

$$SW^t(q) = \sum_{i \in N} u_i(q; t_i) = \sum_{j \in M} v_j(q^j) = V(q). \tag{4}$$

We are interested in examining the social welfare achieved at the equilibria of the resulting game when compared to the optimal social welfare. For a given type profile t we denote with $\mathrm{OPT}(t) = \max_{q \in Q(t)} SW^t(q)$ the maximum welfare.

Equilibria, Existence and Efficiency. We examine both the complete and the incomplete information setting. In the complete information setting, the type (e.g. abilities, budget) of all the players is fixed and common knowledge. We analyze the efficiency of Nash equilibria and of outcomes that arise from no-regret learning strategies of the players when the game is played repeatedly. A Nash equilibrium is a strategy profile where no player can increase his utility by unilaterally deviating. An outcome of a no-regret learning strategy in the limit corresponds to a coarse correlated equilibrium of the game, which is a correlated distribution over strategy profiles, such that no player wants to deviate to some fixed strategy. We note that such outcomes always exist, since no-regret learning algorithms for playing games exist. When the sharing rule induces a game where each players utility is concave with respect to his submitted quality and continuous (e.g. Shapley value) then even a pure Nash equilibrium is guaranteed to exist in our class of games, by the classic result of Rosen [20].

In the incomplete information setting the type t_i of each player is private and is drawn independently from some commonly known distribution F_i on T_i. This defines an incomplete information game where players strategies are mappings $s_i(t_i)$, from types to (possibly randomized) actions, which in our game corresponds to feasible quality vectors. Under this assumption we quantify the efficiency of Bayes-Nash equilibria of the resulting incomplete information game, i.e. strategy profiles where players are maximizing their utility in expectation over other player's types:

$$\mathbb{E}_{t_{-i}}[u_i(s(t))] \geq \mathbb{E}_{t_{-i}}[u_i(s_i', s_{-i}(t_{-i}))] \tag{5}$$

We note that a mixed Bayes-Nash equilibrium in the class of games that we study always exists assuming that the type space is discretized and for a sufficiently small discretization of the strategy space. Even if the strategy and type space is not discretized, a pure Bayes-Nash equilibrium is also guaranteed to exist in the case of soft budget constraints under minimal assumptions (i.e. type space is a convex set and utility share of a player is concave with respect to his submitted quality and is differentiable with bounded slope) as was recently shown by Meirowitz [17].

We quantify the efficiency at equilibrium with respect to the ratio of the optimal social welfare over the worst equilibrium welfare, which is denoted as the *Price of Anarchy*. Equivalently, we quantify what fraction of the optimal welfare is guaranteed at equilibrium.

3   Approximately Efficient Sharing Rules

In this section we analyze a generic class of sharing rules that satisfy the property that locally each player is awarded at least his marginal contribution to the value:

$$u_i^j(q^j) \geq v_j(q^j) - v^j(q_{-i}^j) \tag{6}$$

where q_{-i}^j is the vector of qualities where player i submits 0 and everyone else submits q_i^j.

Several natural and simple sharing rules satisfy the above property, such as sharing proportional to the marginal contribution or according to the local Shapley value.[1] When the value is a concave function of the total quality submitted, then sharing proportional to the quality: i.e. $u_i^j(q^j) = \frac{q_i^j}{\sum_{k \in N_j} q_k^j} v_j(q^j)$, satisfies the marginal contribution property. When the value is the highest quality submission, then just awarding all the value to the highest submission (e.g. only displaying the top response in a Q&A forum) satisfies the marginal contribution property (see full version).

We show that any such sharing rule induces a game that achieves at least a 1/2 approximation to the optimal social welfare, at any no-regret learning outcome and at any Bayes-Nash equilibrium of the incomplete information setting where players' abilities and budgets are private and drawn from commonly known distributions. Our analysis is based on the recently introduced smoothness framework for games of incomplete information by Roughgarden [22] and Syrgkanis [25], which we briefly survey.

Smoothness of Incomplete Information Games. Consider the following class of incomplete information games: Each player i has a type t_i drawn independently from some distribution F_i on some type space T_i, which is common knowledge.

---

[1] The Shapley value corresponds to the expected contribution of a player to the value if we imagine drawing a random permutation and adding players sequentially, attributing to each player his contribution at the time that he was added.

For each type $t_i \in T_i$ each player has a set of available actions $A_i(t_i)$. A players strategy is a function $s_i : T_i \to A_i$ that satisfies $\forall t_i \in T_i : s_i(t_i) \in A_i(t_i)$. A player's utility depends on his type and the action profile: $u_i : T_i \times A \to \mathbb{R}$.

Definition 1 (Roughgarden [22], Syrgkanis [25]). *An incomplete informa- tion game is universally (λ, μ)-smooth if $\forall t \in \times_i T_i$ there exists $a^*(t) \in \times_i A_i(t)$ such that for all $w \in \times_i T_i$ and $a \in \times_i A_i(w)$:*

$$\sum_{i \in N} u_i(a_i^*(t), a_{-i}; t_i) \geq \lambda OPT(t) - \mu \sum_{i \in N} u_i(a; w_i) \tag{7}$$

Theorem 1 (Roughgarden [21, 22], Syrgkanis [25]). *If a game is univer- sally (λ, μ)-smooth then every mixed Bayes-Nash equilibrium of the incomplete information setting and every coarse correlated equilibrium of the complete in- formation setting achieves expected social welfare at least $\frac{\lambda}{1+\mu}$ of the optimal.*

It is easy to observe that our collaboration model, falls into the latter class of incomplete information games, where the action of each player is his submitted quality vector q_i, and the set of feasible quality vectors depend on his private type: $\mathcal{Q}_i(t_i)$ as defined in Equation (2). Last the utility of a player is only a function of the actions of other players and not directly of their types, since it depends only on the qualities that they submitted.

Theorem 2. *The game induced by any sharing rule that satisfies the marginal contribution property is universally $(1, 1)$-smooth.*

Proof. Let t, w be two type profiles, and let $\tilde{q}(t) \in \mathcal{Q}(t)$ be the quality profile that maximizes the social welfare under type profile t, i.e. $\tilde{q}(t) = \arg\max_{q \in \mathcal{Q}(t)} SW(q)$. To simplify presentation we will denote $\tilde{q} = \tilde{q}(t)$, but remind that the vector de- pends on the whole type profile. Consider any quality profile $q \in \mathcal{Q}(w)$. By the fact that $\tilde{q}_i \in \mathcal{Q}_i(t_i)$ is a valid strategy for player i under type profile t_i, we have:

$$\sum_{i \in N} u_i(\tilde{q}_i, q_{-i}; t_i) = \sum_{i \in N} \sum_{j \in M_i} u_i^j(\tilde{q}_i^j, q_{-i}^j)$$

By the marginal contribution property of the sharing rule we have that:

$$\sum_{i \in N} u_i(\tilde{q}_i, q_{-i}; t_i) \geq \sum_{i \in N} \sum_{j \in M_i} \left(v_j(\tilde{q}_i^j, q_{-i}^j) - v_j(q_{-i}^j) \right) = \sum_{j \in M} \sum_{i \in N_j} \left(v_j(\tilde{q}_i^j, q_{-i}^j) - v_j(q_{-i}^j) \right)$$

Following similar analysis as in Vetta [26] for the case of complete information games, by the diminishing marginal returns property of the value functions:

$$v_j(\tilde{q}_i^j, q_{-i}^j) - v_j(q_{-i}^j) \geq v_j(\tilde{q}_{\leq i}^j \vee q_{\leq i}^j, q_{>i}^j) - v_j(\tilde{q}_{<i}^j \vee q_{<i}^j, q_{\geq i}^j)$$

Where it can be seen that the right hand side is the marginal contribution of an extra quality $\tilde{q}_i^j$ added to a larger vector than the vector on the left hand side. Specifically, the left hand side is the marginal contribution of $\tilde{q}_i^j$ to q_{-i}^j, while the right hand side is the marginal contribution of $\tilde{q}_i^j$ to the vector $(q_{<i}^j + \tilde{q}_{<i}^j, q_{\geq i}^j)$. Summing this inequality for every player in N_j we get a telescoping sum:

$$\sum_{i \in N_j} v_j(\tilde{q}_i^j, q_{-i}^j) - v_j(q_{-i}^j) \geq v_j(\tilde{q}^j \vee q^j) - v_j(q^j) \geq v_j(\tilde{q}^j) - v_j(q^j)$$

Combining this with the initial inequality and using the fact that $q \in \mathcal{Q}(w)$, we get the desired universal $(1,1)$-smoothness property:

$$\sum_{i \in N} u_i(\tilde{q}_i, q_{-i}; t_i) \geq \sum_{j \in M} v_j(\tilde{q}^j) - \sum_{j \in M} v_j(q^j) = \text{OPT}(t) - \sum_{i \in N} u_i(q; w_i)$$

$\square$

Corollary 1. *Under a local sharing rule that satisfies the marginal contribution property, every coarse correlated equilibrium of the complete information setting and every mixed Bayes-Nash equilibrium of the incomplete information game achieves at least $1/2$ of the expected optimal social welfare.*

We show that this theorem is tight for the class of games that we study and more specifically, for the proportional sharing rule. The tightness holds even at pure Nash equilibria of the complete information setting, even when all players have the same ability and even when the equilibrium is unique. Intuitively what causes inefficiency is that players prefer to congest a low value project with a high rate of success (i.e. produces almost it's maximal value for a very small quality), e.g. an easy topic, rather than trying their own luck on a hard project that would yield very high value but would require a lot of effort to produce it.

Example 1. Consider the following instance: there are n players and n projects. Every player participates in every project. Each player has a budget of effort of 1 and the quality of his submission at a project is equal to his effort. Project 1 has value function $v_1(x) = 1 - e^{-\alpha x}$, where x is the total submitted quality. The rest of the $n - 1$ projects have value function $\kappa(1 - e^{-\beta x})$. We assume that value is shared proportional to the quality. We show that if we let $\alpha \to \infty$, $\kappa = \frac{n-1}{\beta n^2}$ and $\beta \to 0$, then the unique equilibrium is for all players to put their whole budget on project 1. The optimal on the other hand is for players efforts to be spread out among all the projects. A good approximation to the optimal is for each player to pick a different project and devote his whole effort on it. The ratio of the optimal social welfare and the Nash equilibrium welfare in the limit of the above values will be $1 + (1 - 1/n)^2$, which converges to 2 as the number of players grows large. A detailed equilibrium analysis is given in the full version. $\square$

3.1 Ranking Rules and Approximate Marginal Contribution

An interesting, from both theoretical and practical standpoint, class of sharing rules is that of ranking rules. In a ranking sharing scheme, the mechanism announces a set of fixed portions $a_1^j \geq \ldots \geq a_n^j$, such that $\sum_t a_t^j = 1$. After the players submit their qualities, each player is ranked based on some order that depends on the profile of qualities (e.g. in decreasing quality order or in decreasing marginal contribution order). If a player was ranked at position t then he gets a share of $a_t^j \cdot v_j(q^j)$. Fixed reward rules capture several real world scenarios where the only way of rewarding participants is ordering them in a deterministic manner and the designer doesn't have the freedom to award to the players arbitrary fractions of the produced value.

We show here that although such sharing rules are quite restrictive, they are expressive enough to induce games that achieve only a logarithmic in the number of players loss in efficiency. To prove this we show that by setting the fixed portions inversely proportional to the position, then every player is guaranteed at least an $\log(n)$-fraction of his marginal contribution. It is then easy to generalize our analysis in Theorem 2 to show that sharing rules that award each player a k-fraction of his marginal contribution induce a universally $(1/k, 1/k)$-smooth game and achieve at least $1/(k+1)$ of the optimal welfare at equilibrium.

Lemma 1. *By setting coefficients a_i^j proportional to $\frac{1}{t}$, the game resulting from the ranking sharing rule, where submissions are ranked with respect to the marginal contribution order, achieves a $\Omega(1/log(n))$ fraction of the optimal welfare at every coarse correlated and at every Bayes-Nash equilibrium.*

If the value is a function of the sum of the quality of submissions then a similar guarantee is achieved if submissions are ordered in decreasing quality.

4 Almost Optimality for Uniformly Hard Projects

In this section we identify a subclass of value functions for which the social welfare at equilibrium is a much better approximation to the optimal welfare, achieving almost 95% of the optimal. We start our quest, by observing that the crucial factor that led to the tight lower bound in the previous section, is that different projects have a very different rate of success: the percentage increase in the output for a percentage increase in the input was completely different for different projects and at different qualities within a project. This discrepancy in the output sensitivity was the main force driving the lower bound. In this section we examine a broad class of functions that don't allow for such discrepancies.

The standard economic measure that captures the sensitivity of the output of a function with respect to a change in its input is that of elasticity.

Definition 2. *The elasticity of a function $f(x)$ is defined as: $\epsilon_f(x) = \left| \frac{f'(x)x}{f(x)} \right|$.*

One can show formally that the above parameter of a function has a one-to-one correspondence with the ratio of the percentage change in the output for a percentage of change in the input. Intuitively, projects whose value has the same and constant elasticity have the same and uniform difficulty, though not necessarily the same importance. Based on this reasoning, we examine the setting where all project value functions are functions of the total quality of submissions and have constant elasticity α. It can be easily seen that such functions will take the form $v_j(q^j) = w_j \cdot Q_j^\alpha$, where $Q_j = \sum_{i \in N_j} q_i^j$. The coefficient w_j can be project specific, and will correspond to the importance of a project. For such value functions we prove that the proportional to the quality sharing mechanism achieves social welfare at any pure Nash equilibrium of the complete information setting that is almost optimal.

We point that our class of games always possess a pure Nash equilibrium, since they are games defined on a convex strategy space, with continuous and concave utilities and hence the existence is implied by Rosen [20].

Theorem 3. *Suppose that the project value functions are of the form $v_j(q^j) = w_j \cdot Q_j^\alpha$, for $0 \leq \alpha \leq 1$ and $w_j > 0$. Then, the proportional to the quality sharing rule achieves social welfare at least $\frac{2^{1-\alpha}}{2-\alpha} \geq 0.94$ of the optimal social welfare at every pure Nash equilibrium of the complete information game it defines. (see full version)*

Our analysis is based on the local smoothness framework of Roughgarden and Schoppmann [23], which yields tighter bounds for games with continuous and convex strategy spaces and where utilities are continuous and differentiable. It is easy to see that the strategy spaces in our setting $Q_i(t_i)$ as defined in equation (2) are convex, by the convexity of the functions $x_i^j(\cdot; t_i)$.[2] Additionally, it is easy to check that the utilities of the players under the proportional sharing rule are going to be continuous and differentiable at any point, except potentially at 0.

However, in the full version we show that in equilibrium no project receives 0 total quality with positive probability. We also show that this relaxed condition is sufficient to apply the local smoothness framework to pure Nash equilibria. Alternatively, we can bypass this technicality by assuming there are exclusive players who participate only at a specific project and always invest an ϵ amount of effort. Making the latter assumption, we can use the local smoothness framework in its full generality and our conclusion in Theorem 3 carries over to correlated equilibria of the game (outcomes of no-swap regret learning strategies).

5   Soft Budget Constraints

So far we analyzed the case where players have a hard constraint on their effort, e.g. hard time constraint. In this section we relax this assumption and study the case where instead each player incurs a cost that is a convex function of the total effort he exerts, corresponding to a soft budget constraint on his effort. We exhibit an interesting threshold phenomenon in the inefficiency of the setting: if cost is linear in the total effort then the inefficiency can grow linearly with the number of participants. However, when effort cost is strictly convex, then the inefficiency can be at most a constant independent of the number of participants.

More formally, we will assume that each player has a cost function $c_i(x; t_i)$ that determines his cost when he exerts a total effort of x. This cost function is also dependent on his private type t_i. The total exerted effort can be expressed with respect to the quality of submission as $X_i(q_i; t_i) = \sum_{j \in M_i} x_i^j(q_i^j; t_i)$. Thus a player's utility as a function of the profile of chosen qualities is:

$$u_i(q) = \sum_{j \in M_i} u_i^j(q^j) - c_i\left(X_i(q_i; t_i); t_i\right). \tag{8}$$

Unlike the previous section, the social welfare is not the value produced. Instead:

$$SW^t(q) = \sum_{i \in N} u_i(q; t_i) = \sum_{j \in M} v_j(q^j) - \sum_{i \in N} c_i\left(X_i(q_i; t_i); t_i\right) = V(q) - C^t(q).$$

---

[2] If the effort required to produce two quality vectors $q_i, \hat{q}_i$ is at most $B_i(t_i)$, then the effort required to produce any convex combination of them is also at most $B_i(t_i)$

We refer to $V(q)$ as the production of an outcome q and to $C^t(q)$ as the cost. We first show that when a sharing rule that satisfies the marginal contribution property is used, the production plus the social welfare at equilibrium is at least the value of the optimal social welfare. We then use this result to give bounds on the equilibrium efficiency parameterized by the convexity of the cost functions.

Lemma 2. *Consider the game induced by any sharing rule that satisfies the marginal contribution property and where players have soft budget constraints. Then the expected social welfare plus the expected production at any coarse correlated equilibrium of the complete information setting and at any Bayes-Nash equilibrium of the incomplete information setting, is at least the expected optimal social welfare. (see full version)*

We use the latter result to derive efficiency bounds for both production and social welfare. We assume that the sharing rule used induces a utility share that is a concave function of a players submission quality and such that a player's share at 0 quality is 0. More formally, we assume that: $g(x) = u_i^j(x, q_{-i}^j)$ is concave, continuously differentiable and $g(0) = 0$. We call such sharing rules *concave sharing rules*. It is easy to see that the proportional to quality sharing rule is a concave sharing rule when the value is concave in the total quality. For general value functions, the Shapley sharing rule is also a concave sharing rule.

We show efficiency bounds parameterized by the convexity of the cost functions, using the elasticity of the cost function as the measure of convexity. An increasing convex function that is zero at zero, has an elasticity of at least 1. We will quantify the inefficiency in our game as a function of how far from 1 the elasticity of the cost functions are (e.g. $c_i(x) = \kappa \cdot x^{1+a}$ has elasticity $1 + a$).

Theorem 4. *If a concave sharing rule is used and the elasticity of the cost functions is at least $1 + \mu$ then: i) the expected social welfare at any coarse correlated equilibrium of the complete information setting and at any Bayes-Nash equilibrium of the incomplete information setting is at least $\frac{\mu}{1+2\mu}$ of the optimal, ii) the total value produced in equilibrium is at least $\frac{1}{2}\frac{\mu}{1+\mu}$ of the value produced at the social welfare maximizing outcome. (see full version)*

The proof of the theorem is based on analyzing the first order conditions that an equilibrium strategy must satisfy. Combined with the elasticity of the cost function we manage to relate the expected cost at equilibrium with the expected production. Specifically, we show that at equilibrium and at the optimal strategy profiles the expected production is at least $(1+\mu)$ times the expected cost, where $1+\mu$ is the elasticity of the cost function. Combining, the above with our Lemma 2 we get the result. Thus the theorem uses both a local deviation analyses, via the use of the first order equilibrium conditions and a global deviation analysis, via the smoothness techniques used to prove Lemma 2.

From this theorem, we obtain that as long as $\mu > 0$, the efficiency of any Nash equilibrium, is a constant independent of the number of players. For instance, if the cost is a quadratic function of the total effort then the social welfare at equilibrium is a 3-approximation to the optimal and the produced value is a 4-approximation to the value produced at the welfare-maximizing outcome.

The budget constraint case that we studied in previous sections can be seen as a limit of a family of convex functions that converge to a limit function of the form $c_i(x_i) = 0$ if $x_i < B_i$, and ∞ otherwise. Such a limit function can be thought of as a convex function with infinite elasticity. Taking the limit as $\mu \to \infty$ in the latter theorem, gives that the social welfare at equilibrium is at least half the optimal, which matches our analysis in the previous section.

A corner case is that of linear cost functions, where our Theorem gives no meaningful upper bound. In fact as the following example shows, for linear cost functions the inefficiency can grow linearly with the number of agents.

Example 2. Consider a single project with value $v(Q) = \sqrt{Q}$ and assume that the proportional to the quality sharing rule is used. Moreover, each player pays a cost of 1 per unit of effort, i.e. $c_i(X_i) = X_i$ and where the quality is equal to the effort. The global optimum is the solution to the unconstrained optimization problem: $\max_{Q \in \mathbb{R}_+} \sqrt{Q} - Q$, which leads to $Q^* = 1/4$ and therefore $SW(Q^*) = 1/4$. On the other hand, each player's optimization problem is: $\max_{q_i \in \mathbb{R}_+} q_i \frac{1}{\sqrt{Q}} - q_i$. Symmetry and elementary calculus gives that at the unique Nash equilibrium, the total effort is $Q = \left(\frac{n-1/2}{n}\right)^2$, and the social welfare is $\frac{2n-1}{4n^2} = O(1/n)$.   □

6   Conclusion and Future Work

We analyzed a general model of collaboration under uncertainty, capturing settings such as online social computing and scientific co-authorship. We identified simple value sharing rules that achieve good efficiency in a robust manner with respect to informational assumptions.

Some questions remain open for future research. We showed that ranking rules, which are highly popular [7, 8], achieve a logarithmic approximation, using fixed-prizes independent of the distribution of qualities (prior-free) and of the game instance. Can a constant approximation be achieved if we allow the fixed prizes associated with each position to depend on the distribution of abilities and on the instance of the game? Also, consider a *two-stage* model where in the first stage players choose the projects to participate in and then play our collaboration game in the second stage. Can any efficiency guarantee be given on the welfare achieved at the subgame-perfect equilibria of this two-stage game?

References

1. Anshelevich, E., Hoefer, M.: Contribution Games in Networks. Algorithmica, 1–37 (2011)
2. Archak, N., Sundararajan, A.: Optimal design of crowdsourcing contests. In: ICIS (2009)
3. Bateni, M., Hajiaghayi, M., Immorlica, N., Mahini, H.: The cooperative game theory foundations of network bargaining games. In: Abramsky, S., Gavoille, C., Kirchner, C., Meyer auf der Heide, F., Spirakis, P.G. (eds.) ICALP 2010. LNCS, vol. 6198, pp. 67–78. Springer, Heidelberg (2010)

4. Chawla, S., Hartline, J.D., Sivan, B.: Optimal crowdsourcing contests. In: SODA (2012)
5. DiPalantino, D., Vojnović, M.: Crowdsourcing and all-pay auctions. In: EC (2009)
6. Friedman, E., Moulin, H.: Three methods to share joint costs or surplus. Journal of Economic Theory 87(2), 275–312 (1999)
7. Ghosh, A., Hummel, P.: Implementing optimal outcomes in social computing: a game-theoretic approach. In: WWW (2012)
8. Ghosh, A., Hummel, P.: Learning and incentives in user-generated content: Multi-armed bandits with endogeneous arms. In: ITCS (2013)
9. Ghosh, A., McAfee, P.: Crowdsourcing with endogenous entry. In: WWW (2012)
10. Goemans, M., Li, L.E., Mirrokni, V.S., Thottan, M.: Market sharing games applied to content distribution in ad-hoc networks. In: ACM MobiHoc (2004)
11. Hars, A., Ou, S.: Working for free? - motivations of participating in open source projects. In: HICSS (2001)
12. Hatfield, J.W., Kominers, S.D.: Multilateral Matching. In: ACM EC (2011)
13. Jain, S., Chen, Y., Parkes, D.C.: Designing incentives for online question and answer forums. In: ACM EC (2009)
14. Kleinberg, J., Oren, S.: Mechanisms for (mis)allocating scientific credit. In: STOC (2011)
15. Kleinberg, J., Tardos, E.: Balanced outcomes in social exchange networks. In: STOC (2008)
16. Marden, J., Roughgarden, T.: Generalized efficiency bounds in distributed resource allocation. In: 2010 49th IEEE Conference on Decision and Control (CDC), pp. 2233–2238 (2010)
17. Meirowitz, A.: On the existence of equilibria to bayesian games with non-finite type and action spaces. Economics Letters 78(2), 213–218 (2003)
18. Moulin, H., Watts, A.: Two versions of the tragedy of the commons. Review of Economic Design 2(1), 399–421 (1996)
19. Rashid, A.M., Ling, K., Tassone, R.D., Resnick, P., Kraut, R., Riedl, J.: Motivating participation by displaying the value of contribution. In: Proceedings of the SIGCHI Conference on Human Factors in Computing Systems, CHI 2006, pp. 955–958. ACM, New York (2006)
20. Rosen, J.B.: Existence and uniqueness of equilibrium points for concave n-person games. Econometrica 33(3), 520–534 (1965)
21. Roughgarden, T.: Intrinsic robustness of the price of anarchy. In: STOC (2009)
22. Roughgarden, T.: The price of anarchy in games of incomplete information. In: ACM EC (2012)
23. Roughgarden, T., Schoppmann, F.: Local smoothness and the price of anarchy in atomic splittable congestion games. In: SODA (2011)
24. Shah, S.K.: Motivation, governance, and the viability of hybrid forms in open source software development. Manage. Sci. 52(7), 1000–1014 (2006)
25. Syrgkanis, V.: Bayesian games and the smoothness framework. Arxiv, 1203.5155 (2012)
26. Vetta, A.: Nash equilibria in competitive societies, with applications to facility location, traffic routing and auctions (2002)
27. von Krogh, G., Spaeth, S., Lakhani, K.R.: Community, joining, and specialization in open source software innovation: a case study. Research Policy 32(7), 1217–1241 (2003)
28. Ye, Y., Kishida, K.: Toward an understanding of the motivation of open source software developers. In: ICSE (2003)

Revenue Maximization with Nonexcludable Goods

MohammadHossein Bateni[1], Nima Haghpanah[2,*],
Balasubramanian Sivan[3], and Morteza Zadimoghaddam[4]

[1] Google Research, New York
bateni@google.com
[2] Northwestern University
nima.haghpanah@gmail.com
[3] Microsoft Research
bsivan@microsoft.com
[4] Massachusetts Institute of Technology
morteza@mit.edu

Abstract. We study the design of revenue maximizing mechanisms for selling nonexcludable public goods. In particular, we study revenue maximizing mechanisms in Bayesian settings for facility location problems on graphs where no agent can be excluded from using a facility that has been constructed. We show that the optimization problem involved in implementing the revenue optimal mechanism is hard to approximate within a factor of $\Omega(n^{2-\epsilon})$ (assuming $P \neq NP$) even in star graphs, and that even in expectation over the valuation profiles, the problem is APX-hard. However, in a relevant special case we construct polynomial time truthful mechanisms that approximate the optimal expected revenue within a constant factor. We also study the effect of partially mitigating nonexcludability by collecting tolls for using the facilities. We show that such "posted-price" mechanisms obtain significantly higher revenue, and often approach the optimal revenue obtainable with full excludability.

Keywords: Revenue maximization, nonexcludable goods, hardness of approximation, incentive compatibility.

1 Introduction

How should a seller maximize his revenue while selling nonexcludable services? As a representative example, consider the following facility location problem: firms in an industrial town want to connect their respective warehouses to their outlets with good road links. But the seller (in this case the construction company that lays the roads) cannot enforce exclusivity on the roads it lays: once a firm buys a connection between its warehouse and outlet, the seller cannot exclude others from using those road links. We study the following two questions in this work. Which potential sites should the seller choose for laying roads to maximize his own revenue in such nonexcludable settings? If the seller could enforce partial excludability by collecting tolls for roads,

---

* Part of the work was done while the second and fourth coauthors were visiting Google Research.

Y. Chen and N. Immorlica (Eds.): WINE 2013, LNCS 8289, pp. 40–53, 2013.

where only those firms that pay the toll for a road can use it, how much more revenue can he earn?[1]

The Mechanism Design Problem. We model the problem with a graph where every unordered pair of vertices denotes an agent and every edge denotes the *possibility* of a road to be laid between two vertices. We assume that each agent/firm has a nonnegative private value for getting her pair of vertices connected, drawn independently from a known distribution. We aim for the design of the optimal mechanism which, given the value reports of all the agents, selects a set of edges to build roads on and computes payments to be made by agents so that the expected revenue is maximized (expectation over the distribution of private values) in equilibrium. We invoke Myerson's [14] characterization which states that the expected revenue of any mechanism in a Bayesian Nash Equilibrium (BNE) is equal to the expected *virtual surplus* of the agents *served*, i.e., the sum of virtual values[2] of those agents whose warehouse and outlet are connected in the solution[3]. This reduces the problem of computing the expected revenue maximizing mechanism to a *pointwise* optimization problem: given a profile of values of all agents, compute the set of edges to build roads on so that the sum of the virtual values of the agents served is maximized.

The pointwise optimization problem to maximize virtual values is an interesting graph theoretic problem, which to our knowledge has not been studied before. Given a graph with a weight on every pair of vertices (the weights represent virtual values, and thus, could be negative), design an algorithm that selects a subset of edges that maximizes the sum of the weights of every vertex pair whose endpoints are connected by the edges selected. If the weights were all positive, clearly the optimal choice for the seller would be pick all the edges in the graph. However, weights could be negative. In particular, while aiming to connect some subset of agents (i.e., vertex pairs), another subset of agents with negative weights automatically get connected, and these negative weighted pairs cannot be excluded due to nonexcludability. Deciding on the the set of edges to select for maximizing the sum of weights of the vertex pairs whose endpoints are connected is the nontrivial underlying graph theoretic problem.

The Rooted and Unrestricted Versions. We study two versions of the graph theoretic problem described above. In the unrestricted version, every (unordered) pair of vertices in the graph is an agent. In the rooted version (which is a special case of the unrestricted version), a single vertex is designated as root, and only those unordered pairs having root as one of their vertices are agents. This corresponds to the root being a central location where all warehouses are located, and other nodes being outlet locations.

---

[1] Note that collecting tolls only enforces partial excludability because *every* firm that pays the toll is allowed to use the road, and cannot be excluded.

[2] The virtual value of an agent is a function of her value and the distribution from which her value is drawn; it can be negative. When the virtual value function is not monotone, Myerson [14] applies a fix by considering the ironed virtual value function (see section 2 for more details), and all our results go through with virtual values replaced by ironed virtual values.

[3] Note that Myerson's mechanism remains revenue optimal in all single-parameter settings regardless of whether or not the good is excludable.

Results for Unrestricted Version (Section 3). The natural place to start is to solve the pointwise optimization problem in graphs presented above. Clearly, any α-approximate monotone[4] solution to the pointwise optimization problem implies an α-approximate truthful mechanism to the problem of expected revenue maximization. However, we show that except for very special cases, pointwise optimization is not possible unless $P = NP$. In particular, we show that the pointwise problem is NP-hard to approximate to within a factor $\Omega(n^{2-\epsilon})$ even on star graphs. For the special case where the underlying graph is a path, we give a dynamic program to solve the pointwise optimization. Thus, for paths, we get the expected optimal revenue with nonexcludability for any product distribution over agents' values.

Results for Rooted Version (Section 4). Our results for the rooted version are three-fold.

1. First, we give a simple polynomial time algorithm for pointwise optimization in trees—this contrasts with our result that for the unrestricted version the pointwise problem is hard to approximate beyond a factor of $\Omega(n^{2-\epsilon})$ even for star graphs.
2. Second, as our main result, we give a polynomial time algorithm for optimization in expectation (over the distribution of values, and hence virtual values) for arbitrary graphs, when the agents' valuations are drawn i.i.d.
3. Third, we establish APX-hardness of optimization in expectation for arbitrary graphs, when the valuations of all agents are independent but not necessarily identical.

Our main result, which is the second point above, is that we design an algorithm that guarantees a constant factor approximation to the optimal virtual surplus in expectation. To this end, we extensively use the key (yet simple) property that even though virtual values can be negative, the expected virtual value of an agent is nonnegative. This suggests the following high-level approach to the problem. Partition agents into a constant number of sets. Pick a target set at random, and run an algorithm that is a constant factor approximation for the agents of that set (ignoring other agents). The nonnegativity of the expected virtual value implies that the contribution from nontargeted sets is nonnegative. Since each set is targeted with constant probability, this implies a constant factor approximation. We use this approach to solve the abstract edge-weighted version of the problem (in which edges are agents who derive value from being connected to the root), and then reducing the original problem to the edge-weighted version. In particular, we present an algorithm that partitions the edges of any graph into two sets that, loosely speaking, correspond to edges in well-connected parts of the graph and edges in sparse cuts of the graph. We show that once we contract the edges in a set (that corresponds to ignoring their value in the above high level approach), the remaining graph has a nice structure that allows for constant approximations.

Partial Excludability via Pricing (Section 5). If the seller were allowed to set prices on edges (i.e., collect tolls), can he get close to the optimal revenue when excludability

---

[4] In single-parameter settings, a mechanism's allocation is monotone if fixing the values of other agents and agent i alone reporting a higher value results in agent i getting served with no smaller probability. Monotone allocation is necessary and sufficient for truthfulness, i.e., they alone lend themselves to truthful payments.

is allowed? Typically the optimal revenue with full excludability is much higher than the optimal revenue with nonexcludability, and it is worthwhile for a profit maximizing firm to explore this option. In this problem the seller first has to decide which edges to pick to lay roads on, and decide how to price roads to maximize revenue. We construct a pricing scheme so that when the value distributions are i.i.d., the seller obtains (for the rooted version) the optimal revenue possible when full excludability is allowed. Further, if the distributions are independent but not identical, and satisfy a technical MHR condition[5], we show how to price edges to obtain a $\frac{1}{\log n}$ fraction of the optimal revenue possible when full excludability is allowed (again for the rooted version). An implication of these results is that mitigating externalities via tolls is much more remunerative than aiming for the optimal solution with nonexcludability.

Related Work. Approximating the expected version of a problem that is hard-to-approximate in the worst case, via exploiting the fact that the expected virtual value for any distribution is nonnegative is a relatively new idea. To our knowledge, it has been used only in Haghpanah et al. [10]. Auctions with externality (a notion related to nonexcludability where an agent's utility doesn't just depend on the services he received but also on the outcomes for the other agents) have been studied in multiple flavors before. Settings with positive externality include increased value for having a telephone or a music player or a new technology if more of your neighbors have them [15, 11]. A prime and well-studied example for a setting with negative externality is the sale of contiguous ad slots to two competing businesses [1, 4, 8, 9, 12, 13]. Equilibria which are surprising at first sight, like no allocation equilibria can result in large revenue for the auctioneer in settings with negative externality [7]. Posted pricings have often been a mechanism of choice for settings with externalities [3, 2, 5, 11]. The main difference between these and the posted prices studied in our work (apart from the presence of nonexcludability in our settings) is that these works allow agent-specific prices, whereas our setting is more constrained: we place prices on edges that are common for all agents.

2 Model and Notation

Unrestricted Version. We consider a universe of n potential sites, located on the vertices of a graph, $G = (V, E)$, with m undirected edges. An undirected edge $(i, j) \in E$ means that a link/road connecting sites i and j is allowed (but need not necessarily be constructed). An agent is an unordered pair of sites (i, j), interested in having some path constructed between i and j. Thus, there could be up to $\binom{n}{2}$ agents in a mechanism. An instance $\mathcal{I} = (G, A, F)$ of the problem consists of an undirected graph G, a set A of agents (represented by a set of pairs of vertices), and a distribution F_i associated with agent i (distributions are explained below). Sometimes we use i to denote a single site and sometimes to denote an agent, who is actually a pair of vertices. The context will make it clear whether i is a single site or a pair of sites.

---

[5] Roughly speaking, this means that the tail of the distribution is no heavier than the exponential distribution. Many natural classes of distributions like the uniform, exponential, Gaussian (Normal) distributions satisfy the MHR condition. See Section 2 for a formal definition.

Rooted Version. The rooted version is a special case of the setting described above. A special vertex r is designated as the root. Only vertex pairs of the form (i, r) are agents, i.e., the root r is one of the end points of the path desired for every agent.

An outcome $o \in \Omega = \{0, 1\}^m$ is the set of edges constructed. Agent i has a valuation function $v_i : \Omega \to \mathbf{R}^+ \cup \{0\}$, which maps outcomes to nonnegative real numbers. We study mechanisms in the single-parameter setting where the function $v_i(\cdot)$ takes only two values: v_i and zero. An agent i has a nonzero value v_i if and only if the set of edges selected contains a path between the two sites she represents, and in this case we say that agent i was served. Let $\mathcal{S}$ denote the set of all feasible sets of agents, i.e., the set of all sets of agents that can be simultaneously served. Note that this set system $\mathcal{S}$, is not downward closed: $S \in \mathcal{S}$ does not necessarily mean that $S' \in \mathcal{S}$ for all $S' \subset S$. Clearly, the non-downward closedness stems from the inability to exclude agents from using the roads.

We study mechanisms for this problem in a Bayesian setting, i.e., for every i, the single parameter v_i is assumed to be drawn independently from a publicly known distribution function F_i. Thus $F = F_1 \times F_2 \times \ldots F_n$ denotes the product distribution from which the vector of types $\mathbf{v}$ is drawn. The mechanisms in this paper assume the availability of any expectation defined with respect to F.

A direct revelation mechanism or an auction solicits sealed bids $(b_1, b_2, \ldots, b_n)$ from all the agents, and determines the outcome $\mathbf{x} = (x_1, x_2, \ldots, x_n)$ and payments $\mathbf{p} = (p_1, p_2, \ldots, p_n)$. Each agent i is risk-neutral and has a linear utility $u_i(\mathbf{b}) = v_i \cdot x_i(\mathbf{b}) - p_i(\mathbf{b})$.

Let $(\mathbf{x}(\cdot), \mathbf{p}(\cdot))$ denote a mechanism. When agent i is bidding in the auction, she knows only her own value v_i. A mechanism $(\mathbf{x}(\cdot), \mathbf{p}(\cdot))$ is incentive compatible (IC) if $v_i x_i(v_i, v_{-i}) - p_i(v_i, v_{-i}) \geq v_i x_i(z, v_{-i}) - p_i(z, v_{-i})$ for all i, z. A necessary and sufficient condition on $\mathbf{x}(\cdot)$ for IC payments to exist is monotonicity: for all $i, v_i, v_i' \geq v_i, v_{-i}$, we have $x_i(v_i, v_{-i}) \leq x_i(v_i', v_{-i})$. Since we focus on the class of IC mechanisms, we have $\mathbf{b} = \mathbf{v}$.

Optimal Auctions. To solve for optimal auctions, Myerson [14] defined *virtual valuations* for agents as $\phi_i(v_i) = v_i - \frac{1 - F_i(v_i)}{f_i(v_i)}$ and proved that the expected payment of an agent, $\mathbf{E}_{v_i}[p_i(v_i)]$, in any truthful mechanism, is equal to her expected virtual value $\mathbf{E}_{v_i}[\phi_i(v_i) x_i(v_i)]$ [6]. The distribution F_i is said to be *regular* if the virtual valuation function is monotone. For regular distributions, maximizing virtual values pointwise results in an incentive-compatible allocation rule and therefore the corresponding revenue-optimal auction serves that feasible set of agents who maximize virtual value. For irregular distributions, where maximizing virtual values may result in non-IC allocation rules, Myerson applies a fix by describing a general *ironing* technique. The ironing procedure converts any virtual valuation function $\phi_i(\cdot)$ to an *ironed virtual value function* $\bar{\phi}_i(\cdot)$ such that maximizing $\bar{\phi}_i(\cdot)$ pointwise results in an IC allocation rule.

Theorem 1. *[14] The revenue optimal auction in single-parameter Bayesian settings serves the set S of agents where $S = \mathrm{argmax}_{S \in \mathcal{S}} \sum_{i \in S} \bar{\phi}_i(v_i)$. Further, for all values v_i for which $\bar{\phi}_i(v_i)$ remains the same, agent i's allocation remains the same.*

---

[6] In fact the equality holds even for a specific $\mathbf{v}_{-i}$, i.e., $\mathbf{E}_{v_i}[p_i(\mathbf{v})] = \mathbf{E}_{v_i}[\phi_i(v_i) x_i(\mathbf{v})]$.

As a corollary, when $\mathcal{S}$ contains all possible 2^n sets, the revenue optimal auction puts a price of $\bar{\phi}_i^{-1}(0)$ for agent i and makes a take-it-or-leave-it offer to each of them.

Monotone Hazard Rate. A class of distributions that always satisfy the regularity condition described above are the ones which satisfy the monotone hazard rate condition. A distribution with cdf F satisfies the MHR condition if $\frac{f(x)}{1-F(x)}$ is nondecreasing in x. The MHR condition holds for many natural classes of distributions like the uniform, exponential and normal distributions.

Nonnegative Virtual Valuations. An important property of virtual valuations (and ironed too) is that when the support of the distribution is nonnegative (which is true in our case since values are nonnegative), the virtual valuation in expectation over an agent's distribution is always nonnegative. This property crucially helps us in providing approximations in expectation, for problems which are very hard to approximate for every single realization of values.

3   The Unrestricted Version

In this section, we obtain the following results for the unrestricted version of our problem.

1. A simple polynomial time dynamic program that solves the pointwise problem optimally in paths. (see Appendix A.)
2. A hardness result showing that the pointwise problem is hard to approximate within a factor of $\Omega(n^{2-\epsilon})$ for any $\epsilon > 0$ (assuming $P \neq NP$) even in stars, i.e., we cannot generalize the result on paths even to stars (proof in full version).
3. Given that pointwise approximation is ruled out for arbitrary graphs, the only possible approximation we can hope for is in expectation over values. We show the APX-hardness of this problem in arbitrary graphs. The reduction resembles the one given by Haghpanah et al. [10]. In fact, our hardness result holds even for the rooted version, and hence the proof is presented in Section 4.3 along with other results for the rooted version.

4   The Rooted Version for i.i.d. Agents

Given the hardness results in Section 3 even for undirected graphs, we consider an important special case of our problem in this section: there is a designated root node in an undirected graph, and each nonroot vertex is an agent. Agents values are i.i.d., and an agent derives value only if there is a path constructed from his node to the root node. As before, construction of an edge e can happen only if $e \in G$.

4.1   Pointwise Optimization in Trees

In contrast to the unrestricted version where we showed that the pointwise problem in hard to approximate within a factor of $\Omega(n^{2-\epsilon})$ (assuming $P \neq NP$) even in stars, we show that for the rooted version, we can solve the pointwise problem in polynomial time in trees. The proof is presented in the full version.

4.2   Main Result: Optimization in Expectation in Arbitrary Graphs

Next we present a constant-factor approximation algorithm for optimizing the expected revenue for the rooted, node-weighted version of the problem in arbitrary graphs (the virtual value of an agent is the weight of the nonroot node of that agent). In Section 4.2 we explain how the problem can be solved on edge-weighted graphs, i.e., agents are edges, and they get a value when they are connected to the root. This is later used in Section 4.2 as a subroutine to tackle the vertex-weighted problem.

Edge-Connectivity for the Rooted Version. Given a connected undirected graph, we show how to partition its edges into two parts such that contracting the former edge set yields a 3-edge connected subgraph whereas contracting the latter edge set results in a *roulette* subgraph—a special series-parallel graph to be defined below. We then demonstrate that it is possible to solve the problem (approximately) on each of the two subgraphs, and finally argue that this suffices to obtain a constant-factor approximation for the general rooted edge-weighted case.

Let us define some notations first. For a set S of edges in a graph G, subgraph G/S is obtained from G after contracting all edges S one at a time, where contracting an edge simply refers to removing the edge and identifying its endpoint vertices. We recursively define the class of roulette graphs as follows. A simple cycle is a roulette, and so is a cycle each of whose vertices is replaced by a roulette (with one or two vertices of the inner roulette taking the place of an original vertex of the cycle). Finally, any graph whose 2-edge connected components are roulettes is itself a roulette.

Now we can present the main structural lemma that reduces our general problem into two tractable subproblems.

Lemma 1. *There exists a polynomial-time algorithm that, given a graph $G(V, E)$, partitions the edge set E into two sets S_1 and S_2 such that graphs $G_1 = G/S_1$ and $G_2 = G/S_2$ are respectively 3-edge connected and roulette.*

Proof. Let S_1 be the set of all edges in G that belong to a cut of size at most 2. We note that all cuts of size at most 2 can be found in polynomial time, e.g., using a naïve brute-force search. Since G_1 is obtained by a series of edge contractions, every cut in G_1 represents a cut in G as well. Therefore, since all edges of S_1 are contracted in G_1, no cut of size at most 2 is present in G_1, hence G_1 is 3-edge connected.

On the other hand, every edge in G_2 belongs to a cut of size at most 2 in G, hence in G_2. The bridges in G_2 do not hurt the roulette structure if the 2-edge connected subgraphs of G_2 are roulettes. Thus we assume that the graph G_2 is 2-edge connected. We claim that if each of $\{e_1, e_2\}$ and $\{e_1, e_3\}$ is a cut in G_2, then so is $\{e_2, e_3\}$. Therefore, the edges of G_2 form an equivalence class. To see this equivalence relation, it suffices to focus on the following alternative definition of edge cuts of size 2. In a 2-edge connected graph, two edges e and e' form a cut of size 2 if and only if the set of cycles in the graph that contain e is the same as the set of cycles in the graph that contain e'. Based on this new definition, the two sets of cycles containing e_2 and e_3 respectively are both equal to the set of cycles containing e_1, and therefore they are equal to each other as well which means that e_2 and e_3 form a cut of size 2.

Therefore, the graph looks like a cycle of this equivalence class (the equivalence class of e_1) where some vertices are replaced by another structure; the same argument applies to each of these smaller structures, giving rise to the inductive definition of roulette graphs. We should note that two edges from two of these smaller structures do not belong to the same equivalence class (they cannot form a cut of size 2), and therefore each of the other equivalence classes belong to one smaller structure and is not split between different structures. Thus we can inductively claim each smaller structure is a roulette graph. $\square$

The following decomposition lemma serves as the starting point for our 3-approximation algorithm of the 3-edge connected graph G_1.

Lemma 2. *There exists a polynomial-time algorithm that finds 3 spanning trees T_1, T_2, and T_3 in a 3-edge connected graph G such that every edge of G is missing in at least one of these spanning trees.*

Proof. We replace each edge of G by two parallel edges to get the 6-edge connected graph G^2 with the same vertex set. Catlin et al. [6] show, among other things, that any $2k$-edge connected graph has k edge-disjoint spanning trees, while Roskind and Tarjan [16] show how to find k edge-disjoint spanning trees in a graph (if they exist) in quadratic time. Therefore, we can find 3 edge-disjoint spanning trees T_1, T_2, and T_3 in G^2. Edge-disjointness guarantees that each edge of G can belong to at most two of these spanning trees. $\square$

On the other hand, the problem can be solved optimally for roulette graphs. The intuition behind the algorithm is that roulettes can be shown to have treewidth of at most two, hence, as are many similar problems on bounded-treewidth graphs, our problem can be solved via the dynamic-programming method.

Lemma 3. *There exists a polynomial-time algorithm that finds the optimal solution for roulette graphs.*

Proof. Let us say we have a *cycle decomposition* of our roulette graph, which describes the recursive structure of its 2-edge connected components. Each cycle representing one equivalence class is called an *essential* cycle of the graph. Instead of solving the rooted problem, we consider a slightly more general problem where up to two vertices s, t on an essential cycle are specified, and these vertices should both appear in the connected subgraph of the output. At the beginning we are going to have only one vertex $s = t$ which is the root vertex.

Let us ignore the bridges at this point and assume the graph is 2-edge connected. Focus on the essential cycle, and imagine that all the recursive structures are contracted for now. In the resulting graph, the optimal solution looks like a path or it is the entire cycle. There are polynomially many cases to consider, and we will output the best solution among them. In each case we have a subproblem for each contracted piece if we decide that the optimal solution passes through or ends at the contracted vertex. Since the base cases of the induction (i.e., vertices or cycles) are easily solvable, we can argue inductively that our sligthly more general problem can be solved using a dynamic program.

For each bridge connected to the essential cycle, we compute the best solution for the rooted problem on the other side of the bridge (whose root is the endpoint of the bridge) and add that to the main solution if its weight plus the weight of the bridge turns out positive.   □

We conclude this section by putting together the above ideas to obtain a 4-approximation algorithm for the rooted edge-weighted problem.

Lemma 4. *There exists a polynomial-time monotone algorithm that achieves an approximation factor of 4 for any graph G.*

Proof. If the graph is not connected, we can focus on the connected component that contains the root and disregard the rest of the connected components. Using Lemma 1, we partition the edges of G into two parts S_1 and S_2. We also use Lemma 2 to find three spanning trees T_1, T_2, and T_3 in graph G_1.

We consider four candidate solutions. One solution is the union of S_1 and the edges in tree T_1. Since T_1 is a spanning tree in $G_1 = G/S_1$, the union of T_1 and S_1 is a connected spanning subgraph of G. Serving this connected spanning subgraph allows us to choose any subset of the remaining edges to serve. Among the remaining edges (edges not in $S_1 \cup T_1$), we serve those with positive realized virtual values. In a similar fashion we construct two other candidate solutions based on T_2 and T_3. The fourth and last candidate solution is to contract set S_2 of edges, and solve the problem optimally in the roulette graph G_2 using Lemma 3. The optimal solution we find in G_2 is a connected subgraph of G_2, however, it might not be a connected subgraph of G that includes the root. Nonetheless, it is always possible to serve a subset of edges $S_2' \subseteq S_2$ to make the whole solution not only a connected subgraph of G but also one that includes the root vertex as well. Our fourth candidate solution consists of the edges in S_2' and the optimal soluion for G_2.

We now show that for every instance (i.e., a graph with a designated root and a distribution) one of these candidate solutions that guarantees a 4-approximate solution. Thus picking one at random will guarantee a 4-approximation. For $1 \leq i \leq 3$, if we stick to solution i all the time, our gain from edges in S_1 and T_i is nonnegative (i.e., the sum of their expected virtual values), and in addition we get all edges with positive virtual value outside $S_1 \cup T_i$. Since each edge of S_2 is missing in at least one $S_1 \cup T_i$, the total value we get from the first three solutions is at least the projection of optimal solution on set S_2 of edges. On the other hand in the fourth solution, our expected revenue from edges added from S_2 is nonnegative (once again since the expectation of the virtual values are positive), and our expected revenue from S_1 is at least the amount that the optimal solution gains from them. Therefore, these four solutions together achieve no less than the optimal revenue. Thus, one candidate solution has expected revenue at least a quarter of the optimum.

The algorithm is monotone because firstly it decides which of the four candidate solutions to use independent of the realized values. Further each candidate solution is monotone. The first three solutions are monotone because the edge selection criteria (for the edges they consider) is just non-negative virtual value. The fourth solution is monotone because it is an optimal algorithm for the edges it focuses on.   □

Vertex-Connectivity for the Rooted Version. In this section we provide a constant-factor monotone approximation algorithm for the vertex-connectivity problem using the algorithm for the edge-connectivity version of the problem described above. Let V_2 be the set of degree-2 vertices in graph G. Similar to the edge-connectivity approach, we describe two algorithms (one using a reduction to the edge-connectivity problem) that achieve constant-factor approximations to the problems where the values of $V \setminus V_2$ and V_2 are replaced by zero, respectively. Again, since the expected value of each vertex is nonnegative, this implies a constant approximation for the vertex-connectivity problem.

First consider the instance in which the values of vertices in $V \setminus V_2$ are replaced by zero. Notice that this results in an instance in which any vertex with nonzero value has degree 2. Construct another instance in which each vertex of degree 2 is replaced by an edge with the same value. We can solve this instance using our edge-connectivity algorithm.

Next consider the instance in which the values of vertices in V_2 are replaced by zero. We convert the instance to one without any degree-2 vertices via replacing all paths consisting only of degree-2 vertices by an edge. The following lemma shows that this graph has a spanning tree where at least $\frac{1}{7}$ of its vertices are leaves. The algorithm uses the *internal* nodes (i.e., nonleaves) of this tree to connect the leaves that are positive, to the root. This gives a 7-approximation to the problem because the vertex values are drawn i.i.d.

Putting these two algorithms together, we obtain a constant-factor approximation algorithm for the vertex-connectivity rooted revenue maximization in expectation. In particular, a balancing argument puts a bound of 11 on its approximation ratio.

Monotonicity of the resulting algorithm follows from the monotonicity of the algorithms used for the two subcases. When we use the algorithm for the edge-connected version, the monotonicity of the edge-connected version implies the same here. For the other case, note that a leaf is picked whenever its virtual value is positive thus resulting in a monotone allocation.

Lemma 5. *Given a graph with no degree-2 vertices, a spanning tree can be constructed in polynomial-time where at least $\frac{1}{7}$ of the vertices are leaves.*

Proof. Start from an arbitrary spanning tree T of G. Let T_2 be the set of vertices of degree 2 in T, and let $\hat{T}_2 \subseteq T_2$ be those vertices in T_2 both whose neighbors, too, are in T_2. Modify T as long as any of the following two rules apply.

1. If there exists a vertex $v \in \hat{T}_2$ that has an edge in G to an internal vertex u of T, update T by adding the edge (v, u) to T, and removing the edge incident to v in the unique cycle formed after adding (v, u). Since both neighbors of v in T had degree 2, this process generates a new leaf without removing any of the old leaves.

2. If two vertices $v, u \in \hat{T}_2$ have edges in G to the same leaf l of T, add edges from v and u to that leaf, and remove two edges from T as follows. The addition of edge (u, l) to T produces a cycle that passes through exactly one of the two neighbors of u; call it u'. Note that u' has degree 2 in T by definition of $\hat{T}_2$; let u, u'' be its neighbors. Remove the edge (u', u'') from T, removing the cycle u and maintaining the connectivity of T. We carry out a similar operation, mutatis mutandis, for v. The result will be a tree T on the same set of vertices with one more leaf (increasing the degree of l but turning two other internal vertices into leaves).

The process terminates in a linear number of iterations since the number of leaves increases in each step. We end up with a tree T for which neither of the rules applies. Let T_1 and $T_{\geq 3}$, respectively, denote subsets of vertices of degrees one and at least three in T. We argue below that $|T_1|$ is at least a constant fraction of $|T_2| + |T_{\geq 3}|$. As no vertex of G has degree two, any vertex in $\hat{T}_2$ is bound to have degree at least three in G, hence an edge not in T. This edge cannot be to an internal vertex of T because Rule (1) no longer applies to T. Rule (2), on the other hand, implies that these leaves are distinct for different vertices of $\hat{T}_2$. Therefore, we have

$$|T_1| \geq |\hat{T}_2|. \tag{1}$$

As trees have average degree less than two, we know

$$|T_1| > |T_{\geq 3}|. \tag{2}$$

To bound $|T_2| - |\hat{T}_2|$, if this quantity is not zero, orient T from an arbitrary vertex in $T_2 \setminus \hat{T}_2$ towards the leaves. Assign each vertex $v \in T_2 \setminus \hat{T}_2$ to its closest descendant in the oriented tree that is in $T_1 \cup T_{\geq 3}$. Such an assignment is always possible since no vertex in the former group is a leaf of the (oriented) tree. Each vertex in the latter group is assigned to at most twice, otherwise there should be a path of vertices of degree two with more than two vertices in $T_2 \setminus \hat{T}_2$—a contradiction. As a result, we get

$$|T_2| - |\hat{T}_2| \leq 2(|T_1| + |T_{\geq 3}|) \leq 4|T_1|, \tag{3}$$

where the last inequality is due to (2).

Summing up (1), (2) and (3) with $|T_1| \geq |T_1|$, we obtain $7|T_1| \geq |T_1| + |T_2| + |T_{\geq 3}|$ as desired.   □

4.3   APX-Hardness of Optimization in Expectation in Arbitrary Graphs

In this section we prove the APX-hardness of the rooted version in arbitrary graphs when the valuations of agents are independent but not necessarily identical.

Definition 1. *The prize-collecting set cover problem (PCSCP) consists of a collection of sets $S_1, S_2, \ldots, S_n$ over a universe U. For a collection C of sets, let $Q_C = \cup_{i \in C} S_i$. The goal is to find a collection C^* that maximizes $\alpha|Q_{C^*}| + n - |C^*|$ for some given $\alpha > 0$.*

We first show that there is an approximation preserving reduction from PCSCP to our problem, and then invoke the result from [10] that establishes the APX-hardness of PCSCP, and that its approximation ratio is at least $\frac{530}{529} = 1.002$.

Lemma 6. *There is an approximation preserving reduction from PCSCP to our problem*

See Appendix B for a proof.

5 Posted-Pricing Results

As mentioned in the introduction, one goal of this work to compare two kinds of mechanisms: a) natural mechanisms for nonexcludable public goods, such as a direct revelation mechanism, that have to deal with nonexcludabilities and b) posted price mechanisms which mitigate nonexcludability to a certain extent by ensuring that only those who pay for an edge can use that edge. We show that for the rooted version with i.i.d. agents, a posted price mechanism obtains the optimal revenue possible when full excludability is allowed — i.e., even partially mitigating nonexcludability via tolls gets us the benefit of full excludability. When the agents values are independent but not necessarily identical, we design a pricing scheme that obtains a $O(\frac{1}{\log n})$ fraction of the optimal revenue possible when full excludability is allowed. The details of the prices set, and the proof, are presented in the full version of the paper.

References

[1] Aggarwal, G., Feldman, J., Muthukrishnan, S., Pál, M.: Sponsored search auctions with markovian users. In: Papadimitriou, C., Zhang, S. (eds.) WINE 2008. LNCS, vol. 5385, pp. 621–628. Springer, Heidelberg (2008)

[2] Akhlaghpour, H., Ghodsi, M., Haghpanah, N., Mirrokni, V.S., Mahini, H., Nikzad, A.: Optimal iterative pricing over social networks (Extended abstract). In: Saberi, A. (ed.) WINE 2010. LNCS, vol. 6484, pp. 415–423. Springer, Heidelberg (2010)

[3] Anari, N., Ehsani, S., Ghodsi, M., Haghpanah, N., Immorlica, N., Mahini, H., Mirrokni, V.S.: Equilibrium pricing with positive externalities (Extended abstract). In: Saberi, A. (ed.) WINE 2010. LNCS, vol. 6484, pp. 424–431. Springer, Heidelberg (2010)

[4] Athey, S., Ellison, G.: Position auctions with consumer search. The Quarterly Journal of Economics 126(3), 1213–1270 (2011)

[5] Candogan, O., Bimpikis, K., Ozdaglar, A.: Optimal pricing in the presence of local network effects. In: Saberi, A. (ed.) WINE 2010. LNCS, vol. 6484, pp. 118–132. Springer, Heidelberg (2010)

[6] Catlin, P.A., Lai, H.-J., Shao, Y.: Edge-connectivity and edge-disjoint spanning trees. Discrete Mathematics 309(5), 1033–1040 (2009)

[7] Deng, C., Pekec, S.: Money for nothing: exploiting negative externalities. In: Proceedings of the ACM Conference on Electronic Commerce (EC), pp. 361–370 (2011)

[8] Giotis, I., Karlin, A.R.: On the equilibria and efficiency of the GSP mechanism in keyword auctions with externalities. In: Papadimitriou, C., Zhang, S. (eds.) WINE 2008. LNCS, vol. 5385, pp. 629–638. Springer, Heidelberg (2008)

[9] Gomes, R., Immorlica, N., Markakis, E.: Externalities in keyword auctions: An empirical and theoretical assessment. In: Leonardi, S. (ed.) WINE 2009. LNCS, vol. 5929, pp. 172–183. Springer, Heidelberg (2009)

[10] Haghpanah, N., Immorlica, N., Mirrokni, V., Munagala, K.: Optimal auctions with positive network externalities. In: Proceedings of the 12th ACM Conference on Electronic Commerce, EC 2011, pp. 11–20 (2011)

[11] Hartline, J., Mirrokni, V., Sundararajan, M.: Optimal marketing strategies over social networks. In: Proceedings of the 17th International Conference on World Wide Web, pp. 189–198. ACM (2008)

[12] Jeziorski, P., Segal, I., et al.: What makes them click: Empirical analysis of consumer demand for search advertising. SSRN eLibrary 33 (2009)

[13] Kempe, D., Mahdian, M.: A cascade model for externalities in sponsored search. In: Papadimitriou, C., Zhang, S. (eds.) WINE 2008. LNCS, vol. 5385, pp. 585–596. Springer, Heidelberg (2008)
[14] Myerson, R.B.: Optimal Auction Design. Mathematics of Operations Research 6(1), 58–73 (1981)
[15] Rohlfs, J.: A theory of interdependent demand for a communications service. The Bell Journal of Economics and Management Science, 16–37 (1974)
[16] Roskind, J., Tarjan, R.E.: A note on finding minimum-cost edge-disjoint spanning trees. Mathematics of Operations Research 10(4), 701–708 (1985)

A   The Unrestricted Version

A.1   Pointwise Optimization in Paths

In this section, we show that in paths, a polynomial-time dynamic program is all that is necessary to implement the pointwise optimization involved in implementing Myerson's mechanism.

Consider a path with vertices 1 to n. For $1 \leq i \leq j \leq n$, let $S(i, j)$ be the sum of virtual values of the set of all agents (i.e., vertex pairs) whose both endpoints are in the interval $[i, j]$. We set $S(i, i) = 0$ for all i—assuming that there is no trivial agent whose vertex pairs are the same. Define $OPT(i)$ to be optimal solution when we are restricted to choose only among edges connecting vertices 1 through i. The unrestricted optimal solution OPT is therefore $OPT(n)$. We set $OPT(0) = 0$. The following recursive formula can be used in the dynamic program to solve for OPT.

$$\forall i \leq n, OPT(i) = \max_{0 \leq k < i} OPT(k) + S(k + 1, i).$$

In the above recursion, $k + 1$ is the leftmost vertex that is connected to i. All vertices $k + 1, \ldots, i$ are connected, and $S(k + 1, i)$ is by definition their contribution to the objective. Since the edge connecting k to $k + 1$ is not included, the set of edges chosen among the first k vertices must be equal to $OPT(k)$.

B   The Rooted Version

B.1   Proof of Approximation Preserving Reduction from PCSCP

Proof of Lemma 6. Given an instance of the PCSCP, where the sets are denoted $S_1, S_2, \ldots S_n$ and the elements are denoted $e_1, e_2, \ldots e_m$, we construct an instance of our problem as follows.

Vertices. We start with a root vertex r. For every element e_i, we construct one vertex with the same name e_i. For every set S_i, we construct one vertex denoted by S_i as well.

Edges. For each i, set vertex S_i is connected by an edge to root r and all $e_j \in S_i$.

Agents. The agents (r, S_i) are called set-agents, and (r, e_i) are called element-agents.

Distribution. The value distribution of element agents is deterministic α. For set agents, the value is drawn from the distribution Bernoulli($L - 1, 1/L$)—i.e., the value is equal to $L - 1$ with probability $1/L$, and zero otherwise—where $L \gg mn\alpha$. Thus, the virtual value for these agents is -1 w.p. $1 - 1/L$ and $L - 1$ w.p. $1/L$.

The optimal revenue in our problem, as $L \to \infty$, can be analyzed in two cases.

1. If at least one set-agent has positive virtual valuation (which happens w.p. approximately $n/L \to 0$), the solution chooses all the edges incident on those set agents (with positive virtual valuation) to obtain expected revenue n. The expected revenue from the remaining agents (set-agents with negative virtual valuation, and element-agents) is at most $\alpha mn/L \ll 1$. Therefore, the optimal solution has contribution n from this event as $L \to \infty$, and this solution is trivial to compute.

2. If no set has positive virtual valuation (which happens w.p. $1 - n/L \to 1$), the value of the solution is precisely $\alpha|Q_{C^*}| - |C^*|$. This is because once a set-agent is chosen, clearly all the edge-agents that is contained by this set must be chosen since they all have deterministic positive virtual value α. We should also note that you can obtain the α virtual value of an element agents only if you connect it to root using one of the set agents it belongs to, and for each set you pick you have -1 virtual value in this case.

Therefore the value of the optimal solution is $\alpha|Q_{C^*}| + n - C^*$.   □

On Lookahead Équilibria
in Congestion Games*

Vittorio Bilò[1], Angelo Fanelli[2], and Luca Moscardelli[3]

[1] University of Salento, Italy
vittorio.bilo@unisalento.it
[2] CNRS, (UMR-6211), France
angelo.fanelli@gmail.com
[3] University of Chieti-Pescara, Italy
luca.moscardelli@unich.it

Abstract. We investigate the issues of existence and efficiency of lookahead equilibria in congestion games. Lookahead equilibria, whose study has been initiated by Mirrokni et al. [10], correspond to the natural extension of pure Nash equilibria in which the players, when making use of global information in order to predict subsequent reactions of the other ones, have computationally limited capabilities.

1 Introduction

The definition of the process of interaction among self-interested entities is dependent on the context, and in particular on the set of information available to the players. When they have very little knowledge about each others' costs and strategies, one of the most natural and studied dynamics are *sequential best-responses*, where players play sequentially and each player selects a strategy which is a best-response to the current strategy of the others. In such dynamics, the assumption is that each player has no memory about the past and no knowledge about the available strategies and costs of other players and, thus, myopically responds to the current state, without making any prediction about the consequences of the subsequent responses of the remaining players. One of the basic objective of study of game theory is the concept of equilibrium. An equilibrium can be viewed as a steady state of a dynamics, where no agent has an incentive to unilaterally deviate from. The steady state of a best-response dynamics is known as *pure Nash equilibrium*. It is well known that best-response dynamics do not always lead to a pure Nash equilibrium and that the class of congestion games [13] is a large class of games guaranteeing convergence under best-responses.

In our work, we focus on the settings in which each player has full knowledge of the strategies and costs of the other players, so that, based on such a knowledge,

---

* This work was partially supported by the PRIN 2010–2011 research project ARS TechnoMedia: "Algorithmics for Social Technological Networks" funded by the Italian Ministry of University.

Y. Chen and N. Immorlica (Eds.): WINE 2013, LNCS 8289, pp. 54–67, 2013.

she can make predictions about the others' reactions to her move. We also assume that each player is an entity with limited computational abilities, thus she has the ability of making predictions only on the consequences of a fixed constant number of subsequent consecutive moves. In particular, we study the k-*lookahead dynamics* in which the players sequentially perform k-*lookahead best-responses*. When $k = 1$, the k-lookahead best-response coincides with the best-response. In general, for $k > 1$, the current moving player p evaluates all the possible outcomes resulting from $k - 1$ subsequent moves, by taking into account all the possible orders in which players move and all of their possible strategies. We say that player p has a long-sightedness of k and she makes a prediction by assuming that any player moving $j < k$ steps after her has a long-sightedness of $k - j$. Thus, player p can compute her best move by backward induction starting from the players having long-sightedness of 1, and proceeding backward up to k. When predicting the strategy chosen by any player q having long sightedness $k - j$, it is necessary to make some assumption on which is the next moving player. We take into account two different models: the worst-case and the average-case ones. In the *worst-case model*, player p assumes that the next move after q is performed by a player providing player q the worst possible cost in the final outcome. In the *average-case model*, player p assumes that the next move after q is taken by a player selected uniformly at random. For each of these models, we finally distinguish between the cases of *consecutive* and *non-consecutive moves*, depending on whether player p assumes that the next move after q may be performed by q itself or not.

In our work, we investigate the existence of k-lookahead equilibria and the price of anarchy of 2-lookahead equilibria in congestion games with linear latencies [13]. Congestion games model the settings in which a set of players compete for the usage of a set of common resources. We choose congestion games as representative of a large set of well studied games for which the existence of pure Nash equilibria is always guaranteed.

Results. In Section 3, we discuss our results on the existence of equilibria. We initially focus our attention to the existence of k-lookahead equilibria in strategic games. We are able to show that, in the worst-case model with consecutive moves, for a strategic game, any pure Nash equilibrium is also a k-lookahead equilibrium. This result implicitly shows that the k-lookahead best-responses do not guarantee better performance at equilibrium compared to those achieved by the simple best-responses. In the remainder of Section 3, we focus on the existence of 2-lookahead equilibria in singleton congestion games. We show that in the worst-case model without consecutive moves, any symmetric singleton game always admits 2-lookahead equilibria. For the average-case model, instead, we show that symmetric singleton congestion games do not always admit a 2-lookahead equilibrium regardless of whether consecutive moves are allowed or not.

In Section 4, we present the bounds on the price of anarchy for the 2-lookahead equilibria of linear congestion games, both in the worst-case and in the average-case model. We first show that, in the worst-case model, for any linear congestion game, the price of anarchy is at most 8. For the average-case model, we obtain smaller bounds. In particular, we show that, for any linear congestion game with n players, the price of anarchy is at most $4 + \min\left\{2, \frac{5}{4n-7}\right\}$. This result significantly improves the previous upper bound of $(1 + \sqrt{5})^2 \approx 10.47$ given in [10]. All the bounds mentioned hold either with or without consecutive moves. We also show that, when restricting to singleton strategies, the price of anarchy drops to at most 4 in the worst-case model with or without consecutive moves.

Related Works. The lookahead search was formally proposed by Shannon [14], as a practical heuristic for machines to tackle difficult problems and play games. It is not surprising that Shannon applied the method to chess. More recently, the lookahead search has also been presented by Peral in his book [12] as the most important heuristic used by game-playing programs. Mirrokni et al. [10] initiated the theoretical examination of the consequences of the decision making determined by the use of lookahead search. The authors formally quantify the deterioration of the outcome when players use lookahead search, by bounding the price of anarchy for several games among which are congestion games.

Our work is also related to many papers on congestion games. Congestion games have been introduced by Rosenthal [13] and have been proved to be the only class of games admitting an exact potential function by Monderer and Shapley [11]. There is a long series of works investigating the price of anarchy with respect to the pure Nash equilibria (e.g., [1,3,4,7]), and studying the best-response and approximate improvement dynamics (e.g., [2,5,6,8,9]) for congestion games.

2   Model and Preliminaries

A *congestion game* $\mathcal{G} = (N, E, (\Sigma_i)_{i \in N}, (f_e)_{e \in E}, (c_i)_{i \in N})$ is a non-cooperative strategic game defined by a set E of resources and a set $N = \{1, \ldots, n\}$ of players sharing resources in E.

Any strategy $s_i \in \Sigma_i$ of player i is a non-empty subset of resources, i.e. $\emptyset \neq \Sigma_i \subseteq 2^E$. Given a strategy profile $S = (s_1, \ldots, s_n)$ and a resource e, the number of players using e in S, called the congestion on e, is denoted by $n_e(S) = |\{i \in N : e \in s_i\}|$. A delay function $f_e : \mathbb{N} \mapsto \mathbb{R}_+$ associates to resource e a delay depending on the number of players currently using e, so that the cost of player i for the pure strategy s_i is given by the sum of the delays associated with resources in s_i, i.e. $c_i(S) = \sum_{e \in s_i} f_e(n_e(S))$. We refer to *singleton* congestion games as the games in which all of the players' strategies consist of only a single resource.

In this paper we will focus on *linear* congestion games, that is having linear delay functions with nonnegative coefficients. More precisely, for every resource $e \in E$, $f_e(x) = \alpha_e x + \beta_e$ with $\alpha_e, \beta_e \geq 0$.

Given the strategy profile $S = (s_1, \ldots, s_n)$, the social cost $C(S)$ of S is defined as the sum of all the players' costs, i.e. $C(S) = \sum_{i \in N} c_i(S) = \sum_{e \in E} \left(\alpha_e n_e(S)^2 + \beta_e n_e(S) \right)$. An optimal strategy profile $S^* = (s_1^*, \ldots, s_n^*)$ is one with minimum social cost.

Before introducing the notions of k-lookahead best-response and k-lookahead equilibrium, we briefly define their classical correspondent notions of best-response and Nash equilibrium.

Each player acts selfishly and aims at choosing the strategy lowering her cost. Given a strategy profile S and a strategy $s'_i \in \Sigma_i$, denote with $S \oplus_i s'_i = (s_1, \ldots, s_{i-1}, s'_i, s_{i+1}, \ldots, s_n)$ the strategy profile obtained from S if player i changes her strategy from s_i to s'_i. A *best-response* of player i in S is a strategy $s^b_i \in \Sigma_i$ yielding the minimum possible cost, given the strategic choices of the other players, i.e. $c_i(S \oplus_i s^b_i) \leq c_i(S \oplus_i s'_i)$ for any other strategy $s'_i \in \Sigma_i$.

A (pure) *Nash equilibrium* is a strategy profile in which every player plays a best-response. Given a strategic game $\mathcal{G}$, we denote as $\mathcal{NE}(\mathcal{G})$ the set of its pure Nash equilibria.

We assume that each player, in order to determine her k-lookahead best-response, exploits k-lookahead search, i.e., she predicts $k-1$ consecutive possible re-actions to her move, and selects the best choice according to such a prediction, as shown in the following. More formally, when performing a move starting from a given strategy profile S, player i considers a directed tree game $\mathcal{T} = (V_{\mathcal{T}}^{odd} \cup V_{\mathcal{T}}^{even}, A_{\mathcal{T}}^{odd} \cup A_{\mathcal{T}}^{even})$ of depth $2k-1$ in which odd levels (with the root belonging to level 1) contain *player nodes* belonging to $V_{\mathcal{T}}^{odd}$ and even levels contain *selection nodes* belonging to $V_{\mathcal{T}}^{even}$. Arcs outgoing from nodes in $V_{\mathcal{T}}^{odd}$ ($V_{\mathcal{T}}^{even}$, respectively) belong to $A_{\mathcal{T}}^{odd}$ ($A_{\mathcal{T}}^{even}$, respectively). Each node $v \in V_{\mathcal{T}}^{odd}$ is associated to a player $p(v)$ performing an action, with the root being associated to player i, and each arc a outgoing from node v is associated to her strategy $st(a) \in \Sigma_{p(v)}$; there is an outgoing arc for each strategy of player $p(v)$. Each selection node $v \in V_{\mathcal{T}}^{even}$ is associated to a strategy profile S_v that is obtained in the following way: Initially, S_v is set equal to S. Now, consider the path connecting the root of $\mathcal{T}$ to v; starting from the root, for every arc (u, u') of such a path belonging to $A_{\mathcal{T}}^{odd}$, S_v is updated to $S_v \oplus_{p(u)} st((u, u'))$. In this paper we consider two different settings, depending on whether consecutive moves by a same player are allowed or not in the search tree. In the setting *allowing consecutive moves* by the same player, each selection node has n outgoing arcs, one for each player; in the setting in which they are not allowed, each selection node has $n-1$ outgoing arcs (in this case, the arc (u, u') with $p(u) = p(u')$ is missing).

We assume that, in the k-lookahead search of player i, a player corresponding to a node of level $2j-1$ in $\mathcal{T}$ (for $j = 1, \ldots, k$), has a long-sightedness equal to $k+1-j$ (player i performs a k-lookahead search, the player moving after her a $(k-1)$-lookahead search and so on).

A k-lookahead best-response can be computed by backward induction on the levels of tree $\mathcal{T}$. First of all, it can be computed under two different models:

- The *worst case* model, in which each player assumes that the subsequent move is performed by a player providing her the worst possible cost in the final leaf of tree $\mathcal{T}$.
- The *average case* model, in which the player moving at each step is assumed to be selected uniformly at random.

Notice that for both models we can consider the two settings in which consecutive moves are or are not allowed.

The basis of the induction is the selection of an arc (marked as red) for each node of the last level of $\mathcal{T}$ (being the odd level $2k-1$): for each node v of this last level, the base case reduces to the selection of a 1-lookahead best-response for player $p(v)$ (i.e., a classical best-response to strategy profile S_v); ties are resolved such that player i's cost in the final strategy profile is maximized.

For each $j \geq 1$, given that some outgoing arcs for levels $j+2, \ldots, 2k-1$ have been marked as red, we now show how to mark as red an outgoing arc for each node of level j (being an odd level) and, only in the worst case model, how to mark as red one arc of level $j+1$ (being an even level). In fact, in the average case model, all arcs of level $j+1$ are always marked as red. Given any node v of the odd levels in $\{j+2, \ldots, 2k-1\}$, let $Lf(v)$ be the (maximal) set of leaves of $\mathcal{T}$ such that there exists a path of red arcs going from v to a node in $Lf(v)$. Note that for any node v, $|Lf(v)| = 1$ under the worst case model. Under the worst case model, a $\left(k - \frac{j-1}{2}\right)$-lookahead best-response for player $p(v)$ (with v being a node of level j) is performed by marking as red an arc (v, v') outgoing from v such that the value $c_{p(v)}(S_u)$, with $u \in Lf(v)$ (notice that, under the worst case model, $|Lf(v)| = 1$), is minimized taking into account that the worst case (for player $p(v)$) arc outgoing from v' is also marked as red; ties are resolved such that player i's cost in the final strategy profile is maximized. Under the average model, a $\left(k - \frac{j-1}{2}\right)$-lookahead best-response for player $p(v)$ (with v being a node of level j) is performed by marking as red an arc (v, v') outgoing from v such that the average among values $c_{p(v)}(S_u)$ over all $u \in Lf(v)$ is minimized; again, ties are resolved such that player i's cost in the final strategy profile is maximized.

Suppose that, in a k-lookahead best-response dynamics, player j moves after player i. It is worth noticing that the move performed by j may not be the move anticipated by player i in her own analysis (at the corresponding node of level 3 of $\mathcal{T}$), because in such an analysis of player i, player j was performing a $(k-1)$-lookahead search, while when moving after player i in the "actual" evolution of the game, she is performing a k-lookahead search.

A k-lookahead equilibrium, under the worst or average case model and with or without consecutive moves allowed, is a strategy profile in which every player plays a k-lookahead best-response (under the same setting). Notice that a 1-lookahead best-response corresponds to the classical best-response, and a 1-lookahead equilibrium to a Nash equilibrium.

The k-lookahead price of anarchy of a game $\mathcal{G}$, under the worst or average case model and with or without consecutive moves allowed, is the worst case ratio between the social cost of a k-lookahead Nash equilibrium (under the same setting) and that of an optimal strategy profile, that is, $\mathsf{PoA}(\mathcal{G}) = \max_{S \in \mathcal{LE}_k(\mathcal{G})} \frac{C(S)}{C(S^*)}$, where $\mathcal{LE}_k(\mathcal{G})$ denotes the set of k-lookahead Nash equilibria of $\mathcal{G}$.

3 Existence of Lookahead Equilibria

We show that, in the worst-case model with consecutive moves, the set of pure Nash equilibria of $\mathcal{G}$ is contained in the set of k-lookahead equilibria of $\mathcal{G}$ for any value of k. This result has a double implication when considering the worst-case model with consecutive moves: from one hand, it shows existence of lookahead equilibria in each game admitting pure Nash equilibria and, from the other hand, it tells us that the price of anarchy can only worsen when moving from the classical definition of myopic rationality to the one based on lookahead search.

Theorem 1. *For any strategic game $\mathcal{G}$ and for any index $k \geq 2$, it holds $\mathcal{NE}(\mathcal{G}) \subseteq \mathcal{LE}_k(\mathcal{G})$ in the worst-case model with consecutive moves.*

Proof. First of all, note that, if $\mathcal{G}$ does not possess pure Nash equilibria, then, by definition, $\emptyset = \mathcal{NE}(\mathcal{G}) \subseteq \mathcal{LE}_k(\mathcal{G})$ for any index $k \geq 2$ and we are done. Hence, for the remaining on the proof, assume that $\mathcal{NE}(\mathcal{G}) \neq \emptyset$. The proof is by induction on $k \geq 1$. Note that, the basic case of $k = 1$ holds by definition since the set of 1-lookahead equilibria coincides with that of pure Nash equilibria. Hence, we only need to show the inductive step.

For any index $k \geq 2$ assume, for the sake of induction, that $\mathcal{NE}(\mathcal{G}) \subseteq \mathcal{LE}_j(\mathcal{G})$ for each index j such that $1 \leq j \leq k - 1$. Consider a pure Nash equilibrium $S \in \mathcal{NE}(\mathcal{G})$ and a player i. If i does not change her strategy, then, since S is a $(k-1)$-lookahead Nash equilibrium for $\mathcal{G}$, no player possesses a $(k-1)$-lookahead improving deviation in S and so, the resulting state of i's search tree is S, where i pays $c_i(S)$. If i changes her strategy to s_i', let $S' = S \oplus_i s_i'$ be the resulting state. It holds $c_i(S') \geq c_i(S)$ since S is a pure Nash equilibrium for $\mathcal{G}$. Note that, if the adversary always selects i for the successive $k-1$ moves, the game can never reach a state in which i pays less that $c_i(S)$ (if such a deviation existed, it would contradict the fact that S is a pure Nash equilibrium for $\mathcal{G}$). It follows that, after player i's deviation, the adversary can always select a sequence of player so as to generate a final state S'' such that $c_i(S'') \geq c_i(S)$. Hence, i does not possess any k-lookahead improving deviation from S and the claim is proved. $\square$

For the worst-case model without consecutive moves, we show existence of 2-lookahead Nash equilibria in symmetric singleton congestion games, that is, singleton games in which all players share the same set of strategies.

Theorem 2. *Any symmetric singleton congestion game always admits 2-lookahead Nash equilibria in the worst-case model without consecutive moves.*

Proof. Fix a symmetric singleton game $\mathcal{G}$ and consider the following two cases.

Case 1. $\mathcal{G}$ admits a pure Nash equilibrium S such that there exists two resources with a congestion of at least 2. Consider a player i, using a resource e, whose cost is $c_i(S)$. Since S is a pure Nash equilibrium for $\mathcal{G}$, if i migrates to another resource e', she gets a cost of at least $c_i(S)$. If the adversary selects a player j currently using a resource different than e', the current cost of player i cannot decrease. Since player j always exists under our hypothesis, S has to be a 2-lookahead Nash equilibrium.

Case 2. $\mathcal{G}$ admits a pure Nash equilibrium S such that there exists three re-sources with a congestion of at least 1. With a similar argument as in the previous case, it is possible to show that S has to be a 2-lookahead Nash equilibrium.

If both Cases 1 and 2 do not occur, then, there exists a pure Nash equilibrium S of $\mathcal{G}$ in which there are two resources e and e' with $n_e(S) = 1$ and $n_{e'}(S) = n-1$, $f_e(2) > f_{e'}(n-1)$ and $f_{e''}(1) > f_{e'}(n-1)$ for each $e'' \in E\backslash\{e, e'\}$. Note also that, since S is a pure Nash equilibrium, $f_e(1) \le f_{e''}(1)$ for each $e'' \in E\backslash\{e, e'\}$.

Consider the strategy profile S' such that $n_{e'}(S') = n$. We claim that either S or S' is a 2-lookahead Nash equilibrium.

If S' is a pure Nash equilibrium for $\mathcal{G}$, then it is also a 2-lookahead Nash equilibrium. Hence, we can assume that $f_e(1) < f_{e'}(n)$. Consider any player: If she does not change her strategy, then, no matter which is the other player selected by the adversary, she ends up paying $f_{e'}(n - 1)$. If she changes her strategy, then, no matter which is the other player selected by the adversary, she ends up paying at least $f_e(1)$. Thus, S' is a 2-lookahead Nash equilibrium when $f_{e'}(n - 1) \le f_e(1)$.

On the other hand, since S is a pure Nash equilibrium, player i using resource e in S, ends up paying $f_e(1)$ when not changing her strategy, while any player j using resource e' in S ends up paying $f_{e'}(n - 1)$ when not changing her strategy. If player i changes her strategy, no matter which is the other player selected by the adversary, she ends up paying at least $\min\{f_e(1), f_{e'}(n-1)\}$. If any player j changes her strategy, she ends up paying at least $\min\{f_e(2), f_{e''}(1)\}$. Thus, S is a 2-lookahead Nash equilibrium when $f_{e'}(n - 1) \ge f_e(1)$ and this concludes the proof.   □

For the average-case model, we show that there exists a very simple game $\mathcal{G}$ with 4 symmetric players and 3 singleton strategies admitting no 2-lookahead Nash equilibria independently of whether consecutive moves are allowed or not.

Theorem 3. *In both variants of the average-case model, no 2-lookahead Nash equilibria are guaranteed to exist even in symmetric singleton games.*

4   Bounds on the Price of Anarchy

In this section, we give upper bounds on the price of anarchy of 2-lookahead Nash equilibria of linear congestion games both in the worst-case model and in the average-case model either with or without consecutive moves. To this aim, we use the primal-dual method introduced in [4]. Denoted with $K = (k_1, \ldots, k_n)$ and $O = (o_1, \ldots, o_n)$ the worst 2-lookahead Nash equilibrium and the social optimum, respectively, this method aims at formulating the problem of maximizing the ratio $\frac{C(K)}{C(O)}$ via linear programming. The two strategy profiles K and O play the role of fixed constants, while, for each $e \in E$, the values α_e and β_e defining the delay functions are variables that must be suitably chosen so as to satisfy two constraints: the first, assures that K is a 2-lookahead Nash equilibrium, while the second normalizes to 1 the value of the social optimum $C(O)$.

The objective function aims at maximizing the social value $C(K)$ which, being the social optimum normalized to 1, is equivalent to maximizing the ratio $\frac{C(K)}{C(O)}$. Let us denote with $LP(K, O)$ such a linear program. By the Weak Duality Theorem, each feasible solution to the dual program of $LP(K, O)$ provides an upper bound on the optimal solution of $LP(K, O)$. Hence, by providing a feasible dual solution, we obtain an upper bound on the ratio $\frac{C(K)}{C(O)}$. Anyway, if the provided dual solution is independent on the particular choice of K and O, we obtain an upper bound on the ratio $\frac{C(K)}{C(O)}$ for any possible pair of profiles K and O, which means that we obtain an upper bound on the price of anarchy of 2-lookahead Nash equilibria.

For the sake of brevity, throughout this section, for each $e \in E$, we set $K_e := n_e(K)$ and $O_e := n_e(O)$. Moreover, note that a simplificative argument widely exploited in the literature of linear congestion games states that we do not lose in generality by assuming $\beta_e = 0$ for each $e \in E$ (as long as we are not interested in singleton strategies). Finally, we denote by $c'_i(S, t)$ the cost that player i foresees in her search tree when selecting, in state S, strategy t.

4.1   Worst-Case Model

For the worst-case model without consecutive moves, for any player $i \in N$, strategy profile K and strategy $t \in \Sigma_i$, it holds $c'_i(K, k_i) \geq \sum_{e \in k_i} (\alpha_e K_e) - \sum_{e \in k_i : K_e \geq 2} \alpha_e$ and $c'_i(K, t) \leq \sum_{e \in t} (\alpha_e(K_e + 2))$. In fact, with 2-lookahead best-responses, when selecting strategy k_i, player i has to suffer, for every used resource e for which $K_e \geq 2$, a congestion at least equal to $K_e - 1$, where the decrease of one unit is due to the possibility that the player performing the next move could leave resource e; moreover, when selecting any strategy t, player i can suffer for every used resource e, a congestion at most equal to $K_e + 2$, where the increase of 2 units is due to the fact that player i is selecting e and also the player moving after her could select e.

For the case with consecutive moves, the same inequalities apply as well, since the fact that the adversary can also select again player i can only increase the cost $c'_i(K, k_i)$, whereas the value $\sum_{e \in t} (\alpha_e(K_e + 2))$ is already the maximum possible one that can be suffered by a migrating player in any model of 2-lookahead rationality.

Hence, since K is a 2-lookahead Nash equilibrium, for each player $i \in N$, it holds

$$\sum_{e \in k_i} (\alpha_e K_e) - \sum_{e \in k_i : K_e \geq 2} \alpha_e \leq \sum_{e \in o_i} (\alpha_e(K_e + 2)). \tag{1}$$

Such an inequality was already exploited in [10] in order to study the price of anarchy in the average-case model without consecutive moves. Anyway, as we will see later, in this case a more significant inequality can be derived. When embedded into the primal-dual technique, inequality (1) gives life to the following primal formulation $LP(K, O)$.

$$maximize \sum_{e \in E} \left(\alpha_e K_e^2 \right)$$

$$subject\ to$$

$$\sum_{e \in k_i} \left(\alpha_e K_e \right) - \sum_{e \in k_i : K_e \geq 2} \alpha_e - \sum_{e \in o_i} \left(\alpha_e (K_e + 2) \right) \leq 0, \quad \forall i \in N$$

$$\sum_{e \in E} \left(\alpha_e O_e^2 \right) = 1,$$

$$\alpha_e \geq 0, \qquad\qquad\qquad \forall e \in E$$

The dual program $DLP(K,O)$ is

$$minimize\ \gamma$$

$$subject\ to$$

$$\sum_{i : e \in k_i} \left(x_i (K_e - 1) \right) - \sum_{i : e \in o_i} \left(x_i (K_e + 2) \right) + \gamma O_e^2 \geq K_e^2, \quad \forall e \in E : K_e \geq 2$$

$$\sum_{i : e \in k_i} \left(x_i K_e \right) - \sum_{i : e \in o_i} \left(x_i (K_e + 2) \right) + \gamma O_e^2 \geq K_e^2, \qquad \forall e \in E : K_e < 2$$

$$x_i \geq 0, \qquad\qquad\qquad \forall i \in N$$

Theorem 4. *For any linear congestion game* $\mathcal{G}$, *it holds* $\mathsf{PoA}(\mathcal{G}) \leq 8$ *in the worst-case model.*

Proof. We show the claim by proving that the dual solution such that $x_i = 2$ for each $i \in N$ and $\gamma = 8$ is feasible.

The first dual constraint becomes $f_1(K_e, O_e) \geq 0$ with $f_1(K_e, O_e) := K_e^2 - 2K_e(O_e + 1) + 4O_e(2O_e - 1)$. It holds $f_1(K_e, 0) = K_e^2 - 2K_e$ which implies $f_1(K_e, 0) \geq 0$ for any $K_e \geq 2$. For $O_e \geq 1$, note that the discriminant of the equation $f_1(K_e, O_e) = 0$, when solved for K_e, is $1 + 6O_e - 7O_e^2$ which is always non-positive when $O_e \geq 1$. This implies that $f_1(K_e, O_e) \geq 0$ for each pair of real numbers (K_e, O_e) with $O_e \geq 1$. Hence, it follows that the first dual constraint is always verified for any pair of non-negative integers (K_e, O_e) with $K_e \geq 2$.

The second dual constraint becomes $f_2(K_e, O_e) \geq 0$ with $f_2(K_e, O_e) := K_e^2 - 2K_e O_e + 4O_e(2O_e - 1)$. Note that the discriminant of the equation $f_2(K_e, O_e) = 0$, when solved for K_e, is $4O_e - 7O_e^2$ which is always non-positive when $O_e \geq 0$. This implies that the second dual constraint is always verified for any pair of non-negative reals (K_e, O_e). $\qquad\square$

4.2   Average-Case Model

For the average-case model without consecutive moves, for any player $i \in N$, strategy profile K and strategy $t \in \Sigma_i$, it holds $c_i'(K, k_i) \geq \sum_{e \in k_i} \left(\alpha_e K_e \right) - \sum_{e \in k_i : K_e \geq 2} \frac{\alpha_e (K_e - 1)}{n - 1}$ and $c_i'(K, t) \leq \sum_{e \in t} \left(\alpha_e \left(K_e + 2 - \frac{K_e - 1}{n - 1} \right) \right)$.

In fact, with 2-lookahead best-responses, when selecting strategy k_i, player i has to suffer, for every used resource e for which $K_e \geq 2$, a congestion at least equal to $K_e - 1$, where the decrease of one unit is due to the event, having probability at most $\frac{K_e - 1}{n-1}$, that the player performing the next move leave resource e (because such a player has to belong to the set of players selecting e in K); moreover, when selecting any strategy t, player i can suffer for every used resource e, a congestion at most equal to $K_e + 2$, where the increase of one unit is due to the fact that player i is selecting e and the increase of another unit is due to the event, having probability at most $1 - \frac{K_e - 1}{n-1}$, that also the player moving after i selects e (because such a player has not to belong to the set of players selecting e in K).

Hence, since K is a 2-lookahead Nash equilibrium, for each player $i \in N$, it holds

$$\sum_{e \in k_i} (\alpha_e K_e) - \sum_{e \in k_i : K_e \geq 2} \frac{\alpha_e (K_e - 1)}{n - 1} \leq \sum_{e \in o_i} \left(\alpha_e \left(K_e + 2 - \frac{K_e - 1}{n - 1} \right) \right). \quad (2)$$

When embedded into the primal-dual technique, inequality (2) gives life to the following primal formulation $LP(K, O)$.

$$maximize \sum_{e \in E} (\alpha_e K_e^2)$$

$$subject\ to$$

$$\sum_{e \in k_i} (\alpha_e K_e) - \sum_{e \in k_i : K_e \geq 2} \frac{\alpha_e (K_e - 1)}{n - 1}$$
$$- \sum_{e \in o_i} \left(\alpha_e \left(K_e + 2 - \frac{K_e - 1}{n - 1} \right) \right) \leq 0, \quad \forall i \in N$$
$$\sum_{e \in E} (\alpha_e O_e^2) = 1,$$
$$\alpha_e \geq 0, \qquad\qquad\qquad\qquad \forall e \in E$$

The dual program $DLP(K, O)$ is

$$minimize\ \gamma$$
$$subject\ to$$
$$\sum_{i : e \in k_i} \left(x_i \left(K_e - \frac{K_e - 1}{n - 1} \right) \right)$$
$$- \sum_{i : e \in o_i} \left(x_i \left(K_e + 2 - \frac{K_e - 1}{n - 1} \right) \right) + \gamma O_e^2 \geq K_e^2, \qquad \forall e \in E : K_e \geq 2$$
$$\sum_{i : e \in k_i} (x_i K_e) - \sum_{i : e \in o_i} \left(x_i \left(K_e + 2 - \frac{K_e - 1}{n - 1} \right) \right) + \gamma O_e^2 \geq K_e^2, \quad \forall e \in E : K_e < 2$$
$$x_i \geq 0, \qquad\qquad\qquad\qquad\qquad\qquad\qquad \forall i \in N$$

The following result significantly improves the previous upper bound of $(1 + \sqrt{5})^2 \approx 10.47$ given in [10].

Theorem 5. *For any linear congestion game $\mathcal{G}$, it holds $\mathsf{PoA}(\mathcal{G}) \leq 6$ if $n = 2$ and $\mathsf{PoA}(\mathcal{G}) \leq 4 + \frac{5}{4n-7}$ if $n \geq 3$ in the average-case model without consecutive moves.*

Proof. For $n = 2$, we show that the dual solution such that $x_i = 2$ for each $i \in N$ and $\gamma = 6$ is feasible. The first dual constraint, since $n = 2$ implies $K_e = 2$, becomes $O_e(O_e - 1) \geq 0$ which is always satisfied for any integer value O_e. The second constraint becomes $K_e^2 + 6O_e(O_e - 1) \geq 0$ which is always satisfied for any integer value O_e when $K_e \in \{0, 1\}$.

For $n \geq 3$, we show the claim by proving that the dual solution such that $x_i = \frac{5(n-1)}{4n-7}$ for each $i \in N$ and $\gamma = 4 + \frac{5}{4n-7}$ is feasible.

The first dual constraint becomes $f_1(K_e, O_e) \geq 0$ with $f_1(K_e, O_e) := K_e^2(n - 3) - 5K_e(nO_e - 2O_e - 1) + O_e(16nO_e - 10n - 23O_e + 5)$. It holds $f_1(K_e, 0) = K_e^2(n-3) + 5K_e$ which implies $f_1(K_e, 0) \geq 0$ for any $K_e \geq 2$ and $f_1(K_e, 1) = (K_e^2 - 5K_e)(n-3) + 6n - 18$ which implies $f_1(K_e, 1) \geq 0$ for any integer K_e. The discriminant of the equation $f_1(K_e, O_e) = 0$, when solved for K_e, is $n^2O_e(40 - 39O_e) + 2nO_e(92O_e - 95) - 176O_e^2 + 160O_e + 25$. Note that $-176O_e^2 + 160O_e + 25 \leq 0$ when $O_e \geq 2$ and $n^2O_e(40 - 39O_e) + 2nO_e(92O_e - 95) \leq 0$ when $O_e \geq 2$ and $n \geq 5$. For $n = 4$, the discriminant becomes $-64O_e^2 + 40O_e - 25$ which is always non-positive when $O_e \geq 2$. Finally, for $n = 3$, the dual constraint becomes $(K_e - 5O_e)(1 - O_e) \geq 0$ which is always verified since $O_e \geq 2$ and $K_e \leq n = 3$. Hence, it follows that the first dual constraint is always verified for any pair of non-negative integers (K_e, O_e) with $K_e \geq 2$.

The second dual constraint becomes $f_2(K_e, O_e) \geq 0$ with $f_2(K_e, O_e) := K_e^2(n+2) - 5K_e(nO_e - 2O_e) + O_e(16nO_e - 10n - 23O_e + 5)$. It holds $f_2(0, O_e) = O_e(16nO_e - 10n - 23O_e + 5)$ which is always non-negative for any integer value O_e when $n \geq 3$ and $f_2(1, O_e) = n(16O_e^2 - 15O_e + 1) - 23O_e^2 + 15O_e + 2$ which is always non-negative for any integer value O_e when $n \geq 3$. Hence, it follows that the second dual constraint is always verified for any pair of non-negative integers (K_e, O_e). $\square$

For the average-case model with consecutive moves, for any player $i \in N$, strategy profile K, strategy $t \in \Sigma_i$ and best-response t^* for player i in K, it holds

$$c_i'(K, k_i) \geq \frac{1}{n}c_i(K \oplus_i t^*) + \frac{n-1}{n}\left(\sum_{e \in k_i}(\alpha_e K_e) - \sum_{e \in k_i: K_e \geq 2}\frac{\alpha_e(K_e - 1)}{n - 1}\right)$$

and

$$c_i'(K, t) \leq \frac{1}{n}c_i(K \oplus_i t^*) + \frac{n-1}{n}\sum_{e \in t}\left(\alpha_e\left(K_e + 2 - \frac{K_e - 1}{n - 1}\right)\right).$$

In fact, if consecutive moves are allowed, with probability $\frac{1}{n}$ a 2-lookahead best-response of player i coincides with a classical best-response, and with probability $\frac{n-1}{n}$ the same arguments exploited for the case without repetitions apply.

Hence, the same inequality characterizing the case without repetition occurs also in this case and we can claim the following theorem.

Theorem 6. *For any linear congestion game $\mathcal{G}$, it holds* $\mathsf{PoA}(\mathcal{G}) \leq 6$ *if* $n = 2$ *and* $\mathsf{PoA}(\mathcal{G}) \leq 4 + \frac{5}{4n-7}$ *if* $n \geq 3$ *in the average-case model with consecutive moves.*

4.3   Singleton Strategies

In this subsection, we show that better results can be achieved for the worst-case model when restricting to singleton linear congestion games.

For the worst-case model without consecutive moves, for any player $i \in N$, strategy profile K and strategy $t \in \Sigma_i$, it holds $c_i'(K, k_i) \geq \sum_{e \in k_i} (\alpha_e K_e + \beta_e)$ unless all the players share the same resource in K and $c_i'(K, t) \leq \sum_{e \in t} (\alpha_e(K_e + 2) + \beta_e)$.

In fact, with 2-lookahead best-responses, when selecting strategy k_i consisting of resource e, in the worst-case model the adversary can always select a player not selecting e for the next move (unless all the players share the same resource in K); moreover, when selecting any strategy t consisting of resource e', player i can suffer a congestion at most equal to $K_{e'} + 2$, where the increase of 2 units is due to the fact that player i is selecting e' and also the player moving after her could select e'.

For the case with consecutive moves, the same inequalities apply as well, since the fact that the adversary can also select again player i can only increase the cost $c_i'(K, k_i)$, whereas the value $\sum_{e \in t} (\alpha_e(K_e + 2) + \beta_e)$ is already the maximum possible one that can be suffered by a migrating player in any model of 2-lookahead rationality.

Hence, since K is a 2-lookahead Nash equilibrium, for each player $i \in N$, it holds

$$\sum_{e \in k_i} (\alpha_e K_e + \beta_e) \leq \sum_{e \in o_i} (\alpha_e(K_e + 2) + \beta_e) \tag{3}$$

unless all the players share the same resource in K.

When embedded into the primal-dual technique, inequality (3) allows to formulate the problem similarly to the case of general strategies, and the following theorem holds.

Theorem 7. *For any singleton linear congestion game $\mathcal{G}$, it holds* $\mathsf{PoA}(\mathcal{G}) \leq 4$ *in the worst-case model.*

For the average-case model, no improved bounds, with respect to the ones holding for the case of general strategies, seem possible using our analysis technique. However, the fact that, for singleton strategies, the upper bound on the price of anarchy in the worst-case model is smaller than the one holding for the average-case model might appear counterintuitive and even contradictory. To this aim, in the following example, we show that this is not the case, since there are games with singleton strategies in which the performance of 2-lookahead

Nash equilibria in the worst-case model are better than the one achieved in the average-case model.

Example 1. Let $\mathcal{G}$ be the symmetric singleton game in which there are three players and two resources, namely e_1 and e_2, such that $f_{e_1}(x) = \frac{4}{3}x$ and $f_{e_2}(x) = x$.

Let S be the strategy profile in which two players choose e_1 and one player chooses e_2 and consider the average-case model. In the variant with consecutive moves, the expected cost of any player i choosing e_1 is $\frac{1}{3}\left(2 + \frac{4}{3} + \frac{8}{3}\right) = 2$. If player i switches to resource e_2, her expected cost is 2. Moreover, the expected cost of the player choosing e_2 is $\frac{1}{3}(1 + 2 + 2) = \frac{5}{3}$. If she switches to resource e_1, her expected cost is $\frac{1}{3}\left(2 + \frac{8}{3} + \frac{8}{3}\right) = \frac{22}{9}$. In the variant without consecutive moves, the expected cost of any player i playing e_1 is $\frac{1}{2}\left(\frac{4}{3} + \frac{8}{3}\right) = 2$. If player i switches to resource e_2, her expected cost is 2. Moreover, the expected cost of the player choosing e_2 is $\frac{1}{2}(2 + 2) = 2$. If she switches to resource e_1, her expected cost is $\frac{1}{3}\left(\frac{8}{3} + \frac{8}{3}\right) = \frac{8}{3}$. Thus, in both variants of the average-case model, S is a 2-lookahead Nash equilibrium for $\mathcal{G}$.

Consider now the worst-case model.

First of all, we show that S is not a 2-lookahead Nash equilibrium for $\mathcal{G}$ in both variants of the model. In fact, in both variants, the cost of any player i choosing e_1 is $\frac{8}{3}$. If she switches to resource e_2, her cost is 2. Thus, in both variants, S is not a 2-lookahead Nash equilibrium for $\mathcal{G}$.

Now, let S' be the strategy profile in which one player chooses e_1 and two players choose e_2. In both variants, the cost of the player choosing e_1 is $\frac{4}{3}$. If she switches to resource e_2, her cost is 2. Moreover, in both variants, the cost of any player i choosing e_2 is 2. If she switches to resource e_1, her cost is $\frac{8}{3}$. Thus, in both variants, S' is a 2-lookahead Nash equilibrium for $\mathcal{G}$.

Finally, it is not difficult to see that any profile in which all three players choose the same resource cannot be a 2-lookahead Nash equilibrium for $\mathcal{G}$, again in both variants.

Hence, since S' is the only 2-lookahead Nash equilibrium for $\mathcal{G}$ in the worst-case model, S is a 2-lookahead Nash equilibrium for $\mathcal{G}$ in the average-case model, and $C(S) > C(S')$, we can conclude that the price of anarchy of $\mathcal{G}$ in the average-case model is higher than the one of the worst-case model regardless of whether consecutive moves are allowed or not.

References

1. Aland, S., Dumrauf, D., Gairing, M., Monien, B., Schoppmann, F.: Exact Price of Anarchy for Polynomial Congestion Games. In: Durand, B., Thomas, W. (eds.) STACS 2006. LNCS, vol. 3884, pp. 218–229. Springer, Heidelberg (2006)
2. Awerbuch, B., Azar, Y., Epstein, A., Mirrokni, V.S., Skopalik, A.: Fast convergence to nearly optimal solutions in potential games. In: EC, pp. 264–273 (2008)
3. Bhawalkar, K., Gairing, M., Roughgarden, T.: Weighted Congestion Games: Price of Anarchy, Universal Worst-Case Examples, and Tightness. In: de Berg, M., Meyer, U. (eds.) ESA 2010, Part II. LNCS, vol. 6347, pp. 17–28. Springer, Heidelberg (2010)

4. Bilò, V.: A Unifying Tool for Bounding the Quality of Non-cooperative Solutions in Weighted Congestion Games. In: Erlebach, T., Persiano, G. (eds.) WAOA 2012. LNCS, vol. 7846, pp. 215–228. Springer, Heidelberg (2013)
5. Caragiannis, I., Fanelli, A., Gravin, N., Skopalik, A.: Efficient Computation of Approximate Pure Nash Equilibria in Congestion Games. In: FOCS, pp. 532–541 (2011)
6. Chien, S., Sinclair, A.: Convergence to approximate Nash equilibria in congestion games. In: SODA, pp. 169–178 (2007)
7. Christodoulou, G., Koutsoupias, E.: The price of anarchy of finite congestion games. In: STOC, pp. 67–73 (2005)
8. Fanelli, A., Flammini, M., Moscardelli, L.: The speed of convergence in congestion games under best-response dynamics. ACM Transactions on Algorithms 8(3), 25 (2012)
9. Fanelli, A., Moscardelli, L.: On best response dynamics in weighted congestion games with polynomial delays. Distributed Computing 24(5), 245–254 (2011)
10. Mirrokni, V., Thain, N., Vetta, A.: A Theoretical Examination of Practical Game Playing: Lookahead Search. In: Serna, M. (ed.) SAGT 2012. LNCS, vol. 7615, pp. 251–262. Springer, Heidelberg (2012)
11. Monderer, D., Shapley, L.S.: Potential games. Games and Economic Behavior 14(1), 124–143 (1996)
12. Pearl, J.: Heuristics: Intelligent search strategies for computer problem solving. Addison-Wesley (1984)
13. Rosenthal, R.W.: A class of games possessing pure-strategy Nash equilibria. International Journal of Game Theory 2, 65–67 (1973)
14. Shannon, C.: Programming a computer for playing chess. Philosophical Magazine, 7th Series 41(314), 256–275 (1950)

Trading Agent Kills Market Information
Evidence from Online Social Lending

Rainer Böhme[1] and Jens Grossklags[2]

[1] Department of Information Systems, University of Münster, Germany
[2] College of Information Sciences and Technology,
Pennsylvania State University, USA

Abstract. The proliferation of Internet technology has created numerous new markets as social coordination mechanisms, including those where human decision makers and computer algorithms interact. Because humans and computers differ in their capabilities to emit and process complex market signals, there is a need to understand the determinants of the provision of market information. We tackle the general research question from the perspective of new electronic credit markets. On online social lending platforms, loan applications typically contain detailed personal information of prospective borrowers next to hard facts, such as credit scores. We investigate whether a change of the market mechanism in the form of the introduction of an automated trading agent shifts the dynamics of information revelation from a high-effort norm to a low-effort information equilibrium. We test our hypothesis with a natural experiment on Smava.de and find strong support for our proposition.

1 Introduction

Credit markets are envisioned to serve as efficient social coordination mechanisms between lenders and borrowers [1]. The idea of online social lending (also known as peer-to-peer lending) is to provide a marketplace for unsecured personal loans. An electronic platform lists borrowers' loan applications so that individual lenders can review this information and decide in which projects they want to invest. Each lender contributes a small fraction of the financed amount. This distributes the credit risk in loan-specific pools of lenders. As compensation for taking risk, lenders receive interest payments, whereas platforms charge fixed (risk-free) fees [2,3,4].

Traditional institutional lending relies on a number of information sources including hard facts such as requested amount, interest rate, credit rating information and past repayment performance as well as soft facts that consider the wider context of a potential transaction. In online social lending, soft facts find typically consideration in the form of credit profiles that may include an essay description of the project complemented with a picture and other personal information. The careful evaluation of the profile may enable lenders to differentiate between borrowers and to eventually reduce the risk of loan defaults. Additional information typically *allows* for, but not necessarily leads to more efficient contracts [5].

Y. Chen and N. Immorlica (Eds.): WINE 2013, LNCS 8289, pp. 68–81, 2013.
© Springer-Verlag Berlin Heidelberg 2013

Due to the novelty of online social lending, previous research has focused on the negotiation phase rather than long-term consequences. For example, Böhme and Pötzsch report that references to outside options in the traditional banking sector, even if unverified, are rewarded with better financing conditions. However, statements targeted at arousing pity are penalized [2]. The credit profile helps to reduce the information asymmetry between borrowers and lenders. While requesters of funds might conceal information that would make them appear less desirable [6], they will also pro-actively signal to lenders their credit-worthiness [7]. In summary, in credit markets, information is more important compared to many other financial markets that price more standardized goods [8].

However, the wealth of informally provided information and the growth in popularity of social lending also pose challenges to the efficiency of these marketplaces. In particular, lenders must find ways to overcome the information overload originating from the abundance of loan applications. One option is the creation of reputation schemes to favor well-established and reliable borrowers. A different approach is the consideration of alternative market designs and changes in trading rules [9].

In July 2009, *Smava.de*, a popular German social lending platform, changed its market mechanism by introducing *immediate loans*. Instead of waiting for a posted loan application to be funded, a borrower may consult an automated agent that is suggesting an interest rate high enough so that the loan can be financed instantaneously by lenders who pre-committed offers, resulting in a form of order book. In this paper, we investigate whether the introduction of this automated trading agent shifts the dynamics of information revelation from a high-effort information norm to a low-effort equilibrium. We scrutinize our hypothesis in the form of a natural experiment on *Smava.de* and find strong support for our proposition.

As to the organization of this paper, Section 2 reviews theoretical approaches to the research question, which also include our analysis of the strategic options of market providers, lenders, and borrowers. Section 3 describes our empirical strategy to study the research question with a natural experiment observed on *Smava.de*. The results, presented in Section 4, support our theory both descriptively and, more specifically, by regression analyses of disaggregated data. We offer concluding remarks and present trajectories for future work in Section 5.

2   Incentives of Market Providers and Participants

We are not aware of research on the impact of agents on human-populated online social lending markets, and related work for financial markets is surprisingly sparse. Lin and Kraus survey research on the question whether software agents can successfully negotiate with humans on a variety of commodity markets [10], and Duffy reviews research on markets populated with automated traders in comparison to similar work in experimental economics [11]. In the context of financial markets, software agents are expected to improve market efficiency because they follow predefined rules and do not make mistakes with respect to their algorithms. In addition, software agents can process more data in a given time span and interact faster with the market via APIs than human traders are able to utilize any user interface [12,13].

We focus the following theoretical discussion on the incentive structure in online social lending considering the different stakeholders: market providers, lenders, and borrowers.

2.1   Rationale of the Market Provider

Online social lending markets are two-sided with significant positive cross-side network effects, i.e., lenders prefer to have a larger group of borrowers to choose from, and vice versa. The intermediary is interested in the overall growth of the platform to reap first-mover advantages to, amongst other factors, erect barriers to entry. To enhance long-term viability, the platform can support matching that will lead to low loan default rates by excluding, for example, untenable risks. Other concerns include the transfer of credit risk to non-banks as well as adherence to financial regulations (e.g., the Dodd–Frank Wall Street Reform and Consumer Protection Act in the United States [14]).

The market provider's profit is derived from closing and late fees and potential future opportunities that may result from the growth of the platform. As a financial intermediary, it is critically important for the market provider to foster an image of professionalism and reliability [15]. One important implication of such an evaluation is the trend towards *uniformity*. In finance, "a preference for uniformity is consistent with a preference for strong uncertainty avoidance leading to a concern for law and order and rigid codes of behaviour, a need for written rules and regulations, [and] a respect for *conformity*" [16]. Such consistency is primarily driven by the evaluation of borrower profiles and can be guided through the default format of these profiles.

The intermediary can further influence the appeal of the platform via market design [9]. A banking report argues that automatic bidding and secondary markets (i.e., the trading of existing loan notes) "inject new professionalism," but also shift attention from humans to artificial agents [17]. As a result, the comprehensiveness of borrower profiles decreases in importance for negotiations that are mediated by automatic agents.

2.2   Rationale of the Market Participants

Lenders. Non-bank lenders may understand online social lending as a viable alternative for portfolio diversification, for example, to complement low-risk/low-return certificates of deposit, and stock market portfolios that promise higher expected returns, but come with a significant degree of uncertainty in the short term. The inherent trade-off for online social lenders is the expectation of a relatively high rate of return weighted against the default risk associated with a particular group of borrowers.

However, due to its novelty, we cannot expect a high degree of domain-specific financial literacy within the lender population and, therefore, sufficient expertise to independently avoid borrowers with default risks [18]. The potentially unjustified reliance on soft information in borrowers' profiles might further exacerbate the asymmetric selection problem. In contrast, lenders may derive immaterial benefits from investing in real individuals' aspirations and plans, and learning about them in their self-descriptions.

The crisis in mortgage lending and institutional finance has reopened the discussion about effective protection of non-professional market participants [19]. Further, while online social lending acts as an instrument to escape credit scarcity, borrowers who are not served by traditional banking may also pose additional risks. Taken together, lenders will benefit from marketplace designs that limit overlending as well as contribute to the selection of appropriate credit terms for manageable risks [20].

The existence of an automatic lending agent addresses some of these problems. It limits the search costs that arise from the need to investigate a large amount of soft information and sharpens the focus on verified information. The interaction with the recommendation features of the agent also reduces the likelihood of significant misquoting of interest rates.

Borrowers. Borrowers' prime objective is to gain access to financing at reasonable conditions and without other unattractive contractual obligations. Further, the unbureaucratic and innovative nature of online social lending might appeal to individuals with unsuccessful interactions with the traditional banking sector.

Borrowers aim for a favorable evaluation of their loan applications through a number of factors. Borrowers publish a desired amount and purpose, they provide verifiable information about themselves including the credit grade. In addition, a customizable profile allows them to personalize their funding appeal.

The accessibility of personal profiles to every potential lender might, however, be perceived as a privacy risk by borrowers [2]. For example, details about personal finances or unfortunate circumstances could, when used outside of the platform context, cause ridicule by acquaintances or colleagues. The typical interactions with institutional banks and credit bureaus are not immune to privacy concerns [21], however, the open and social nature of online social lending amplifies these worries.

Further, borrowers might also attempt to conceal relevant information [6]. For example, on *Smava.de* over 30 % of all loans are awarded to small business owners or self-employed professionals. It follows that personal credit ratings may not accurately reflect inherent business-related risks [17].

The availability of an artificial agent allows borrowers to choose between automated matching and the human-driven process. The former contributes to an amelioration of privacy concerns, but a weakening of success prospects for individuals with low credit ratings.

2.3   Information Revelation as a Coordination Problem

Borrowers are not only affected by lenders' behavior, and vice versa, but also by the actions of others within their group. For example, past studies have presented evidence for herding behavior with respect to bidding on *Prosper.com*, a US-based online social lending platform [22,23]. Similarly, the presentation of the personal profile is subject to mimicry. Borrowers copy information from their peers' profiles. Interestingly, they do not seem to copy from successful recent applications more often than from pending or unsuccessful applications [24].

The presence of an automatic agent introduces an unprecedented speed into the process of social lending, as well as an increased focus on verified information. The reliance on the agent may trigger a desire to decrease the provisioning of comprehensive personal profiles to reduce signaling costs. Further, recent behavioral research suggests that the mere exposure to indicators of instant gratification (e.g., fast food symbols) may contribute to a shift of preferences towards economic impatience [25]. It follows that even borrowers who are not directly utilizing the artificial agent may change their behaviors.

The resulting net impact on signaling is far from obvious. In the early years, online social lending platforms have emphasized the social aspect of lending. For example, *Smava.de* advertised its services with the slogan "loans from human to human." This has contributed to a *norm of comprehensive textual signaling* in the form of long personal profiles with the expectation to adhere as a matter of proper conduct [26].

At the same time, our discussion shows that restricted information focused on verifiable facts is unlikely to be inferior from an economic perspective, in particular, considering humans' innate bounded ability of information gathering and processing [27,28].

Considering the information revelation of borrowers as a coordination problem is helpful to understand the dynamics of their behavior on *Smava.de*. We argue that different jointly chosen degrees of soft information revelation can be equilibria, and it depends on exogenous coordinating factors which outcome is reached [29]. For example, the user interface design for personal profiles as well as *Smava.de*'s framing as a human-to-human lending platform jointly served as a focal point for a high degree of information revelation. In contrast, the introduction of the artificial agent is a strong driver for brevity.

More specifically, we can describe the coordination between borrowers where lenders react as follows. When lenders make their funding decisions, they cannot know the true value of soft information (compared to hard verifiable information). Hence, they estimate it by observing the usage of soft information in the marketplace (i.e., the average of all soft information revealed). If the majority of borrowers reveal no information then lenders would reckon that such soft information is of no value. In contrast, a market in which borrowers heavily utilize soft information would suggest to lenders that such information has value.

From this basic premise at least two potential outcomes may result. At the one extreme, if none of the borrowers reveals any soft information, then it follows that none of the borrowers could improve his position by revealing soft information as the cost of revelation is positive and the value of revelation is zero. In contrast, if all borrowers reveal soft information, then any borrower would harm his position by not including soft information. While on the one hand, the borrower could reduce his cost by omitting soft information, the loss of apparent creditworthiness (from the perspective of lenders) outweighs this benefit. In that sense, soft information is productive.

The introduction of the trading agent has the potential to change the focal point since those borrowers who use loan matching by the trading agent know that adding soft information does not influence the lending decision. Lenders, however, have no direct means to tell loans matched with a trading agent and conventional loans apart. Hence, their estimated value of soft information is impacted by the mixture of the two regimes. As the share of automatically matched loans exceeds a certain threshold, the market will tip towards low information revelation.

The distribution of individually heterogeneous costs for information revelation and lenders' belief structure influence the strength of the described processes and the threshold of the tipping point which motivates an empirical analysis. We hypothesize that the amount of soft information provided on *Smava.de* decreases after the introduction of the trading agent independent of whether the trading agents has been used by an individual borrower.

3   Empirical Strategy

3.1   Institutional Background

Smava.de, established in February 2007, is the largest online social lending platform in Germany handling a total of € 77 million allocated to about 9000 loans (as of July 2013). Unlike *Prosper.com*, the dominant platform in the US, *Smava.de* does not use an auction mechanism. Instead, borrowers post loan applications including amount, interest rate, and maturity along with verified demographic information (age, occupation, state of residence) and a credit grade between A (best, nominal default risk < 1.3 %) and H (worst, default risk 17 %). These applications serve as take-it-or-leave-it offers for lenders, who decide if and how much (in units of € 250) they want to contribute to financing each pending loan. Loan applications are settled when they are fully funded or after two weeks. Borrowers have the option to revise the interest rate upwards if the loan does not receive funding as quickly as desired. They may also complement their loan application by unverified information, such as textual descriptions, motivation statements, or custom pictures. We use this voluntary provision of information as indicator for revelation behavior.

Figure 1 depicts a typical profile on the platform. In this example, a potential male borrower applies for an amount equivalent to $7000 to finance education expenses to become a certified optician.

Smava.de introduced and gradually extended automatic trading agents to assist their lenders. A more substantial change was the introduction of an automatic loan placement agent in July 2009. This agent assists borrowers in finding the currently relevant interest rate such that a loan application would immediately be approved by the lenders' trading agents. In other words, the new

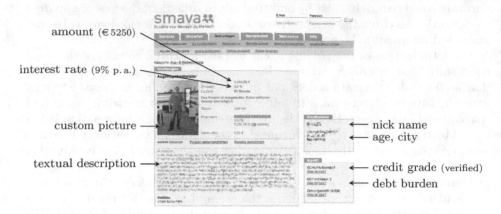

amount (€ 5250)

interest rate (9% p. a.)

custom picture

textual description

nick name

age, city

credit grade (verified)

debt burden

Fig. 1. Example credit profile of a male applicant to finance certifications to become a professional optician. (Contents obfuscated by the authors for borrower privacy.)

agent reinterprets the parameterization of the lenders' trading agents—all controlled by the platform—as an order book, and replaces the take-it-or-leave-it mechanism by a matching mechanism.[1]

Since July 15th, 2009 both mechanisms coexist. This forms a unique natural experiment to study not only the influence of trading agents on information revelation in the part of the market served by the agents, but also on the rest of the loans which continue to use the old mechanism.

3.2 Data

Our study uses public information only. We downloaded all $N = 931$ loan applications listed on *Smava.de* between April and October 2009. This sample has been split into contrast groups consisting of 380 loan applications before and 551 applications after the intervention. We remove the month of July to exclude all loan applications that overlap the intervention date (see Figure 2).

Our *independent* variable is the presence of the trading agent for borrowers. We measure our *dependent* variable, information revelation, by two proxies. First, we follow Herzenstein *et al.* [4] and measure the length of all unverified descriptions of a loan application and the attached borrower profile in characters. Within each contrast group, this variable can be reasonably approximated with a Gaussian distribution after taking logs (see Figure 3). The second indicator of information revelation is inspired by Pope and Syndor [30]. We take the binary fact whether or not borrowers illustrate their loan applications by uploading custom pictures which replace the default icon defined by *Smava.de*.

We do not try to measure the semantics of the description or the picture, i. e., whether they contain any relevant information or valence. As even in the "before" condition, only one quarter of loan applications is illustrated with a

---

[1] The basic process is similar to Priceline.com counteroffers.

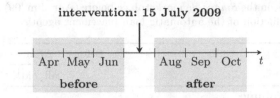

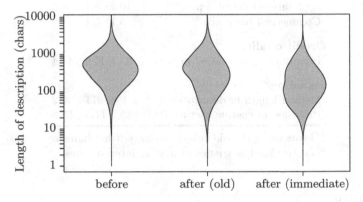

Fig. 2. Design of the natural experiment analysis

Fig. 3. Length of description: Violin plots of smoothed empirical distribution (left) and log normal fit (right) for contrast groups compared in this study (Gaussian kernel, bandwidth 0.5, $N = 931$)

custom picture, it is fair to assume that borrowers will only upload carefully selected pictures which they believe help their cases. Likewise, writing longer descriptions is associated with opportunity costs and privacy loss.

Moreover, we collected a number of *control* variables which might interact with the hypothesized relationship. Most importantly, we try to identify whether a loan has been granted using the old or new mechanism in the "after" condition. This information is not directly visible on the platform and has been inferred from the succession of bid times, which are available at a resolution of one minute. Agent-matched ("immediate") loans are characterized by complete funding whereby no two bids differ by more than one minute. All other loans are classified as take-it-or-leave-it ("old"). In addition, we collected the amount, interest rate, credit grade, and the assignment to one of 19 credit categories[2] for every loan application in the sample.

---

[2] The categories on *Smava.de* are: debt restructuring; liquidity; home, gardening & do-it-yourself; cars & motorbikes; events; education & training; family & education; antiques & art; collection & rarity; electronics; health & lifestyle; sports & leisure; travel; pets & animals; volunteering; commercial; business investment; business extension; miscellaneous.

Table 1. Activities on the *Smava.de* marketplace before (Apr–Jun '09) and after (Aug–Oct '09) the introduction of the automatic loan placement agent

	Before	After	
		all	old[1]
Volume			
Number of loans	380	551	378
Financed amount (€ millions)	2.9	4.7	3.5
Credit conditions			
Avg. interest rate (% p. a.)	10.2	8.7	8.5
Commercial bank rate[2]	5.1	5.1	—
Credit quality			
Investment grade (A–C in %)	43.7	46.6	43.4
Signaling			
Median length of description	456	271	332
Provision of custom picture (%)	25.5	11.3	15.3

[1] loans using the old take-it-or-leave-it mechanism
[2] central bank statistics of market interest rates

4   Results

4.1   Descriptive Analysis

Table 1 shows aggregated statistics broken down by the contrast groups before and after the intervention. The "after" condition is further refined by a separate column for loans using the old take-it-or-leave-it mechanism. One can observe three major effects to be discussed in the following.

The number of loans grew by 45 %. At first sight, it looks like immediate loans tap into a new segment, as the number of loans using the old mechanism is almost constant. The average loan amount grew by 12 % with a tendency for larger loans to use the old mechanism while smaller loans are matched through the trading agents. The observed development is in line with our analysis of the market provider's strategy (Sect. 2.1). But the evidence is relatively weak because the general growth path of the platform impedes a direct causal attribution to the intervention.

The average interest rate dropped by 1.5 %-pts with immediate loans being marginally more expensive than take-it-or-leave-it loans. The latter discrepancy can be explained by time preferences. This development is remarkable because we can rule out third factors such as general trends in consumer credit interest rates. The official statistics of comparable loans to consumers of traditional banks report stable and significantly lower interest rates. The level shift is due to higher quality requirements in the banking sector compared to *Smava.de*. The drop in interest rates after the intervention cannot be explained by a significant change in average credit quality, either. The fact that immediate loans exhibit slightly higher credit quality may in fact be due to inverse causality: high-quality (i. e., low risk) borrowers have an outside option in the banking sector, which becomes comparably less attractive if the interest rates on *Smava.de* decline.

Note that we must not interpret this result as support for the hypothesis that trading agents improve market efficiency. Unless we observe actual default rates, it is too early to tell if the borrower-friendly low risk premium is in fact closer to the equilibrium price of risk on *Smava.de* [31].

Both indicators of information revelation, length of description and provision of custom picture, show a substantial decline after the intervention. Interestingly, this is not limited to the immediate loans (where the effect is most pronounced because the agents do not evaluate unverified information). So the presence and visibility of agent-matched deals appears to spill over and change the information revelation conventions on the *entire* marketplace.

Superficially, these numbers already tell a story. But the evidence for this interpretation from Table 1 alone is weak. Market expansion, borrower-friendly conditions, and other effects might interact with each other and lead to spurious results in the aggregated numbers. For example, an alternative explanation could be that lower interest rates have attracted better risks with more self-explanatory credit projects. To gain more robust insights, we conduct a disaggregated analysis on individual loans for the phenomenon of disappearing information.

4.2 Regression Analysis

To isolate the effect of the introduction of a trading agent on information revelation from other shifts in the market conditions, a series of multivariate regression models has been estimated. First, we explain the length of description (ℓ) with the following equation,

$$\log_2 \ell_i = \beta_1 A_i + \beta_2 R_i + \beta_3 T_i + \beta_4 I_i + cC_{(i,\cdot)} + gG_{(i,\cdot)} + \varepsilon_i, \tag{1}$$

where A_i is the log amount, and R_i is the interest rate in percent p. a. of loan i. T_i is a dummy variable taking value 1 if the loan has been listed *after* the introduction of the trading agent, 0 otherwise. I_i takes value 1 if the loan is an "immediate loan", i. e., it has been financed by using the trading agent. Matrices C and G contain a series of dummy variables as fixed effects for 19 credit categories and 8 credit grades, respectively. Equation (1) is estimated using ordinary least squares, i. e.,

$$(\hat{\beta}_1, \hat{\beta}_2, \hat{\beta}_3, \hat{\beta}_4, \hat{c}, \hat{g}) = \arg \min_{(\beta_1, \ldots, g)} \sum_{i=1}^{N} \varepsilon_i^2. \tag{2}$$

Table 2 reports the estimated coefficients for a stepwise inclusion of the T_i and I_i terms along with statistical significance tests of the null hypothesis $\beta = 0$.

M1 is the default model over both periods together. It identifies a highly significant positive correlation between the length of description and the amount (both in logs): borrowers who ask for more money are willing to explain their project

Table 2. Results of regression analyses: Effect of presence and use of trading agent on information revelation while controlling for credit volume, conditions, and quality

	Length of description ($\log_2 \ell$)		
Terms	M1	M2	M3
Amount [log]	0.24 ***	0.35 ***	0.30 ***
Interest rate [%-pts]	−0.07 *	−0.21 ***	−0.16 ***
Trading agent present		−1.14 ***	−0.91 ***
Trading agent used			−0.49 **
Category fixed effects	yes	yes	yes
Credit grade fixed effects	yes	yes	yes
Adjusted R^2 [%]	3.7	13.2	14.0

Sig. levels: [*]$p < 0.05$, [**]$p < 0.01$, [***]$p < 0.001$; $N = 931$

better. We also find a significant negative correlation between interest rate and length of description suggesting that borrowers who are less verbose are penalized ceteris paribus with (slightly) worse credit conditions. All predictors in M1 explain less than 4 % of the variance of the dependent variable. This is because the hidden heterogeneity—the regime change—is not reflected in this specification.

Models M2 and M3 include a term for the presence of the trading agent, which adds another 10 %-pts of explained variance. The coefficient is negative—indicating disappearing information—and highly significant. This supports our above hypothesis with strong evidence on the micro-level and after controlling for third variables. The effect of the intervention can be further decomposed on the individual loan level to isolate contributions from the mere presence of a trading agent and the fact that the trading agent was actually used to settle a particular loan. This is realized in M3. Interestingly, the platform-wide effect is responsible for the lion's share in the decline of signaling whereas the actual use of the trading agent is of subordinate importance. We interpret this as support for a switch in the equilibrium situation stimulated by the option to use the new mechanism.

Regression diagnosis via inspection of the residual distribution and fixed effects coefficients revealed nothing surprising or worrying. For example, categories with positive significant fixed effects include events, volunteering, and business extensions; arguably the least self-explanatory ventures. Post-hoc ANOVA checks between M1 and M2, as well as M2 and M3, respectively, indicate highly significant differences in explained variance.

A remaining doubt is that detailed information might have disappeared due to a gradual shift in the conventions on *Smava.de*, which would be confounded with our natural experiment. To test this, we re-estimated M3 including a linear time trend as additional term. The coefficient (-0.001, $p = 0.44$) indicated no prevalence of a persistent time trend between April and October 2009. This strengthens the evidence that the observed differences before and after the introduction of the trading agent were indeed caused by this intervention.

Table 3. Effect of trading agent on provision of custom picture

Terms	Log odds ratio of custom picture		
	Model 4	Model 5	Model 6
Amount [log]	0.1	0.3 *	0.2
Interest rate [%-pts]	−0.1	−0.3 ***	−0.2 *
Trading agent present		−1.5 ***	−1.1 ***
Trading agent used			−1.7 **
Category fixed effects	yes	yes	yes
Credit grade fixed effects	yes	yes	yes
Pseudo-R^2 [%]	8.8	16.8	18.8

Sig. levels: $^*p < 0.05$, $^{**}p < 0.01$, $^{***}p < 0.001$; $N = 931$

A second indicator of information revelation is the provision of a custom picture. This is a binary indicator, and we use logistic regression analysis to regress the predictors of Equation (1) on the log odds ratio for the provision of a custom picture. The resulting coefficients, as reported in Table 3, have to be transformed to the probability domain to interpret their absolute magnitudes. Nevertheless, it is straightforward to interpret their sign and relative size.

Provision of a custom picture is a cruder indicator. Hence, the terms for amount and interest rate are barely significant, yet estimated with plausible signs. Again, after controlling for third variables, the intervention has a strongly significant negative effect on the willingness to provide custom pictures. Note that model M6 attributes a larger contribution to the actual use of the trading agent than to its mere presence.

4.3   Limitations

Natural experiments with a single intervention date suffer from the difficulty to exclude unobserved third variables as causes. Therefore, they do not permit causal inference in a strict sense. Although, we controlled for observable factors and linear time trends, there may be non-linear dynamics of growth or overlapping interventions we are not aware of.

We intentionally avoided conjectures about efficiency or welfare aspects of information revelation regimes. Reliable empirical statements on market efficiency and long-term costs or benefits of signaling in this marketplace depend on the availability of actual default rates. These cannot be observed before the 3–5 year maturity of the outstanding loans has been reached.

5   Concluding Remarks

To the best of our knowledge, this work is the first attempt to study the effect of automatic trading on information revelation behavior in marketplaces where humans and computers interact. We have theorized how voluntary disclosure of unverified information forms a coordination problem with at least two equilibria for high, and respectively low information regimes. A natural experiment in the context of online social lending, an information-rich market, enables us to test our hypothesis empirically and study the effects of the introduction of an optional trading agent on information revelation. The latter was measured by two quantitative indicators. While controlling for third variables, both were found to be negatively affected by the introduction of the trading agent.

Generally speaking, our results illustrate how changes in the market mechanism, even if limited to parts of the market, may reset focal points and cause spillovers to rebalance the equilibria in the initially unaffected segments of the market. If this logic is transferred to other markets, or more generally to coordination games (e.g., real-name policies in virtual communities), then utmost care should be taken when introducing automated agents. Even if the automation is optional and affects only part of the market or community, an avalanche effect might follow and its precise consequences are difficult to predict in advance.

We believe that this opens an interesting and relevant direction with many research questions. Obvious next steps include the differentiation of signals on a semantic level [2], or the interpretation of the temporary shut-down of the US social lending platform *Prosper.com* as a natural experiment [32].

References

1. Hayek, F.: The use of knowledge in society. American Economic Review 35(4), 519–530 (1945)
2. Böhme, R., Pötzsch, S.: Privacy in online social lending. In: Proc. of AAAI Spring Symposium on Intelligent Privacy Management, Palo Alto, CA, pp. 23–28 (2010)
3. Berger, S., Gleisner, F.: Emergence of financial intermediaries in electronic markets: The case of online P2P lending. BuR - Business Research 2(1), 39–65 (2009)
4. Herzenstein, M., Andrews, R., Dholakia, U., Lyandres, E.: The democratization of personal consumer loans? Determinants of success in online peer-to-peer lending communities (June 2008)
5. Varian, H.: Computer mediated transactions. American Economic Review 100(2), 1–10 (2010)
6. Posner, R.: An economic theory of privacy. Regulation, 19–26 (1978)
7. Spence, M.: Job market signaling. The Quarterly Journal of Economics 87(3), 355–374 (1973)
8. Malkiel, B.: The efficient market hypothesis and its critics. Journal of Economic Perspectives 17(1), 59–82 (2003)
9. Chen, N., Ghosh, A., Lambert, N.: Social lending. In: Proc. of ACM EC, Palo Alto, CA, pp. 335–344 (2009)
10. Lin, R., Kraus, S.: Can automated agents proficiently negotiate with humans? Communications of the ACM 53(1), 78–88 (2010)
11. Duffy, J.: Agent-based models and human-subject experiments. In: Judd, K., Tesfatsion, L. (eds.) Handbook of Computational Economics, vol. 2, pp. 949–1012. North Holland (2006)

12. Das, R., Hanson, J., Kephart, J., Tesauro, G.: Agent-human interactions in the continuous double auction. In: Proc. of IJCAI 2001, Seattle, WA, pp. 1169–1187 (2001)
13. Grossklags, J., Schmidt, C.: Software agents and market (in)efficiency - A human trader experiment. IEEE Transactions on System, Man, and Cybernetics: Part C 36(1), 56–67 (2006)
14. Brill, A.: Peer-to-peer lending: Innovative access to credit and the consequences of Dodd–Frank (December 2010), http://www.aei.org/article/102873
15. Howcroft, B., Lavis, J.: Image in retail banking. International Journal of Bank Marketing 4(4), 3–13 (1986)
16. Gray, S.: Towards a theory of cultural influence on the development of accounting systems internationally. Abacus: A Journal of Accounting, Finance and Business Studies 24(1), 1–15 (1988)
17. Meyer, T.: Innovations in P2P lending may put computers over people: Welcome to the machine (November 2009), http://www.dbresearch.de/
18. Klafft, M.: Online peer-to-peer lending: A lenders' perspective. In: Proceedings of the International Conference on E-Learning, E-Business, Enterprise Information Systems, and E-Government, pp. 371–375 (2008)
19. Bar-Gill, O., Warren, E.: Making credit safer. University of Pennsylvania Law Review 157(1), 1–101 (2008)
20. de Meza, D.: Overlending? The Economic Journal 112(477), 17–31 (2002)
21. Jentzsch, N.: Financial Privacy. Springer, Heidelberg (2007)
22. Ceyhan, S., Shi, X., Leskovec, J.: Dynamics of bidding in a P2P lending service: Effects of herding and predicting loan success. In: Proc. of ACM WWW (2011)
23. Shen, D., Krumme, C., Lippman, A.: Follow the profit or the herd? Exploring social effects in peer-to-peer lending. In: Proc. of IEEE SocialCom, pp. 137–144 (2010)
24. Böhme, R., Pötzsch, S.: Collective exposure: Peer effects in voluntary disclosure of personal data. In: Danezis, G. (ed.) FC 2011. LNCS, vol. 7035, pp. 1–15. Springer, Heidelberg (2012)
25. Zhong, C., DeVoe, S.: You are how you eat: Fast food and impatience. Psychological Science 21(5), 619–622 (2011)
26. Young, P.: Social norms. In: Durlauf, S., Blume, L. (eds.) The New Palgrave Dictionary of Economics, Second Edition, 2nd edn. Palgrave Macmillan (2008)
27. Selten, R.: What is bounded rationality? In: Gigerenzer, G., Selten, R. (eds.) Bounded Rationality: The Adaptive Toolbox, pp. 13–36. MIT Press (2001)
28. Weiss, G., Pelger, K., Horsch, A.: Mitigating adverse selection in P2P lending – Empirical evidence from Prosper.com (July 2010)
29. Schelling, T.: The Strategy of Conflict. Oxford University Press, Oxford (1965)
30. Pope, D., Syndor, J.: What's in a picture? Evidence of discrimination from Prosper.com. Journal of Human Resources 46(1), 53–92 (2011)
31. Karlan, D., Zinman, J.: Observing unobservables: Identifying information asymmetries with a consumer credit field experiment. Econometrica 77(6), 1993–2008 (2009)
32. Eisenbeis, H.: The government crackdown on peer-to-peer lending. The Big Money (March 2009)

Designing Markets for Daily Deals

Yang Cai[1], Mohammad Mahdian[2], Aranyak Mehta[2], and Bo Waggoner[3]

[1] MIT
ycai@csail.mit.edu
[2] Google
{mahdian,aranyak}@google.com
[3] Harvard
bwaggoner@fas.harvard.edu

Abstract. Daily deals platforms such as Amazon Local, Google Offers, GroupOn, and LivingSocial have provided a new channel for merchants to directly market to consumers. In order to maximize consumer acquisition and retention, these platforms would like to offer deals that give good value to users. Currently, selecting such deals is done manually; however, the large number of submarkets and localities necessitates an automatic approach to selecting good deals and determining merchant payments.

We approach this challenge as a market design problem. We postulate that merchants already have a good idea of the attractiveness of their deal to consumers as well as the amount they are willing to pay to offer their deal. The goal is to design an auction that maximizes a combination of the revenue of the auctioneer (platform), welfare of the bidders (merchants), and the positive externality on a third party (the consumer), despite the asymmetry of information about this consumer benefit. We design auctions that truthfully elicit this information from the merchants and maximize the social welfare objective, and we characterize the consumer welfare functions for which this objective is truthfully implementable. We generalize this characterization to a very broad mechanism-design setting and give examples of other applications.

1 Introduction

Daily deals websites such as Amazon Local, Google Offers, GroupOn, and LivingSocial have provided a new channel of direct marketing for merchants. In contrast to standard models of advertising such as television ads and web search results, the daily deals setting provides two new challenges to platforms.

First, in models of advertising such as web search, the advertisement is shown on the side of the main content; in contrast, daily deals websites offer consumers web pages or emails that contain only advertisements (*i.e.*, coupons). Therefore, for the long-term success of a platform, the decision of which coupons to show to the user must depend heavily on the benefit these coupons provide to consumers.

Second, the merchant often has significantly more information than the advertising platform about this consumer benefit. This benefit depends on many

Y. Chen and N. Immorlica (Eds.): WINE 2013, LNCS 8289, pp. 82–95, 2013.
© Springer-Verlag Berlin Heidelberg 2013

things: how much discount the coupon is offering, how the undiscounted price compares with the price of similar goods at the competitors, the price elasticity of demand for the good, the fine prints of the coupon, and so on. These parameters are known to the merchants, who routinely use such information to optimize their pricing and their inventory, but not to the platform provider who cannot be expected to be familiar with all markets and would need to invest significant resources to learn these parameters. Furthermore, unlike standard advertising models where an ad is displayed over a time period to a number of users and its value to the user (often measured using proxies like click-through rate or conversion rate) can be estimated over time, the structure of the daily deals market does not permit much experimentation: A number of deals must be selected at the beginning of each day to be sent to the subscribers all at once, and the performance of previous coupons, if any, by the same advertiser is not a good predictor of the performance of the current coupon, as changing any of the terms of the coupons can significantly affect its value.

These challenges pose a novel *market design* problem: How can we select deals with good benefit to the consumer in the presence of strongly asymmetric information about this benefit? This is precisely our goal in this paper. We postulate that merchants hold, as private information, two parameters: A *valuation* equalling the overall utility the merchant gains from being selected (as in a standard auction); and a *quality* that represents the attractiveness of their deal to a user. The task is to design an auction mechanism that incentivizes the merchants to reveal their private information about both their valuation and quality, then picks deals that maximize a combination of platform, merchant, and consumer values. We show that, if consumer welfare is a convex function of quality, then we can design a truthful auction that maximizes total social welfare; furthermore, we show that the convexity condition is necessary. We give negative results for another natural goal, achieving a constant-fraction welfare objective subject to a quality threshold guarantee. The main idea behind our positive results is to design a mechanism where bidders' total payment is contingent (in a carefully chosen way) upon whether the consumer purchases the coupon. Not surprisingly, the theory of proper scoring rules comes in handy here.

We then extend these results to characterize incentive-compatible mechanisms for social welfare maximization in a very general auction setting, where the type of each bidder has both a valuation and a quality component. Quality is modeled as a distribution over possible states of the world; a *consumer welfare function* maps these distributions to the welfare of some non-bidding party. We design truthful welfare-maximizing mechanisms for this setting and characterize implementable consumer welfare functions with a convexity condition that captures expected welfare and, intuitively, risk-averse preferences. We give a number of example applications demonstrating that our framework can be applied in a broad range of mechanism design settings, from network design to principal agent problems.

The rest of this paper is organized as follows: In the next section, we formally define the setting and the problem. In Section 3, we give a mechanism for maximizing social welfare when consumer welfare is a convex function the quality. In Section 4, we show that no truthful mechanism even approximates the objective of maximizing the winner's value subject to a minimum quality; we also show that the convexity assumption in Section 3 is necessary. Finally, in Section 5, we extend our mechanisms and characterization to a much more general setting.

Related work. To the best of our knowledge, our work is the first to address mechanism design in a market for daily deals. There has been unrelated work on other aspects of daily deals (*e.g.* impact on reputation) [1,2,3]. A related, but different line of work deals with mechanism design for pay-per-click (PPC) advertising. In that setting, as in ours, each ad has a value and a quality (representing click-through rate for PPC ads and the probability of purchasing the deal in our setting). The objective is often to maximize the combined utility of the advertisers and the auctioneer [4,5], but variants where the utility of the user is also taken into account have also been studied [6]. The crucial difference is that in PPC advertising, the auctioneer holds the quality parameter, whereas in our setting, this parameter is only known to the merchant and truthful extraction of the parameter is an important part of the problem. Other work on auctions with a quality component [7,8] assume that a quality level may be assigned by the mechanism to the bidder (who always complies), in contrast to our setting where quality is fixed and private information.

We make use of proper scoring rules, an overview of which appears in [9]; to our knowledge, proper scoring rules have been used in auctions only to incentivize agents to guess others' valuations [10]. Our general setting is related to an extension of proper scoring rules, decision rules and decision markets [11,12]. There, a mechanism designer elicits agents' predictions of an event conditional on which choice she makes. She then selects an outcome, observes the event, and pays the agents according to the accuracy of their predictions. Unlike our setting, agents are assumed not to have preferences over the designer's choice, except in [13], which (unlike us) assumes that the mechanism has partial knowledge of these preferences and does not attempt to elicit preferences. Our general model may be interpreted as a fully general extension to the decision-rule setting in which we introduce the novel challenge of *truthfully eliciting* these preferences and incorporate them into the objective. However, we focus on deterministic mechanisms, while randomized mechanisms have been shown to have nice properties in a decision-rule setting [14].

Another related line of work examines when a proper scoring rule might incentivize an agent to take undesirable actions in order to improve his prediction's accuracy. When the mechanism designer has preferences over different states, scoring rules that incentivize beneficial actions are termed *principal-aligned* scoring rules [15]. A major difference is that the mechanism designer in the principal-aligned setting, unlike in ours, does not select between outcomes of any mechanism, but merely observes a state of the world and makes payments.

2   The Model

In this section, we formulate the problem in its simplest form: when an auctioneer has to select just one of the interested merchants to display her coupon to a single consumer.[1] In Section 5, our model and results will be generalized to a much broader setting.

There are m bidders, each with a single coupon. We also refer to the bidders as *merchants* and to coupons as *deals*. An auctioneer selects at most one of these coupons to display. For each bidder i, there is a probability $p_i \in [0, 1]$ that if i's coupon is displayed to a consumer, it will be purchased by the consumer. We refer to p_i as the *quality* of coupon i. Furthermore, for each bidder i, there is a value $v_i \in \mathbb{R}$ that represents the expected value that i gets if her coupon is chosen to be displayed to the advertiser. Both v_i and p_i are private information of the bidder i, and are unknown to the auctioneer.[2] We refer to (v_i, p_i) as bidder i's *type*. We assume that the bidders are expected utility maximizers and their utility is quasilinear in payment.

Note that v_i is i's total expected valuation for being selected; in particular, it is *not* a value-per-purchase (as in *e.g.* search advertisement). Rather, v_i is the maximum amount i would be willing to pay to be selected (before observing the consumer's purchasing decision). Also, we allow v_i and p_i to be related in an arbitrary manner. If, for instance, i derives value a_i from displaying the coupon plus an additional c_i if the consumer purchases the coupon, then i would compute $v_i = a_i + p_i c_i$ and submit her true type (v_i, p_i). For our results, we do not need to assume any particular model of how v_i is computed or of how it relates to p_i.

An auction mechanism functions as follows. It asks each bidder i to reveal her private type (v_i, p_i). Let $(\hat{v}_i, \hat{p}_i)$ denote the type reported by bidder i. Based on these reports, the mechanism chooses one bidder i^* as the *winner* of the auction, i.e., the merchant whose deal is shown. Then, a consumer arrives; with probability p_{i^*}, she decides to purchase the deal. Let $\omega \in \{0, 1\}$ denote the consumer's decision (where 1 is a purchase). The mechanism observes the consumer's decision and then charges the bidders according to a payment rule, which may depend on ω.

We require the mechanism to be *truthful*, which means that it is, first, *incentive compatible*: for every merchant i and every set of types reported by the other merchants, i's expected utility is maximized if she reports her true type (v_i, p_i); and second, *interim individually rational*: each merchant receives a non-negative utility in expectation (over the randomization involved in the consumer's purchasing decision) if she reports her true type.

The goal of the auctioneer is to increase some combination of the welfare of all the parties involved. If we ignore the consumer, this can be modeled by the sum of the utilities of the merchants and the auctioneer, which, by quasi-linearity of

---

[1] Our mechanisms for this model can be immediately extended to the case of many consumers by scaling.

[2] In Section 5.3, we will briefly discuss extensions in which both parties have quality information.

the utilities, is precisely v_{i*}. To capture the welfare of the user, we suppose that a reasonable proxy is the quality p_{i*} of the selected deal. We study two natural ways to combine the merchant/auctioneer welfare v_{i*} with the consumer welfare p_{i*}. One is to maximize v_{i*} subject to the deal quality p_{i*} meeting a minimum threshold α. Another is to model the consumer's welfare as a function $g(p_{i*})$ of quality and seek to maximize total welfare $v_{i*} + g(p_{i*})$. In the latter case, when g is a convex function, we construct in the next section a truthful mechanism that maximizes this social welfare function (and we show in Section 4 that, when g is not convex, there is no such mechanism). For the former case, in Section 4, we prove that it is not possible to achieve the objective, even approximately.

3   A Truthful Mechanism via Proper Scoring Rules

In this section, we show that for every *convex* function g, there is an incentive-compatible mechanism that maximizes the social welfare function $v_{i*} + g(p_{i*})$. A convex consumer welfare g function may be natural in many settings. Most importantly, it includes the natural special case of a linear function; and it also intuitively models *risk aversion*, because (by definition of convexity) the average welfare of taking a guaranteed outcome, which is $pg(1) + (1 - p)g(0)$, is larger than the welfare $g(p)$ of facing a lottery over those outcomes.[3,4]

We will make use of *binary scoring rules*, which are defined as follows.

Definition 1. *A binary scoring rule $S : [0, 1] \times \{0, 1\} \mapsto \mathbb{R}$ is a function that assigns a real number $S(\hat{p}, \omega)$ to each probability report $\hat{p} \in [0, 1]$ and state $\omega \in \{0, 1\}$. The expected value of $S(\hat{p}, \omega)$, when ω is drawn from a Bernoulli distribution with probability p, is denoted by $S(\hat{p}; p)$. A scoring rule S is (strictly) proper if, for every p, $S(\hat{p}; p)$ is (uniquely) maximized at $\hat{p} = p$.*

Traditionally, proper binary scoring rules are used to truthfully extract the probability of an observable binary event from an agent who knows this probability: It is enough to pay the agent $S(\hat{p}, \omega)$ when the agent reports the probability $\hat{p}$ and the state turns out to be ω. In our setting, obtaining truthful reports is not so straightforward: A bidder's report affects whether or not they win the auction as well as any scoring rule payment. However, the following theorem shows that, when the consumer welfare function g is convex, then a careful use of proper binary scoring rules yields an incentive-compatible auction mechanism.

---

[3] To see this, suppose 100 consumers arrive, and the welfare of each is the convex function $g(p) = p^2$. If 50 consumers see a deal with $p = 0$ and 50 see a deal with $p = 1$, the total welfare is $50(0) + 50(1) = 50$. If all 100 see a deal with $p = 0.5$, the total welfare is $100 \left(0.5^2\right) = 25$. Under this welfare function, the "sure bet" of 50 purchases is preferable to the lottery of 100 coin flips.

[4] Note that risk aversion is often associated with *concave* functions. These are unrelated as they do *not* map probability distributions to welfare; they are functions $u : \mathbb{R} \rightarrow \mathbb{R}$ that map wealth to welfare. Concavity represents risk aversion in that setting because the welfare of a guaranteed payoff x, which is $u(x)$, is larger than the welfare of facing a draw from a distribution with probability x, which is $xu(1) + (1 - x)u(0)$.

Theorem 1. *Let $g : \mathbb{R} \to \mathbb{R}$ be a convex function. Then there is a truthful auction that picks the bidder i^* that maximizes $v_{i^*} + g(p_{i^*})$ as the winner.*

The proof of this theorem relies on the following lemma about proper binary scoring rules, which is well known and fully proven, for example, in [9].

Lemma 1. *Let $g : [0,1] \to \mathbb{R}$ be a (strictly) convex function. Then there is a (strictly) proper binary scoring rule S_g such that for every p, $S_g(p; p) = g(p)$.*

The proof of Lemma 1 proceeds by checking the claims (omitted here) after constructing S_g. To do so, letting $g'(p)$ be a subgradient of g at point p (that is, the slope of any tangent line to g at p, *e.g.* equalling the derivative if g is differentiable at p), we take $S_g(p,1) = g(p) + (1 - p)g'(p)$ and $S_g(p,0) = g(p) - pg'(p)$.

Proof (Theorem 1). Let h be the following "adjusted value" function: $h(\hat{v}, \hat{p}) = \hat{v} + g(\hat{p})$. For convenience, rename the bidders so that bidder 1 has the highest adjusted value, bidder 2 the next highest, and so on. The mechanism deterministically gives the slot to bidder $1 = i^*$. All bidders except bidder 1 pay zero. Bidder 1 pays $h(\hat{v}_2, \hat{p}_2) - S_g(\hat{p}_1, \omega)$, where S_g is a proper binary scoring rule satisfying $S_g(p; p) = g(p)$ and ω is 1 if the customer purchases the coupon and 0 otherwise. The existence of this binary scoring rule is guaranteed by Lemma 1.

We now show that the auction is truthful. If i bids truthfully and does not win, i's utility is zero. If i bids truthfully and wins, i's expected utility is

$$v_i - h(\hat{v}_2, \hat{p}_2) + S_g(p_i; p_i)$$
$$= h(v_i, p_i) - h(\hat{v}_2, \hat{p}_2).$$

This expected utility is always at least 0 because i is selected as winner only if $h(v_i, p_i) \geq h(v_2, p_2)$. This shows that the auction is interim individually rational.

Now suppose that i reports $(\hat{v}_i, \hat{p}_i)$. If i does not win the auction with this report, then i's utility is zero, but a truthful report always gives at least zero. So we need only consider the case where i wins the auction with this report. Then, i's expected utility is

$$v_i - h(\hat{v}_2, \hat{p}_2) + S_g(\hat{p}_i; p_i)$$
$$\leq v_i - h(\hat{v}_2, \hat{p}_2) + S_g(p_i; p_i)$$
$$= h(v_i, p_i) - h(\hat{v}_2, \hat{p}_2).$$

using the properness of S_g and the definition of $h(v_i, p_i)$. There are two cases. First, if $h(v_i, p_i) < h(\hat{v}_2, \hat{p}_2)$, then $U(\hat{v}_i, \hat{p}_i) < 0$. But, if i had reported truthfully, i would have gotten a utility of zero (having not have been selected as the winner). Second, if $h(v_i, p_i) \geq h(\hat{v}_2, \hat{p}_2)$, then $U(\hat{v}_i, \hat{p}_i) \leq h(v_i, p_i) - h(\hat{v}_2, \hat{p}_2)$. But, if i had reported truthfully, i would have gotten an expected utility of $h(v_i, p_i) - h(\hat{v}_2, \hat{p}_2)$. This shows incentive compatibility.

4 Impossibility Results

An alternative way to combine consumer welfare with the advertiser/auctioneer welfare is to ask for an outcome that maximizes the advertiser/auctioneer welfare subject to the winner's quality parameter meeting a minimum threshold. It is not hard to show that achieving such "discontinuous" objective functions is impossible.[5] A more reasonable goal is to obtain an incentive-compatible mechanism with the following property: for two given thresholds α and β with $\alpha < \beta$, the mechanism always selects a winner i^* with quality p_{i^*} at least α, and with a value v_{i^*} that is at least $v^* := \max_{i:p_i \geq \beta}\{v_i\}$ (or an approximation of v^*).

One approach to solving this problem is to use the result of the previous section (Theorem 1) with an appropriate choice of the function g. Indeed, if we assume the values are from a bounded range $[0, V_{\max})$ and use the auction mechanism from Theorem 1 with a function g defined as follows,

$$g(p) = \begin{cases} 0 & \text{if } p < \alpha \\ \frac{p-\alpha}{\beta-\alpha}\cdot V_{\max} & \text{if } p \geq \alpha \end{cases}$$

then if there is at least one bidder with quality parameter at least β, then the mechanism is guaranteed to pick a winner with quality at least α. This is easy to see: the adjusted bid of the bidder with quality at least β is at least $V_{\max}$, while the adjusted bid of any bidder with quality less than α is less than $V_{\max}$. In terms of the value, however, this mechanism cannot provide any multiplicative approximation guarantee, as it can select a bidder with quality 1 and value 0 over a bidder with quality β and any value less than $\frac{1-\alpha}{\beta-\alpha}V_{\max}$.

Unfortunately, as we show in Theorem 2, this is unavoidable: unless $\beta = 1$ (that is, unless welfare is compared only against bidders of "perfect" quality), there is no deterministic, truthful mechanism that can guarantee a bounded multiplicative approximation guarantee in the above setting.

Theorem 2. *For a given $0 \leq \alpha < \beta \leq 1$ and $\lambda \geq 1$, suppose that a deterministic truthful mechanism satisfies that, if there is some bidder i with $p_i \geq \beta$:*

1. The winner has $p_{i^*} > \alpha$;
2. The winner has value $v_{i^*} \geq v^*/\lambda$, where $v^* := \max_{i:p_i \geq \beta}\{v_i\}$.
Then $\beta = 1$. This holds even if valuations are upper-bounded by a constant V_{max}.

Due to space constraints, the proof is deferred to the full version of the paper [16]. The main idea is as follows. Fix all bids except i's and suppose v^* is the highest bid of any other agent whose quality exceeds α. First, suppose i has quality below α: i must not want to win (so expected payment must be at least V_{max}). Second, suppose i has perfect quality: i must not want to win if v_i is below v^*/λ (so expected payment must be at least v^*/λ). Third, suppose $v_i \geq \lambda v^*$ and $p_i \geq \beta$: i must want to win (so expected payment must be less than λv_j). Combining these inequalities and finding that i's expected payment is convex in p_i (taking into account i's optimal misreport if necessary!), we get that

$$\lambda v^* \geq V_{max}\left(\frac{1-\beta}{1-\alpha}\right) + \frac{v^*}{\lambda}\left(\frac{\beta-\alpha}{1-\alpha}\right),$$

---

[5] Intuitively, the reason is that it is impossible to distinguish between a coin whose probability of heads is α and one whose probability is $\alpha - \epsilon$, when ϵ can be arbitrarily small, by the result of a single flip.

which will hold for all v^* only if $\beta = 1$; otherwise, it fails for $v^* \ll V_{max}$. We can extend the techniques used in the above proof to show that the convexity assumption in Theorem 1 is indeed necessary. The proof is deferred to the full version of the paper [16].

Theorem 3. *Assume $g : \mathbb{R} \to \mathbb{R}$ is a function for which there exists a deterministic truthful auction that always picks the bidder i^* that maximizes $v_{i^*} + g(p_{i^*})$ as the winner. Then g is a convex function.*

5   A General Framework

Daily deals websites generally offer many deals simultaneously, and to many consumers. A more realistic model of this scenario must take into account complex *valuation functions* as well as general *quality reports*. Merchants' valuations may depend on which slot (top versus bottom, large versus small) or even *subset* of slots they win; they may also change depending on which competitors are placed in the other slots. Meanwhile, merchants might like to report quality in different units than purchase probability, such as (for example) total number of coupon sales in a day, coupon sales relative to those of competitors, or so on.

In this section, we develop a general model that can cover these cases and considerably more. As in a standard multidimensional auction, bidders have a valuation for each outcome of the mechanism (for instance, each assignment of slots to bidders). For quality reports, our key insight is that they may be modeled by a *belief* or *prediction* over possible states of the world, where each state has some verifiable quality. This naturally models many scenarios where the designer would like to make a social choice (such as allocating goods) based not only on the valuations of the agents involved, but also on the likely externality on some non-bidding party; however, this externality can be best estimated by the bidders. We model this externality by a function, which we call the *consumer welfare function*, that maps probability distributions to a welfare value. A natural consumer welfare function is the expected value of a distribution.

When this consumer welfare satisfies a convexity condition, we construct truthful mechanisms for welfare maximization in this general setting; we also prove matching negative results. This allows us to characterize implementable welfare functions in terms of *component-wise convexity*, which includes the special case of expected value and can also capture intuitively risk-averse preferences.

We start with a definition of the model in Section 5.1, and then give a truthful mechanism as well as a matching necessary condition for implementability in this model in Section 5.2. In Section 5.3 we give a number of applications and extensions of our general framework.

5.1   Model

We now define the general model, using the multi-slot daily deals problem as a running example to illustrate the definition.

There are m *bidders* (also called *merchants*) indexed 1 through m, and a finite set $\mathcal{O}$ of possible *outcomes* of the mechanism. Each bidder has as private information a valuation function $v_i : \mathcal{O} \to \mathbb{R}$ that assigns a value $v_i(o)$ to each outcome o. For instance, each outcome o could correspond to an assignment of merchants to the available slots, and $v_i(o)$ is i's expected value for this assignment, taking into account the slot(s) assigned to i as well as the coupons in the other slots.

For each $o \in \mathcal{O}$ and each bidder i, there is a finite set of observable disjoint *states* of interest $\Omega_{i,o}$ representing different events that could occur when the mechanism's choice is o. For example, if merchant i is awarded a slot under outcome o, then $\Omega_{i,o}$ could be the possible total numbers of sales of i's coupon when the assignment is o, *e.g.* $\Omega_{i,o} = \{$fewer than 1000, 1000 to 5000, more than 5000$\}$.

Given an outcome o chosen by the mechanism, nature will select at random one of the states ω in $\Omega_{i,o}$ for each bidder i.[6] In the running example, some number of consumers choose to purchase i's coupon, so perhaps $\omega = $ "1000 to 5000".

We let $\Delta_{\Omega_{i,o}}$ denote the probability simplex over the set $\Omega_{i,o}$, i.e., $\Delta_{\Omega_{i,o}} = \{p \in [0,1]^{\Omega_{i,o}} : \sum_{\omega \in \Omega_{i,o}} p_\omega = 1\}$. Each bidder i holds as private information a set of beliefs (or predictions) $p_i : \mathcal{O} \to \Delta_{\Omega_{i,o}}$. For each outcome o, $p_i(o) \in \Delta_{\Omega_{i,o}}$ is a probability distribution over states $\omega \in \Omega_{i,o}$. Thus, under outcome o where i is assigned a slot, $p_i(o)$ would give the probability that i sells fewer than 1000 coupons, that i sells between 1000 and 5000 coupons, and that i sells more than 5000 coupons. We denote the vector of predictions $(p_1(o), \ldots, p_m(o))$ at outcome o by $\boldsymbol{p}(o) \in \times_{i=1}^{m} \Delta_{\Omega_{i,o}}$.

The goal of the mechanism designer is to pick an outcome that maximizes a notion of welfare. The combined welfare of the bidders and the auctioneer can be represented by $\sum_{i=1}^{m} v_i(o)$. If this was the goal, then the problem could have been solved by ignoring the $p_i(o)$'s and using the well-known Vickrey-Clarke-Groves mechanism [17,18,19]. In our setting, however, there is another component in the welfare function, which for continuity with the daily deals setting we call the *consumer welfare*. This component, which depends on the probabilities $p_i(o)$, represents the welfare of a non-bidding party that the auctioneer wants to keep happy (which could even be the auctioneer herself!). The consumer welfare when the mechanism chooses outcome o is given by an arbitrary function $g_o : \times_{i=1}^{m} \Delta_{\Omega_{i,o}} \to \mathbb{R}$ which depends on the bidders' predictions $\boldsymbol{p}(o)$. The goal of the mechanism designer is then to pick an outcome o that maximizes

$$\left(\sum_{i=1}^{m} v_i(o) \right) + g_o(\boldsymbol{p}(o)).$$

For example, in the multi-slot problem, consumer welfare at the outcome o could be defined as the sum of the expected number of clicks of the deals that are allocated a slot in o.

---

[6] These choices do not have to be independent across bidders; indeed, all bidders could be predicting the same event, in which case $\Omega_{i,o} = \Omega_{i',o}$ for all i, i' and nature selects the same state for each i.

A mechanism in this model elicits bids $(\hat{v}_i, \hat{p}_i)$ from each bidder i and picks an outcome o based on these bids. Then, for each i, the mechanism observes the state ω_i picked by nature from $\Omega_{i,o}$ and charges i an amount that can depend on the bids as well as the realized state ω_i. This mechanism is *truthful* (incentive compatible and individually rational) if, for each bidder i, and for any set of reports of other bidders $(\hat{v}_{-i}, \hat{p}_{-i})$, bidder i can maximize her utility by bidding her true type (v_i, p_i), and this utility is non-negative.

5.2 Characterization of Truthful Mechanisms

We begin by defining the convexity property we will use in our characterization.

Definition 2. *A function $f : \Delta_\Omega \mapsto \mathbb{R}$ is convex if and only if for each $x, y \in \Delta_\Omega$ and each $\alpha \in [0, 1]$,*

$$f(\alpha x + (1 - \alpha)y) \le \alpha f(x) + (1 - \alpha)f(y).$$

We call a function $g_o : \times_{i=1}^m \Delta_{\Omega_{i,o}} \mapsto \mathbb{R}$ component-wise convex if for each i and for each vector $\boldsymbol{p}_{-i}(o) \in \times_{j:j \ne i} \Delta_{\Omega_{j,o}}$ of predictions of bidders other than i, $g_o(p_i(o), \boldsymbol{p}_{-i}(o))$ is a convex function of $p_i(o)$.

Component-wise convexity includes the important special case of expected value, and can also capture an intuitive notion of risk aversion with respect to each bidder's prediction, as it requires that the value of taking a draw from some distribution gives lower utility than the expected value of that draw (see footnotes 3 and 4). It also includes functions such as $g(p_1, p_2) = p_1 p_2$ that are component-wise convex, but not convex.

We now state our results for the general model:

Theorem 4. *There is a deterministic truthful mechanism that selects an outcome o maximizing $\sum_{i=1}^m v_i(o) + g_o(\boldsymbol{p}(o))$ if and only if, for each o, the consumer welfare function g_o is component-wise convex.*

As in the simple model, our mechanism uses proper scoring rules, defined for the general setting below. We also need a generalization of Lemma 1.

Definition 3. *A scoring rule $S : \Delta_\Omega \times \Omega \to \mathbb{R}$ is a function that assigns a real number $S(p, \omega)$ to each probability report $p \in \Delta_\Omega$ and state $\omega \in \Omega$. The expected value of $S(\hat{p}, \omega)$ when ω is drawn according to the distribution $p \in \Delta_\Omega$ is denoted by $S(\hat{p}; p)$. A scoring rule S is (strictly) proper if, for every p, $S(\hat{p}; p)$ is (uniquely) maximized at $\hat{p} = p$.*

Lemma 2 ([9,20]). *For every convex function $g : \Delta_\Omega \to \mathbb{R}$ there is a proper scoring rule S_g such that for every p, $S_g(p; p) = g(p)$.*

The full proof of Theorem 4 is given in the full version of the paper [16]; here, we define the mechanism and sketch the main idea behind showing truthfulness.

The proof that component-wise convexity is necessary uses similar ideas to that of Theorem 3.

To define the mechanism, note that, because each g_o is component-wise convex, we can use Lemma 2 to construct, for each outcome o, bidder i, and set of fixed reports $\mathbf{p}_{-i}$ of other bidders, a proper scoring rule $S_{o,i,\mathbf{p}_{-i}(o)}(p_i(o), \omega)$. This scoring rule takes $p_i(o)$, which is i's prediction conditional on choice o, along with the state $\omega \in \Omega_{i,o}$ observed by the mechanism. The expected value for a truthful report is $g_o(p_i, \mathbf{p}_{-i})$.

Let $W^o = \sum_{i=1}^{m} v_i(o) + g_o(\mathbf{p}(o))$. The mechanism chooses the outcome o^* with maximum value W^{o^*}. Let W_{-i} be the value of the choice of the mechanism (that is, what W^{o^*} would be) if i had not participated; then bidder i's payment when outcome o is selected and the state $\omega \in \Omega_{i,o}$ is realized is

$$W_{-i} - \sum_{i' \neq i} v_{i'}(o^*) - S_{o^*,i,\mathbf{p}_{-i}(o^*)}(p_i(o^*), \omega).$$

The proof of truthfulness follows by showing that bidder i's expected utility for reporting truthfully is $W^{o^*} - W_{-i}$, whereas i's expected utility for a misreport that results in the mechanism choosing o' is at most $W^{o'} - W_{-i}$ by properness of the scoring rule, and $W^{o'} \leq W^{o^*}$ by construction of the mechanism.

5.3   Applications

In this section, we present a few sample applications and extensions of our general framework. This demonstrates that the results of Section 5.2 can be used to characterize achievable objective functions and design truthful mechanisms in a very diverse range of settings.

Daily Deals with Both Merchant and Platform Information. In some cases, it might be reasonable in a daily deals setting to suppose that the platform, as well as the merchant, has some relevant private information about deal quality. For example, perhaps the merchant has specific information about his particular deal, while the auctioneer has specific information about typical consumers under particular circumstances (days of the week, localities, and so on). Many such extensions are quite straightforward; intuitively, this is because we solve the difficult problem: incentivizing merchants to truthfully reveal quality information.

To illustrate, consider a simple model where merchant i gets utility a_i from displaying a deal to a consumer and an additional c_i if the user purchases it. For every assignment of slots o containing the merchant's deal, its quality (probability of purchase) is a function $f_{o,i}$ of two pieces of private information: x_i, held by the merchant, and y_i, held by the platform. Each merchant is asked to submit (a_i, c_i, x_i). The platform computes, for each slot assignment o, $p_i(o) = f_{o,i}(x_i, y_i)$, then sets $v_i(o) = a_i + p_i(o)c_i$ for all o that include i's coupon ($v_i(o) = 0$ otherwise). Then, the platform runs the auction defined in Theorem 4, setting i's bid equal to (v_i, p_i). By Theorem 4, bidder i maximizes expected utility when v_i is her true valuation for winning and p_i is her true deal quality; therefore, she can maximize expected utility by truthfully submitting (a_i, c_i, x_i), as this allows the mechanism to correctly compute v_i and p_i.

Reliable Network Design. Consider a graph G, where each edge is owned by a different agent. The auctioneer wants to buy a path from a source node s to a destination node t. Each edge has a cost for being used in the path, and also a probability of failure. Both of these parameters are private values of the edge. The goal of the mechanism designer is to buy a path from s to t that minimizes the total cost of the edges plus the cost of failure, which is a fixed constant times the probability that at least one of the edges on the path fails.

It is easy to see that the above problem fits in our general framework: each bidder's value is the negative of the cost of the edge; each "outcome" is a path from s to t; for each edge i on a path, the corresponding "states" are fail and succeed; the consumer welfare function g_o for an outcome o is the negative of the failure cost of that path. For each edge, fixing all other reports, g_o is a linear function of failure probability. Therefore, g_o is component-wise convex, and Theorem 4 gives a truthful mechanism for this problem.

We can also model a scenario where each edge has a probabilistic delay instead of a failure probability. When edge i is included in the path, the possible states $\Omega_{i,o}$ correspond to the possible delays experienced on that edge. A natural objective function is to minimize the total cost of the path from s to t plus its expected delay, which is a linear function of probability distributions. We can also implement costs that are concave functions of the delay on each edge (as welfare, the negative of cost, is then convex). These model risk aversion, as, intuitively, the cost of a delay drawn from a distribution is higher than the cost of the expected delay of that distribution. (Note that our results imply that a *concave* objective function is *not* implementable!)

The exact same argument shows that other network design problems fit in our framework. For example, the goal can be to pick a k-flow from s to t, or a spanning tree in the graph. The "failure" function can also be more complicated, although we need to make sure the convexity condition is satisfied.

Principal-Agent Models with Probabilistic Signals. Another application of our mechanism is in a *principal-agent* setting, where a principal would like to incentivize agents to exert an optimal level of effort, but can only observe a probabilistic signal of this effort. Suppose the principal wishes to hire a set of agents to complete a project; the principal only observes whether each agent succeeds or fails at his task, but the probability of each's success is influenced by the amount of effort he puts in. More precisely, let $c_i(e)$ denote the cost of exerting effort e for agent i and $p_i(e)$ denote the probability of the agent's success if this agent is hired and exerts effort e. The welfare generated by the project is modeled by a component-wise convex function of the agents' probabilities of success (for instance, a constant times the probability that all agents succeed).

At the first glance, it might seem that this problem does not fit within our framework, since each agent can affect its success probability by exerting more or less effort. However, suppose we define an outcome of the mechanism as selecting both a set of agents and an assignment of effort levels to these agents. Each agent submits as his type the cost $c_i(e)$ and probability of success $p_i(e)$ for each possible effort level e. Theorem 4 then gives a welfare-maximizing mechanism that truthfully elicits the $c_i(e)$ and $p_i(e)$ values from each agent and selects the agents to hire, the effort levels they should exert, and the payment each receives conditional on whether his component of the project succeeds. Agents maximize expected utility by declaring their true types and exerting the amount of effort they are asked to.[7]

6   Conclusion

Markets for daily deals present a challenging new mechanism-design setting, in which a mechanism designer (the platform) wishes to pick an outcome (merchant and coupon to display) that not only gives good bidder/auctioneer welfare, but also good welfare for a third party (the consumer); however, this likely consumer welfare is private information of the bidders.

Despite the asymmetry of information, we show that, when the consumer welfare function is a convex function of bidders' quality, we can design truthful mechanisms for social welfare maximization in this setting. We give a matching negative result showing that no truthful, deterministic mechanism exists when consumer welfare is not convex. Another natural objective, approximating welfare subject to meeting a quality threshold, also cannot be achieved in this setting.

Extending the daily deals setting to a more general domain yields a rich setting with many potential applications. We model this setting as an extension to traditional mechanism design: Now, agents have both preferences over outcomes and probabilistic beliefs conditional on those outcomes. The goal is to maximize social welfare including the welfare of a non-bidding party, modeled by a consumer welfare function taking probability distributions over states of the world to welfare.

A truthful mechanism must incentivize bidders to reveal their true preferences *and* beliefs, even when these revealed beliefs influence the designer to pick a less favorable outcome for the bidders. We demonstrate that this is possible if and only if the consumer welfare function is component-wise convex, and when it is, we explicitly design mechanisms to achieve the welfare objective. Component-wise convexity includes expected-welfare maximization and intuitively can capture risk averse preferences. Finally, we demonstrate the generality of our results with a number of example extensions and applications.

References

1. Byers, J., Mitzenmacher, M., Zervas, G.: Daily deals: Prediction, social diffusion, and reputational ramifications. In: Proceedings of the Fifth ACM International Conference on Web Search and Data Mining, pp. 543–552. ACM (2012)

---

[7] The proof is the same as that of truthfulness: If the agent deviates and exerts some other effort level, his expected utility will be bounded by if he had reported the truth and the mechanism had assigned him that effort level; but by design, this is less than his utility under the choice actually made by the mechanism.

2. Byers, J., Mitzenmacher, M., Zervas, G.: The groupon effect on yelp ratings: a root cause analysis. In: Proceedings of the 13th ACM Conference on Electronic Commerce, pp. 248–265. ACM (2012)
3. Lu, T., Boutilier, C.: Matching models for preference-sensitive group purchasing. In: Proceedings of the 13th ACM Conference on Electronic Commerce, pp. 723–740. ACM (2012)
4. Varian, H.: Position auctions. International Journal of Industrial Organization 25(6), 1163–1178 (2007)
5. Edelman, B., Ostrovsky, M., Schwarz, M.: Internet advertising and the generalized second-price auction: Selling billions of dollars worth of keywords. American Economic Review 97(1), 242–259 (2007)
6. Abrams, Z., Schwarz, M.: Ad auction design and user experience. In: Deng, X., Graham, F.C. (eds.) WINE 2007. LNCS, vol. 4858, pp. 529–534. Springer, Heidelberg (2007)
7. Che, Y.K.: Design competition through multidimensional auctions. RAND Journal of Economics 24(4), 668–680 (1993)
8. Espinola-Arredondo, A.: Green auctions: A biodiversity study of mechanism design with externalities. Ecological Economics 67(2), 175–183 (2008)
9. Gneiting, T., Raftery, A.: Strictly proper scoring rules, prediction, and estimation. Journal of the American Statistical Association 102(477), 359–378 (2007)
10. Azar, P., Chen, J., Micali, S.: Crowdsourced bayesian auctions. In: Proceedings of the 3rd Innovations in Theoretical Computer Science Conference, pp. 236–248. ACM (2012)
11. Hanson, R.: Decision markets. IEEE Intelligent Systems 14(3), 16–19 (1999)
12. Othman, A., Sandholm, T.: Decision rules and decision markets. In: Proceedings of the 9th International Conference on Autonomous Agents and Multiagent Systems (AAMAS), pp. 625–632 (2010)
13. Boutilier, C.: Eliciting forecasts from self-interested experts: scoring rules for decision makers. In: Proceedings of the 11th International Conference on Autonomous Agents and Multiagent Systems. International Foundation for Autonomous Agents and Multiagent Systems, vol. 2, pp. 737–744 (2012)
14. Chen, Y., Kash, I., Ruberry, M., Shnayder, V.: Decision markets with good incentives. In: Chen, N., Elkind, E., Koutsoupias, E. (eds.) WINE 2011. LNCS, vol. 7090, pp. 72–83. Springer, Heidelberg (2011)
15. Shi, P., Conitzer, V., Guo, M.: Prediction mechanisms that do not incentivize undesirable actions. In: Leonardi, S. (ed.) WINE 2009. LNCS, vol. 5929, pp. 89–100. Springer, Heidelberg (2009)
16. Cai, Y., Mahdian, M., Mehta, A., Waggoner, B.: Designing markets for daily deals. Available online at arxiv.org (2013)
17. Vickrey, W.: Counterspeculation, auctions, and competitive sealed tenders. The Journal of Finance 16(1), 8–37 (2012)
18. Clarke, E.: Multipart pricing of public goods. Public Choice 11(1), 17–33 (1971)
19. Groves, T.: Incentives in teams. Econometrica: Journal of the Econometric Society, 617–631 (1973)
20. Savage, L.: Elicitation of personal probabilities and expectations. Journal of the American Statistical Association 66(336), 783–801 (1971)

The Exact Computational Complexity
of Evolutionarily Stable Strategies

Vincent Conitzer

Departments of Computer Science and Economics
Duke University
Durham, NC, USA
conitzer@cs.duke.edu

Abstract. While the computational complexity of many game-theoretic solution concepts, notably Nash equilibrium, has now been settled, the question of determining the exact complexity of computing an evolutionarily stable strategy has resisted solution since attention was drawn to it in 2004. In this paper, I settle this question by proving that deciding the existence of an evolutionarily stable strategy is Σ_2^P-complete.

Keywords: Algorithmic game theory, equilibrium computation, evolutionarily stable strategies.

1 Introduction

Game theory provides ways of formally representing strategic interactions between multiple players, as well as a variety of *solution concepts* for the resulting games. The best-known solution concept is that of Nash equilibrium [Nash, 1950], where each player plays a best response to all the other players' strategies. The computational complexity of, given a game in strategic form, computing a (any) Nash equilibrium, remained open for a long time and was accorded significant importance [Papadimitriou, 2001]. An elegant algorithm for the two-player case, the Lemke-Howson algorithm [Lemke and Howson, 1964], was proved to require exponential time on some game families [Savani and von Stengel, 2006]. Finally, in a breakthrough series of papers, the problem was established to be PPAD-complete, even in the two-player case [Daskalakis *et al.*, 2009; Chen *et al.*, 2009].[1]

Not all Nash equilibria are created equal; for example, one can Pareto-dominate another. Moreover, generally, the set of Nash equilibria does not satisfy *interchangeability*. That is, if player 1 plays her strategy from one Nash equilibrium, and player 2 plays his strategy from another Nash equilibrium, the result is not guaranteed to be a Nash equilibrium. This leads to the dreaded *equilibrium selection problem*: if one plays a game for the first time, how is

---

[1] Depending on the precise formulation, the problem can actually be FIXP-complete for more than 2 players [Etessami and Yannakakis, 2010].

Y. Chen and N. Immorlica (Eds.): WINE 2013, LNCS 8289, pp. 96–108, 2013.

one to know according to which equilibrium to play? This problem is arguably exacerbated by the fact that determining whether equilibria with particular properties, such as placing probability on a particular pure strategy or having at least a certain level of social welfare, exist is NP-complete in two-player games (and associated optimization problems are inapproximable unless P=NP) [Gilboa and Zemel, 1989; Conitzer and Sandholm, 2008]. In any case, equilibria are often seen as a state to which play could reasonably converge, rather than an outcome that can necessarily be arrived at immediately by deduction. Many other solution concepts have been studied from a computational perspective, including refinements of Nash equilibrium [Hansen et al., 2010; Sørensen, 2012], coarsenings of Nash equilibrium (such as correlated equilibrium [Papadimitriou and Roughgarden, 2008; Jiang and Leyton-Brown, 2013] and equilibria of repeated games [Littman and Stone, 2005; Borgs et al., 2010; Kontogiannis and Spirakis, 2008; Andersen and Conitzer, 2013]), and incomparable concepts such as Stackelberg equilibrium [Conitzer and Sandholm, 2006; von Stengel and Zamir, 2010; Conitzer and Korzhyk, 2011].

In this paper, we consider the concept of *evolutionarily stable strategies*, a solution concept for symmetric games with two players. s will denote a pure strategy and σ a mixed strategy, where $\sigma(s)$ denotes the probability that mixed strategy σ places on pure strategy s. $u(s, s')$ is the utility that a player playing s obtains when playing against a player playing s', and

$$u(\sigma, \sigma') = \sum_{s,s'} \sigma(s)\sigma'(s')u(s, s')$$

is the natural extension to mixed strategies.

Definition 1 (Price and Smith [1973]). *Given a symmetric two-player game, a mixed strategy σ is said to be an* evolutionarily stable strategy (ESS) *if both of the following properties hold.*

1. *(Symmetric Nash equilibrium property) For any mixed strategy σ', we have $u(\sigma, \sigma) \geq u(\sigma', \sigma)$.*
2. *For any mixed strategy σ' ($\sigma' \neq \sigma$) for which $u(\sigma, \sigma) = u(\sigma', \sigma)$, we have $u(\sigma, \sigma') > u(\sigma', \sigma')$.*

The intuition behind this definition is that a population of players playing σ cannot be successfully "invaded" by a small population of players playing some $\sigma' \neq \sigma$, because they will perform *strictly* worse than the players playing σ and therefore they will shrink as a fraction of the population. They perform strictly worse either because (1) $u(\sigma, \sigma) > u(\sigma', \sigma)$, and because σ has dominant presence in the population this outweighs performance against σ'; or because (2) $u(\sigma, \sigma) = u(\sigma', \sigma)$ so the second-order effect of performance against σ' becomes significant, but in fact σ' performs worse against itself than σ performs against it, that is, $u(\sigma, \sigma') > u(\sigma', \sigma')$.

Example (Hawk-Dove game [Price and Smith, 1973]). Consider the following symmetric two-player game:

	Dove	Hawk
Dove	1,1	0,2
Hawk	2,0	-1,-1

The unique symmetric Nash equilibrium σ of this game is 50% Dove, 50% Hawk. For any σ', we have $u(\sigma, \sigma) = u(\sigma', \sigma) = 1/2$. That is, everything is a best reponse to σ. We also have $u(\sigma, \sigma') = 1.5\sigma'(\text{Dove}) - 0.5\sigma'(\text{Hawk}) = 2\sigma'(\text{Dove}) - 0.5$, and $u(\sigma', \sigma') = 1\sigma'(\text{Dove})^2 + 2\sigma'(\text{Hawk})\sigma'(\text{Dove}) + 0\sigma'(\text{Dove})\sigma'(\text{Hawk}) - 1\sigma'(\text{Hawk})^2 = -2\sigma'(\text{Dove})^2 + 4\sigma'(\text{Dove}) - 1$. The difference between the former and the latter expression is $2\sigma'(\text{Dove})^2 - 2\sigma'(\text{Dove}) + 0.5 = 2(\sigma'(\text{Dove}) - 0.5)^2$. The latter is clearly positive for all $\sigma' \neq \sigma$, implying that σ is an ESS.

Intuitively, the problem of computing an ESS appears significantly harder than that of computing a Nash equilibrium, or even a Nash equilibrium with a simple additional property such as those described earlier. In the latter type of problem, while it may be difficult to find the solution, once found, it is straightforward to verify that it is in fact a Nash equilibrium (with the desired simple property). This is not so for the notion of ESS: given a candidate strategy, it does not appear straightforward to figure out whether there exists a strategy that successfully invades it. However, appearances can be deceiving; perhaps there is a not entirely obvious, but nevertheless fast and elegant way of checking whether such an invading strategy exists. Even if not, it is not immediately clear whether this makes the problem of *finding* an ESS genuinely harder. Computational complexity provides the natural toolkit for answering these questions.

The complexity of computing whether a game has an evolutionarily stable strategy (for an overview, see Chapter 29 of the Algorithmic Game Theory book [Suri, 2007]) was first studied by Etessami and Lochbihler [2008], who proved that the problem is both NP-hard and coNP-hard, as well as that the problem is contained in Σ_2^P (the class of decision problems that can be solved in nondeterministic polynomial time when given access to an NP oracle). Nisan [2006] subsequently[2] proved the stronger hardness result that the problem is $\text{co}D^P$-hard. He also observed that it follows from his reduction that the problem of determining whether a given strategy is an ESS is coNP-hard (and Etessami and Lochbihler [2008] then pointed out that this also follows from their reduction). Etessami and Lochbihler [2008] also showed that the problem of determining the existence of a *regular* ESS is NP-complete. As was pointed out in both papers, all of this still leaves the main question of the exact complexity of the general ESS problem open. In this paper, this is settled: the problem is in fact Σ_2^P-complete.

The proof is structured as follows. Lemma 1 shows that the slightly more general problem of determining whether an ESS exists whose support is restricted to a subset of the strategies is Σ_2^P-hard. This is the main part of the proof.

---

[2] An early version of Etessami and Lochbihler [2008] appeared in 2004.

Then, Lemma 2 points out that if two pure strategies are exact duplicates, neither of them can occur in the support of any ESS. By this, we can disallow selected strategies from taking part in any ESS simply by duplicating them. Combining this with the first result, we arrive at the main result, Theorem 1.

One may well complain that Lemma 2 is a bit of a cheat; perhaps we should just consider duplicate strategies to be "the same" strategy and merge them back into one. As the reader probably suspects, such a hasty and limited patch will not avoid the hardness result. Even something a little more thorough, such as iterated elimination of very weakly dominated strategies (in some order), will not suffice: in Appendix A I show, with additional analysis and modifications, that the result holds even in games where each pure strategy is the unique best response to some mixed strategy.

2   Hardness with Restricted Support

Definition 2. *In ESS-RESTRICTED-SUPPORT, we are given a symmetric two-player normal-form game G with strategies S, and a subset $T \subseteq S$. We are asked whether there exists an evolutionarily stable strategy of G that places positive probability only on strategies in T (but not necessarily on all strategies in T).*

Definition 3 (MINMAX-CLIQUE). *We are given a graph $G = (V, E)$, sets I and J, a partition of V into subsets V_{ij} for $i \in I$ and $j \in J$, and a number k. We are asked whether it is the case that for every function $t : I \to J$, there is a clique of size (at least) k in the subgraph induced on $\bigcup_{i \in I} V_{i,t(i)}$. (Without loss of generality, we will require $k > 1$.)*

Example. Figure 1 shows a tiny MINMAX-CLIQUE instance (let $k = 2$). The answer to this instance is "no" because for $t(1) = 2, t(2) = 1$, the graph induced on $\bigcup_{i \in I} V_{i,t(i)} = V_{12} \cup V_{21} = \{v_{12}, v_{21}\}$ has no clique of size at least 2.

Recall that $\Pi_2^P = \text{co}\Sigma_2^P$.

Known Theorem 1 ([Ko and Lin, 1995]). *MINMAX-CLIQUE is Π_2^P-complete.*

Lemma 1. *ESS-RESTRICTED-SUPPORT is Σ_2^P-hard.*

Proof: We reduce from the complement of MINMAX-CLIQUE. That is, we show how to transform any instance of MINMAX-CLIQUE into a symmetric two-player normal-form game with a distinguished subset T of its strategies, so that this game has an ESS with support in T if and only if the answer to the MINMAX-CLIQUE instance is "no."

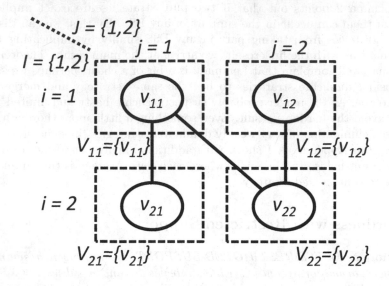

Fig. 1. An example MINMAX-CLIQUE instance (with $k = 2$), for which the answer is "no."

The Reduction. For every $i \in I$ and every $j \in J$, create a strategy s_{ij}. For every $v \in V$, create a strategy s_v. Finally, create a single additional strategy s_0.

- For all $i \in I$ and $j \in J$, $u(s_{ij}, s_{ij}) = 1$.
- For all $i \in I$ and $j, j' \in J$ with $j \neq j'$, $u(s_{ij}, s_{ij'}) = 0$.
- For all $i, i' \in I$ with $i \neq i'$ and $j, j' \in J$, $u(s_{ij}, s_{i'j'}) = 2$.
- For all $i \in I$, $j \in J$, and $v \in V$, $u(s_{ij}, s_v) = 2 - 1/|I|$.
- For all $i \in I$ and $j \in J$, $u(s_{ij}, s_0) = 2 - 1/|I|$.
- For all $i \in I$, $j \in J$, and $v \in V_{ij}$, $u(s_v, s_{ij}) = 2 - 1/|I|$.
- For all $i \in I$, $j, j' \in J$ with $j \neq j'$, and $v \in V_{ij}$, $u(s_v, s_{ij'}) = 0$.
- For all $i, i' \in I$ with $i \neq i'$, $j, j' \in J$, and $v \in V_{ij}$, $u(s_v, s_{i'j'}) = 2 - 1/|I|$.
- For all $v \in V$, $u(s_v, s_v) = 0$.
- For all $v, v' \in V$ with $v \neq v'$ where $(v, v') \notin E$, $u(s_v, s_{v'}) = 0$.
- For all $v, v' \in V$ with $v \neq v'$ where $(v, v') \in E$, $u(s_v, s_{v'}) = (k/(k-1))(2 - 1/|I|)$.
- For all $v \in V$, $u(s_v, s_0) = 0$.
- For all $i \in I$ and $j \in J$, $u(s_0, s_{ij}) = 2 - 1/|I|$.
- For all $v \in V$, $u(s_0, s_v) = 0$.
- $u(s_0, s_0) = 0$.

We are asked whether there exists an ESS that places positive probability only on strategies s_{ij} with $i \in I$ and $j \in J$. That is, $T = \{s_{ij} : i \in I, j \in J\}$.

Example. Consider again the MINMAX-CLIQUE instance from Figure 1. The game to which the reduction maps this instance is:

	s_{11}	s_{12}	s_{21}	s_{22}	$s_{v_{11}}$	$s_{v_{12}}$	$s_{v_{21}}$	$s_{v_{22}}$	s_0
s_{11}	1	0	2	2	3/2	3/2	3/2	3/2	3/2
s_{12}	0	1	2	2	3/2	3/2	3/2	3/2	3/2
s_{21}	2	2	1	0	3/2	3/2	3/2	3/2	3/2
s_{22}	2	2	0	1	3/2	3/2	3/2	3/2	3/2
$s_{v_{11}}$	3/2	0	3/2	3/2	0	0	3	3	0
$s_{v_{12}}$	0	3/2	3/2	3/2	0	0	0	3	0
$s_{v_{21}}$	3/2	3/2	3/2	0	3	0	0	0	0
$s_{v_{22}}$	3/2	3/2	0	3/2	3	3	0	0	0
s_0	3/2	3/2	3/2	3/2	0	0	0	0	0

It has an ESS σ with weight $1/2$ on each of s_{12} and s_{21}. In contrast, (for example) σ' with weight $1/2$ on each of s_{11} and s_{21} is invaded by the strategy σ'' with weight $1/2$ on each of $s_{v_{11}}$ and $s_{v_{21}}$, because $u(\sigma'', \sigma') = u(\sigma', \sigma') = 3/2$ and $u(\sigma'', \sigma'') = u(\sigma', \sigma'') = 3/2$.

Proof of Equivalence. Suppose there exists a function $t : I \to J$ such that every clique in the subgraph induced on $\bigcup_{i \in I} V_{i,t(i)}$ has size strictly less than k. We will show that the mixed strategy σ that places probability $1/|I|$ on $s_{i,t(i)}$ for each $i \in I$ (and 0 everywhere else) is an ESS.

First, we show that σ is a best response against itself. For any s_{ij} in the support of σ, we have $u(s_{ij}, \sigma) = (1/|I|) \cdot 1 + (1 - 1/|I|) \cdot 2 = 2 - 1/|I|$, and hence we also have $u(\sigma, \sigma) = 2 - 1/|I|$. For s_{ij} not in the support of σ, we have $u(s_{ij}, \sigma) = (1/|I|) \cdot 0 + (1 - 1/|I|) \cdot 2 = 2 - 2/|I| < 2 - 1/|I|$. For all $i \in I$, for all $v \in V_{i,t(i)}$, we have $u(s_v, \sigma) = (1/|I|) \cdot (2 - 1/|I|) + (1 - 1/|I|) \cdot (2 - 1/|I|) = 2 - 1/|I|$. For all $i \in I$, $j \in J$ with $j \neq t(i)$, and $v \in V_{ij}$, we have $u(s_v, \sigma) = (1/|I|) \cdot 0 + (1 - 1/|I|) \cdot (2 - 1/|I|) = (1 - 1/|I|)(2 - 1/|I|) < 2 - 1/|I|$. Finally, $u(s_0, \sigma) = 2 - 1/|I|$. So σ is a best response to itself.

It follows that if there were a strategy $\sigma' \neq \sigma$ that could successfully invade σ, then σ' must put probability only on best responses to σ. Based on the calculations in the previous paragraph, these best responses are s_0, and, for any i, $s_{i,t(i)}$ and, for all $v \in V_{i,t(i)}$, s_v. The expected utility of σ against any of these is $2 - 1/|I|$ (in particular, for any i, we have $u(\sigma, s_{i,t(i)}) = (1/|I|) \cdot 1 + (1 - 1/|I|) \cdot 2 = 2 - 1/|I|$). Hence, $u(\sigma, \sigma') = 2 - 1/|I|$, and to successfully invade, σ' must attain $u(\sigma', \sigma') \geq 2 - 1/|I|$.

We can write $\sigma' = p_0 s_0 + p_1 \sigma_1' + p_2 \sigma_2'$, where $p_0 + p_1 + p_2 = 1$, σ_1' only puts positive probability on the $s_{i,t(i)}$ strategies, and σ_2' only puts positive probability on the s_v strategies with $v \in V_{i,t(i)}$. The strategy that results from conditioning σ' on σ_1' not being played may be written as $(p_0/(p_0+p_2))s_0 + (p_2/(p_0+p_2))\sigma_2'$, and thus we may write $u(\sigma', \sigma') = p_1^2 u(\sigma_1', \sigma_1') + p_1(p_0 + p_2)u(\sigma_1', (p_0/(p_0 + p_2))s_0 + (p_2/(p_0 + p_2))\sigma_2') + (p_0 + p_2)p_1 u((p_0/(p_0 + p_2))s_0 + (p_2/(p_0 + p_2))\sigma_2', \sigma_1') + (p_0 + p_2)^2 u((p_0/(p_0 + p_2))s_0 + (p_2/(p_0 + p_2))\sigma_2', (p_0/(p_0 + p_2))s_0 + (p_2/(p_0 + p_2))\sigma_2')$. Now, if we shift probability mass from s_0 to σ_2', i.e., we decrease p_0 and increase

p_2 by the same amount, this will not affect any of the coefficients in the previous expression; it will not affect any of $u(\sigma_1', \sigma_1')$, $u(\sigma_1', (p_0/(p_0 + p_2))s_0 + (p_2/(p_0 + p_2))\sigma_2')$ (because $u(s_{ij}, v) = u(s_{ij}, s_0) = 2 - 1/|I|$), and $u((p_0/(p_0 + p_2))s_0 + (p_2/(p_0 + p_2))\sigma_2', \sigma_1')$ (because $u(s_0, s_{ij}) = u(s_v, s_{ij}) = 2 - 1/|I|$ when $v \in V_{ij}$ or $v \in V_{i'j'}$ with $i' \neq i$); and it will not decrease $u((p_0/(p_0 + p_2))s_0 + (p_2/(p_0 + p_2))\sigma_2', (p_0/(p_0 + p_2))s_0 + (p_2/(p_0 + p_2))\sigma_2')$ (because for any $v \in V$, $u(s_0, s_0) = u(s_0, s_v) = u(s_v, s_0) = 0$). Therefore, we may assume without loss of generality that $p_0 = 0$, and hence $\sigma' = p_1\sigma_1' + p_2\sigma_2'$.

It follows that we can write $u(\sigma', \sigma') = p_1^2 u(\sigma_1', \sigma_1') + p_1 p_2 u(\sigma_1', \sigma_2') + p_2 p_1 u(\sigma_2', \sigma_1') + p_2^2 u(\sigma_2', \sigma_2')$. We first note that $u(\sigma_1', \sigma_1')$ can be at most $2 - 1/|I|$. Specifically, $u(\sigma_1', \sigma_1') = (\sum_i \sigma_1'(s_{i,t(i)})^2) \cdot 1 + (1 - \sum_i \sigma_1'(s_{i,t(i)})^2) \cdot 2$, and this expression is uniquely maximized by setting each $\sigma_1'(s_{i,t(i)})$ to $1/|I|$. $u(\sigma_1', \sigma_2')$ is easily seen to also be $2 - 1/|I|$, and $u(\sigma_2', \sigma_1')$ is easily seen to be at most $2 - 1/|I|$ (in fact, it is exactly that). Thus, to obtain $u(\sigma', \sigma') \geq 2 - 1/|I|$, we must have either $p_1 = 1$ or $u(\sigma_2', \sigma_2') \geq 2 - 1/|I|$. However, in the former case, we would require $u(\sigma_1', \sigma_1') = 2 - 1/|I|$, which can only be attained by setting each $\sigma_1'(s_{i,t(i)})$ to $1/|I|$—but this would result in $\sigma' = \sigma$. Thus, we can conclude $u(\sigma_2', \sigma_2') \geq 2 - 1/|I|$. But then σ_2' would also successfully invade σ. Hence, we can assume without loss of generality that $\sigma' = \sigma_2'$, i.e., $p_0 = p_1 = 0$ and $p_2 = 1$.

That is, we can assume that σ' only places positive probability on strategies s_v with $v \in \bigcup_{i \in I} V_{i,t(i)}$. For any $v, v' \in V$, we have $u(s_v, s_{v'}) = u(s_{v'}, s_v)$. Specifically, $u(s_v, s_{v'}) = u(s_{v'}, s_v) = (k/(k-1))(2 - 1/|I|)$ if $v \neq v'$ and $(v, v') \in E$, and $u(s_v, s_{v'}) = u(s_{v'}, s_v) = 0$ otherwise. Now, suppose that $\sigma'(s_v) > 0$ and $\sigma'(s_{v'}) > 0$ for $v \neq v'$ with $(v, v') \notin E$. We can write $\sigma' = p_0\sigma'' + p_1 s_v + p_2 s_{v'}$, where p_0, $p_1 = \sigma'(s_v)$, and $p_2 = \sigma'(s_{v'})$ sum to 1. We have $u(\sigma', \sigma') = p_0^2 u(\sigma'', \sigma'') + 2p_0 p_1 u(\sigma'', s_v) + 2p_0 p_2 u(\sigma'', s_{v'})$ (because $u(s_v, s_v) = u(s_{v'}, s_{v'}) = u(s_v, s_{v'}) = 0$). Suppose, without loss of generality, that $u(\sigma'', s_v) \geq u(\sigma'', s_{v'})$. Then, if we shift all the mass from $s_{v'}$ to s_v (so that the mass on the latter becomes $p_1 + p_2$), this can only increase $u(\sigma', \sigma')$, and it reduces the size of the support of σ' by 1. By repeated application, we can assume without loss of generality that the support of σ' corresponds to a clique of the induced subgraph on $\bigcup_{i \in I} V_{i,t(i)}$. We know this clique has size c where $c < k$. $u(\sigma', \sigma')$ is maximized if σ' randomizes uniformly over its support, in which case $u(\sigma', \sigma') = ((c - 1)/c)(k/(k-1))(2 - 1/|I|) < ((k-1)/k)(k/(k-1))(2 - 1/|I|) = 2 - 1/|I|$. But this contradicts that σ' would successfully invade σ. It follows that σ is indeed an ESS.

Conversely, suppose that there exists an ESS σ that places positive probability only on strategies s_{ij} with $i \in I$ and $j \in J$. We must have $u(\sigma, \sigma) \geq 2 - 1/|I|$, because otherwise s_0 would be a better response to σ. First suppose that for every $i \in I$, there is at most one $j \in J$ such that σ places positive probability on s_{ij} (we will shortly show that this must be the case). Let $t(i)$ denote the $j \in J$ such that $\sigma(s_{ij}) > 0$ (if there is no such j for some i, then choose an arbitrary j to equal $t(i)$). Then, $u(\sigma, \sigma)$ is uniquely maximized by setting $\sigma(s_{i,t(i)}) = 1/|I|$ for all $i \in I$, resulting in $u(\sigma, \sigma) = (1/|I|) \cdot 1 + (1 - 1/|I|) \cdot 2 = 2 - 1/|I|$. Hence, this is the only way to ensure that $u(\sigma, \sigma) \geq 2 - 1/|I|$, under the assumption that for every $i \in I$, there is at most one $j \in J$ such that σ places positive probability on s_{ij}.

Now, let us consider the case where there exists an $i \in I$ such that there exist $j, j' \in J$ with $j \neq j'$, $\sigma(s_{ij}) > 0$, and $\sigma(s_{ij'}) > 0$, to show that such a strategy cannot obtain a utility of $2 - 1/|I|$ or more against itself. We can write $\sigma = p_0\sigma' + p_1 s_{ij} + p_2 s_{ij'}$, where σ' places probability zero on s_{ij} and $s_{ij'}$. We observe that $u(\sigma', s_{ij}) = u(s_{ij}, \sigma')$ and $u(\sigma', s_{ij'}) = u(s_{ij'}, \sigma')$, because when the game is restricted to these strategies, each player always gets the same payoff as the other player. Moreover, $u(\sigma', s_{ij}) = u(\sigma', s_{ij'})$, because σ' does not place positive probability on either s_{ij} or $s_{ij'}$. Hence, we have that $u(\sigma, \sigma) = p_0^2 u(\sigma', \sigma') + 2p_0(p_1 + p_2)u(\sigma', s_{ij}) + p_1^2 + p_2^2$. But then, if we shift all the mass from $s_{ij'}$ to s_{ij} (so that the mass on the latter becomes $p_1 + p_2$) to obtain strategy σ'', it follows that $u(\sigma'', \sigma'') > u(\sigma, \sigma)$. By repeated application, we can find a strategy σ''' such that $u(\sigma''', \sigma''') > u(\sigma, \sigma)$ and for every $i \in I$, there is at most one $j \in J$ such that σ''' places positive probability on s_{ij}. Because we showed previously that the latter type of strategy can obtain expected utility at most $2 - 1/|I|$ against itself, it follows that it is in fact the *only* type of strategy (among those that randomize only over the s_{ij} strategies) that can obtain expected utility $2 - 1/|I|$ against itself. Hence, we can conclude that the ESS σ must have, for each $i \in I$, exactly one $j \in J$ (to which we will refer as $t(i)$) such that $\sigma(s_{i,t(i)}) = 1/|I|$, and that σ places probability 0 on every other strategy.

Finally, suppose, for the sake of contradiction, that there exists a clique of size k in the induced subgraph on $\bigcup_{i \in I} V_{i,t(i)}$. Consider the strategy σ' that places probability $1/k$ on each of the corresponding strategies s_v. We have that $u(\sigma, \sigma) = u(\sigma, \sigma') = u(\sigma', \sigma) = 2 - 1/|I|$. Moreover, $u(\sigma', \sigma') = (1/k) \cdot 0 + ((k - 1)/k) \cdot (k/(k-1))(2 - 1/|I|) = 2 - 1/|I|$. It follows that σ' successfully invades σ—but this contradicts σ being an ESS. It follows, then, that t is such that every clique in the induced graph on $\bigcup_{i \in I} V_{i,t(i)}$ has size strictly less than k. ∎

3   Hardness without Restricted Support

Lemma 2 (No duplicates in ESS). *Suppose that strategies s_1 and s_2 ($s_1 \neq s_2$) are duplicates, i.e., for all s, $u(s_1, s) = u(s_2, s)$.[3] Then no ESS places positive probability on s_1 or s_2.*

Proof: For the sake of contradiction, suppose σ is an ESS that places positive probability on s_1 or s_2 (or both). Then, let $\sigma' \neq \sigma$ be identical to σ with the exception that $\sigma'(s_1) \neq \sigma(s_1)$ and $\sigma'(s_2) \neq \sigma(s_2)$ (but it must be that $\sigma'(s_1) + \sigma'(s_2) = \sigma(s_1) + \sigma(s_2)$). That is, σ' redistributes some mass between s_1 and s_2. Then, σ cannot repel σ', because $u(\sigma, \sigma) = u(\sigma', \sigma)$ and $u(\sigma, \sigma') = u(\sigma', \sigma')$. ∎

---

[3] It is fine to require $u(s, s_1) = u(s, s_2)$ as well, and we will do so in the proof of Theorem 1, but it is not necessary for this lemma to hold.

Definition 4. *In ESS, we are given a symmetric two-player normal-form game G. We are asked whether there exists an evolutionarily stable strategy of G.*

Theorem 1. *ESS is Σ_2^P-complete.*

Proof: Etessami and Lochbihler [2008] proved membership in Σ_2^P. We prove hardness by reduction from ESS-RESTRICTED-SUPPORT, which is hard by Lemma 1. Given the game G with strategies S and subset of strategies $T \subseteq S$ that can receive positive probability, construct a modified game G' by duplicating all the strategies in $S \setminus T$. (At this point, for duplicate strategies s_1 and s_2, we require $u(s, s_1) = u(s, s_2)$ as well as $u(s_1, s) = u(s_2, s)$.) If G has an ESS σ that places positive probability only on strategies in T, this will still be an ESS in G', because any strategy that uses the new duplicate strategies will still be repelled, just as its equivalent strategy that does not use the new duplicates was repelled in the original game. (Here, it should be noted that the equivalent strategy in the original game cannot turn out to be σ, because σ does not put any probability on a strategy that is duplicated.) On the other hand, if G' has an ESS, then by Lemma 2, this ESS can place positive probability only on strategies in T. This ESS will still be an ESS in G (all of whose strategies also exist in G'), and naturally it will still place positive probability only on strategies in T.   ∎

A   Hardness without Duplication

In this appendix, it is shown that with some additional analysis and modifications, the result holds even in games where each pure strategy is the unique best response to some mixed strategy. That is, the hardness is not simply an artifact of the introduction of duplicate or otherwise redundant strategies.

Definition 5. *In the MINMAX-CLIQUE problem, say vertex v dominates vertex v' if they are in the same partition element V_{ij}, there is no edge between them, and the set of neighbors of v is a superset (not necessarily strict) of the set of neighbors of v'.*

Lemma 3. *Removing a dominated vertex does not change the answer to a MINMAX-CLIQUE instance.*

Proof: In any clique in which dominated vertex v' participates (and therefore its dominator v does not), v can participate in its stead.   ∎

Modified Lemma 1. *ESS-RESTRICTED-SUPPORT is Σ_2^P-hard, even if every pure strategy is the unique best response to some mixed strategy.*

Proof: We use the same reduction as in the proof of Lemma 1. We restrict our attention to instances of the MINMAX-CLIQUE problem where $|I| \geq 2$, $|J| \geq 2$, there are no dominated vertices, and every vertex is part of at least one edge. Clearly, the problem remains Π_2^P-complete when restricting attention to these instances. For the games resulting from these restricted instances, we show that every pure strategy is the unique best response to some mixed strategy. Specifically:

- s_{ij} is the unique best response to the strategy that distributes $1 - \epsilon$ mass uniformly over the $s_{i'j'}$ with $i' \neq i$, and ϵ mass uniformly over the $s_{ij'}$ with $j' \neq j$. (This is because only pure strategies $s_{ij'}$ will get a utility of 2 against the part with mass $1 - \epsilon$, and among these only s_{ij} will get a utility of 1 against the part with mass ϵ.)
- s_v (with $v \in V_{ij}$) is the unique best response to the strategy that places $(1/|I|)(1 - \epsilon)$ probability on s_{ij} and $(1/(|I||J|))(1 - \epsilon)$ probability on every $s_{i'j'}$ with $i' \neq i$, and that distributes the remaining ϵ mass uniformly over the vertex strategies corresponding to neighbors of v. (This is because s_v obtains an expected utility of $2 - 1/|I|$ against the part with mass $1 - \epsilon$, and an expected utility of $(k/(k - 1))(2 - 1/|I|)$ against the part with mass ϵ; strategies $s_{v'}$ with $v' \notin V_{ij}$ obtain utility strictly less than $2 - 1/|I|$ against the part with mass $1 - \epsilon$; and strategies $s_{i''j''}$, s_0, and $s_{v'}$ with $v' \in V_{ij}$ obtain utility at most $2 - 1/|I|$ against the part with mass $1 - \epsilon$, and an expected utility of strictly less than $(k/(k - 1))(2 - 1/|I|)$ against the part with mass ϵ. (In the case of $s_{v'}$ with $v' \in V_{ij}$, this is because by assumption, v' does not dominate v, so either v has a neighbor that v' does not have, which gets positive probability and against which $s_{v'}$ gets a utility of 0; or, there is an edge between v and v', so that $s_{v'}$ gets positive probability and $s_{v'}$ gets utility 0 against itself.))
- s_0 is the unique best response to the strategy that randomizes uniformly over all the s_{ij}. (This is because it obtains utility $2 - 1/|I|$ against that strategy, and all the other pure strategies obtain utility strictly less against that strategy, due to getting utility 0 against at least one pure strategy in its support.)

∎

The following lemma is a generalization of Lemma 2.

Modified Lemma 2. *Suppose that subset $S' \subseteq S$ satisfies:*

- *for all $s \in S \backslash S'$ and $s', s'' \in S'$, we have $u(s', s) = u(s'', s)$ (that is, strategies in S' are interchangeable when they face a strategy outside S');[4] and*
- *the restricted game where players must choose from S' has no ESS.*

Then no ESS of the full game places positive probability on any strategy in S'.

Proof: Consider a strategy σ that places positive probability on S'. We can write $\sigma = p_1\sigma_1 + p_2\sigma_2$, where $p_1 + p_2 = 1$, σ_1 places positive probability only on $S \backslash S'$, and σ_2 places positive probability only on S'. Because no ESS exists

---

[4] Again, it is fine to require $u(s, s') = u(s, s'')$ as well, and we will do so in the proof of Modified Theorem 1, but it is not necessary for the lemma to hold.

in the game restricted to S', there must be a strategy σ'_2 (with $\sigma'_2 \neq \sigma_2$) whose support is contained in S' that successfully invades σ_2, so either (1) $u(\sigma'_2, \sigma_2) > u(\sigma_2, \sigma_2)$ or (2) $u(\sigma'_2, \sigma_2) = u(\sigma_2, \sigma_2)$ and $u(\sigma'_2, \sigma'_2) \geq u(\sigma_2, \sigma'_2)$. Now consider the strategy $\sigma' = p_1\sigma_1 + p_2\sigma'_2$; we will show that it successfully invades σ. This is because $u(\sigma', \sigma) = p_1^2 u(\sigma_1, \sigma_1) + p_1 p_2 u(\sigma_1, \sigma_2) + p_2 p_1 u(\sigma'_2, \sigma_1) + p_2^2 u(\sigma'_2, \sigma_2) = p_1^2 u(\sigma_1, \sigma_1) + p_1 p_2 u(\sigma_1, \sigma_2) + p_2 p_1 u(\sigma_2, \sigma_1) + p_2^2 u(\sigma'_2, \sigma_2) \geq p_1^2 u(\sigma_1, \sigma_1) + p_1 p_2 u(\sigma_1, \sigma_2) + p_2 p_1 u(\sigma_2, \sigma_1) + p_2^2 u(\sigma_2, \sigma_2) = u(\sigma, \sigma)$, where the second equality follows from the property assumed in the lemma. If case (1) above holds, then the inequality is strict and σ is not a best response against itself. If case (2) holds, then we have equality; moreover, $u(\sigma', \sigma') = p_1^2 u(\sigma_1, \sigma_1) + p_1 p_2 u(\sigma_1, \sigma'_2) + p_2 p_1 u(\sigma'_2, \sigma_1) + p_2^2 u(\sigma'_2, \sigma'_2) = p_1^2 u(\sigma_1, \sigma_1) + p_1 p_2 u(\sigma_1, \sigma'_2) + p_2 p_1 u(\sigma_2, \sigma_1) + p_2^2 u(\sigma'_2, \sigma'_2) \geq p_1^2 u(\sigma_1, \sigma_1) + p_1 p_2 u(\sigma_1, \sigma'_2) + p_2 p_1 u(\sigma_2, \sigma_1) + p_2^2 u(\sigma_2, \sigma'_2) = u(\sigma, \sigma')$, where the second equality follows from the property assumed in the lemma. So in this case too, σ' successfully invades σ.   ∎

Modified Theorem 1. *ESS is Σ_2^P-complete, even if every pure strategy is the unique best response to some mixed strategy.*

Proof: Again, Etessami and Lochbihler [2008] proved membership in Σ_2^P. For hardness, we use a similar proof strategy as in Theorem 1, again reducing from ESS-RESTRICTED-SUPPORT, which is hard even if every pure strategy is the unique best response to some mixed strategy, by Modified Lemma 1. Given the game G with strategies S and subset of strategies $T \subseteq S$ that can receive positive probability, construct a modified game G' by replacing each pure strategy $s \in S \setminus T$ by three new pure strategies, s^1, s^2, s^3. For each $s' \notin \{s^1, s^2, s^3\}$, we will have $u(s^i, s') = u(s, s')$ (the utility of the original s) and $u(s', s^i) = u(s', s)$ for all $i \in \{1, 2, 3\}$; for all $i, j \in \{1, 2, 3\}$, we will have $u(s^i, s^j) = u(s, s) + \rho(i, j)$, where ρ gives the payoffs of rock-paper-scissors (with -1 for a loss, 0 for a tie, and 1 for a win).

If G has an ESS that places positive probabilities only on strategies in T, this will still be an ESS in G' because any strategy σ' that uses new strategies s^i will still be repelled, just as the corresponding strategy σ'' that put the mass on the corresponding original strategies s (i.e., $\sigma''(s) = \sigma'(s^1) + \sigma'(s^2) + \sigma'(s^3)$) was repelled in the original game. (Unlike in the proof of the original Theorem 1, here it is perhaps not immediately obvious that $u(\sigma'', \sigma'') = u(\sigma', \sigma')$, because the right-hand side involves additional terms involving ρ. But ρ is a symmetric zero-sum game, and any strategy results in an expected utility of 0 against itself in such a game.) On the other hand, if G' has an ESS, then by Modified Lemma 2 (letting $S' = \{s^1, s^2, s^3\}$ and using the fact that rock-paper-scissors has no ESS), this ESS can place positive probability only on strategies in T. This ESS will still be an ESS in G (for any potentially invading strategy in G there would be an equivalent such strategy in G', for example replacing s by s^1 as needed), and naturally it will still place positive probability only on strategies in T.

Finally it remains to be shown that in G' each pure strategy is the unique best response to some mixed strategy, using the fact that this is the case for G. For a pure strategy in T, we can simply use the same mixed strategy as we use for that pure strategy in G, replacing mass placed on each $s \notin T$ in G with a uniform mixture over s^1, s^2, s^3 where needed. (By using a uniform mixture, we guarantee that each s^i obtains the same expected utility against the mixed strategy as the corresponding s strategy in G.) For a pure strategy $s^i \notin T$, we cannot simply use the same mixed strategy as we use for the corresponding s in G (with the same uniform mixture trick), because s^1, s^2, s^3 would all be equally good responses. But because these three would be the *only* best responses, we can mix in a sufficiently small amount of s^{i+1} (mod 3) (where i beats $i+1$ (mod 3) in ρ) to make s^i the unique best response. ∎

Acknowledgments. I thank ARO and NSF for support under grants W911NF-12-1-0550, W911NF-11-1-0332, IIS-0953756, and CCF-1101659. The compendium by Schaefer and Umans [2008] guided me to the MINMAX-CLIQUE problem.

References

Andersen, G., Conitzer, V.: Fast equilibrium computation for infinitely repeated games. In: Proceedings of the Twenty-Seventh AAAI Conference on Artificial Intelligence, Bellevue, WA, USA, pp. 53–59 (2013)

Borgs, C., Chayes, J., Immorlica, N., Kalai, A.T., Mirrokni, V., Papadimitriou, C.: The myth of the Folk Theorem. Games and Economic Behavior 70(1), 34–43 (2010)

Chen, X., Deng, X., Teng, S.-H.: Settling the complexity of computing two-player Nash equilibria. Journal of the ACM 56(3) (2009)

Conitzer, V., Korzhyk, D.: Commitment to correlated strategies. In: Proceedings of the National Conference on Artificial Intelligence (AAAI), San Francisco, CA, USA, pp. 632–637 (2011)

Conitzer, V., Sandholm, T.: Computing the optimal strategy to commit to. In: Proceedings of the ACM Conference on Electronic Commerce (EC), Ann Arbor, MI, USA, pp. 82–90 (2006)

Conitzer, V., Sandholm, T.: New complexity results about Nash equilibria. Games and Economic Behavior 63(2), 621–641 (2008)

Daskalakis, C., Goldberg, P., Papadimitriou, C.H.: The complexity of computing a Nash equilibrium. SIAM Journal on Computing 39(1), 195–259 (2009)

Etessami, K., Lochbihler, A.: The computational complexity of evolutionarily stable strategies. International Journal of Game Theory 37(1), 93–113 (2004); an earlier version was made available as ECCC tech report TR04-055 (2004)

Etessami, K., Yannakakis, M.: On the complexity of Nash equilibria and other fixed points. SIAM Journal on Computing 39(6), 2531–2597 (2010)

Gilboa, I., Zemel, E.: Nash and correlated equilibria: Some complexity considerations. Games and Economic Behavior 1, 80–93 (1989)

Hansen, K.A., Miltersen, P.B., Sørensen, T.B.: The computational complexity of trembling hand perfection and other equilibrium refinements. In: Kontogiannis, S., Koutsoupias, E., Spirakis, P.G. (eds.) SAGT 2010. LNCS, vol. 6386, pp. 198–209. Springer, Heidelberg (2010)

Jiang, A.X., Leyton-Brown, K.: Polynomial-time computation of exact correlated equilibrium in compact games. Games and Economic Behavior (2013)

Ko, K.-I., Lin, C.-L.: On the complexity of min-max optimization problems and their approximation. In: Du, D.-Z., Pardalos, P.M. (eds.) Minimax and Applications, pp. 219–239. Kluwer Academic Publishers, Boston (1995)

Kontogiannis, S.C., Spirakis, P.G.: Equilibrium points in fear of correlated threats. In: Papadimitriou, C., Zhang, S. (eds.) WINE 2008. LNCS, vol. 5385, pp. 210–221. Springer, Heidelberg (2008)

Lemke, C., Howson, J.: Equilibrium points of bimatrix games. Journal of the Society of Industrial and Applied Mathematics 12, 413–423 (1964)

Littman, M.L., Stone, P.: A polynomial-time Nash equilibrium algorithm for repeated games. Decision Support Systems 39, 55–66 (2005)

Nash, J.: Equilibrium points in n-person games. Proceedings of the National Academy of Sciences 36, 48–49 (1950)

Nisan, N.: A note on the computational hardness of evolutionary stable strategies. Electronic Colloquium on Computational Complexity (ECCC) 13(076) (2006)

Papadimitriou, C.H., Roughgarden, T.: Computing correlated equilibria in multi-player games. Journal of the ACM 55(3) (2008)

Papadimitriou, C.H.: Algorithms, games and the Internet. In: Proceedings of the Annual Symposium on Theory of Computing (STOC), pp. 749–753 (2001)

Price, G., Smith, J.M.: The logic of animal conflict. Nature 246, 15–18 (1973)

Savani, R., von Stengel, B.: Hard-to-solve bimatrix games. Econometrica 74, 397–429 (2006)

Schaefer, M., Umans, C.: Completeness in the polynomial-time hierarchy: A compendium (2008)

Sørensen, T.B.: Computing a proper equilibrium of a bimatrix game. In: Proceedings of the ACM Conference on Electronic Commerce (EC), Valencia, Spain, pp. 916–928 (2012)

Suri, S.: Computational evolutionary game theory. In: Nisan, N., Roughgarden, T., Tardos, E., Vazirani, V. (eds.) Algorithmic Game Theory, vol. 29. Cambridge University Press (2007)

von Stengel, B., Zamir, S.: Leadership games with convex strategy sets. Games and Economic Behavior 69, 446–457 (2010)

The Price of Anarchy of the Proportional Allocation Mechanism Revisited

José R. Correa[1], Andreas S. Schulz[2], and Nicolás E. Stier-Moses[3,4]

[1] Universidad de Chile, Santiago, Chile
correa@uchile.cl
[2] Massachusetts Institute of Technology, Cambridge, MA, USA
schulz@mit.edu
[3] Columbia University, New York, NY, USA
stier@gsb.columbia.edu
[4] Universidad Torcuato Di Tella, Buenos Aires, Argentina

Abstract. We consider the proportional allocation mechanism first studied by Kelly (1997) in the context of congestion control algorithms for communication networks. A single infinitely divisible resource is to be allocated efficiently to competing players whose individual utility functions are unknown to the resource manager. If players anticipate the effect of their bids on the price of the resource and their utility functions are concave, strictly increasing and continuously differentiable, Johari and Tsitsiklis (2004) proved that the price of anarchy is 4/3. The question was raised whether there is a relationship between this result and that of Roughgarden and Tardos (2002), who had earlier shown exactly the same bound for nonatomic selfish routing with affine-linear congestion functions. We establish such a relationship and show, in particular, that the efficiency loss can be characterized by precisely the same geometric quantity. We also present a new variational inequality characterization of Nash equilibria in this setting, which enables us to extend the price-of-anarchy analysis to important classes of utility functions that are not necessarily concave.

1 Introduction

In a pioneering paper[1], Roughgarden and Tardos [9] established that the loss of efficiency caused by selfish behavior in a multicommodity-flow network with affine-linear latency functions is at most 33%, if compared to the cost of a system-optimal solution. Shortly thereafter, in another widely cited paper, Johari and Tsitsiklis [5] observed virtually the same price of anarchy in a completely different context, in which a finite number of bidders with concave, strictly increasing and continuously differentiable utility functions compete for a single divisible resource, which is allocated in proportion to the bids, as suggested by Kelly [6].

---

[1] http://www.acm.org/press-room/news-releases/2012/goedel-prize-2012

Y. Chen and N. Immorlica (Eds.): WINE 2013, LNCS 8289, pp. 109–120, 2013.
© Springer-Verlag Berlin Heidelberg 2013

From the get-go, the question arose whether the sameness of the two bounds was just a coincidence.[2] For instance, Johari and Tsitsiklis [5, Page 418] write:

> However, it remains an open question whether a relationship can be drawn between the two games; in particular, we note that while [our main theorem] holds even if the utility functions are nonlinear, Roughgarden and Tardos have shown that the price of anarchy in traffic routing may be arbitrarily high if link latency functions are nonlinear.

In this paper, we offer an explanation as to why the upper bound on the price of anarchy is the same in both situations. We show that the Johari-Tsitsiklis bound follows from the same geometric quantity that describes the price of anarchy in the selfish-routing game that was analyzed by Roughgarden and Tardos and, for more general latency functions, by Roughgarden [8]. In earlier work [1], the authors of this paper had shown that the price of anarchy in the latter setting is bounded by $1/(1 - \beta(\mathcal{U}))$, where $\mathcal{U}$ is the class of (latency) functions considered and

$$\beta(\mathcal{U}) = \sup_{u \in \mathcal{U}} \ \sup_{0 \le x \le y \le 1} \frac{x\big(u(y) - u(x)\big)}{yu(y)} \ ,$$

as long as the functions in $\mathcal{U}$ are nonnegative, nondecreasing and continuous.[3] For a specific function u and specific values $x \le y$, it is easy to see that the numerator in the expression above is equal to the area of the shaded rectangle in Figure 1, while the denominator corresponds to the area of the big rectangle. In particular, it follows immediately from elementary geometric arguments that this ratio never exceeds $1/4$ for affine-linear functions, leading to a bound of $4/3$ on the price of anarchy. In fact, it was already noted in [1] that this remains true for more general classes of functions, including concave functions.

We show that the price of anarchy of the proportional allocation mechanism of Kelly [6] is bounded by $1/(1 - \beta(\mathcal{U}))$ as well. Here, we assume that all $u \in \mathcal{U}$ are concave, strictly increasing and continuously differentiable, as did Johari and Tsitsiklis [5].[4] In particular, we get the same bound of $4/3$. In addition, our proof is considerably simpler than Johari and Tsitsiklis' original proof, and it follows

---

[2] Koutsoupias and Papadimitriou [7] introduced the price of anarchy as the worst-case ratio of the cost of an equilibrium to that of an optimum. In particular, in the minimization context of Roughgarden and Tardos, the price of anarchy is $4/3$. Johari and Tsitsiklis considered a maximization problem and established a worst-case efficiency loss of 25%. For reasons of consistency, we assume that the price of anarchy in a maximization setting is defined as the worst-case ratio of the value of an optimum to that of an equilibrium. In particular, for us, the price of anarchy is always greater than or equal to one.

[3] As noted in [1], $1/(1 - \beta(\mathcal{U}))$ is equal to the price-of-anarchy value $\alpha(\mathcal{U})$ presented in [8], but the use of $\beta(\mathcal{U})$ allows for the inclusion of more general functions and the geometric interpretation first pointed out in [2].

[4] This assumption ensures the existence and uniqueness of a Nash equilibrium, as long as there are at least two players; see [4,5] for details.

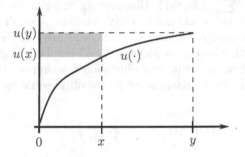

Fig. 1. The geometric interpretation of $\beta(\mathcal{U})$ as the supremum of the shaded area over the area of the big rectangle defined by the origin and the point $(y, u(y))$

the same steps as our earlier proof for the price of anarchy in selfish routing: The key inequality is delivered by a variational inequality derived from the optimality conditions of a concave program. Roughgarden [4, Proof of Theorem 21.4] used the same idea for the proportional allocation mechanism, but relied on a different concave program, which does not lend itself to the same quantity $1/(1 - \beta(\mathcal{U}))$ that arises in the context of selfish routing. Moreover, both the original proof by Johari and Tsitsiklis and Roughgarden's proof make explicit use of the concavity of the utility functions, whereas ours does not, allowing us to extend the analysis of the proportional allocation mechanism to situations in which utility functions are not necessarily concave. For this, we present a characterization of equilibria by a new variational inequality, which holds true as long as the second derivative of the utility functions does not become too positive. This condition encompasses certain convex functions, allowing us to capture some aspects of economies of scale. Corresponding functions include specific polynomials, exponential functions and queueing delay functions, which all give rise to a constant price of anarchy.

2   The Proportional Allocation Mechanism

Johari and Tsitsiklis [5] consider the following model. There is a single divisible resource shared by a set N of players.[5] Each player $i \in N$ has a concave, strictly increasing and continuously differentiable utility function $U_i : [0, 1] \to \mathbb{R}_+$, so that her utility from receiving a fraction x_i of the resource equals $U_i(x_i)$.[6] In a utilitarian setting, the resource would ideally be allocated so as to maximize the total utility:

$$\max_{x \in \Delta} \sum_{i \in N} U_i(x_i),$$

---

[5] To exclude pathological cases, we assume throughout the paper that $|N| \geq 2$.
[6] For convenience, it is assumed that utility is measured in monetary units.

where $\Delta := \{x \in \mathbb{R}_+^N : \sum_{i \in N} x_i \le 1\}$. However, in order to implement this solution, the resource manager would need to have knowledge of all utility functions, which is often not realistic. Alternatively, one could use Kelly's proportional allocation mechanism [6], where each player $i \in N$ bids a nonnegative amount v_i, which is going to be her payment, and obtains a fraction of the resource equal to $x_i = v_i / \sum_{j \in N} v_j$. In particular, player i's payoff is equal to

$$U_i(x_i) - v_i = U_i\left(\frac{v_i}{\sum_{j \in N} v_j}\right) - v_i,$$

and we assume that her payoff is zero if all players bid zero. Player i wants to maximize this expression, and an equilibrium is a vector of bids such that each player bids optimally, given the bids of the other players. Hajek and Gopalakrishnan [3] proved that there exists a unique equilibrium. Moreover, it is not hard to see (e.g., [5, Proof of Theorem 2]) that a vector $v \in \mathbb{R}^N$ is an equilibrium if and only if, for all $i \in N$,

$$(1 - x_i^{\mathrm{NE}})U_i'(x_i^{\mathrm{NE}}) = V \text{ if } v_i > 0, \text{ and } U_i'(0) \le V \text{ if } v_i = 0. \qquad (1)$$

Here, $V = \sum_{j \in N} v_j$ and $x_i^{\mathrm{NE}} = v_i / V$.

3   A New Proof for the Price of Anarchy

We now introduce a new concave program and show that the equilibrium allocation x^{NE} is an optimal solution. Consider

$$\max_{x \in \Delta} \quad \sum_{i \in N} (1 - x_i^{\mathrm{NE}})U_i(x_i), \qquad (2)$$

where x_i^{NE} is fixed in the objective function. The optimality conditions are:

$$(1 - x_i^{\mathrm{NE}})U_i'(x_i) = \lambda + \mu_i \quad \text{for all } i \in N,$$
$$\sum_{i \in N} x_i \le 1,$$
$$\mu_i x_i = 0 \qquad \text{for all } i \in N,$$
$$\mu_i \le 0 \qquad \text{for all } i \in N,$$
$$x_i \ge 0 \qquad \text{for all } i \in N,$$
$$\lambda \ge 0.$$

By taking $\lambda = V$ and $\mu_i = U_i'(0) - V$ when $x_i^{\mathrm{NE}} = 0$, it follows that x^{NE} satisfies the Karush-Kuhn-Tucker conditions and, thus, is a maximum of (2).

Using the optimality of x^{NE} for (2), the monotonicity of the utility functions and the definition of $\beta(\mathcal{U})$, we are ready to prove the desired bound on the price of anarchy of the proportional allocation mechanism by Kelly. Here, $x^* \in \Delta$ is an arbitrary feasible vector, and the result follows when x^* is taken to be the socially optimal assignment:

$$\sum_{i \in N} U_i(x_i^{\mathrm{NE}}) \geq \sum_{i \in N} \left(x_i^{\mathrm{NE}} U_i(x_i^{\mathrm{NE}}) + (1 - x_i^{\mathrm{NE}}) U_i(x_i^*) \right)$$

$$\geq \sum_{i \in N} U_i(x_i^*) - \sum_{i \in N : x_i^* > x_i^{\mathrm{NE}}} x_i^{\mathrm{NE}} \left(U_i(x_i^*) - U_i(x_i^{\mathrm{NE}}) \right)$$

$$\geq \sum_{i \in N} U_i(x_i^*) - \sum_{i \in N : x_i^* > x_i^{\mathrm{NE}}} \beta(\mathcal{U}) x_i^* U_i(x_i^*)$$

$$\geq (1 - \beta(\mathcal{U})) \sum_{i \in N} U_i(x_i^*).$$

We have given a new proof of the upper bound in the following theorem and, at the same time, established a connection to the quantity $\beta(\mathcal{U})$, which plays a similar role in the price of anarchy of selfish routing, as discussed in the introduction.

Theorem 1 (Johari and Tsitsiklis 2004). *The price of anarchy for the proportional allocation mechanism is 4/3 when utility functions are strictly increasing, continuously differentiable and concave.*

4   A Variational Inequality for the Nonconcave Case

In the derivation of the price of anarchy for the proportional allocation mechanism in Section 3 we made use of a new variational inequality obtained easily from the optimality conditions of the concave program (2):

$$\sum_{i \in N} (1 - x_i^{\mathrm{NE}}) \left(U_i(x_i^{\mathrm{NE}}) - U_i(x_i) \right) \geq 0 \text{ for all } x \in \Delta. \tag{3}$$

We will now derive another variational inequality that continues to characterize equilibria even if the players' utility functions are not concave anymore. This allows us to extend the price-of-anarchy analysis to settings that include economies of scale and other situations in which players' utilities may not be concave.

Johari and Tsitsiklis [5] characterized equilibria as optimal solutions to the following nonlinear program:

$$\max_{x \in \Delta} \sum_{i \in N} \left((1 - x_i) U_i(x_i) + \int_0^{x_i} U_i(z) dz \right). \tag{4}$$

The partial derivative of the objective function in direction x_i is $(1 - x_i) U_i'(x_i)$, giving exactly the equilibrium conditions shown in (1). The first-order optimality conditions of Problem (4) can be written as a variational inequality: An equilibrium allocation x^{NE} is characterized by

$$\sum_{i \in N} (1 - x_i^{\mathrm{NE}}) U_i'(x_i^{\mathrm{NE}})(x_i^{\mathrm{NE}} - x_i) \geq 0 \quad \text{for all } x \in \Delta. \tag{5}$$

This is exactly the variational inequality used by Roughgarden [4, Proof of Theorem 21.4] in his proof of the price-of-anarchy result of Johari and Tsitsiklis [5]. In their setting, concavity and monotonicity of the utility functions U_i guarantee that the objective function in (4) is concave, making sure that the optimality conditions of the nonlinear program characterize globally optimal solutions. We will exploit the fact that concavity of this objective can still be guaranteed by less stringent assumptions on U_i. For convenience, we now assume that each U_i is twice-differentiable (and we continue to assume that it is strictly increasing). Then, for the objective function of (4) to be concave over the feasible region Δ, it suffices if, for all $i \in N$,

$$\frac{U_i''(x)}{U_i'(x)} \leq \frac{1}{1 - x} \quad \text{for } 0 \leq x < 1. \tag{6}$$

If this is the case, an equilibrium continues to exist, it is unique, and it is still characterized by the first-order optimality conditions of (4). Notice that (6) is indeed a relaxation of concavity, because for strictly increasing, concave functions, the left-hand side is not positive. In fact, Condition (6) is satisfied by certain convex functions, such as some polynomials, or exponential functions, or wait functions of queueing networks, such as $(c - x)^{-1}$ for $c \geq 2$.

Assuming (6), the following nonlinear program is concave and has exactly the same first-order optimality conditions as (4), as one can easily check by taking the derivative of the objective function:

$$\max_{x \in \Delta} \quad \sum_{i \in N} \left(U_i(x_i) - \int_{U_i(0)}^{U_i(x_i)} U_i^{-1}(z) dz \right).$$

Here, U_i^{-1} is the inverse function of U_i. In turn, after a change of variables, $y_i = U_i(x_i)$, this problem is equivalent to

$$\max_{y : y_i \geq U_i(0), \sum_i U_i^{-1}(y_i) \leq 1} \sum_{i \in N} \left(y_i - \int_0^{y_i} U_i^{-1}(z) dz \right). \tag{7}$$

The optimal solution y^{NE} of (7) is equal to the vector $\left(U_i(x_i^{\text{NE}}) \right)_{i \in N}$, because x^{NE} is the unique solution to (4), and all the nonlinear programs above are equivalent. If all utility functions are concave, y^{NE} satisfies the first-order optimality conditions

$$\sum_{i \in N} \left(U_i^{-1}(y_i^{\text{NE}}) - 1 \right) \left(y_i - y_i^{\text{NE}} \right) \geq 0$$

for all vectors y that are feasible for the constraints of (7). Undoing the change of variables, this variational inequality is equivalent to (3), and we have provided an alternative way of deriving variational inequality (3). In case of nonconcave utility functions that satisfy (6), we can use the following variational inequality to characterize equilibria, which follows directly from (7):

$$\sum_{i \in N} (y_i^{\text{NE}} - y_i) - \left(\sum_{i \in N} \int_{y_i}^{y_i^{\text{NE}}} U_i^{-1}(z) dz \right) \geq 0 \tag{8}$$

for all vectors y that are feasible for the constraints of (7). We will use this variational inequality in the next section to derive price-of-anarchy bounds for non-concave utility functions satisfying (6).

5   The Price of Anarchy in the Nonconcave Case

From now on, we will work with the relaxed concavity assumption (6), which suffices for the variational inequality (8) to hold. We will compute the price of anarchy in this more general setting. For this, we introduce a new constant "beta" that depends again just on the class of utility functions considered:

$$\hat{\beta}(\mathcal{U}) := \sup_{u \in \mathcal{U}} \sup_{u(0) \leq y \leq u(1)} \frac{\int_{u(0)}^{y} u^{-1}(z)dz}{y} .$$

With this definition and variational inequality (8) in place, we bound the price of anarchy of the proportional allocation mechanism using an approach similar to that in Section 3, only that we have one less inequality in the derivation:[7]

$$\sum_{i \in N} U_i(x_i^{\mathrm{NE}}) \geq \sum_{i \in N} U_i(x_i^*) - \sum_{i \in N : x_i^* > x_i^{\mathrm{NE}}} \int_{U_i(x_i^{\mathrm{NE}})}^{U_i(x_i^*)} U_i^{-1}(z)dz$$

$$\geq \left(1 - \hat{\beta}(\mathcal{U})\right) \sum_{i \in N} U_i(x_i^*).$$

Theorem 2. *The price of anarchy for the proportional allocation mechanism is at most $1/(1 - \hat{\beta}(\mathcal{U}))$, when all utility functions belong to a family U of strictly increasing and twice differentiable functions that satisfy* (6).

It remains to compute the actual value of $\hat{\beta}$ for concrete families $\mathcal{U}$ of this kind, which is the content of the next section.

6   Computing $\hat{\beta}$

For the calculation of $\hat{\beta}$, we will parameterize the family of nonnegative, strictly increasing, twice differentiable utility functions that satisfy (6) as follows: For $c \geq 1$, we let $\mathcal{U}^c$ be the set of all such functions U_i for which

$$\left(1 - \frac{x}{c}\right) U_i''(x) \leq U_i'(x) \text{ for } 0 \leq x \leq 1.$$

---

[7] This happens because, compared to the definition of β, the product in the denominator is replaced by a single value. Interestingly, had we defined $\hat{\beta}$ with $y\,u^{-1}(y)$ in the denominator, its geometric interpretation would have been that we are seeking an upper bound on the ratio of the area defined by the integral in the numerator to the area of the rectangle defined by the origin and the point $(u^{-1}(y), y)$, which is easily seen to be at most $1/2$ for concave functions (see Figure 2). While this would suffice to replicate the "easy [...] bound" of Johari and Tsistsiklis [5, Page 415], $\hat{\beta}$ as defined here has the potential to lead to stronger results.

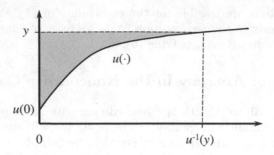

Fig. 2. The geometric meaning of the numerator in the definition of $\hat{\beta}(\mathcal{U})$

It is straightforward to see that $\mathcal{U}^1$ is equal to the set of all functions that satisfy (6), while $\mathcal{U}^\infty$ contains the functions that satisfy

$$U_i''(x) \leq U_i'(x) \quad \text{for } 0 \leq x \leq 1. \tag{9}$$

We begin with the computation of $\hat{\beta}(\mathcal{U}^\infty)$, which amounts to computing

$$\sup_{u:u(0)\geq 0,u'\geq 0,u''\leq u'} \sup_{u(0)\leq y\leq u(1)} \frac{\int_{u(0)}^{y} u^{-1}(z)dz}{y}.$$

Note that for any fixed function u the supremum over y is attained at $y = u(1)$ since the derivative of the argument equals

$$\frac{u^{-1}(y)y - \int_{u(0)}^{y} u^{-1}(z)dz}{y^2},$$

which is greater than zero as u^{-1} is strictly increasing. Therefore we have that

$$\hat{\beta}(\mathcal{U}^\infty) = \sup_{u:u(0)\geq 0,u'\geq 0,u''\leq u'} \frac{\int_{u(0)}^{u(1)} u^{-1}(z)dz}{u(1)}.$$

Note that we may assume $u(0) = 0$; otherwise subtracting $u(0)$ from a given function will maintain feasibility and only increase the supremum. Also, we may assume $u(1) = 1$ since, when dividing a function u by this quantity, the areas that represent the denominator and the numerator will shrink by the same factor, and, of course, feasibility will not be affected. We conclude that:

$$\hat{\beta}(\mathcal{U}^\infty) = \sup_{u:u(0)=0,u(1)=1,u'\geq 0,u''\leq u'} \int_0^1 u^{-1}(z)dz.$$

Therefore, solving the differential equation $u''(z) = u'(z)$ with the initial values $u(0) = 0$ and $u(1) = 1$ provides a feasible solution to our problem, given by $u(z) = (e^z - 1)/(e - 1)$. We now prove that this is optimal.

To maximize the area in the supremum we need to find a function u with the required properties that is as small as possible for any given $0 < z < 1$. However, in principle this function may not exist, as for different values of z we may have different functions being the smallest, so we define the function $g : [0,1] \to \mathbb{R}_+$ by

$$g(z) := \inf_{u:u(0)=0,u(1)=1,u'\geq 0,u''\leq u'} u(z).$$

Clearly $\hat{\beta}(\mathcal{U}^\infty) \leq \int_0^1 g^{-1}(z)dz$, so it remains to compute g to provide a matching upper bound. Note that we may interpret $u(\cdot)$ as a cumulative distribution function (cdf). Calling its pdf h, we have that

$$g(z) = \inf_{h \text{ density s.t. } h'\leq h} \int_0^z h(t)dt.$$

We complete the argument by proving that $g(z) = (e^z - 1)/(e - 1)$. To get a contradiction, suppose that $g(z) < (e^z - 1)/(e - 1)$ and let us consider $f(z) := e^z/(e - 1)$. Then, there exists a density h, such that $h' \leq h$, satisfying that

$$\int_0^z h(t)dt < \frac{e^z - 1}{e - 1} = f(z) - f(0) = \int_0^z f(t)dt.$$

Thus, there exists a point $z_1 < z$ for which $h(z_1) < f(z_1)$, and, as both h and f are density functions, $\int_0^1 h(t)dt = \int_0^1 f(t)dt$, so that we may consider z_2 as the smallest point larger than z_1 for which $h(z_2) = f(z_2)$. Thus h is smaller than f in the interval $[z_1, z_2)$, and they are equal at z_2. Since in this interval h grows more than f, it is immediate that there exists a point $x \in [z_1, z_2)$ for which $h'(x) > f'(x)$. But on the other hand $f(x) = f'(x)$ and $h(x) \geq h'(x)$, implying that $h(x) > f(x)$. A contradiction follows.

We conclude that $g(z) = (e^z - 1)/(e - 1)$ and thus

$$\hat{\beta}(\mathcal{U}^\infty) = 1 - \int_0^1 \frac{e^z - 1}{e - 1}dz = \frac{1}{e - 1} \approx 0.581977,$$

yielding an upper bound of 2.392213 on the price of anarchy.

We now use a similar approach to compute $\hat{\beta}(\mathcal{U}^c)$. Note first that the solution to the differential equation $(1 - z/c)u''(z) = u'(z)$ with initial values $u(0) = 0$ and $u(1) = 1$, for an arbitrary $c > 1$, is given by

$$\bar{u}(z) = \frac{(c - z)^{1-c} - c^{1-c}}{(c - 1)^{1-c} - c^{1-c}} .$$

Therefore $\hat{\beta}(\mathcal{U}^c) \geq \int_0^1 \bar{u}^{-1}(z)dz$. We now prove that this is actually an equality. Again we consider the function $g : [0,1] \to \mathbb{R}_+$ defined by

$$g(z) := \inf_{u:u(0)=0,u(1)=1,u'\geq 0,(1-z/c)u''\leq u'} u(z),$$

which allows us to interpret u as a cdf. Calling its pdf h, we have that

$$g(z) = \inf_{h \text{ density s.t. } (1-z/c)h'\leq h} \int_0^z h(t)dt.$$

To get a contradiction, suppose that $g(z) < \bar{u}(z)$ and let us consider the density function

$$f(z) := \bar{u}'(z) = \frac{(c-1)(c-z)^{-c}}{(c-1)^{1-c} - c^{1-c}}.$$

Then, there exists a density h, such that $(1 - z/c)h' \leq h$, satisfying that

$$\int_0^z h(t)dt < \bar{u}(z) = \int_0^z f(t)dt.$$

Thus, there exists a point $z_1 < z$ for which $h(z_1) < f(z_1)$, and, as both h and f are density functions, $\int_0^1 h(t)dt = \int_0^1 f(t)dt$. We may therefore consider z_2 as the smallest point larger than z_1 for which $h(z_2) = f(z_2)$. Thus, h is smaller than f in the interval $[z_1, z_2)$, and they are equal at z_2. Since in this interval h grows more than f, it is immediate that there exists a point $x \in [z_1, z_2)$ for which $h'(x) > f'(x)$. On the other hand, $f(x) = f'(x)(1 - x/c)$ and $h(x) \geq h'(x)(1 - x/c)$, implying that $h(x) > f(x)$. A contradiction follows.

We conclude that $g(z) = \bar{u}(z)$ and thus

$$\hat{\beta}(\mathcal{U}^c) = 1 - \int_0^1 \bar{u}(z)dz,$$

leading to an upper bound on the price of anarchy of

$$\left(1 - c + \frac{c-1}{c-2} \cdot \frac{(c-1)^{2-c} - c^{2-c}}{(c-1)^{1-c} - c^{1-c}}\right)^{-1}.$$

The concrete numerical value as a function of c can be seen in Table 1 and Figure 3.

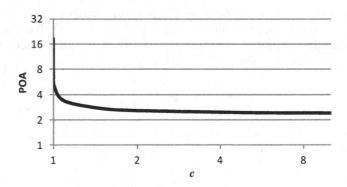

Fig. 3. The new bound on the price of anarchy as a function of $c \geq 1$ (log-log scale)

Table 1. The new bound on the price of anarchy as a function of $c \geq 1$.

c	POA
1	∞
1.000000001	20.72326624
1.00000001	18.42068566
1.0000001	16.11813311
1.000001	13.81578405
1.00001	11.51480886
1.0001	9.222259832
1.001	6.974091719
1.01	4.907853238
1.1	3.345367672
1.5	2.732050808
2	2.587079623
5	2.449115044
10	2.418357995
20	2.404781333
50	2.397126363
100	2.394650571

7   Concluding Remarks

We have added a link between the price-of-anarchy analysis of the nonatomic selfish routing game with concave latency functions and that of the proportional allocation mechanism with concave utility functions. In both cases, the price of anarchy is governed by the same geometric quantity. We have also presented two new variational inequalities characterizing equilibria in the proportional allocation mechanism; one for the case of concave utility functions, and one that works for certain classes of nonconcave utility functions. This allowed us to give the first price-of-anarchy analysis of the proportional allocation mechanism with nonconcave utility functions. Even though we worked under the assumption that, for any $x \in (0, 1)$, the ratio between the second and the first derivative of each utility function is at most $1/(1 - x)$, a closer look at this game reveals that it would actually suffice if that ratio were at most $2/(1 - x)$, which is something to be exploited in the future. We also leave as an open problem to determine tight bounds for the price of anarchy for the nonconcave classes of utility functions considered here. Finally, it is also worth mentioning that the concave problem (7) does not require the differentiability of the U_i's. One can actually prove that the optimal solutions to this problem always coincide with equilibria (a standard application of subdifferentials). Then variational inequality (3), which can be derived from (7), leads to the same price of anarchy in the concave, but not necessarily differentiable case.

Acknowledgements. This work was partially supported by Nucleo Milenio Información y Coordinación en Redes ICM/FIC P10-024F and by CONICET Argentina through grant PICT-2012-1324.

References

1. Correa, J.R., Schulz, A.S., Stier-Moses, N.E.: Selfish routing in capacitated networks. Mathematics of Operations Research 29, 961–976 (2004)
2. Correa, J.R., Schulz, A.S., Stier-Moses, N.E.: A geometric approach to the price of anarchy in nonatomic congestion games. Games and Economic Behavior 64, 457–469 (2008)
3. Hajek, B., Gopalakrishnan, G.: Do greedy autonomous systems make for a sensible Internet? Presented at the Conference on Stochastic Networks, Stanford University, CA (2002)
4. Johari, R.: The price of anarchy and the design of scalable resource allocation mechanisms. In: Nisan, N., Roughgarden, T., Tardos, É., Vazirani, V.V. (eds.) Algorithmic Game Theory. Cambridge University Press (2007)
5. Johari, R., Tsitsiklis, J.N.: Efficiency loss in a network resource allocation game. Mathematics of Operations Research 29, 407–435 (2004)
6. Kelly, F.: Charging and rate control for elastic traffic. European Transactions on Telecommunications 8, 33–37 (1997)
7. Koutsoupias, E., Papadimitriou, C.: Worst-case equilibria. In: Meinel, C., Tison, S. (eds.) STACS 1999. LNCS, vol. 1563, pp. 404–413. Springer, Heidelberg (1999)
8. Roughgarden, T.: The price of anarchy is independent of the network topology. Journal of Computer and System Sciences 67, 341–364 (2003)
9. Roughgarden, T., Tardos, É.: How bad is selfish routing? Journal of the ACM 49, 236–259 (2002)

Can Credit Increase Revenue?

Nishanth Dikkala[1,*] and Éva Tardos[2,**]

[1] Indian Institute of Technology, Bombay, India
nishanth@cse.iitb.ac.in
[2] Dept. of Computer Science, Cornell University, Ithaca, NY, USA
eva.tardos@cornell.edu

Abstract. Classical models of private value auctions assume that bidders know their own private value for the item being auctioned. We explore games where players have a private value, but can only learn this value through experimentation, a scenario that is typical in AdAuctions. We consider this question in a repeated game context, where early participation in the auction can help the bidders learn their own value. We consider what is a good bidding strategy for a player in this game, and show that with low enough competition new bidders will enter and experiment, but with a bit higher level of competition, initial credit offered by the platform can encourage experimentation, and hence ultimately can increase revenue.

Keywords: auction, learning, revenue.

1 Introduction

In on-line advertisement systems, such as sponsored search auction systems, advertisers are repeatedly involved in auctions to acquire advertisement spaces. Standard models of auctions assume that bidders have a private value v_i for the item under auction, which they know. Alternately, models consider a common (or affiliated) value model, where bidders get signals about the common value of an item, such as the classical paper of Milgrom and Weber [6], or the recent work of Abraham et al [1] and Kempe at al [7] motivated by AdAuctions. However, in the context of AdAuctions bidders are not often well-informed about the value of the item they are bidding on (such as an ad spot), and common or affiliated value is just one of the reasons for this. In this paper we will focus on private value auctions, where bidders are not fully informed about their own value. In the context of a single shot auction, such a not fully informed bidder will use her expected value for the item in her bid. However, AdAuctions are best modeled as a repeated game, repeatedly selling identical items (such as ads displayed related to one keyword in search), as advertisers typically have large budgets

* Research was done while visiting Cornell university, supported by NSF grant CCF-0910940.
** Supported in part by NSF grants CCF-0910940 and CCF-0729006, ONR grant N00014-08-1-0031, a Yahoo! Research Alliance Grant, and a Google Research Grant.

Y. Chen and N. Immorlica (Eds.): WINE 2013, LNCS 8289, pp. 121–133, 2013.

compared to the magnitude of their bids. In a repeated auction, winning early runs can be helpful not only for the inherent value of the item won, but also as a means of learning about the value, and hence allowing better bidding behavior in future runs. We explore possible bidding strategies, trading off the instant value of winning with the value of learning, which was also considered by Iyer et al [2] and Gummadi et al [4] using the notion of mean-field equilibria.

The main result of this paper is understanding the effect of offering credit to such participants on the revenue of the auctioneer. Work of Bulow and Klemperer [3] shows that recruiting additional bidders in auctions is better than setting the optimal reserve price. One practical strategy for recruiting additional bidders is to offer special "deals" for new bidders, such as free or discounted bidding for an initial time period. In this paper we develop a simple model to study the value of early credit in such repeated auctions. We'll show that it can be in the platform's direct financial interest to allow bidders, who may not be fully informed about their own value, free experimentation. With low competition in the current auction, a new bidder will want to enter and experiment, and credit isn't needed to entice them. With extremely high competition, the dominant effect of credit to new entrants is the loss of revenue (from auctions when credit was used). We show that there is a middle range of competition levels, where the initial revenue loss of such credit is compensated for by recruiting bidders who learn, through such experimentation, of the high value of participation. We show that in addition to just continued participation beyond the initial credit, such bidders will also learn faster their best bidding strategy, and hence start bidding aggressively earlier, further enhancing revenue.

We will consider auction with bidders who are not fully aware of their own value in the context of AdAuctions. Standard models of AdAuctions assume that the advertiser has a private value v_i for each click (that she knows), and there is a known probability γ_i, possibly depending on the advertiser i, of getting a click when the ad is displayed. This then would result in a value of $v_i\gamma_i$ for the impression. In the standard pay-per-click auction Advertisers then bid their maximum willingness to pay for a click, and the platform needs to learn the click-through-rate γ_i to be able to rank advertisers for their declared value of an impression. However, most of the value for a click is coming from conversions, i.e., events when the viewer not only clicked to look at the advertiser's web page, but also purchased something, or subscribed etc. In this paper, we will assume that advertisers have a known value w_i for conversion, but they need to learn what is the probability q_i that a click will lead to a conversion. Alternate variants of AdAuctions suggest charging advertisers by impression, that is every time the ad is displayed, or only charging by conversion, that is, only when the click resulted in a desired purchase by the viewer. From this perspective, the standard pay-per-click auction has many advantages, as clicks are easy to monitor both by the platform and by the advertiser. However, the pay-per-click auction leaves both the advertiser and the platform with an uncertainty. The platform needs to learn the click-through rate γ_i for each advertiser to be able to estimate the claimed value of the impression $\gamma_i b_i$ for a bidder with bid (claimed value) of b_i.

The advertiser is faced with a similar estimation problem in trying to find her own expected value $v_i = w_i q_i$ for a click. Lahaie and McAfee [5] explore the effect on the platform of not knowing the click-through-rates γ_i, and suggest a ranking rule of ads that trades off the immediate value of displaying ads against the added value of learning of the click-through rates γ_i. Iyer at al [2] and Gummadi et al [4] consider the optimal bidding strategy of such bidders. In this paper, we initiate a similar study of understanding how the bidder's uncertainty about her own value effects the platform.

We analyze the problem faced by a single advertiser competing in a series of second-price auctions. Analogous to the work of [2] and Gummadi et al [4] we focus on a decisions by a single new bidder, and will assume that the other bidders are in steady state, the highest bids of the opponents are independent and identically distributed over different auctions. To simplify the presentation, we will assume that the click-through-rates γ_i are known, and we will further assume that γ_i the equal to 1 for all bidders throughout most of the paper, but with known click-through-rates γ_i our results extend to the general case. We model the value of the advertiser in two steps, we assume that each advertiser has a value w_i for a conversion, which is known both to her and to the platform, and there is a probability q_i that a click will lead to a conversion. It is this probability that is typically not known to an advertiser who is new to on-line advertising. We will assume that q_i is drawn from a known probability distribution. We assume that every time the ad of advertiser i is clicked on, the click results in a conversion of value w_i with probability q_i, and different clicks are independent. Under this assumption, a click will not only result in expected value $v_i = w_i q_i$ for the advertiser, but will also help her in gaining information about the value q_i, and as a result, help to get a better estimate of her value $v_i = w_i q_i$.

Our Results: The main results of this paper are to characterize the optimal bidding strategy for an advertiser, and show that credit offered to new participants have the potential to improve the auctioneer's revenue depending on the level of competition. In particular, we propose a simple model to study this phenomena, and find the optimal strategy of bidders, (note that rational bidders need to bid more aggressively than suggested by the expected value $E(v_i) = w_i E(q_i)$, as winning early auctions helps improve the bidders' estimate their own value). Further, we'll show that depending on the level of current competition, it can be in the platform's financial interest to allow new bidders free experimentation, as free experimentation leads to faster learning, which helps not only the bidders, but also the auctioneer.

2 Preliminaries

The Auction Model: Our action model is a simplified model of repeated auctions for advertisement space, similar to the model used by Iyer et al [2]. In each time step a single item (ad slot) is sold on a second price auction. Each item is identical. We assume that the auction already has steady participation from a number of well established bidders and we focus on a new entrant into the auction. We assume that the bids of the established bidders are identically distributed, and this distribution is not effected by the new entrant. The new bidder's bid is denoted by b, the highest opponent's bid by b', and the second highest opponents' bid by b''. The resulting revenue is b' if the bidder in focus won the auction and $\max(b, b'')$ is she did not win.

We start in Section 3 by analyzing the full information version of the auction with b' and b'' known and fixed. For the full information case, we assume that a bidder who knows that he won't win, simply won't participate in the auction. This assumption reflects our focus aiming to understand to what extent credit recruits valuable new bidders. In more realistic scenarios bidders are not usually this well informed, and may not know to withdraw from the auction. In Section 4 we consider the Bayesian version, assuming only the distribution of the highest and second highest bids is known, and is identical across time steps. With the Bayesian uncertainty, new bidders may want to stay in with a low bid, even if they rarely win.

Conversion Rate and Bidder's Value: We focus on a single bidder, and hence will drop the index i in what follows. The AdAuctions requires the bidder to enter a bid b (claimed value) for a click. The advertiser has a value w for a conversion and has a probability q that a click results in a conversion. For simplicity of presentation, without loss of generality, we assume throughout most of the paper that winning the auction is guaranteed to result a click ($\gamma = 1$ using the notation from the introduction). A click will then result in a conversion with probability q, and the bidder has value $v = wq$ for each conversion.

We model uncertainty about the value by focusing on the uncertainty about the conversion rate q. We assume that w is known both to the bidder as well as to the auctioneer. To simplify notation, we will assume, without loss of generality, that the new entrant's value for conversion is $w = 1$. The bidder does not know her conversion rate q, which is drawn at random uniformly from $[0, 1]$. With each click she wins, she learns more about q, using Bayesian rule, described below. The utility of the bidder in each round is computed by taking the difference between the value gained and amount spent to win the auction. The utility gained in a single round is defined below:

$$U = \begin{cases} 0 - b' & \text{if the bidder wins the auction but does not convert} \\ 1 - b' & \text{if the bidder wins the auction and converts} \\ 0 & \text{if the bidder doesn't win the auction .} \end{cases} \tag{1}$$

Hence if the value of the conversion rate is q, then the expected utility given that the bidder wins the auction is $q(1 - b') + (1 - q)(0 - b') = q - b'$.

Note that we assume that advertisers have no budget constraint, and their goal is simply to maximize the total expected utility over time, using the exponential discounting as described above.

Temporal Discounting: We'll use a model with infinite time horizon and temporal exponential discounting on both the utility gained by the bidder and the revenue gained by the auctioneer with the factor $d(\geq 0)$, i.e utility gained by the bidder due to a conversion at time t is $(1 - b') \times e^{-td}$. Therefore, if the bidder gains a utility of U in every round starting from the kth round, her total expected utility at the beginning would be $Ue^{-kd} + Ue^{-(k+1)d} + Ue^{-(k+2)d} + \cdots = \dfrac{Ue^{-kd}}{1 - e^{-d}}$. We use the same temporal exponential discounting for the revenue of the auctioneer.

Modeling Credit: The main goal of this paper is to understand the effect on the auctioneer's revenue of credit given to the bidder for initial experimentation. We will model credit as a single free bid, and assume that with the credit the bidder can bid high enough to win the auction.

The Learning Model: The bidder in focus has a single parameter q, the conversion rate, which she needs to learn. Winning a round of the auction, doesn't only result in added utility for the winner, but also helps her learn the value of the conversion rate q.

- At every stage of the repeated auction process, the bidder's knowledge about the value q is a distribution from which the value of q is chosen.
- We assume that the bidder starts with a uniform distribution for $q \in [0, 1]$ initially. Given a prior distribution in a round, the posterior is computed using a Bayesian rule as follows: If the density function of the prior is f, it is unchanged if the bidder doesn't win, as no new information is available. If the bidder wins, the posterior is either f_w, if he converts, and f_l if he doesn't convert, where f_w and f_l are:

$$f_w(q) = \frac{qf(q)}{\int_0^1 qf(q)\,dq} = \frac{qf(q)}{E[q]}. \tag{2}$$

$$f_l(q) = \frac{(1-q)f(q)}{\int_0^1 (1-q)f(q)\,dq} = \frac{(1-q)f(q)}{1 - E[q]}. \tag{3}$$

- If the bidder in focus does not participate or does not win the auction in the current round, her posterior is the same as her prior. She does not gain any knowledge in this round and must wait for the next round to do so.

We will focus on a model where bidders only use a single time step to improve their estimate of the conversion rate q, as this single time step learning already demonstrates the important features of this model.

Note that initially we have $E(q) = 1/2$ as q is drawn uniformly at random from $[0, 1]$. Using the above formulas, we get that after winning a single click, the expectation is $2/3$ if the click results in a conversion, and $1/3$ if it does not.

3   Full Information Model

We first consider the full information model, when b' and b'' are fixed. Without a learning opportunity, the new bidder has to evaluate her expected value in participating the auction. When the value for a conversion is $w = 1$, and the expected value of the conversion rate q is $1/2$, expected value of a click is $wE(q) = 1/2$. So in a second price auction, the dominant strategy for the new bidder is to bid her expected value $b = 1/2$, and win if and only if $b' < 1/2$.

Lemma 1. *If a new bidder with value w and unknown conversion rate, has to bid a single value b that will be used through the auction, then bidding $wE(q)$ is dominant strategy.*

We can summarize this, as the new bidder will participate if and only if $b' < 1/2$, and drop out when $b' < 1/2$.

With a learning opportunity, a bidder has more strategy options. Although the bidder has a choice of bidding any value between 0 and 1, since the value of b' is known, the only choice the bidder makes in each round is whether to win the auction (by bidding above b') or not take part in the auction at all. Also, since she is allowed to learn only in the first round, from the second round onwards the choice she needs to make is whether to continue or quit and once this choice is made it never changes in future rounds. The options are summarized by the following strategies:

1. Never participate in the auction and withdraw permanently from the first round itself.
2. Win the first round, i.e $t = 0$ and then do the following:
   (a) If conversion occurs in the first round continue winning the auction forever.
   (b) If conversion does not occur in the first round withdraw from the auction permanently.
3. Always participate and win in the auction by bidding a value above b' in all rounds.

If b' is very low, the cost of winning is low, and hence even after a click with no conversion, the expected value of winning exceeds the cost. Recall that the expected value of q conditioned on the first click not resulting in a conversion is $1/3$, so if $b' < 1/3$, the bidder's dominant strategy is the first one. Its not hard to see that when $b' > 2/3$, then never participating in the auction is dominant strategy for the bidder, but we'll see that when b' is close to $2/3$ the expected revenue cannot make up for the cost of winning. As b' goes higher the bidder's best strategy changes from the first, the second, and then to the last. To decide what is the best strategy, we need to evaluate the expected value for each option.

Lemma 2. *The expected utility gained by the bidder with each strategy is given below:*

- *Option 1: has utility 0.*
- *Option 2: has utility*

$$-b' + \frac{1}{2}\left(1 + \left(\frac{2}{3} - b'\right)\frac{e^{-d}}{1 - e^{-d}}\right). \tag{4}$$

- *Option 3: has utility $\left(\frac{1}{2} - b'\right)\frac{1}{1-e^{-d}}$.*

Proof. In the case when we never participate the utility gained is clearly 0.

When the optimal strategy is unconditional participation, the expected value of the utility in any given round is $E[q] - b'$. Since the distribution of q is uniform, $E[q] = \frac{1}{2}$ and hence the total utility is computed along with time discounting to be equal to $(\frac{1}{2} - b')(1 + e^{-d} + e^{-2d} + \cdots) = \frac{(\frac{1}{2} - b')}{1 - e^{-d}}$.

When the optimal strategy is to experiment for a round, the bidder initially has $E[q] = \frac{1}{2}$ and hence believes she will get a conversion in the first round with a probability of $\frac{1}{2}$, and so has value $\frac{1}{2}$ in expectation. Given a conversion, the expectation goes up to $\frac{2}{3}$ and hence from the second round onwards she believes that she will win with a probability of $\frac{2}{3}$ in expectation. There is always a payment of b' to be made in the first round which is the cost of experimentation. Hence the expected utility turns out to be equal to $-b' + \frac{1}{2} + \frac{1}{2}(\frac{2}{3} - b')\frac{e^{-d}}{1 - e^{-d}}$.

Given the utilities resulting from each strategy option, we can now decide what the best strategy is for the bidder, depending on the value of b'. We denote the point (value of b') where the second strategy begins to dominate the first by β and the point where the third strategy takes over the second by α.

Corollary 1. *The value of α and β are given by:*

$$\alpha = \frac{3 - e^{-d}}{3(2 - e^{-d})},$$
(5)

$$\beta = \frac{1}{3}.$$
(6)

Remark: Note that the value of α is above $\frac{1}{2}$, so with the opportunity to withdraw after a round, bidders experiment, even with higher values of b'. On the other hand, α is always less than $\frac{2}{3}$, the expected value of q after a positive outcome of the experiment. So in the first round, the bidder is trading off the cost of the experiment b' which could be above the expected value of the outcome in this round, against the future value of learning that q may be more likely high.

Next, we will explore how credit given in the form of a single free-start affects the strategies, and outcomes. With a free start, there is no cost and a possible benefit in participating in the first round, so the strategies available to the bidder are:

1. Always participate and win in the auction by bidding a value above b'.
2. Win the first round, i.e $t = 0$ and do the following:
   (a) If conversion occurs in the first round continue winning the auction forever.
   (b) If conversion does not occur in the first round withdraw from the auction permanently.
3. Win the first round and withdraw permanently thereafter no matter what happens in the first round.

As before, depending on the value of b' the best strategy changes from the first (when b' is low), to the second and third options as b' gets higher. For the version with credit, we denote the point (value of b') where the second strategy begins to dominate the first by β' and the point where the third strategy takes over the second by α'.

Lemma 3. *The expected utility in the fixed b' case when credit in the form of a single free-start is given is:*

$$E[U(b')] = \begin{cases} \frac{1}{2} & \text{if } b' \geq \alpha' \\ \frac{1}{2}\left(1 + (\frac{2}{3} - b')\frac{e^{-d}}{1-e^{-d}}\right) & \text{if } \alpha' \geq b' \geq \beta \\ (\frac{1}{2} - b')\frac{1}{1-e^{-d}} + b' & \text{if } b' < \beta', \end{cases} \tag{7}$$

where the values of α' and β are the following:

$$\alpha' = \frac{2}{3}, \tag{8}$$

$$\beta' = \frac{1}{3}. \tag{9}$$

Proof. Since the bidder no longer has to pay a cost of b' in the first round, there is no cost to the first round of participation. After the first round, the expected value of the bidder is either $\frac{1}{3}$ or $\frac{2}{3}$ depending on the outcome of the first round, so by Lemma 1, she will continue in either case if $b' \leq \beta' = \frac{1}{3}$ and continue only on positive outcome if $\frac{1}{3} \leq b' \leq \alpha' = \frac{2}{3}$, and otherwise withdraw in either case.

Remark: Notice that $\beta' = \beta$, as after the first round of experiments, the decision to continue participating is no longer affected by the cost of the first round (which is now sunk cost). On the other hand, $\alpha' > \alpha$. With costly experimentation, the cost of the experiment is traded off against the potential gain. There is a range of $\alpha' > b' > \alpha$, where the new bidder will remain in the auction with probability $\frac{1}{2}$ (if the first win results in a conversion), but with costly experimentation she would have never tried.

Now considering revenue, Corollary 1 and Lemma 3 shows that giving credit when $b' \leq \alpha'$ or if $b' > \alpha$, will only cost the auctioneer one round of income, without effecting the probability of recruiting a new bidder. If competition isn't too high, a new bidder will join, or at least experiment, even without credit. With $b' \leq \beta = \beta'$ the new bidder will remain in the auction in any case, with $\beta < b' \leq \alpha$, he will experiment, even on his own cost, and remain in the auction with probability $\frac{1}{2}$ (if he converts in the first round). If competition is already very high, even free initial experimentation cannot recruit a new bidder, if $b' \geq \alpha'$, the new bidder will withdraw after a round, even if he wins. But if $\alpha < b' < \alpha'$, free credit changes the behavior of the new bidder. To see whether this change is positive for the auctioneer's revenue, we need to trade off the lost revenue from the free round, with the possible gain of an extra bidder. The region where credit can increase revenue is shown on Figure 1.

Theorem 1. *In the full information model with fixed b', and single-shot learning, the auctioneer has a possibility of recruiting an additional bidder via giving credit in the form of a single free-start when $\frac{2}{3} \geq b' \geq \frac{3-e^{-d}}{3(2-e^{-d})}$. This free credit will increase revenue in expectation if $b' > (2e^d - 1)b''$.*

Proof. Giving credit for the first round results in a loss of revenue of b'' for the first round, but a possible gain if the new entrant participates in future auctions, while she would not have participated without the credit. The range in which credit possibly effects future participation is $\frac{2}{3} \geq b' \geq \frac{3-e^{-d}}{3(2-e^{-d})}$. The loss of revenue is b''. With probability $\frac{1}{2}$ the credit will lead to no conversion, and hence not lead to future participation, but with probability $\frac{1}{2}$, credit will lead to conversion, and hence the new entrant remains in the auction. This will raise the revenue in each round from b'' to b', resulting in a total of $(b' - b'')\frac{e^{-d}}{1-e^{-d}}$ increase in revenue. Therefore credit is beneficial when $b'' < \frac{1}{2}(b' - b'')\frac{e^{-d}}{1-e^{-d}}$, which is $(2e^{-d} - 1)b'' < b'$, as claimed.

We have so far assumed that the CTR γ equals 1 for all bidders. However, this assumption is not crucial for our finding. If the CTR's of different bidders are known, but not all equal to 1, let $b'\gamma'$ be the highest product of the bid and CRT of bidders already in the auction, while let $b\gamma$ be the product of bid and CRT of the new bidder. The bidder wins if $b\gamma \geq b'\gamma'$, and pays $p' = \frac{b'\gamma'}{\gamma}$ for each click. If free credit is offered for one display of the ad, the tradeoff in utility and revenue is analogous to that of Corollary 1 and Lemma 3 with b' replaced by the price p', and scaled with the probability γ that the click will happen (and hence the cost and utility and learning materialized). Similarly, if free credit is offered for the first click, again assuming that the CTRs are known, the cost and benefit is $1/\gamma$ times those with just one free display, and the same conclusion holds.

Corollary 2. *In the full information model with fixed b', and single-shot learning, the auctioneer has a possibility of recruiting an additional bidder via giving credit in the form of a single free display or single free click of the ad when $\frac{2}{3} \geq \gamma p' \geq \frac{3-e^{-d}}{3(2-e^{-d})}$. This free credit will increase revenue in expectation if $\gamma p' > (2e^d - 1)b''$.*

Remark: To simplify the presentation, we focused on a single round of learning. However, similar conclusion also holds for multiple rounds of learning. The region where revenue increase can take place is no longer a single contiguous region but a number of non-contiguous regions. This is showed in Figure 2, where the green regions are those where an increase in revenue is possible.

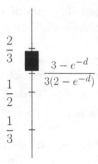

Fig. 1. Region of revenue increase possibility. (single shot learning model with fixed b').

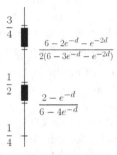

Fig. 2. Regions of revenue increase possibility. (2-shot learning model with fixed b').

4   Bayesian Model

In this section we consider a model where the maximum bid b' comes from a known distribution. We assume that each time step is independent, and bids come from the same distribution. Analogous to Lemma 1 if the entering bidder has no opportunity to learn and has to provide a bid b used throughout the auctions, then her dominant strategy is to bid her expected value $b = wE(q) = \frac{1}{2}$ even against variable opponents.

Similarly, if credit is offered for the first bid, she should bid high enough to win the first auction (for free). If this results in conversion, the resulting conditional expectation for q is $\frac{2}{3}$, so her dominant strategy is to bid $\frac{2}{3}$ in all further auctions; while if the first win doesn't result in a conversion then her dominant strategy is to bid $\frac{1}{3}$ in all further auctions.

Lemma 4. *With no further opportunity to learn, it is dominant strategy for the bidder to bid his expected value for a click. So with no learning opportunity, it is dominant strategy to bid $\frac{1}{2}$. With credit given in the first round, it is dominant strategy for the bidder to bid high enough to win the first round, and then bid 2/3 or 1/3 depending whether the first win resulted in a conversion.*

It is more challenging to evaluate what the best strategy is without credit. We understand the model with a single learning step to mean that after the first time period that the bidder wins, she has the chance to withdraw from further rounds. Under this model, the best bid in initial rounds may be some value b between 0 and 1 (the smallest and largest possible value for the bidder depending on the value of the conversion probability q). A smaller value of b will take longer to result in a win, while a higher value of b results in higher b' in expectation. So the cost of experimentation is now traded off against delay, and hence lost value due in the initial rounds. After this first win, the bidder has a dominant strategy of bidding $\frac{2}{3}$ or $\frac{1}{3}$ depending on whether conversion occurred at the first win.

Next we claim that the optimal strategy for the bidder would be to bid a constant value till the first win, that depends on the distributions b' and q.

Lemma 5. *The optimal strategy for the bidder given the distribution of b' and the distribution of q is to bid a constant bid b, the value of which depends on the distributions b' and q.*

Proof. The expected utility of a bid is a function of the bid b and the distributions b' and q only. Hence, given distributions b' and q, there will exist a value of b for which the expected utility is maximum, and this is the optimal bid until he gets new information about the distribution of q, by winning an auction and experiencing if this win translated into a conversion.

Therefore, for the repeated auction scenario the strategy space of the bidder looks as follows:

- Make a bid $b > \frac{1}{2}$ in initial rounds till she wins an auction.
- Make a bid $b_w = \frac{2}{3}(\geq b)$ after winning a first auction if we convert in the first round.
- Make a bid $b_l = \frac{1}{3}(\leq b)$ after winning a first auction if we do not convert in the first round.

We'll use F to denote the cumulative distribution function of the random variable b', the maximum bid without the new bidder, so $F(b)$ is the probability that $b' < b$ for a given value b. The expected utility is dependent on the bid b made in the initial rounds, and the bids $b_w = 3/2$ and $b_l = 1/3$ which would be made after winning a round depending on whether conversion occurred. Let $E[U]$ denote the expected utility made and $E[U|w]$ and $E[U|l]$ denote the expected utilities gained given a conversion in the first round and no conversion in the first round respectively. The expression for the expected utility $E[U]$ is:

$$E[U] = F(b)\left[-E[b'|b' \leq b] + \frac{1}{2}\left(1 + e^{-d}E[U|w]\right) + \frac{1}{2}\left(e^{-d}E[U|l]\right)\right]$$

$$+ \left[1 - F(b)\right]e^{-d}E[U].$$

The second term $[1 - F(b)]e^{-d}E[U]$ is the utility we would gain if we did not win the auction in the first round by continuing to try from the second round.

The expressions for $E[U|w]$ and $E[U|l]$ are given below using the fact that each step yields the same expected utility:

$$E[U|w] = \frac{F(2/3)(\frac{2}{3} - E[b'|b' \leq 2/3])}{1 - e^{-d}}. \tag{10}$$

$$E[U|l] = \frac{F(1/3)(\frac{1}{3} - E[b'|b' \leq 1/3])}{1 - e^{-d}}. \tag{11}$$

While solving explicitly for the optimal bid b is less easy, this more complex scenario results in qualitatively the same conclusion as we have seen in the previous section: initial credit can improve revenue by making the entering bidder more aggressive initially and hence helping her find out her own value (or conversion rate). Not being as aggressive without the credit now manifests itself by using a lower bid value b and hence taking longer before learning about her conversion rate.

Recall that with credit for one round of auction, the bidder can bid high enough to win the first round, and hence the expression for the expected utility becomes:

$$E[U] = \frac{1}{2}(1 + e^{-d}E[U|w]) + \frac{1}{2}(e^{-d}E[U|l]). \tag{12}$$

Theorem 2. *Depending on the distribution of the highest and second highest bids b' and b'', allowing a round of free experimentation for a new bidder, can increase profit for the auctioneer.*

Proof. We will show that the condition for profit in a variable b' case is $e^{-d}(1 - F(b)) \left(\frac{\lambda(2/3) + \lambda(1/3)}{2} \right) > \lambda(b)$, where

1. $\lambda(b) = F(b)E[b'|b' \leq b] + (1 - F(b))E[max(b, b'')|b' > b]$
2. b is the optimal bid of the bidder in the initial rounds.

To see this let R_{nc} denote the revenue of the auctioneer in the case when credit is not given, and R_c denote the revenue of the auctioneer in the case when credit is given (in the form of a free-start). The expressions for the two revenues are:

$$R_{nc} = F(b)[E[b'|b' \leq b] + \frac{e^{-d}}{2(1 - e^{-d})}[F(2/3)E[b'|b' \leq 2/3] +$$

$$+F(1/3)E[b'|b' \leq 1/3] + (1 - F(2/3))max(2/3, b'') +$$

$$+(1 - F(1/2))max(1/3, b'')]] + (1 - F(b))[max(b, 2/3) + e^{-d}R_{nc}].$$

$$R_c = 1[0 + \frac{e^{-d}}{2(1 - e^{-d})}[F(2/3)E[b'|b' \leq 2/3] +$$

$$+F(1/3)E[b'|b' \leq 1/3] + (1 - F(2/3))max(b_w, b'')$$

$$(1 - F(1/3))max(1/3, b'')].$$

$$=> R_{nc} = \frac{F(b)(E[b'|b' \leq b] + R_c) + (1 - F(b))max(b, b'')}{1 - e^{-d} + e^{-d}F(b)}.$$

The condition for profit would be $R_c > R_{nc}$. To simplify the evaluation of the above condition we define

$$\lambda(b) = F(b)E[b'|b' \le b] + (1 - F(b))b. \qquad (13)$$

Now the condition for profit in terms of $\lambda(.)$ is

$$e^{-d}(1 - F(b)) \left(\frac{\lambda(b_w) + \lambda(b_l)}{2} \right) > \lambda(b). \qquad (14)$$

The function λ can be viewed as the expected revenue gained by the auctioneer in one round of the auction as a function of the bidder's utility maximizing bid given her knowledge at that time. From the above inequality the following remarks can be made:

1. If b the optimal bid in the first round is 1, i.e the bidder always participates in the auction irrespective of the outcome, a profit cannot be made.
2. The effect of the discounting parameter d is two-fold, it hampers the possibility of a profit by reducing the value of future revenue gained by the auctioneer but it also can have an increasing effect on the value of b.

5   Conclusion

We have shown that credit to encourage experimentation by a new entrant can increase revenue in repeated auctions. To simplify the presentation, we assumed that the unknown conversion rate q that the bidder is learning while experimenting is drawn from the uniform distribution, and that learning takes place by only a single experimentation. While these assumptions simplify the presentation, neither is crucial for the conclusion of the paper.

References

1. Abraham, I., Athey, S., Babaioff, M., Grubb, M.: Peaches, lemons, and cookies: Designing auction markets with dispersed information. In: ACM Conference on Economic Commerce (2013)
2. Iyer, K., Johari, R., Sundararajan, M.: Mean Field Equilibria of Dynamic Auctions with Learning. In: ACM Conference on Economic Commerce (2011)
3. Bulow, J., Klemperer, P.: Auctions Versus Negotiations. American Economic Review (1996)
4. Gummadi, R., Key, P., Proutiere, A.: Optimal Bidding Strategies and Equilibria in Repeated Auctions with Budget Constraints. In: Ad Auctions Workshop (2012)
5. Lahaie, S., McAfee, R.P.: Efficient Ranking in Sponsored Search. In: Chen, N., Elkind, E., Koutsoupias, E. (eds.) WINE 2011. LNCS, vol. 7090, pp. 254–265. Springer, Heidelberg (2011)
6. Milgrom, P.R., Weber, R.J.: A theory of auctions and competitive bidding. Econometrica 50(5), 1089–1122 (1982)
7. Syrgkanis, V., Kempe, D., Tardos, E.: Information Asymmetries in Common-Value Auctions with Discrete Signals. In: Ad Auctions Workshop (2013)

Mechanism Design for Aggregating Energy Consumption and Quality of Service in Speed Scaling Scheduling

Christoph Dürr[1], Łukasz Jeż[2,3], and Óscar C. Vásquez[4]

[1] CNRS, LIP6, Université Pierre et Marie Curie, Paris, France
[2] Institute of Computer Science, University of Wrocław, Poland
[3] DIAG, Sapienza University of Rome, Italy
[4] LIP6 and Industrial Engineering Department, University of Santiago of Chile

Abstract. We consider a strategic game, where players submit jobs to a machine that executes all jobs in a way that minimizes energy while respecting the jobs' deadlines. The energy consumption is then charged to the players in some way. Each player wants to minimize the sum of that charge and of their job's deadline multiplied by a priority weight. Two charging schemes are studied, the *proportional cost share* which does not always admit pure Nash equilibria, and the *marginal cost share*, which does always admit pure Nash equilibria, at the price of overcharging by a constant factor.

1 Introduction

In many computing systems, minimizing energy consumption and maximizing quality of service are opposed goals. This is also the case for the speed scaling scheduling model considered in this paper. It has been introduced in [9], and triggered a lot of work on offline and online algorithms; see [1] for an overview.

The online and offline optimization problem for minimizing flow time while respecting a maximum energy consumption has been studied for the single machine setting in [14,2,5,8] and for the parallel machines setting in [3]. For the variant where an aggregation of energy and flow time is considered, polynomial approximation algorithms have been presented in [7,4,11].

In this paper we propose to study this problem from a different perspective, namely as a strategic game. In society many ecological problems are either addressed in a centralized manner, like forcing citizens to sort household waste, or in a decentralized manner, like tax incentives to enforce ecological behavior. This paper proposes incentives for a scheduling game, in form of an energy cost charging scheme.

Consider a scheduling problem for a single processor, that can run at variable speed, such as the modern microprocessors Intel SpeedStep, AMD PowerNow! or IBM EnergyScale. Higher speed means that jobs finish earlier at the price of a higher energy consumption. Each job has some workload, representing a number of instructions to execute, and a release time before which it cannot be

Y. Chen and N. Immorlica (Eds.): WINE 2013, LNCS 8289, pp. 134–145, 2013.

scheduled. Every user submits a single job to a common processor, declaring the jobs parameters, together with a deadline, that the player chooses freely.

The processor will schedule the submitted jobs preemptively, so that all release times and deadlines are respected and the overall energy usage is minimized. The energy consumed by the schedule needs to be charged to the users. The individual goal of each user is to minimize the sum of the energy cost share and of the requested deadline weighted by the user's priority, which represents a quality of service coefficient. This individual priority weight implies a conversion factor that allows of aggregation of deadline and energy.

In a companion paper [15] we study this game from the point of view of the game regulator, and compare different ways to organize the game which would lead to truthfulness. In this paper we focus on a particular game setting, described in the next section.

2   The Model

Formally, we consider a non-cooperative game with n players and a regulator. The regulator manages the machine where the jobs are executed. Each player has a job i with a workload w_i, a release time r_i and a priority p_i, representing a quality of service coefficient. The player submits its job together with a deadline $d_i > r_i$ to the regulator. Workloads, release times and deadlines are public information known to all players, while quality of service coefficients can be private.

The regulator implements some *cost sharing mechanism*, which is known to all users. This mechanism defines a cost share function b_i specifying how much player i is charged. The penalty of player i is the sum of two values: his *energy cost share* $b_i(w, r, d)$ defined by the mechanism, where $w = (w_1, \ldots, w_n)$, $r = (r_1, \ldots, r_n)$ and $d = (d_1, \ldots, d_n)$, and his *waiting cost*, which can be either $p_i d_i$ or $p_i(d_i - r_i)$; we use the former waiting cost throughout the article but all our results apply to both. The sum of all player's penalties, i.e., energy cost shares and waiting costs will be called the *utilitarian social cost*.

The regulator computes a minimum energy schedule for a single machine in the speed scaling model, which stipulates that at any point in time t the processor can run at arbitrary speed $s(t) \geq 0$; for a time interval I, the workload executed in I is $\int_{t \in I} s(t)dt$, while the energy consumed is $\int_{t \in I} s(t)^\alpha dt$ for some fixed physical constant $\alpha \in [2, 3]$ characteristic for a device [6]. The sum of the energy used by this optimum schedule and of all the players' waiting costs will be called the *effective social cost*.

The minimum energy schedule can be computed in time $O(n^2 \log n)$ [10] and has (among others) the following properties [16]. The jobs in the schedule are executed by preemptive earliest deadline first order (EDF), and the speed $s(t)$ at which they are processed is piecewise linear. Preemptive EDF means that at every time point among all jobs which are already released and not yet completed, the job with the smallest deadline is executed, using job indices to break ties.

The cost sharing mechanism defines the game completely. Ideally, we would like the game and the mechanism to have the following properties.

Existence of Pure Nash Equilibria. This means that there is a strategy profile vector d such that no player can unilaterally deviate from their strategy d_i while strictly decreasing their penalty.

Budget Balance. The mechanism is c-budged balanced, when the sum of the cost shares is no smaller than the total energy consumption and no larger than c times the energy consumption.

In the sequel we introduce and study two different cost sharing mechanisms, namely PROPORTIONAL COST SHARING where every player pays exactly the cost generated during the execution of his job, and MARGINAL COST SHARING where every player pays the increase of energy cost generated by adding this player to the game.

3　Proportional Cost Sharing

The proportional cost sharing is the simplest budget balanced cost sharing scheme one can think of. Every player i is charged exactly the energy consumed during the execution of his job. Unfortunately this mechanism does not behave well as we show in Theorem 1.

Fact 1. *In a single player game, the player's penalty is minimized by the deadline*

$$r_1 + w_1(\alpha - 1)^{1/\alpha}p_1^{-1/\alpha}.$$

Proof. If player 1 chooses deadline $d_1 = r_1 + x$ then the schedule is active between time r_1 and $r_1 + x$ at speed w_1/x. Therefore his penalty is

$$p_1(r_1 + x) + x^{1-\alpha}w_1^\alpha.$$

Deriving this expression in x, and using the fact that the penalty is concave in t for any $x > 0$ and $\alpha > 0$, we have that the optimal x for the player will set to zero the derivative. This implies the claimed deadline.　　　　□

If there are at least two players however, the game does not have nice properties as we show now.

Theorem 1. *The* PROPORTIONAL COST SHARING *does not always admit a pure Nash equilibrium.*

The proof consists of a very simple example: there are 2 identical players with identical jobs, say $w_1 = w_2 = 1$, $r_1 = r_2 = 0$ and $p_1 = p_2 = 1$. First we determine the best response of player 1 as a function of player 2, then we conclude that there is no pure Nash equilibrium.

Table 1. The local minimum in the range of f corresponding to f_i is a function of α and d_2, which we denote by $d_1^{(i)}$. The value at such local minimum is again a function of α and d_2, which we denote by $g_i(d_2)$. These are only potential minima: they exist if and only if the condition given in the last column is satisfied.

argument	value	applicable range
$d_1^{(1)} = (\alpha - 1)^{1/\alpha}$	$g_1(d_2) = \alpha(\alpha - 1)^{1/\alpha - 1}$	$d_2 \geq 2(\alpha - 1)^{1/\alpha}$
$d_1^{(2)} = \frac{d_2}{2}$	$g_2(d_2) = d_2/2 + (d_2/2)^{1-\alpha}$	$d_2 \leq 2(\alpha - 1)^{1/\alpha}$
$d_1^{(3)} = 2\left(\frac{\alpha-1}{2}\right)^{1/\alpha}$	$g_3(d_2) = \alpha\left(\frac{\alpha-1}{2}\right)^{1/\alpha - 1}$	$\left(\frac{\alpha-1}{2}\right)^{1/\alpha} \leq d_2 \leq 2\left(\frac{\alpha-1}{2}\right)^{1/\alpha}$
$d_1^{(4)} = d_2 + (\alpha - 1)^{1/\alpha}$	$g_4(d_2) = d_2 + \alpha(\alpha - 1)^{1/\alpha - 1}$	$d_2 \leq (\alpha - 1)^{1/\alpha - 1}$

Lemma 1. *Given the second player's choice d_2, the penalty of the first player as a function of his choice d_1 is given by*

$$f(d_1) = \begin{cases} f_1(d_1) = d_1 + d_1^{1-\alpha} & \text{if } d_1 \leq \frac{d_2}{2} \\ f_2(d_1) = d_1 + (\frac{d_2}{2})^{1-\alpha} & \text{if } \frac{d_2}{2} \leq d_1 \leq d_2 \\ f_3(d_1) = d_1 + (\frac{d_1}{2})^{1-\alpha} & \text{if } d_2 \leq d_1 \leq 2d_2 \\ f_4(d_1) = d_1 + (d_1 - d_2)^{1-\alpha} & \text{if } d_1 \geq 2d_2 \end{cases} \qquad (1)$$

The local minima of $f(d_1)$ are summarized in Table 1, and the penalties corresponding to player 1 picking these minima are illustrated in Figure 1.

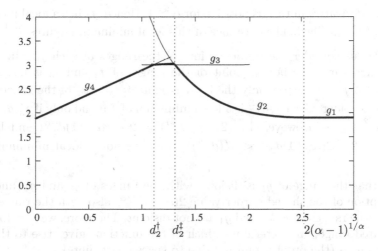

Fig. 1. First player's penalty (in bold) when choosing his best response as a function of second player's strategy d_2, here for $\alpha = 3$

Proof. Formula (1) follows by a straightforward case inspection. Then, to find all the local minima of f, we first look at the behavior of each of f_i, finding their local minima in their respective intervals, and afterwards we inspect the border points of these intervals.

Range of f_1: The derivative of f_1 is

$$f_1'(d_1) = 1 - (\alpha - 1)d_1^{-\alpha} \ ,$$

whose derivative in turn is positive for $\alpha > 1$. Therefore, f_1 has a local minimum at $d_1^{(1)}$ as specified. Since we require that this local minimum is within the range where f coincides with f_1, the necessary and sufficient condition is $d_1^{(1)} \le \frac{d_2}{2}$.

Range of f_2: f_2 is an increasing function, and therefore it attains a minimum value only at the lower end of its range, $d_1^{(3)}$. However, if $d_1^{(2)}$ is to be a local minimum of f, there can be no local minimum of f in the range of f_1 (immediately to the left), so the applicable range of $d_1^{(2)}$ is the complement of that of $d_1^{(1)}$.

Range of f_3: The derivative of f_3 is

$$f_3'(d_1) = 1 - \frac{\alpha - 1}{2}(d_1/2)^{-\alpha} \ , \tag{2}$$

whose derivative in turn is positive for $\alpha > 1$. Hence, f_3 has a local minimum at $d_1^{(3)}$ as specified. The existence of this local minimum requires $d_2 \le d_1^{(3)} \le 2d_2$, which is equivalent to $\frac{d_1^{(3)}}{2} \le d_2 \le d_1^{(3)}$.

Range of f_4: The derivative of f_4 is

$$f_4'(d_1) = 1 - (\alpha - 1)(d_1 - d_2)^{-\alpha} \ , \tag{3}$$

whose derivative in turn is positive for $\alpha > 1$. Hence, f_4 has a local minimum at $d_1^{(4)}$ as specified. The existence of this local minimum requires $d_1^{(4)} \ge 2d_2$.

Now let us consider the border points of the ranges of each f_i. Since f_2 is strictly increasing, the border point of the ranges of f_2 and f_3 is not a local minimum of f. This leaves only the border point $d_1^{(2)} = 2d_2$ of the ranges of f_3 and f_4 to consider. Clearly, $d_1^{(2)}$ is a local minimum of f if and only if $f_3'(d_1^{(2)}) \le 0$ and $f_4'(d_1^{(2)}) \ge 0$. However, by (2), $f_3'(d_1^{(2)}) = 2 - (\alpha - 1)d_2^{-\alpha}$, and by (3), $f_4'(d_1^{(2)}) = 2 - 2(\alpha - 1)d_2^{-\alpha} < f_3'(d_1^{(2)})$, so $d_1^{(2)}$ is not a local minimum of f either. $\qquad\square$

Note that the range of g_1 is disjoint with the ranges of g_3 and g_4, and with the exception of the shared border value $2(\alpha - 1)^{1/\alpha}$, also with the range of g_2. However, the ranges of g_2, g_3 and g_4 are not disjoint. Therefore, we now focus on their shared range, and determine which of the functions gives rise to the true local minimum (the proof is omitted due to space constraints).

Lemma 2. *The function $g_3(d_2)$ is constant, the function $g_4(d_2)$ is an increasing linear function, and the function $g_2(d_2)$ is decreasing for $d_2 < d_1^{(3)}$. Moreover, there exist two unique values*

$$d_2^\dagger = \alpha(\alpha - 1)^{1/\alpha - 1}(2^{1 - 1/\alpha} - 1) \quad \text{such that} \quad g_4(d_2^\dagger) = g_3(d_2^\dagger) \ , \tag{4}$$

$$d_2^\ddagger \in \left(d_2^\dagger, d_1^{(3)}\right) \qquad\qquad\qquad \text{such that} \quad g_2(d_2^\ddagger) = g_3(d_2^\ddagger) \ . \tag{5}$$

With Lemma 1 and Lemma 2, whose statements are summarized in Table 1 and Figure 1, we can finally determine what is the best response of the first player as a function of d_2.

Lemma 3. *The best response for player 1 as function of d_2 is*

$$d_1^{(4)} = d_2 + (\alpha - 1)^{1/\alpha} \quad if \qquad\qquad 0 < d_2 \leq d_2^\dagger ,$$

$$d_1^{(3)} = 2\left(\frac{\alpha - 1}{2}\right)^{1/\alpha} \quad if \qquad\qquad d_2^\dagger < d_2 \leq d_2^\ddagger ,$$

$$d_1^{(2)} = \frac{d_2}{2} \quad\qquad if \qquad\qquad d_2^\ddagger < d_2 \leq 2(\alpha - 1)^{1/\alpha} ,$$

$$d_1^{(1)} = (\alpha - 1)^{1/\alpha} \quad if \quad 2(\alpha - 1)^{1/\alpha} < d_2 .$$

Proof. The proof consists in determining which of the applicable local minima of f is the global minimum for each range of d_2. Again, the cases are depicted in Figure 1.

case (i) $0 < d_2 \leq d_2^\dagger$: In this case, we claim that the best response of player 1 is
$$d_1^{(4)} = d_2 + (\alpha - 1)^{1/\alpha} .$$

First we prove that
$$d_2^\dagger \in \left(\left(\frac{\alpha - 1}{2}\right)^{1/\alpha} , (\alpha - 1)^{1/\alpha - 1}\right) .$$

The upper bound hold since
$$\alpha(\alpha - 1)^{1/\alpha - 1}(2^{1 - 1/\alpha} - 1) < (\alpha - 1)^{1/\alpha - 1}$$
$$\alpha(2^{1 - 1/\alpha} - 1) < 1,$$

holds for $\alpha \geq 2$.
The lower bound holds since,

$$\alpha(\alpha - 1)^{1/\alpha - 1}(2^{1 - 1/\alpha} - 1) > \left(\frac{\alpha - 1}{2}\right)^{1/\alpha}$$
$$\frac{\alpha}{\alpha - 1}(2^{1 - 1/\alpha} - 1) > 2^{-1/\alpha}$$
$$\frac{\alpha}{\alpha - 1}(2 - 2^{1/\alpha}) > 1$$

holds for $\alpha \geq 2$.
In fact, both inequalities are true even for $\alpha > 1$, but as we require $\alpha \geq 2$ due to Lemma 2, we settle for simpler proofs.
These bounds imply that in case (i) player 1 chooses the minimum among the 3 local minima $d_1^{(2)}$, $d_1^{(3)}$, and $d_1^{(4)}$, where the middle one is only an option for $\left(\frac{\alpha - 1}{2}\right)^{1/\alpha} \leq d_2 \leq d_2^\dagger$. It follows from Lemma 2 that the last option always

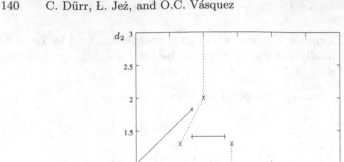

Fig. 2. Best response of player 1 as function of d_2, and best response of player 2 as function of d_1. Here for $\alpha = 3$.

dominates: by (5), for every $\left(\frac{\alpha-1}{2}\right)^{1/\alpha} \leq d_2 < d_2^{\ddagger}$, we have $g_3(d_2) < g_2(d_2)$, and by (4), for every $\left(\frac{\alpha-1}{2}\right)^{1/\alpha} \leq d_2 \leq d_2^{\dagger}$, we have $g_4(d_2) < g_3(d_2)$. This concludes the analysis for case (i).

case (ii) $d_2^{\dagger} < d_2 \leq d_2^{\ddagger}$: In this case, we claim that the best response of player 1 is

$$d_1^{(3)} = 2\left(\frac{\alpha - 1}{2}\right)^{1/\alpha} .$$

First we observe that by Lemma 2 (5),

$$d_2^{\ddagger} < d_1^{(4)} ,$$

which rules out $d_1^{(1)}$ as a choice for player 1, leaving only $d_1^{(2)}$, $d_1^{(3)}$, and $d_1^{(4)}$. Again, Lemma 2 implies that $d_1^{(4)}$ dominates other choices: by (5), we have $g_3(d_2) < g_2(d_2)$ for all $\left(\frac{\alpha-1}{2}\right)^{1/\alpha} \leq d_2 < d_2^{\ddagger}$, and by (4), we have $g_3(d_2) < g_4(d_2)$ for all $d_2 > d_2^{\dagger}$.

Note that for $\alpha = 2$, the range of this case is empty.

case (iii) $d_2^{\ddagger} < d_2 \leq 2(\alpha - 1)^{1/\alpha}$: For this range, only $d_1^{(2)}$ and $d_1^{(3)}$ are viable choices for player 1, and Lemma 2 (5) implies that $d_1^{(2)}$ dominates. Therefore first player's best response is

$$d_1^{(2)} = \frac{d_2}{2} .$$

case (iv) $2(\alpha - 1)^{1/\alpha} < d_2$: For this range, the only viable choice for player 1 is

$$d_1^{(1)} = (\alpha - 1)^{1/\alpha} ,$$

which is therefore his best response.

This concludes the proof of the lemma.   □

By the symmetry of the players, the second player's best response is in fact an identical function of d_1 as the one stated in Lemma 3. By straightforward inspection it follows that there is no fix point (d_1, d_2) to this game, which implies the following theorem, see figure 2 for illustration.

4 Marginal Cost Sharing

In this section we propose a different cost sharing scheme, that improves on the previous one in the sense that it admits pure Nash equilibria, however for the price of overcharging by at most a constant factor.

Before we give the formal definition we need to introduce some notations. Let $\mathrm{OPT}(d)$ be the energy minimizing schedule for the given instance, and $\mathrm{OPT}(d_{-i})$ be the energy minimizing schedule for the instance where job i is removed. We denote by $E(S)$ the energy cost of schedule S.

In the marginal cost sharing scheme, player i pays the penalty function

$$p_i d_i + E(\mathrm{OPT}(d)) - E(\mathrm{OPT}(d_{-i})).$$

This scheme defines an exact potential game by construction [12]. Formally, let n be the number of players, $D = \{d | \forall j : d_j > r_j\}$ be the set of action profiles (deadlines) over the action sets D_i of each player.

Let us denote the effective social cost corresponding to a strategy profile d by $\Phi(d)$. Then

$$\Phi(d) = \sum_{i=1}^{n} p_i d_i + E(\mathrm{OPT}(d)).$$

Clearly, if a player i changes its strategy d_i and his penalty decreases by some amount Δ, then the effective social cost decreases by the same amount Δ, because $E(\mathrm{OPT}(d_{-i}))$ remains unchanged.

4.1 Existence of Equilibria

While the best response function is not continuous in the strategy profile, precluding the use of Brouwer's fixed-point theorem, existence of pure Nash equilibria can nevertheless be easily established.

To this end, note that the global minimum of the effective social cost, if it exists, is a pure Nash equilibrium. Its existence follows from (1) compactness of a non-empty sub-space of strategies with bounded social cost and (2) continuity of Φ.

For (2), note that $\sum_i p_i d_i$ is clearly continuous in d, and hence $\Phi(d)$ is continuous if $E(\mathrm{OPT}(d))$ is. The continuity of the latter is clear once considers all possible relations of the deadlines chosen by the players.

For (1), let d' be any (feasible) strategy profile such that $d_i > r_i$ for each player i. The subspace of strategy profiles d such that $\Phi(d) \leq \Phi(d')$ is clearly closed, and bounded due to the $p_i d_i$ terms. Thus it is a compact subspace that contains the global minimum of Φ.

4.2   Convergence Can Take Forever

In this game the strategy set is infinite. Moreover, the convergence time can be infinite as we demonstrate below in Theorem 2. Notice that this also proves that in general there are no dominant strategies in this game.

Theorem 2. *For the game with the marginal cost sharing mechanism, the convergence time to reach a pure Nash equilibrium can be unbounded.*

Proof. The proof is by exhibiting again the same small example, with 2 players, release times 0, unit weights, unit penalty factors, and $\alpha > 2$.

For this game there are two pure Nash equilibria, the first one is

$$d_1 = \left(\frac{\alpha - 1}{2}\right)^{1/\alpha}, \quad d_2 = d_1 + (\alpha - 1)^{1/\alpha},$$

while the second one is symmetric for players 1 and 2.

In the reminder of the proof, we assume that player 1 chooses a deadline which is close to the pure Nash equilibrium above. By analyzing the best responses of the players, we conclude that after a best response of player 2, and then of player 1 again, he chooses a deadline which is even closer to the pure Nash equilibrium above but different from it, leading to an infinite convergence sequence of best responses. The proofs of the following two lemmas are omitted.

Now suppose $d_1 = \delta \left(\frac{\alpha-1}{2}\right)^{1/\alpha}$ for some $1 < \delta < 2^{1/\alpha}$. What is the best response for player 2?

Lemma 4. *Given the first player's choice d_1, the penalty of the second player as a function of his choice d_2 is given by*

$$h(d_2, d_1) = \begin{cases} h_1(d_2, d_1) = d_2 + d_2^{1-\alpha} + (d_1 - d_2)^{1-\alpha} - d_1^{1-\alpha} & \text{if } d_2 \leq \frac{d_1}{2} \\ h_2(d_2, d_1) = d_2 + (2^\alpha - 1)d_1^{1-\alpha} & \text{if } \frac{d_1}{2} \leq d_2 \leq d_1 \\ h_3(d_2, d_1) = d_2 + 2^\alpha d_2^{1-\alpha} - d_1^{1-\alpha} & \text{if } d_1 \leq d_2 \leq 2d_1 \\ h_4(d_2, d_1) = d_2 + (d_2 - d_1)^{1-\alpha} & \text{if } d_2 \geq 2d_1, \end{cases}$$

and the best response for player 2 as function of d_1 is

$$d_1 + (\alpha - 1)^{1/\alpha} = (\alpha - 1)^{1/\alpha}(1 + 2^{-1/\alpha}\delta) \tag{6}$$

From now on we assume that player 2 chooses $d_2 = d_1 + (\alpha - 1)^{1/\alpha} = (\alpha - 1)^{1/\alpha}(1 + 2^{-1/\alpha}\delta)$. What is the best response for player 1?

Lemma 5. *Given the second player's choice d_2, the penalty of the first player as a function of his choice d_1 is given by $h(d_1, d_2)$ and the best response for player 1 is*

$$d_1 = \delta' \left(\frac{\alpha - 1}{2}\right)^{1/\alpha},$$

for some $\delta' \in (1, \delta)$.

This concludes the proof of Theorem 2.   □

4.3   Bounding Total Charge

In this section we bound the total cost share for the MARGINAL COST SHARING SCHEME, by showing that it is at least $E(\mathrm{OPT}(d))$ and at most α times this value. In fact we show a stronger claim for individual cost shares.

Theorem 3. *For every player i, its marginal costshare is at least its proportional costshare and at most α times the proportional costshare.*

Proof. Fix a player i, and denote by S_{-i} the schedule obtained from $\mathrm{OPT}(d)$ when all executions of i are replaced by idle times. Clearly we have the following inequalities.

$$E(\mathrm{OPT}(d_{-i})) \le E(S_{-i}) \le E(\mathrm{OPT}(d))$$

Then the marginal cost share of player i can be lower bounded by

$$E(\mathrm{OPT}(d)) - E(\mathrm{OPT}(d_{-i})) \ge E(\mathrm{OPT}(d)) - E(S_{-i}).$$

According to [16] the schedule OPT can be obtained by the following iterative procedure. Let S be the support of a partial schedule. For every interval $[t, t')$ we define its domain $I_{t,t'} := [t, t') \backslash S$, the set of included jobs $J_{t,t'} := \{j : [r_j, d_j) \subseteq [t, t')\}$, and the density $\sigma_{t,t'} := \sum_{j \in J_{t,t'}} w_j / |I_{t,t'}|$. The procedure starts with $S = \emptyset$, and while not all jobs are scheduled, selects an interval $[t, t')$ with maximal density, and schedules all jobs from $J_{t,t'}$ in earliest deadline order in $I_{t,t'}$ at speed $\sigma_{t,t'}$ adding $I_{t,t'}$ to S.

For the upper bound, let $t_1 < t_2 < \ldots < t_\ell$ be the sequence of all release times and deadlines for some $\ell \le 2n$. Clearly both schedules S run at uniform speed in every interval $[t_{k-1}, t_k)$. For every $1 \le k \le n$ let s_k be the speed of S in $[t_{k-1}, t_k)$, and s'_k the speed of S' in the same interval.

From the algorithm above it follows that every job is scheduled at constant speed, so let s_a be the speed at which job i is scheduled in $\mathrm{OPT}(d)$. It also follows that if $s_k > s_a$, then $s'_k = s_k$, and if $s_k \le s_a$, then $s'_k \le s_k$.

We establish the following upper bound.

$$
\begin{aligned}
E(\mathrm{OPT}(d)) - E(\mathrm{OPT}(d_{-i})) &= \sum_{k=1}^{\ell} s_k^\alpha (t_k - t_{k-1}) - s_k'^\alpha (t_k - t_{k-1}) \\
&= \sum (t_k - t_{k-1})(s_k^\alpha - (s_k - (s_k - s'_k))^\alpha) \\
&= \sum (t_k - t_{k-1}) s_k^\alpha \left(1 - \left(1 - \frac{s_k - s'_k}{s_k}\right)^\alpha\right) \\
&\le \sum (t_k - t_{k-1}) s_k^\alpha \left(1 - \left(1 - \alpha \frac{s_k - s'_k}{s_k}\right)\right) \\
&= \sum (t_k - t_{k-1}) \alpha s_k^{\alpha-1}(s_k - s'_k) \\
&\le \alpha s_a^{\alpha-1} \sum (t_k - t_{k-1})(s_k - s'_k) \\
&= \alpha s_a^{\alpha-1} w_i \\
&= \alpha(E(\mathrm{OPT}(d)) - E(S_{-i})).
\end{aligned}
$$

The first inequality uses the generalized Bernoulli inequality, and the last one the fact that for all k with $s_k \neq s'_k$ we have $s_k \leq s_a$.

The theorem follows from the fact that $s_a^{\alpha-1} w_i$ is precisely the proportional cost share of job i in OPT(d). ☐

A tight example is given by n jobs, each with workload $1/n$, release time 0 and deadline 1. Clearly the optimal energy consumption is 1 for this instance. The marginal cost share for each player is $1 - (1 - 1/n)^\alpha$. Finally we observe that the total marginal cost share tends to α, i.e.

$$\lim_{n \to +\infty} n - n(1 - 1/n)^\alpha = \alpha.$$

5   A Note on Cross-Monotonicity

We conclude this paper by a short note on *cross-monotonicity*. This is a property of cost sharing games, stating that whenever new players enter the game, the cost share of any fixed player does not increase. This property is useful for stability in the game, and is the key to the Moulin carving algorithm [13], which selects a set of players to be served for specific games.

In the game that we consider, the minimum energy of an optimal schedule for a set S of jobs contrasts with many studied games, where serving more players becomes more cost effective, because the used equipment is better used.

Consider a very simple example of two identical players, submitting their respective jobs with the same deadline 1. Suppose the workload of each job is w, then the minimum energy necessary to schedule one job is w^α, while the cost to serve both jobs is $(2w)^\alpha$, meaning that the cost share increase whenever a second player enters the game. Therefore the marginal cost sharing scheme is not cross-monotonic.

Acknowledgements. We would like to thank anonymous referees for remarks and suggestions on an earlier version of this paper.

Christoph Dürr and Oscar C. Vásquez were partially supported by grant ANR-11-BS02-0015. Łukasz Jeż was partially supported by MNiSW grant N N206 368839, 2010-2013, EU ERC project 259515 PAAl, and FNP Start scholarship.

References

1. Albers, S.: Energy-efficient algorithms. Communications of the ACM 53(5), 86–96 (2010)
2. Albers, S., Fujiwara, H.: Energy-efficient algorithms for flow time minimization. ACM Transactions on Algorithms (TALG) 3(4), 49 (2007)
3. Angel, E., Bampis, E., Kacem, F.: Energy aware scheduling for unrelated parallel machines. In: 2012 IEEE International Conference on Green Computing and Communications (GreenCom), pp. 533–540. IEEE (2012)

4. Bansal, N., Chan, H.-L., Katz, D., Pruhs, K.: Improved bounds for speed scaling in devices obeying the cube-root rule. Theory of Computing 8, 209–229 (2012)
5. Bansal, N., Kimbrel, T., Pruhs, K.: Speed scaling to manage energy and temperature. Journal of the ACM (JACM) 54(1), 3 (2007)
6. Brooks, D.M., Bose, P., Schuster, S.E., Jacobson, H., Kudva, P.N., Buyukto-sunoglu, A., Wellman, J., Zyuban, V., Gupta, M., Cook, P.W.: Power-aware microarchitecture: Design and modeling challenges for next-generation micropro-cessors. IEEE Micro 20(6), 26–44 (2000)
7. Carrasco, R.A., Iyengar, G., Stein, C.: Energy aware scheduling for weighted completion time and weighted tardiness. Technical report, arxiv.org (2011)
8. Chan, S.-H., Lam, T.-W., Lee, L.-K.: Non-clairvoyant speed scaling for weighted flow time. In: de Berg, M., Meyer, U. (eds.) ESA 2010, Part I. LNCS, vol. 6346, pp. 23–35. Springer, Heidelberg (2010)
9. Irani, S., Pruhs, K.R.: Algorithmic problems in power management. ACM SIGACT News 36(2), 63–76 (2005)
10. Li, M.G., Yao, A.C., Yao, F.F.: Discrete and continuous min-energy schedules for variable voltage processor. Proceedings of the National Academy of Sciences of the United States of America, PNAS 2006 103, 3983–3987 (2006)
11. Megow, N., Verschae, J.: Dual techniques for scheduling on a machine with varying speed. In: Fomin, F.V., Freivalds, R., Kwiatkowska, M., Peleg, D. (eds.) ICALP 2013, Part I. LNCS, vol. 7965, pp. 745–756. Springer, Heidelberg (2013)
12. Monderer, D., Shapley, L.S.: Potential games. Games and Economic Behavior 14, 124–143 (1996)
13. Moulin, H., Shenker, S.: Strategyproof sharing of submodular costs: budget balance versus efficiency. Economic Theory 18(3), 511–533 (2001)
14. Pruhs, K., Uthaisombut, P., Woeginger, G.: Getting the best response for your erg. ACM Transactions on Algorithms (TALG) 4(3), 38 (2008)
15. Vasquez, O.C.: Energy in computing systems with speed scaling: optimization and mechanisms design. Technical report, arxiv.org (2012)
16. Yao, F., Demers, A., Shenker, S.: A scheduling model for reduced cpu energy. In: Proceedings of the 36th Annual Symposium on Foundations of Computer Science, FOCS 1995, pp. 374–382. IEEE Computer Society, Washington, DC (1995)

Valuation Compressions in VCG-Based Combinatorial Auctions

Paul Dütting[1,*], Monika Henzinger[2,**], and Martin Starnberger[2,**]

[1] Department of Computer Science, Cornell University
136 Hoy Road, Ithaca, NY 14850, USA
paul.duetting@cornell.edu
[2] Faculty of Computer Science, University of Vienna
Währinger Straße 29, 1090 Wien, Austria
{monika.henzinger,martin.starnberger}@univie.ac.at

Abstract. The focus of classic mechanism design has been on truthful direct-revelation mechanisms. In the context of combinatorial auctions the truthful direct-revelation mechanism that maximizes social welfare is the VCG mechanism. For many valuation spaces computing the allocation and payments of the VCG mechanism, however, is a computationally hard problem. We thus study the performance of the VCG mechanism when bidders are forced to choose bids from a subspace of the valuation space for which the VCG outcome can be computed efficiently. We prove improved upper bounds on the welfare loss for restrictions to additive bids and upper and lower bounds for restrictions to non-additive bids. These bounds show that the welfare loss increases in expressiveness. All our bounds apply to equilibrium concepts that can be computed in polynomial time as well as to learning outcomes.

1 Introduction

An important field at the intersection of economics and computer science is the field of mechanism design. The goal of mechanism design is to devise mechanisms consisting of an outcome rule and a payment rule that implement desirable outcomes in strategic equilibrium. A fundamental result in mechanism design theory, the so-called *revelation principle*, asserts that any equilibrium outcome of any mechanism can be obtained as a truthful equilibrium of a direct-revelation mechanism. However, the revelation principle says nothing about the computational complexity of such a truthful direct-revelation mechanism.

In the context of combinatorial auctions the truthful direct-revelation mechanism that maximizes welfare is the *Vickrey-Clarke-Groves (VCG) mechanism* [29,4,10]. Unfortunately, for many valuation spaces computing the VCG allocation and payments is a computationally hard problem. This is, for example, the case for subadditive, fractionally subadditive, and submodular valuations [16].

* Research supported by an SNF Postdoctoral Fellowship.
** This work was funded by the Vienna Science and Technology Fund (WWTF) through project ICT10-002, and by the University of Vienna through IK I049-N.

Y. Chen and N. Immorlica (Eds.): WINE 2013, LNCS 8289, pp. 146–159, 2013.

We thus study the performance of the VCG mechanism in settings in which the bidders are forced to use bids from a subspace of the valuation space for which the allocation and payments can be computed efficiently. This is obviously the case for additive bids, where the VCG-based mechanism can be interpreted as a separate second-price auction for each item. But it is also the case for the syntactically defined bidding space OXS, which stands for ORs of XORs of singletons, and the semantically defined bidding space GS, which stands for gross substitutes. For OXS bids polynomial-time algorithms for finding a maximum weight matching in a bipartite graph such as the algorithms of [28] and [8] can be used. For GS bids there is a fully polynomial-time approximation scheme due to [15] and polynomial-time algorithms based on linear programming [30] and convolutions of $M^{\#}$-concave functions [21,20,22].

One consequence of restrictions of this kind, that we refer to as *valuation compressions*, is that there is typically no longer a truthful dominant-strategy equilibrium that maximizes welfare. We therefore analyze the *Price of Anarchy*, i.e., the ratio between the optimal welfare and the worst possible welfare at equilibrium. We focus on equilibrium concepts such as correlated equilibria and coarse correlated equilibria, which can be computed in polynomial time [24,13], and naturally emerge from learning processes in which the bidders minimize external or internal regret [7,11,17,2].

Our Contribution. We start our analysis by showing that for restrictions from subadditive valuations to additive bids deciding whether a pure Nash equilibrium exists is $\mathcal{NP}$-hard. This shows the necessity to study other bidding functions or other equilibrium concepts.

We then define a smoothness notion for mechanisms that we refer to as *relaxed smoothness*. This smoothness notion is weaker in some aspects and stronger in another aspect than the weak smoothness notion of [27]. It is weaker in that it allows an agent's deviating bid to depend on the distribution of the bids of the other agents. It is stronger in that it disallows the agent's deviating bid to depend on his own bid. The former gives us more power to choose the deviating bid, and thus has the potential to lead to better bounds. The latter is needed to ensure that the bounds on the welfare loss extend to coarse correlated equilibria and minimization of external regret.

We use relaxed smoothness to prove an upper bound of 4 on the Price of Anarchy with respect to correlated and coarse correlated equilibria. Similarly, we show that the average welfare obtained by minimization of internal and external regret converges to $1/4$-th of the optimal welfare. The proofs of these bounds are based on an argument similar to the one in [6]. Our bounds improve the previously known bounds for these solution concepts by a logarithmic factor. We also use relaxed smoothness to prove bounds for restrictions to non-additive bids. For subadditive valuations the bounds are $O(\log(m))$ resp. $\Omega(1/\log(m))$, where m denotes the number of items. For fractionally subadditive valuations the bounds are 2 resp. $1/2$. The proofs require novel techniques as non-additive bids lead to non-additive prices for which most of the techniques developed in prior work fail. The bounds extend the corresponding bounds of [3,1] from additive to non-additive bids.

Table 1. Summary of our results (bold) and the related work (regular) for coarse correlated equilibria and minimization of external regret through repeated play. The range indicates upper and lower bounds on the Price of Anarchy.

		valuations	
		less general	subaddtive
bids	additive	[2,2]	[2,4]
	more general	[2, 2]	[2.4,O(log(m))]

Finally, we prove lower bounds on the Price of Anarchy. By showing that VCG-based mechanisms satisfy the *outcome closure property* of [19] we show that the Price of Anarchy with respect to pure Nash equilibria weakly increases with expressiveness. We thus extend the lower bound of 2 from [3] from additive to non-additive bids. This shows that our upper bounds for fractionally subadditive valuations are tight. We prove a lower bound of 2.4 on the Price of Anarchy with respect to pure Nash equilibria that applies to restrictions from subadditive valuations to OXS bids. Together with the upper bound of 2 of [1] for restrictions from subadditive valuations to additive bids this shows that the welfare loss can strictly increase with expressiveness.

Our analysis leaves a number of interesting open questions, both regarding the computation of equilibria and regarding improved upper and lower bounds. Interesting questions regarding the computation of equilibria include whether or not mixed Nash equilibria can be computed efficiently for restrictions from subadditive to additive bids or whether pure Nash equilibria can be computed efficiently for restrictions from fractionally subadditive valuations to additive bids. A particularly interesting open problem regarding improved bounds is whether the welfare loss for computable equilibrium concepts and learning outcomes can be shown to be strictly larger for restrictions to non-additive, say OXS, bids than for restrictions to additive bids. This would show that additive bids are not only sufficient for the best possible bound but also necessary.

Related Work. The Price of Anarchy of restrictions to additive bids is analyzed in [3,1,6] for second-price auctions and in [12,6] for first price auctions. The case where all items are identical, but additional items contribute less to the valuation and agents are forced to place additive bids is analyzed in [18,14]. Smooth games are defined and analyzed in [25,26]. The smoothness concept is extended to mechanisms in [27].

Organization. We describe our model in Section 2. We give the hardness result in Section 3, and define relaxed smoothness in Section 4. The upper and lower bounds can be found in Sections 5 to 8. All proofs omitted from this extended abstract are given in the full version of the paper.

2 Preliminaries

Combinatorial Auctions. In a *combinatorial auction* there is a set N of n *agents* and a set M of m *items*. Each agent $i \in N$ employs preferences over bundles of items, represented by a *valuation function* $v_i : 2^M \to \mathbb{R}_{\geq 0}$. We use V_i for the *class of valuation functions* of agent i, and $V = \prod_{i \in N} V_i$ for the class of joint valuations. We write $v = (v_i, v_{-i}) \in V$, where v_i denotes agent i's valuation and v_{-i} denotes the valuations of all agents other than i. We assume that the valuation functions are *normalized* and *monotone*, i.e., $v_i(\emptyset) = 0$ and $v_i(S) \leq v_i(T)$ for all $S \subseteq T$.

A mechanism $M = (f, p)$ is defined by an *allocation rule* $f : B \to \mathcal{P}(M)$ and a *payment rule* $p : B \to \mathbb{R}_{\geq 0}^n$, where B is the *class of bidding functions* and $\mathcal{P}(M)$ denotes the *set of allocations* consisting of all possible partitions X of the set of items M into n sets $X_1, \ldots, X_n$. As with valuations we write b_i for agent i's bid, and b_{-i} for the bids by the agents other than i. We define the *social welfare* of an allocation X as the sum $\text{SW}(X) = \sum_{i \in N} v_i(X_i)$ of the agents' valuations and use $\text{OPT}(v)$ to denote the maximal achievable social welfare. We say that an allocation rule f is *efficient* if for all bids b it chooses the allocation $f(b)$ that maximizes the sum of the agent's bids, i.e., $\sum_{i \in N} b_i(f_i(b)) = \max_{X \in \mathcal{P}(M)} \sum_{i \in N} b_i(X_i)$. We assume *quasi-linear preferences*, i.e., agent i's *utility* under mechanism M given valuations v and bids b is $u_i(b, v_i) = v_i(f_i(b)) - p_i(b)$.

We focus on the *Vickrey-Clarke-Groves (VCG)* mechanism [29,4,10]. Define $b_{-i}(S) = \max_{X \in \mathcal{P}(S)} \sum_{j \neq i} b_j(X_j)$ for all $S \subseteq M$. The VCG mechanisms starts from an efficient allocation rule f and computes the payment of each agent i as $p_i(b) = b_{-i}(M) - b_{-i}(M \setminus f_i(b))$. As the payment $p_i(b)$ only depends on the bundle $f_i(b)$ allocated to agent i and the bids b_{-i} of the agents other than i, we also use $p_i(f_i(b), b_{-i})$ to denote agent i's payment.

If the bids are additive then the VCG prices are additive, i.e., for every agent i and every bundle $S \subseteq M$ we have $p_i(S, b_{-i}) = \sum_{j \in S} \max_{k \neq i} b_k(j)$. Furthermore, the set of items that an agent wins in the VCG mechanism are the items for which he has the highest bid, i.e., agent i wins item j against bids b_{-i} if $b_i(j) \geq \max_{k \neq i} b_k(j) = p_i(j)$ (ignoring ties). Many of the complications in this paper come from the fact that these two observations do *not* apply to non-additive bids.

Valuation Compressions. Our main object of study in this paper are *valuation compressions*, i.e., restrictions of the class of bidding functions B to a strict subclass of the class of valuation functions V.[1] Specifically, we consider valuations and bids from the following hierarchy due to [16],

$$\text{OS} \subset \text{OXS} \subset \text{GS} \subset \text{SM} \subset \text{XOS} \subset \text{CF} \ ,$$

where OS stands for additive, GS for gross substitutes, SM for submodular, and CF for subadditive.

The classes OXS and XOS are syntactically defined. Define OR ($\vee$) as $(u \vee w)(S) = \max_{T \subseteq S}(u(T) + w(S \setminus T))$ and XOR ($\otimes$) as $(u \otimes w)(S) = \max(u(S), w(S))$. Define XS as the class of valuations that assign the same value to all bundles that contain a specific item and zero otherwise. Then OXS is the class of valuations that can be described as ORs of XORs of XS valuations and

---

[1] This definition is consistent with the notion of *simplification* in [19,5].

XOS is the class of valuations that can be described by XORs of ORs of XS valuations.

Another important class is the class β-XOS, where $\beta \geq 1$, of β-fractionally subadditive valuations. A valuation v_i is β-fractionally subadditive if for every subset of items T there exists an additive valuation a_i such that (a) $\sum_{j \in T} a_i(j) \geq v_i(T)/\beta$ and (b) $\sum_{j \in S} a_i(j) \leq v_i(S)$ for all $S \subseteq T$. It can be shown that the special case $\beta = 1$ corresponds to the class XOS, and that the class CF is contained in $O(\log(m))$-XOS (see, e.g., Theorem 5.2 in [1]). Functions in XOS are called *fractionally subadditive*.

Solution Concepts. We use game-theoretic reasoning to analyze how agents interact with the mechanism, a desirable criterion being stability according to some solution concept. In the *complete information* model the agents are assumed to know each others' valuations, and in the *incomplete information* model the agents' only know from which distribution the valuations of the other agents are drawn. In the remainder we focus on complete information. The definitions and our results for incomplete information are given in the full version of the paper.

The static solution concepts that we consider in the complete information setting are:

$$\text{DSE} \subset \text{PNE} \subset \text{MNE} \subset \text{CE} \subset \text{CCE} ,$$

where DSE stands for dominant strategy equilibrium, PNE for pure Nash equilibrium, MNE for mixed Nash equilibrium, CE for correlated equilibrium, and CCE for coarse correlated equilibrium.

In our analysis we only need the definitions of pure Nash and coarse correlated equilibria. Bids $b \in B$ constitute a *pure Nash equilibrium (PNE)* for valuations $v \in V$ if for every agent $i \in N$ and every bid $b_i' \in B_i$, $u_i(b_i, b_{-i}, v_i) \geq u_i(b_i', b_{-i}, v_i)$. A distribution $\mathcal{B}$ over bids $b \in B$ is a *coarse correlated equilibrium (CCE)* for valuations $v \in V$ if for every agent $i \in N$ and every pure deviation $b_i' \in B_i$, $\mathrm{E}_{b \sim \mathcal{B}}[u_i(b_i, b_{-i}, v_i)] \geq \mathrm{E}_{b \sim \mathcal{B}}[u_i(b_i', b_{-i}, v_i)]$.

The dynamic solution concept that we consider in this setting is regret minimization. A sequence of bids $b^1, \ldots, b^T$ incurs *vanishing average external regret* if for all agents i, $\sum_{t=1}^{T} u_i(b_i^t, b_{-i}^t, v_i) \geq \max_{b_i'} \sum_{t=1}^{T} u_i(b_i', b_{-i}^t, v_i) - o(T)$ holds, where $o(\cdot)$ denotes the little-oh notation. The empirical distribution of bids in a sequence of bids that incurs vanishing external regret converges to a coarse correlated equilibrium (see, e.g., Chapter 4 of [23]).

Price of Anarchy. We quantify the welfare loss from valuation compressions by means of the *Price of Anarchy (PoA)*.

The PoA with respect to PNE for valuations $v \in V$ is defined as the worst ratio between the optimal social welfare $\text{OPT}(v)$ and the welfare $\text{SW}(b)$ of a PNE $b \in B$,

$$\text{PoA}(v) = \max_{b:\ \text{PNE}} \frac{\text{OPT}(v)}{\text{SW}(b)} .$$

Similarly, the PoA with respect to MNE, CE, and CCE for valuations $v \in V$ is the worst ratio between the optimal social welfare $\text{SW}(b)$ and the expected welfare $\mathrm{E}_{b \sim \mathcal{B}}[\text{SW}(b)]$ of a MNE, CE, or CCE $\mathcal{B}$,

$$\text{PoA}(v) = \max_{\mathcal{B}: \text{ MNE, CE or CCE}} \frac{\text{OPT}(v)}{E_{b\sim\mathcal{B}}[\text{SW}(b)]} .$$

We require that the bids b_i for a given valuation v_i are *conservative*, i.e., $b_i(S) \leq v_i(S)$ for all bundles $S \subseteq M$. Similar assumptions are made and economically justified in the related work [3,1,6].

3   Hardness Result for PNE with Additive Bids

Our first result is that deciding whether there exists a pure Nash equilibrium for restrictions from subadditive valuations to additive bids is $\mathcal{NP}$-hard. The proof of this result is by reduction from 3-PARTITION [9] and uses an example with no pure Nash equilibrium from [1]. The same decision problem is simple for $V \subseteq \text{XOS}$ because pure Nash equilibria are guaranteed to exist [3].

Theorem 1. *Suppose that $V = CF$, $B = OS$, that the VCG mechanism is used, and that agents bid conservatively. Then it is $\mathcal{NP}$-hard to decide whether there exists a PNE.*

4   Smoothness Notion and Extension Results

Next we define a smoothness notion for mechanisms. It is weaker in some aspects and stronger in another aspect than the weak smoothness notion in [27]. It is weaker because it allows agent i's deviating bid a_i to depend on the marginal distribution $\mathcal{B}_{-i}$ of the bids b_{-i} of the agents other than i. This gives us more power in choosing the deviating bid, which might lead to better bounds. It is stronger because it does *not* allow agent i's deviating bid a_i to depend on his own bid b_i. This allows us to prove bounds that extend to coarse correlated equilibria and not just correlated equilibria.

Definition 1. *A mechanism is relaxed (λ, μ_1, μ_2)-smooth for $\lambda, \mu_1, \mu_2 \geq 0$ if for every valuation profile $v \in V$, every distribution over bids $\mathcal{B}$, and every agent i there exists a bid $a_i(v, \mathcal{B}_{-i})$ such that*

$$\sum_{i\in N} \mathop{E}_{\mathcal{B}_{-i}} [u_i((a_i, b_{-i}), v_i)] \geq \lambda OPT(v) - \mu_1 \sum_{i\in N} \mathop{E}_{\mathcal{B}}[p_i(X_i(b), b_{-i})] - \mu_2 \sum_{i\in N} \mathop{E}_{\mathcal{B}}[b_i(X_i(b))].$$

Theorem 2. *If a mechanism is relaxed (λ, μ_1, μ_2)-smooth, then the Price of Anarchy under conservative bidding with respect to coarse correlated equilibria is at most $(\max\{\mu_1, 1\} + \mu_2)/\lambda$.*

Proof. Fix valuations v. Consider a coarse correlated equilibrium $\mathcal{B}$. For each b from the support of $\mathcal{B}$ denote the allocation for b by $X(b) = (X_1(b), \ldots, X_n(b))$. Let $a = (a_1, \ldots, a_n)$ be defined as in Definition 1. Then,

$$\mathop{E}_{b\sim\mathcal{B}}[\text{SW}(b)] = \sum_{i\in N} \mathop{E}_{b\sim\mathcal{B}}[u_i(b, v_i)] + \sum_{i\in N} \mathop{E}_{b\sim\mathcal{B}}[p_i(X_i(b), b_{-i})]$$

$$\geq \sum_{i\in N} \underset{b_{-i}\sim\mathcal{B}_{-i}}{\mathrm{E}} [u_i((a_i, b_{-i}), v_i)] + \sum_{i\in N} \underset{b\sim\mathcal{B}}{\mathrm{E}} [p_i(X_i(b), b_{-i})]$$

$$\geq \lambda\, \mathrm{OPT}(v) - (\mu_1 - 1) \sum_{i\in N} \underset{b\sim\mathcal{B}}{\mathrm{E}} [p_i(X_i(b), b_{-i})] - \mu_2 \sum_{i\in N} \underset{b\sim\mathcal{B}}{\mathrm{E}} [b_i(X_i(b))],$$

where the first equality uses the definition of $u_i(b, v_i)$ as the difference between $v_i(X_i(b))$ and $p_i(X_i(b), b_{-i})$, the first inequality uses the fact that $\mathcal{B}$ is a coarse correlated equilibrium, and the second inequality holds because $a = (a_1, \ldots, a_n)$ is defined as in Definition 1.

Since the bids are conservative this can be rearranged to give

$$(1 + \mu_2) \underset{b\sim\mathcal{B}}{\mathrm{E}} [\mathrm{SW}(b)] \geq \lambda\, \mathrm{OPT}(v) - (\mu_1 - 1) \sum_{i\in N} \underset{b\sim\mathcal{B}}{\mathrm{E}} [p_i(X_i(b), b_{-i})].$$

For $\mu_1 \leq 1$ the second term on the right hand side is lower bounded by zero and the result follows by rearranging terms. For $\mu_1 > 1$ we use that $\mathrm{E}_{b\sim\mathcal{B}}[p_i(X_i(b), b_{-i})] \leq \mathrm{E}_{b\sim\mathcal{B}}[v_i(X_i(b))]$ to lower bound the second term on the right hand side and the result follows by rearranging terms. $\square$

Theorem 3. *If a mechanism is relaxed* (λ, μ_1, μ_2)*-smooth and* $(b^1, \ldots, b^T)$ *is a sequence of conservative bids with vanishing external regret, then*

$$\frac{1}{T} \sum_{t=1}^{T} SW(b^t) \geq \frac{\lambda}{\max\{\mu_1, 1\} + \mu_2} \cdot OPT(v) - o(1).$$

Proof. Fix valuations v. Consider a sequence of bids $b^1, \ldots, b^T$ with vanishing average external regret. For each b^t in the sequence of bids denote the corresponding allocation by $X(b^t) = (X_1(b^t), \ldots, X_n(b^t))$. Let $\delta_i^t(a_i) = u_i(a_i, b_{-i}^t, v_i) - u_i(b^t, v_i)$ and let $\Delta(a) = \frac{1}{T}\sum_{t=1}^{T}\sum_{i=1}^{n} \delta_i^t(a_i)$. Let $a = (a_1, \ldots, a_n)$ be defined as in Definition 1, where $\mathcal{B}$ is the empirical distribution of bids. Then,

$$\frac{1}{T} \sum_{t=1}^{T} \mathrm{SW}(b^t) = \frac{1}{T} \sum_{t=1}^{T}\sum_{i=1}^{n} u_i(b_i^t, b_{-i}^t, v_i) + \frac{1}{T} \sum_{t=1}^{T}\sum_{i=1}^{n} p_i(X_i(b^t), b_{-i}^t)$$

$$= \frac{1}{T} \sum_{t=1}^{T}\sum_{i=1}^{n} u_i(a_i, b_{-i}^t, v_i) + \frac{1}{T} \sum_{t=1}^{T}\sum_{i=1}^{n} p_i(X_i(b^t), b_{-i}^t) - \Delta(a)$$

$$\geq \lambda\, \mathrm{OPT}(v) - (\mu_1 - 1)\frac{1}{T} \sum_{t=1}^{T}\sum_{i=1}^{n} p_i(X_i(b^t, b_{-i}^t))$$

$$- \mu_2 \frac{1}{T} \sum_{t=1}^{T}\sum_{i=1}^{n} b_i(X_i(b^t)) - \Delta(a),$$

where the first equality uses the definition of $u_i(b_i^t, b_{-i}^t, v_i)$ as the difference between $v_i(X_i(b^t))$ and $p_i(X_i(b^t), b_{-i}^t)$, the second equality uses the definition of $\Delta(a)$, and the third inequality holds because $a = (a_1, \ldots, a_n)$ is defined as in Definition 1.

Since the bids are conservative this can be rearranged to give

$$(1 + \mu_2) \frac{1}{T} \sum_{t=1}^{T} \mathrm{SW}(b^t) \geq \lambda\, \mathrm{OPT}(v) - (\mu_1 - 1)\frac{1}{T} \sum_{t=1}^{T} \sum_{i=1}^{n} p_i(X_i(b^t), b^t_{-i}) - \Delta(a).$$

For $\mu_1 \leq 1$ the second term on the right hand side is lower bounded by zero and the result follows by rearranging terms provided that $\Delta(a) = o(1)$. For $\mu_1 > 1$ we use that $\frac{1}{T} \sum_{t=1}^{T} \sum_{i=1}^{n} p_i(X_i(b^t), b^t_{-i}) \leq \frac{1}{T} \sum_{t=1}^{T} \sum_{i=1}^{n} v_i(X_i(b^t))$ to lower bound the second term on the right hand side and the result follows by rearranging terms provided that $\Delta(a) = o(1)$.

The term $\Delta(a)$ is bounded by $o(1)$ because the sequence of bids $b^1, \ldots, b^T$ incurs vanishing average external regret and, thus,

$$\Delta(a) \leq \frac{1}{T} \sum_{i=1}^{n} \left[\max_{b'_i} \sum_{t=1}^{T} u_i(b'_i, b^t_{-i}, v_i) - \sum_{t=1}^{T} u_i(b^t, v_i) \right] \leq \frac{1}{T} \sum_{i=1}^{n} o(T). \qquad \square$$

5   Upper Bounds for CCE and Minimization of External Regret for Additive Bids

We conclude our analysis of restrictions to additive bids by showing how the argument of [6] can be adopted to show that for restrictions from $V = CF$ to $B = OS$ the VCG mechanism is relaxed $(1/2, 0, 1)$-smooth. Using Theorem 2 we obtain an upper bound of 4 on the Price of Anarchy with respect to coarse correlated equilibria. Using Theorem 3 we conclude that the average social welfare for sequences of bids with vanishing external regret converges to at least $1/4$ of the optimal social welfare. We thus improve the best known bounds by a logarithmic factor.

Proposition 1. *Suppose that $V = CF$ and that $B = OS$. Then the VCG mechanism is relaxed $(1/2, 0, 1)$-smooth under conservative bidding.*

To prove this result we need two auxiliary lemmata.

Lemma 1. *Suppose that $V = CF$, that $B = OS$, and that the VCG mechanism is used. Then for every agent i, every bundle Q_i, and every distribution $\mathcal{B}_{-i}$ on the bids b_{-i} of the agents other than i there exists a conservative bid a_i such that*

$$\operatorname*{E}_{b_{-i} \sim \mathcal{B}_{-i}} [u_i((a_i, b_{-i}), v_i)] \geq \frac{1}{2} \cdot v_i(Q_i) - \operatorname*{E}_{b_{-i} \sim \mathcal{B}_{-i}} [p_i(Q_i, b_{-i})] .$$

Proof. Consider bids b_{-i} of the agents $-i$. The bids b_{-i} induce a price $p_i(j) = \max_{k \neq i} b_k(j)$ for each item j. Let T be a maximal subset of items from Q_i such that $v_i(T) \leq p_i(T)$. Define the *truncated* prices q_i as follows:

$$q_i(j) = \begin{cases} p_i(j) & \text{for } j \in Q_i \setminus T, \text{ and} \\ 0 & \text{otherwise.} \end{cases}$$

The distribution $\mathcal{B}_{-i}$ on the bids b_{-i} induces a distribution $\mathcal{C}_i$ on the prices p_i as well as a distribution $\mathcal{D}_i$ on the truncated prices q_i.

We would like to allow agent i to draw his bid b_i from the distribution $\mathcal{D}_i$ on the truncated prices q_i. For this we need that (1) the truncated prices are additive and that (2) the truncated prices are conservative. The first condition is satisfied because additive bids lead to additive prices. To see that the second condition is satisfied assume by contradiction that for some set $S \subseteq Q_i \setminus T$, $q_i(S) > v_i(S)$. As $p_i(S) = q_i(S)$ it follows that

$$v_i(S \cup T) \leq v_i(S) + v_i(T) \leq p_i(S) + p_i(T) = p_i(S \cup T),$$

which contradicts our definition of T as a maximal subset of Q_i for which $v_i(T) \leq p_i(T)$.

Consider an arbitrary bid b_i from the support of $\mathcal{D}_i$. Let $X_i(b_i, p_i)$ be the set of items won with bid b_i against prices p_i. Let $Y_i(b_i, q_i)$ be the subset of items from Q_i won with bid b_i against the truncated prices q_i. As $p_i(j) = q_i(j)$ for $j \in Q_i \setminus T$ and $p_i(j) \geq q_i(j)$ for $j \in T$ we have $Y_i(b_i, q_i) \subseteq X_i(b_i, p_i) \cup T$. Thus, using the fact that v_i is subadditive, $v_i(Y_i(b_i, q_i)) \leq v_i(X_i(b_i, p_i)) + v_i(T)$. By the definition of the prices p_i and the truncated prices q_i we have $p_i(Q_i) - q_i(Q_i) = p_i(T) \geq v_i(T)$. By combining these inequalities we obtain

$$v_i(X_i(b_i, p_i)) + p_i(Q_i) \geq v_i(Y_i(b_i, q_i)) + q_i(Q_i).$$

Taking expectations over the prices $p_i \sim \mathcal{C}_i$ and the truncated prices $q_i \sim \mathcal{D}_i$ gives

$$\operatorname*{E}_{p_i \sim \mathcal{C}_i}[v_i(X_i(b_i, p_i)) + p_i(Q_i)] \geq \operatorname*{E}_{q_i \sim \mathcal{D}_i}[v_i(Y_i(b_i, q_i)) + q_i(Q_i)].$$

Next we take expectations over $b_i \sim \mathcal{D}_i$ on both sides of the inequality. Then we bring the $p_i(Q_i)$ term to the right and the $q_i(Q_i)$ term to the left. Finally, we exploit that the expectation over $q_i \sim \mathcal{D}_i$ of $q_i(Q_i)$ is the same as the expectation over $b_i \sim \mathcal{D}_i$ of $b_i(Q_i)$ to obtain

$$\operatorname*{E}_{b_i \sim \mathcal{D}_i}\left[\operatorname*{E}_{p_i \sim \mathcal{C}_i}[v_i(X_i(b_i, p_i))]\right] - \operatorname*{E}_{b_i \sim \mathcal{D}_i}[b_i(Q_i)]$$
$$\geq \operatorname*{E}_{b_i \sim \mathcal{D}_i}\left[\operatorname*{E}_{q_i \sim \mathcal{D}_i}[v_i(Y_i(b_i, q_i))]\right] - \operatorname*{E}_{p_i \sim \mathcal{C}_i}[p_i(Q_i)] \quad (1)$$

Now, using the fact that b_i and q_i are drawn from the same distribution $\mathcal{D}_i$, we can lower bound the first term on the right-hand side of the preceding inequality by

$$\operatorname*{E}_{b_i \sim \mathcal{D}_i}\left[\operatorname*{E}_{q_i \sim \mathcal{D}_i}[v_i(Y_i(b_i, q_i))]\right] = \frac{1}{2} \cdot \operatorname*{E}_{b_i \sim \mathcal{D}_i}\left[\operatorname*{E}_{q_i \sim \mathcal{D}_i}[v_i(Y_i(b_i, q_i)) + v_i(Y_i(q_i, b_i))]\right]$$
$$\geq \frac{1}{2} \cdot v_i(Q_i), \quad (2)$$

where the inequality in the last step comes from the fact that the subset $Y_i(b_i, q_i)$ of Q_i won with bid b_i against prices q_i and the subset $Y_i(q_i, b_i)$ of Q_i won with bid q_i against prices b_i form a partition of Q_i and, thus, because v_i is subadditive, it must be that $v_i(Y_i(b_i, q_i)) + v_i(Y_i(q_i, b_i)) \geq v_i(Q_i)$.

Note that agent i's utility for bid b_i against bids b_{-i} is given by his valuation for the set of items $X_i(b_i, p_i)$ minus the price $p_i(X_i(b_i, p_i))$. Note further that the price $p_i(X_i(b_i, p_i))$ that he faces is at most his bid $b_i(X_i(b_i, p_i))$. Finally note that his bid $b_i(X_i(b_i, p_i))$ is at most $b_i(Q_i)$ because b_i is drawn from $\mathcal{D}_i$. Together with inequality (1) and inequality (2) this shows that

$$\mathop{\mathrm{E}}_{b_i \sim \mathcal{D}_i} \left[\mathop{\mathrm{E}}_{b_{-i} \sim \mathcal{B}_{-i}} [u_i((b_i, b_{-i}), v_i)] \right] \geq \mathop{\mathrm{E}}_{b_i \sim \mathcal{D}_i} \left[\mathop{\mathrm{E}}_{p_i \sim \mathcal{C}_i} [v_i(X_i(b_i, p_i)) - b_i(Q_i)] \right]$$

$$\geq \frac{1}{2} \cdot v_i(Q_i) - \mathop{\mathrm{E}}_{p_i \sim \mathcal{C}_i} [p_i(Q_i)].$$

Since this inequality is satisfied in expectation if bid b_i is drawn from distribution $\mathcal{D}_i$ there must be a bid a_i from the support of $\mathcal{D}_i$ that satisfies it.   □

Lemma 2. *Suppose that $V = CF$, that $B = OS$, and that the VCG mechanism is used. Then for every partition $Q_1, \ldots, Q_n$ of the items and all bids b,*

$$\sum_{i \in N} p_i(Q_i, b_{-i}) \leq \sum_{i \in N} b_i(X_i(b)).$$

Proof. For every agent i and each item $j \in Q_i$ we have $p_i(j, b_{-i}) = \max_{k \neq i} b_k(j) \leq \max_k b_k(j)$. Hence an upper bound on the sum $\sum_{i \in N} p_i(Q_i, b_{-i})$ is given by $\sum_{i \in N} \max_k b_k(j)$. The VCG mechanisms selects allocation $X_1(b), \ldots, X_n(b)$ such that $\sum_{i \in N} b_i(X_i(b))$ is maximized. The claim follows.   □

Proof of Proposition 1. The claim follows by applying Lemma 1 to every agent i and the corresponding optimal bundle O_i, summing over all agents i, and using Lemma 2 to bound $\mathrm{E}_{b_{-i} \sim \mathcal{B}_{-i}}[\sum_{i \in N} p_i(O_i, b_{-i})]$ by $\mathrm{E}_{b \sim \mathcal{B}}[\sum_{i \in N} b_i(X_i(b))]$.   □

An important observation is that the proof of the previous proposition requires that the class of price functions, which is induced by the class of bidding functions via the formula for the VCG payments, is contained in B. While this is the case for additive bids that lead to additive (or "per item") prices this is *not* the case for more expressive bids. In fact, as we will see in the next section, even if the bids are from OXS, the least general class from the hierarchy of [16] that strictly contains the class of additive bids, then the class of price functions that is induced by B is no longer contained in B. This shows that the techniques that led to the results in this section *cannot* be applied to the more expressive bids that we study next.

6   A Lower Bound for PNE with Non-additive Bids

We start our analysis of non-additive bids with the following separation result: While for restrictions from subadditive valuations to additive bids the bound is 2 for pure Nash equilibria [1], we show that for restrictions from subadditive valuations to OXS bids the corresponding bound is at least 2.4. This shows that more expressiveness can lead to strictly worse bounds.

Theorem 4. *Suppose that $V = CF$, that $OXS \subseteq B \subseteq XOS$, and that the VCG mechanism is used. Then for every $\delta > 0$ there exist valuations v such that the PoA with respect to PNE under conservative bidding is at least $2.4 - \delta$.*

7   Upper Bounds for CCE and Minimization of External Regret for Non-additive Bids

Our next group of results concerns upper bounds for the PoA for restrictions to non-additive bids. For β-fractionally subadditive valuations we show that the VCG mechanism is relaxed $(1/\beta, 1, 1)$-smooth. By Theorem 2 this implies that the Price of Anarchy with respect to coarse correlated equilibria is at most 2β. By Theorem 3 this implies that the average social welfare obtained in sequences of repeated play with vanishing external regret converges to $1/(2\beta)$ of the optimal social welfare. For subadditive valuations, which are $O(\log(m))$-fractionally subadditive, we thus obtain bounds of $O(\log(m))$ resp. $\Omega(1/\log(m))$. For fractionally subadditive valuations, which are 1-fractionally subadditive, we thus obtain bounds of 2 resp. $1/2$. We thus extend the results of [3,1] from additive to non-additive bids.

Proposition 2. *Suppose that $V \subseteq \beta\text{-}XOS$ and that $OS \subseteq B \subseteq XOS$, then the VCG mechanism is relaxed $(1/\beta, 1, 1)$-smooth under conservative bidding.*

We will prove that the VCG mechanism satisfies the definition of relaxed smoothness point-wise. For this we need two auxiliary lemmata.

Lemma 3. *Suppose that $V \subseteq \beta\text{-}XOS$, that $OS \subseteq B \subseteq XOS$, and that the VCG mechanism is used. Then for all valuations $v \in V$, every agent i, and every bundle of items $Q_i \subseteq M$ there exists a conservative bid $a_i \in B_i$ such that for all conservative bids $b_{-i} \in B_{-i}$, $u_i(a_i, b_{-i}, v_i) \geq \frac{v_i(Q_i)}{\beta} - p_i(Q_i, b_{-i})$.*

Proof. Fix valuations v, agent i, and bundle Q_i. As $v_i \in \beta\text{-}XOS$ there exists a conservative, additive bid $a_i \in OS$ such that $\sum_{j \in X_i} a_i(j) \leq v_i(X_i)$ for all $X_i \subseteq Q_i$, and $\sum_{j \in Q_i} a_i(j) \geq \frac{v_i(Q_i)}{\beta}$. Consider conservative bids b_{-i}. Suppose that for bids (a_i, b_{-i}) agent i wins items X_i and agents $-i$ win items $M \setminus X_i$. As VCG selects outcome that maximizes the sum of the bids,

$$a_i(X_i) + b_{-i}(M \setminus X_i) \geq a_i(Q_i) + b_{-i}(M \setminus Q_i).$$

We have chosen a_i such that $a_i(X_i) \leq v_i(X_i)$ and $a_i(Q_i) \geq v_i(Q_i)/\beta$. Thus,

$$v_i(X_i) + b_{-i}(M \setminus X_i) \geq a_i(X_i) + b_{-i}(M \setminus X_i)$$
$$\geq a_i(Q_i) + b_{-i}(M \setminus Q_i) \geq \frac{v_i(Q_i)}{\beta} + b_{-i}(M \setminus Q_i).$$

Subtracting $b_{-i}(M)$ from both sides gives

$$v_i(X_i) - p_i(X_i, b_{-i}) \geq \frac{v_i(Q_i)}{\beta} - p_i(Q_i, b_{-i}).$$

As $u_i((a_i, b_{-i}), v_i) = v_i(X_i) - p_i(X_i, b_{-i})$ this shows that $u_i((a_i, b_{-i}), v_i) \geq v_i(Q_i)/\beta - p_i(Q_i, b_{-i})$ as claimed. □

Lemma 4. *Suppose that $OS \subseteq B \subseteq XOS$ and that the VCG mechanism is used. For every allocation $Q_1, \ldots, Q_n$ and all conservative bids $b \in B$ and corresponding allocation $X_1, \ldots, X_n$, $\sum_{i=1}^{n}[p_i(Q_i, b_{-i}) - p_i(X_i, b_{-i})] \leq \sum_{i=1}^{n} b_i(X_i)$.*

Proof. We have $p_i(Q_i, b_{-i}) = b_{-i}(M) - b_{-i}(M \setminus Q_i)$ and $p_i(X_i, b_{-i}) = b_{-i}(M) - b_{-i}(M \setminus X_i)$ because the VCG mechanism is used. Thus,

$$\sum_{i=1}^{n}[p_i(Q_i, b_{-i}) - p_i(X_i, b_{-i})] = \sum_{i=1}^{n}[b_{-i}(M \setminus X_i) - b_{-i}(M \setminus Q_i)]. \tag{3}$$

We have $b_{-i}(M \setminus X_i) = \sum_{k \neq i} b_k(X_k)$ and $b_{-i}(M \setminus Q_i) \geq \sum_{k \neq i} b_k(X_k \cap (M \setminus Q_i))$ because $(X_k \cap (M \setminus Q_i))_{i \neq k}$ is a feasible allocation of the items $M \setminus Q_i$ among the agents $-i$. Thus,

$$\sum_{i=1}^{n}[b_{-i}(M \setminus X_i) - b_{-i}(M \setminus Q_i)] \leq \sum_{i=1}^{n}[\sum_{k \neq i} b_k(X_k) - \sum_{k \neq i} b_k(X_k \cap (M \setminus Q_i))]$$

$$\leq \sum_{i=1}^{n}[\sum_{k=1}^{n} b_k(X_k) - \sum_{k=1}^{n} b_k(X_k \cap (M \setminus Q_i))]$$

$$= \sum_{i=1}^{n}\sum_{k=1}^{n} b_k(X_k) - \sum_{i=1}^{n}\sum_{k=1}^{n} b_k(X_k \cap (M \setminus Q_i)). \tag{4}$$

The second inequality holds due to the monotonicity of the bids. Since XOS = 1-XOS for every agent k, bid $b_k \in$ XOS, and set X_k there exists a bid $a_{k,X_k} \in$ OS such that $b_k(X_k) = a_{k,X_k}(X_k) = \sum_{j \in X_k} a_{k,X_k}(j)$ and $b_k(X_k \cap (M \setminus Q_i)) \geq a_{k,X_k}(X_k \cap (M \setminus Q_i)) = \sum_{j \in X_k \cap (M \setminus Q_i)} a_{k,X_k}(j)$ for all i. As $Q_1, \ldots, Q_n$ is a partition of M every item is contained in exactly one of the sets $Q_1, \ldots, Q_n$ and hence in $n - 1$ of the sets $M \setminus Q_1, \ldots, M \setminus Q_n$. By the same argument for every agent k and set X_k every item $j \in X_k$ is contained in exactly $n - 1$ of the sets $X_k \cap (M \setminus Q_1), \ldots, X_k \cap (M \setminus Q_n)$. Thus, for every fixed k we have that $\sum_{i=1}^{n} b_k(X_k \cap (M \setminus Q_i)) \geq (n - 1) \cdot \sum_{j \in X_k} a_{k,X_k}(j) = (n - 1) \cdot a_{k,X_k}(X_k) = (n - 1) \cdot b_k(X_k)$. It follows that

$$\sum_{i=1}^{n}\sum_{k=1}^{n} b_k(X_k) - \sum_{i=1}^{n}\sum_{k=1}^{n} b_k(X_k \cap (M \setminus Q_i))$$

$$\leq n \cdot \sum_{k=1}^{n} b_k(X_k) - (n - 1) \cdot \sum_{k=1}^{n} b_k(X_k) = \sum_{i=1}^{n} b_k(X_k). \tag{5}$$

The claim follows by combining inequalities (3), (4), and (5). □

Proof of Proposition 2. Applying Lemma 3 to the optimal bundles $O_1, \ldots, O_n$ and summing over all agents i,

$$\sum_{i \in N} u_i(a_i, b_{-i}, v) \geq \frac{1}{\beta} \mathrm{OPT}(v) - \sum_{i \in N} p_i(O_i, b_{-i}).$$

Applying Lemma 4 we obtain

$$\sum_{i \in N} u_i(a_i, b_{-i}, v) \geq \frac{1}{\beta} \mathrm{OPT}(v) - \sum_{i \in N} p_i(X_i(b), b_{-i}) - \sum_{i \in N} b_i(X_i(b)). \qquad \square$$

8   More Lower Bounds for PNE with Non-additive Bids

We conclude by proving matching lower bounds for the VCG mechanism and restrictions from fractionally subadditive valuations to non-additive bids. We prove this result by showing that the VCG mechanism satisfies the *outcome closure property* of [19], which implies that when going from more general bids to less general bids no new pure Nash equilibria are introduced. Hence the lower bound of 2 for pure Nash equilibria and additive bids of [3] translates into a lower bound of 2 for pure Nash equilibria and non-additive bids.

Theorem 5. *Suppose that $OXS \subseteq V \subseteq CF$, that $OS \subseteq B \subseteq XOS$, and that the VCG mechanism is used. Then the PoA with respect to PNE under conservative bidding is at least 2.*

It should be noted that the previous result applies even if valuation and bidding space coincide, and the VCG mechanism has an efficient, dominant-strategy equilibrium. This is because the VCG mechanism also admits other, non-efficient equilibria (and the Price of Anarchy metric does not restrict to dominant-strategy equilibria if they exist).

References

1. Bhawalkar, K., Roughgarden, T.: Welfare guarantees for combinatorial auctions with item bidding. In: Proc. of 22nd SODA, pp. 700–709 (2011)
2. Cesa-Bianchi, N., Mansour, Y., Stoltz, G.: Improved second-order bounds for prediction with expert advice. Machine Learning 66(2-3), 321–352 (2007)
3. Christodoulou, G., Kovács, A., Schapira, M.: Bayesian combinatorial auctions. In: Aceto, L., Damgård, I., Goldberg, L.A., Halldórsson, M.M., Ingólfsdóttir, A., Walukiewicz, I. (eds.) ICALP 2008, Part I. LNCS, vol. 5125, pp. 820–832. Springer, Heidelberg (2008)
4. Clarke, E.H.: Multipart pricing of public goods. Public Choice 11, 17–33 (1971)
5. Dütting, P., Fischer, F., Parkes, D.C.: Simplicity-expressiveness tradeoffs in mechanism design. In: Proc. of 12th EC, pp. 341–350 (2011)
6. Feldman, M., Fu, H., Gravin, N., Lucier, B.: Simultaneous auctions are (almost) efficient. In: Proc. of 45th STOC, pp. 201–210 (2013)
7. Foster, D., Vohra, R.: Calibrated learning and correlated equilibrium. Games and Economic Behavior 21, 40–55 (1997)

8. Fredman, M.L., Tarjan, R.E.: Fibonacci heaps and their uses in improved network optimization algorithms. Journal of the ACM 34(3), 596–615 (1987)
9. Garey, M.R., Johnson, D.S.: Computers and Intractability: A Guide to the Theory of NP-Completeness. W. H. Freeman and Company, New York (1979)
10. Groves, T.: Incentives in teams. Econometrica 41, 617–631 (1973)
11. Hart, S., Mas-Colell, A.: A simple adaptive procedure leading to correlated equilibrium. Econometrica 68, 1127–1150 (2000)
12. Hassidim, A., Kaplan, H., Mansour, Y., Nisan, N.: Non-price equilibria in markets of discrete goods. In: Proc. of 12th EC, pp. 295–296 (2011)
13. Jiang, A.X., Leyton-Brown, K.: Polynomial-time computation of exact correlated equilibrium in compact games. In: Proc. of 12th EC, pp. 119–126 (2011)
14. de Keijzer, B., Markakis, E., Schäfer, G., Telelis, O.: On the inefficiency of standard multi-unit auctions, pp. 1–16 (2013), http://arxiv.org/abs/1303.1646
15. Kelso, A.S., Crawford, V.: Job matching, coalition formation, and gross substitutes. Econometrica 50, 1483–1504 (1982)
16. Lehmann, B., Lehmann, D., Nisan, N.: Combinatorial auctions with decreasing marginal utilities. Games and Economic Behavior 55, 270–296 (2005)
17. Littlestone, N., Warmuth, M.K.: The weighted majority algorithm. Information and Computation 108(2), 212–261 (1994)
18. Markakis, E., Telelis, O.: Uniform price auctions: Equilibria and efficiency. In: Serna, M. (ed.) SAGT 2012. LNCS, vol. 7615, pp. 227–238. Springer, Heidelberg (2012)
19. Milgrom, P.: Simplified mechanisms with an application to sponsored-search auctions. Games and Economic Behavior 70, 62–70 (2010)
20. Murota, K.: Valuated matroid intersection II: Algorithm. SIAM Journal of Discrete Mathematics 9, 562–576 (1996)
21. Murota, K.: Matrices and Matroids for Systems Analysis. Springer, Heidelberg (2000)
22. Murota, K., Tamura, A.: Applications of m-convex submodular flow problem to mathematical economics. In: Eades, P., Takaoka, T. (eds.) ISAAC 2001. LNCS, vol. 2223, pp. 14–25. Springer, Heidelberg (2001)
23. Nisan, N., Roughgarden, T., Tardos, E., Vazirani, V.: Algorithmic Game Theory. Cambridge University Press, New York (2007)
24. Papadimitriou, C., Roughgarden, T.: Computing correlated equilibria in multi-player games. Journal of the ACM 55, 14 (2008)
25. Roughgarden, T.: Intrinsic robustness of the price of anarchy. In: Proc. of 41st STOC, pp. 513–522 (2009)
26. Roughgarden, T.: The price of anarchy in games of incomplete information. In: Proc. of 13th EC, pp. 862–879 (2012)
27. Syrgkanis, V., Tardos, É.: Composable and efficient mechanisms. In: Proc. of 45th STOC, pp. 211–220 (2013)
28. Tarjan, R.E.: Data structures and network algorithms. Society for Industrial and Applied Mathematics, Philadelphia (1983)
29. Vickrey, W.: Counterspeculation, auctions, and competitive sealed tenders. J. of Finance 16(1), 8–37 (1961)
30. de Vries, S., Bikhchandani, S., Schummer, J., Vohra, R.V.: Linear programming and Vickrey auctions. In: Dietrich, B., Vohra, R.V. (eds.) Mathematics of the Internet: E-Auctions and Markets, IMA Volumes in Mathematics and its Applications, pp. 75–116. Springer, Heidelberg (2002)

Limits of Efficiency in Sequential Auctions

Michal Feldman[1], Brendan Lucier[2], and Vasilis Syrgkanis[3]

[1] Hebrew Univsersity
mfeldman@huji.ac.il
[2] Microsoft Research
brlucier@microsoft.com
[3] Dept of Computer Science, Cornell University
vasilis@cs.cornell.edu

Abstract. We study the efficiency of sequential first-price item auctions at (subgame perfect) equilibrium. This auction format has recently attracted much attention, with previous work establishing positive results for unit-demand valuations and negative results for submodular valuations. This leaves a large gap in our understanding between these valuation classes. In this work we resolve this gap on the negative side. In particular, we show that even in the very restricted case in which each bidder has either an additive valuation or a unit-demand valuation, there exist instances in which the inefficiency at equilibrium grows linearly with the minimum of the number of items and the number of bidders. Moreover, these inefficient equilibria persist even under iterated elimination of weakly dominated strategies. Our main result implies linear inefficiency for many natural settings, including auctions with gross substitute valuations, capacitated valuations, budget-additive valuations, and additive valuations with hard budget constraints on the payments. For capacitated valuations, our results imply a lower bound that equals the maximum capacity of any bidder, which is tight following the upper-bound technique established by Paes Leme et al. [20].

1 Introduction

Consider the following natural auction setting. An auction house has a number of items that will go up for auction on a particular day. To orchestrate this, the auction house publishes a list of the items to be sold and the order in which they will be auctioned off. The items are then sold one at a time in the given order. A group of bidders attends this session of auctions, with each bidder being allowed to participate in any or all of the single-item auctions that will be run throughout the day. Since the auctions are run one at a time, in sequence, this format is referred to as a sequential auction.

This way of auctioning multiple items is prevalent in practice, due to its relative simplicity and transparency. This model is also related to electronic markets, such as eBay, due to the asynchronous nature of the multiple single-item auctions that are executed on the platform. A natural question, then, is how well such a sequential auction performs in practice. While the auction of a single

Y. Chen and N. Immorlica (Eds.): WINE 2013, LNCS 8289, pp. 160–173, 2013.
© Springer-Verlag Berlin Heidelberg 2013

item is relatively simple, equilibria of the larger game may be significantly more complex. For instance, a bidder who views two of the items as substitutes might prefer to win whichever sells at the lower price, and hence when bidding on the first item he must look ahead to the anticipated outcome of the second auction. What's more, the sequential nature of the mechanism implies that the outcome of one auction can influence the behavior of bidders in subsequent auctions. This gives rise to complex reasoning about the value of individual outcomes, with the potential to undermine the efficiency of the overall auction.

In this work we study the efficiency of sequential single-item first-price auctions, where items are sold sequentially using some predefined order and each item is sold by means of a first-price auction. We study the efficiency of outcomes at subgame perfect equilibrium, which is the natural solution concept for a dynamic, sequential game. Theoretical properties of these sequential auctions have been long studied in the economics literature starting from the seminal work of Weber [23]. However, most of the prior literature has focused on very restricted settings, such as unit-demand valuations, identical items, and symmetrically distributed player valuations. The few exceptions that have attempted to study equilibria when bidders have more complex valuations tend to have other restrictions, such as a very limited number of players or items [12,10,3,2]. Much of the difficulty in studying these auctions under complex environments and/or valuations stems from the inherent complexity of the equilibrium structure, which (as alluded to above) can involve complex reasoning about future auction outcomes.

Paes Leme et al. [20] and Syrgkanis and Tardos [21] circumvented this difficulty by performing an indirect analysis on efficiency using the price of anarchy framework. They showed that when bidders have unit-demand valuations (UD), items are heterogeneous, and bidders' valuations are arbitrarily asymmetrically distributed, then the social welfare at every equilibrium is a constant fraction of the optimal welfare. Syrgkanis and Tardos [22] extended this result to no-regret learning outcomes and to settings with budget constraints. On the negative side, Paes Leme et al. [20] showed that this result does not extend to submodular valuations (SM): there exists an instance where the unique "natural" subgame perfect equilibrium has inefficiency that scales linearly with the number of items.

The above results leave a large gap between the positive regime (unit-demand bidders) and the negative (submodular bidders). Many natural and heavily-studied classes of valuations fall in the range between UD and SM valuations. Among them are the following, arranged roughly from most to least general:

- *Gross-substitutes valuations (GS):* Whenever the cost of one item increases, this cannot reduce the demand for another item whose price did not increase.
- *k-capacitated valuations (k-CAP):* Each player i has a capacity $k_i \leq k$ and a value for each item; the value for a set of items is then the value of the k_i highest-valued items in the set.
- *Budget-additive valuations (BA):* The value of a player i is additive up to a player-specific budget B_i and then remains constant.

The class of GS valuations is motivated by the fact that it is (in a certain sense) the largest class of valuations for which a Walrasian equilibrium is guaranteed to exist [13], and a Walrasian equilibrium, if exists, is always efficient (see, e.g., [6]). It is known that every k-capacitated valuation satisfies gross substitutes [9]. Moreover, every gross substitutes valuation is submodular [16], and it is easy to see that unit-demand valuations are precisely 1-capacitated valuations. We therefore have UD $\subset$ k-CAP $\subset$ GS $\subset$ SM. The set of budget-additive valuations is incomparable to UD, k-CAP, and GS, but it is known that BA $\subset$ SM.

We ask: *for which of the above classes does the sequential first-price auction obtain a constant fraction of the optimal social welfare at equilibrium?* In this work we show that the answer to the above question is *none of them*.

Specifically, we show that for the case of gross substitutes valuations and for budget additive valuations, the inefficiency of equilibrium can grow linearly with the number of items and the number of players. Thus, even for settings in which a Walrasian equilibrium is guaranteed to exist, an auction that handles items sequentially cannot find an approximately optimal outcome at equilibrium. For the case of k-capacitated valuations, we show that the inefficiency can be as high as k. This bound of k is tight, following the upper bound established by [20].

To prove these lower bounds we consider a different, conceptually more restrictive, class of valuations: the union of unit-demand and additive valuations. We construct an instance in which every bidder has either a unit-demand valuation or an additive valuation, then show that the unique "natural" equilibrium for this instance has extremely poor social efficiency. We then adapt this construction to provide a lower bound for the valuation classes described above.

We also extend our lower bound to apply to one other setting: additive valuations when players have hard budget-constraints on their payments. This setting falls outside the quasi-linear regime, but is very relevant in the sequential auction setting: for instance, each bidder may arrive at an auction session with only a certain fixed amount of money to spend. Note that this is different from the BA valuation class, since it does not restrict the value of a player for a set of items, but rather limits the total payment that a player can make. For this setting, it is known that maximizing welfare is not an achievable goal in most auction settings, as a participant with low budget is necessarily ineffective at maximizing the value of the item(s) she obtains. Instead, the natural notion of social efficiency is the "effective welfare," in which the contribution of each participant to the welfare is capped by her budget [22]. We show that, even comparing against the benchmark of effective welfare, our negative result also applies to this setting: for additive valuations with hard budget constraints, the inefficiency can grow linearly with the number of items or players. This is in stark contrast to the setting of *simultaneous* first-price auctions, where it is known that a constant fraction of the optimal effective social welfare occurs at equilibrium for bidders with hard budget constraints, even when valuations are fractionally subadditive [22] (where this class falls between submodular and subadditive valuations).

Sequential auctions with additive bidders and hard budget constraints have been studied in only very limited settings in the economics literature and have recently begun to attract the attention of the computer science community [15].

The negative results described above rely heavily on the fact that items can be sold in an arbitrary order. This leads naturally to the following *design* question: does there always exists an order on the items that results in better outcomes at a subgame perfect equilibrium? We conjecture that a concrete class of item orders (that we propose) always contains a good order that leads to the VCG outcome at equilibrium, for the class of *single-valued unit-demand* valuations. We leave the resolution of this conjecture as an open problem.

1.1 Related Work

Sequential auctions have been long studied in the economics literature. Weber [23] and Milgrom and Weber [18] analyzed first- and second-price sequential auctions with identical items and unit-demand bidders in an incomplete-information setting and showed that the unique symmetric equilibrium is efficient and the prices have an upward drift. The behavior of prices in sequential studies was subsequently studied in [1,17]. Boutilier el al. [7] studies first-price auctions in a setting with uncertainty, and devised a dynamic-programming algorithm for finding the optimal strategies (assuming stationary distribution of others' bids).

The setting of multi-unit demand has also been studied under the complete-information model. Several papers studied the two-bidder case, where there is a unique subgame perfect equilibrium that survives the iterated elimination of weakly dominated strategies (IEWDS) [12,10]. Bae et al. [3,2] studied the case of sequential second-price auctions of identical items with two bidders with concave valuations and showed that the unique outcome that survives IEWDS achieves a social welfare at least $1 - e^{-1}$ of the optimum. Here we consider more than two bidders and heterogeneous items.

Recently, Paes Leme et al. [20] analyzed sequential first- and second-price auctions for heterogeneous items and multi-unit demand valuations in the complete-information setting. For sequential first-price auctions they showed that when bidders are unit-demand, every subgame perfect equilibrium achieves at least $1/2$ of the optimal welfare, while for submodular bidders the inefficiency can grow with the number of items, even with a constant number of bidders. The positive results were later extended to the incomplete-information setting in [21] and to no-regret outcomes and budget-constrained bidders in [22]. In this work we close the gap between positive and negative results and show that inefficiency can grow linearly with the minimum of the number of items and bidders even when bidders are either additive or unit-demand.

This work can be seen as part of the recent interest line of research on simple auctions. The closest literature to our work is that of simultaneous item-bidding auctions [5,8,4,14,11,22], which is the simultaneous counterpart of sequential auctions. In contrast to sequential auctions, in simultaneous item auctions constant efficiency guarantees have been established for general complement-free valuations, even under incomplete-information settings or outcomes that emerge from learning behavior. We refer to [19] for a recent survey on the efficiency of simultaneous and sequential item-auctions.

2 Model and Preliminaries

We consider settings with n bidders and m items, where every bidder $i \in [n]$ has a valuation function $v_i : 2^{[m]} \to \mathbb{R}_+$, associating a non-negative real value with every subset of items. We denote the set of bidders by $[n]$ and the set of items by $[m]$. The valuation function is assumed to be monotone (i.e., $v_i(T) \leq v_i(S)$ for every $T \subseteq S$). An *allocation* is a vector $x = (x_1, \ldots, x_n)$, where x_i denotes the set of items allocated to bidder i, and such that $x_i \cap x_j = \emptyset$ for every $i \neq j$.

Sequential item auctions. The auction proceeds in steps, where a single item is sold in every step using a first-price auction. In every step $t = 1, \ldots, m$, every bidder i offers a bid $b_i(t)$, and the item is allocated to the agent with the highest bid for a payment that equals his bid. Each bid in each step can be a function of the history of the game, which is assumed to be visible to all bidders. More formally, a strategy of bidder i is a function that, for every step t, associates a bid as a function of the sequence of the bidding profiles in all periods $1, \ldots, t-1$. The *utility* of an agent is defined, as standard, to be his value for the items he won minus the total payment he made throughout the auction (i.e., quasi-linear utility). We will also assume that the bid space is discretized in small negligible δ-increments, and for ease of presentation we will use b^+ to denote the bid $b + \delta$.

This setting is captured by the framework of extensive-form games (see, e.g., [20]), where the natural solution concept is that of a *subgame-perfect equilibrium* (SPE). In an SPE, the bidding strategy profiles of the players constitute a Nash equilibrium in every subgame. That is, at every step t and for every possible partial bidding profile $b(1), b(2), \ldots, b(t-1)$ up to (but not including) step t, the strategy profile in the subgame that begins in step t constitutes a Nash equilibrium in the induced (i.e., remaining) game.

Elimination of Weakly Dominated Strategies. We wish to further restrict our attention to "natural" equilibria, that exclude (for example) dominated over-bidding strategies. We therefore consider a natural and well-studied refinement of the set of subgame perfect equilibria: those that survive iterated elimination of weakly dominated strategies (IEWDS). A strategy s is *weakly dominated* by a strategy s' if, for every profile of other players' strategies s_{-i}, we have $u_i(s, s_{-i}) \leq u_i(s', s_{-i})$, and moreover there exists some s_{-i} such that $u_i(s, s_{-i}) < u_i(s', s_{-i})$. Roughly speaking, under IEWDS, each player removes from her strategy space the set of all weakly dominated strategies. This removal may cause new strategies to become weakly dominated for a player, which are then removed from her strategy space, and so on until no weakly dominated strategies remain. A formal definition appears in the full version of the paper.

Price of anarchy. The price of anarchy (PoA) measures the inefficiency that can arise in strategic settings. The PoA for subgame perfect equilibria is defined as the worst (i.e., largest) possible ratio between the welfare obtained in the optimal allocation and the welfare obtained in any subgame perfect equilibrium of the game. We note that all of our lower bounds on the price of anarchy will involve "natural" equilibria that survive IEWDS.

3   A Simple Example

To develop some intuition regarding the strategic considerations that might take place in sequential auctions, we give a simple example in which one bidder has value for many items (i.e., wholesale buyer) and another bidder has value for only one item (i.e., retail buyer).

In particular, consider a sequence of two auctions for two identical items and two buyers, A and B. Buyer A is a "wholesale" buyer, having an additive valuation with a value of 9 for each of the two items. Buyer B is a "retail" buyer, who wants only one item (unit-demand) and has a value of 5 for either of the two. The items are sold sequentially using a first-price auction for each item.

Consider the situation from the perspective of the additive buyer A. Thinking strategically and farsightedly, he reasons that if he wins the first auction, then in the second auction he will have to compete with buyer B and will therefore have to pay 5 dollars to win the second item. If, however, he lets buyer B win the first item, then buyer B will have no value for the second item and hence the only undominated strategy for buyer B will be to bid 0 in the second auction, and hence buyer A will win the second item for free. What must buyer A pay in order to win the first item? Buyer B knows that if the first item goes to buyer A, then buyer B will certainly lose the second item as well; therefore buyer B is willing to pay up to 5 for the first item. Therefore, in order to win the first item, buyer A will have to bid at least 5 in the first auction.

Thus bidder A needs to choose between the following options: he can win both auctions for a price of 5 each, or let bidder B win the first auction and win only the second, but pay nothing. The first option gives bidder A a utility of 8 ($= 2 \cdot (9 - 5)$) while the second option gives him a utility of 9 ($= 1 \cdot (9 - 0)$). Consequently, bidder A will choose to forego the first item in order to improve his situation in the second one. Interestingly, this equilibrium outcome is socially suboptimal, since the efficient outcome is for bidder A to win both items.

One can modify this example by taking bidder A's value to be $10 - \epsilon$ for each item. In this case the unique subgame perfect equilibrium that survives elimination of dominated strategies is a $4/3$ approximation to the optimal welfare, even though the items are identical (and therefore the inefficiency is irrespective of item ordering). In the next section we demonstrate that with heterogeneous items, the social welfare of sequential item auctions at subgame perfect equilibrium can be as low as an $O(m)$ fraction of the optimal social welfare.

4   Lower Bound for Additive and Unit-demand Valuations

We now present our main result by providing an instance of a sequential first price auction with unit-demand and additive bidders, where the social welfare at a subgame-perfect equilibrium that survives IEWDS[1] achieves social welfare

---

[1] The equilibrium that we describe is, in some sense, the unique natural equilibrium: if we were to ask players to submit bids sequentially within each auction, rather than simultaneously, then there would be a unique equilibrium (solvable by backward induction), which is the equilibrium that we describe.

that is only an $O(\min\{n, m\})$-fraction of the optimal welfare. Therefore, our example shows that inefficiency can arise at equilibrium in a robust manner.

Theorem 1. *The price of anarchy of the sequential first-price item auctions with additive and unit-demand bidders is $\Omega(\min\{n, m\})$. Moreover, this result persists even if we consider only equilibria that survive IEWDS.*

Informal Description. Before we delve into the details of the proof of Theorem 1, we give a high-level idea of the type of strategic manipulations that lead to inefficiency and compare to the corresponding simultaneous auction.

Consider an auction instance where two additive bidders have identical values for most of the items for sale, but their valuations differ only on the last few items that are sold. Specifically, assume that there are two items Z_1 and Z_2, auctioned last, such that only player 1 has value for Z_1 and only player 2 has value for Z_2. We will refer to these items as the *non-competitive items* and to all other items as the *competitive items*. The additive bidders know that it is hopeless to try to achieve any positive utility from the competitive items on which they have identical interests. The only utility they can ever derive is from the last, non-competitive items on which they don't compete with each other. If these were the only two players in the auction, then we would obtain the optimal outcome: the two bidders would simply compete on each of the competitive items, with one of them acquiring each competitive item at zero utility.[2]

We now imagine adding unit-demand bidders to the auction in order to perturb the optimality. Specifically, suppose there is a unit-demand bidder that has value for the two *non-competitive items*, with the value for item Z_i being slightly less than player i's value for Z_i, $i \in \{1, 2\}$. This endangers the additive bidders' hopes of getting non-negligible utility, since competition from the unit-demand player may drive up the prices of Z_1 and Z_2. The only hope that the additive bidders have is that the unit-demand bidder will have his demand satisfied prior to these final two auctions, in which case the unit-demand bidder would not bother to bid on them. Hence, the two additive bidders would do anything in their power to guide the auction to such an outcome, even if that means sacrificing all the competitive items! This is exactly the effect that we achieve in our construction. Specifically, we create an instance where this competing unit-demand bidder has his demand satisfied prior to the auctions for Z_1 and Z_2 if and only if a very specific outcome occurs: the additive bidders don't bid at all on all the competitive items, but rather other small-valued bidders acquire the competitive items instead. These small-valued bidders contribute almost nothing to the welfare, and therefore all of the welfare from the competitive items is lost.

It is useful to compare this example with what would happen if the auctions were run simultaneously, rather than sequentially. This uncovers the crucial property of sequential auctions that leads to inefficiency: the *ability to respond to deviations*. If all auctions happened simultaneously, then the behavior of the additive bidders that we described above could not possibly be an equilibrium:

---

[2] In fact, optimality is always achieved when all bidders are additive, in general.

one additive bidder, knowing that his additive competitor bids 0 on all the competitive items, would simply deviate to outbid him on the competitive items and get a huge utility. However, because the items are sold sequentially, this deviation cannot be undertaken without consequence: the moment one of the additive bidders deviates to bidding on the competitive items, in all subsequent auctions the competitor will respond by bidding on subsequent competitive items, leading to zero utility for the remainder of the auctions. Moreover, this response need not be punitive, but is rather the only rational response once the auction has left the equilibrium path (since the additive bidders know that there is no way to obtain positive utility in subsequent auctions). Thus, in a sequential auction, an additive player can only extract utility from at most one competitive item, which is not sufficient to counterbalance the resulting utility-loss due to the increased competition on the last non-competitive item.

The Lower Bound. We now proceed with a formal proof of Theorem 1. Consider an instance with 2 additive players, $k + 1$ unit-demand players and $k + 3$ items. Denote with $\{a, b\}$ the two additive players and with $\{p_0, p_1, \ldots, p_k\}$ the $k + 1$ unit-demand players. Also denote the items with $\{I_1, \ldots, I_k, Y, Z_1, Z_2\}$. The valuations of the additive players are represented by the following table of v_{ij}, where $\epsilon > 0$ is an arbitrarily small constant:

	I_k	$\ldots$	I_1	Y	Z_1	Z_2
a	$1 + \epsilon$	$\ldots$	$1 + \epsilon$	0	10	0
b	1	$\ldots$	1	0	0	10

In addition, the unit-demand valuations for the remaining $k + 1$ players are given by the table of v_{ij} that follows (an empty entry corresponds to a 0 valuation), though now a valuation of a player when getting a set S is $\max_{j \in S} v_{ij}$:

	I_k	I_{k-1}	I_{k-2}	$\ldots$	I_2	I_1	Y	Z_1	Z_2
p_0				$\ldots$			$10 - \epsilon$	$10 - \epsilon$	$10 - \epsilon$
p_1				$\ldots$		δ_1	10		
p_2				$\ldots$	δ_2	δ_2			
$\ldots$									
p_{k-1}		δ_{k-1}	δ_{k-1}	$\ldots$					
p_k	δ_k	δ_k		$\ldots$					

The constants $\delta_1, \ldots, \delta_k$ are chosen to satisfy the following condition:

$$\delta_k > \delta_{k-1} > \ldots > \delta_2 > \delta_1 > \epsilon \tag{1}$$

Note that, by taking ϵ to be arbitrarily small, we can take each δ_i to be arbitrarily small as well.

In the optimal allocation, player a gets all the items $I_1, \ldots, I_k$ and Z_1, player b gets Z_2 and player p_1 gets Y. The resulting social welfare is $k(1 + \epsilon) + 30$. We assume that the auctions take place in the order depicted in the valuation tables: $\{I_k, \ldots, I_1, Y, Z_1, Z_2\}$. We will show that there is a subgame perfect equilibrium for this auction instance such that the unit-demand players win all the items

$I_1, \ldots, I_k$. Specifically, player p_i wins item I_i, player a wins Z_1, player b wins Z_2, and player p_0 wins Y, resulting in a social welfare of $30 - \epsilon + \sum_{i=1}^{k} \delta_i$. Taking δ sufficiently small, this welfare is at most 31. This will establish that the price of anarchy for this instance is at least $\frac{k(1+\epsilon)+30}{31} = O(k)$, establishing Theorem 1. Furthermore, we will show that this subgame perfect equilibrium is *natural*, in the sense that it survives iterated deletion of weakly dominated strategies.

The intuition is the following: after the first k auctions have been sold, player p_0 has to decide if he will target (and win) item Y, or if he will instead target items Z_1 and/or Z_2. If he targets item Y, he competes with player p_1 and afterwards lets players a and b win items Z_1, Z_2 for free. This decision of player p_0 depends on whether player p_1 has won item I_1, which in turn depends on the outcomes of the first $k-1$ auctions. In particular, player p_1 can win item I_1 only if player p_2 has won item I_2. In turn, p_2 can win I_2 only if p_3 has won item I_3 and so on. Hence, it will turn out that in order for p_0 to want to target item Y, it must be that each item I_i is sold to bidder p_i. Thus, if either player a or b acquires any of the items $I_1, \ldots, I_k$, they will be guaranteed to obtain low utility on items Z_1 and Z_2. This will lead them to bidding truthfully on all subsequent I_i auctions, leading to a severe drop in utility gained from future auctions.

In the remainder of this section, we provide a more formal analysis of the equilibrium in this auction instance. We begin by examining what happens in the last three auctions of Y, Z_1 and Z_2, conditional on the outcomes of the first k auctions. We first examine the outcome of auctions Y, Z_1, Z_2 conditional on the outcome of auction I_1:

Case 1: p_1 has won I_1. Player p_1 has marginal value of $10 - \delta_1$ for item Y. Hence, he is willing to bid at most $10 - \delta_1$ on item Y. Player p_0 knows that if he loses Y then in the subgame perfect equilibrium in that subgame he will bid $10 - \epsilon$ on Z_1 and Z_2 and lose. Thus he expects no utility from the future if he loses Y. Thus he is willing to pay at most $10 - \epsilon$ for item Y. Since by assumption (1) $\delta_1 > \epsilon$, player p_0 will win Y at a price of $10 - \delta_1$. Then players a, b will win Z_1 and Z_2 for free. Thus the utilities in this case from this subgame are: $u(a) = 10$, $u(b) = 10$, $u(p_0) = \delta_1 - \epsilon$, $u(p_1) = 0$.

Case 2: p_1 has lost I_1. Player p_1 has marginal value of 10 for item Y. Hence, he is willing to bid at most 10 on item Y. Player p_0 performs the exact same thinking as in the previous case and thereby is willing to bid at most $10 - \epsilon$ for item Y. Thus in this case p_1 will win item Y at a price of $10 - \epsilon$. Then, as predicted, p_0 will bid $10 - \epsilon$ on Z_1 and Z_2 and lose. Thus the utilities of the players in this case are: $u(a) = \epsilon$, $u(b) = \epsilon$, $u(p_0) = 0$, $u(p_1) = \epsilon$.

Now we focus on the auction of item I_1. As was explained in Paes Leme et al. [20] this auction will be an auction with externalities where each player has a different utility for each different winner outcome. These utilities can be concisely expressed in a table of v_{ij}'s where v_{ij} is the value of player i when player j wins. The only players that potentially have any incentive to bid on item I_1 are a, b, p_0, p_1, p_2. The following table summarizes their values for each possible winner outcome of auction I_1 as was calculated in the previous case-analysis (we point that in the diagonal we also add the actual value that a player acquires from item I_1 to his future utility conditional on winning I_1) .

$$[v_{ij}] = \begin{array}{c|ccccc} & a & b & p_0 & p_1 & p_2 \\ \hline a & 1+2\epsilon & \epsilon & \epsilon & 10 & \epsilon \\ b & \epsilon & 1+\epsilon & \epsilon & 10 & \epsilon \\ p_0 & 0 & 0 & 0 & \delta_1 - \epsilon & 0 \\ p_1 & \epsilon & \epsilon & \epsilon & \delta_1 & \epsilon \\ p_2 & 0 & 0 & 0 & 0 & \delta_2 \cdot 1_{\text{hasn't won } I_2} \end{array}$$

For example, player a obtains utility 10 if player p_1 wins item I_1. We see from the table that, at this auction, everyone except p_2 achieves their maximum value when p_1 wins the auction. Player p_2 has value for winning the auction only if he hasn't won I_2. In addition, since $\delta_2 > \delta_1$, if p_2 hasn't won I_2 then he can definitely outbid p_1 on I_1 and therefore p_1 has no chance of winning the auction of I_1. As we now show, this implies that there is a unique equilibrium of the auction conditioning on whether or not p_2 has won I_2:

Case 1: If p_2 has won I_2 then he has no value for I_1. There exists an equilibrium in undominated strategies where all players a, b, p_0, p_2 will bid 0, while p_1 bids 0^+. In fact this is in some sense the most natural equilibrium since it yields the highest utility for a and b. In this case the utility of the players from auctions I_1 and onward will be: $u(a) = 10$, $u(b) = 10$, $u(p_0) = \delta_1 - \epsilon$, $u(p_1) = \delta_1$, $u(p_2) = 0$.

Case 2: If p_2 has lost I_2, then he has value of $\delta_2 > \delta_1$ for I_1. Hence, p_1 has no chance of winning item I_1. Thus, the unique equilibrium that survives elimination of weakly dominated strategies in this case is for player a to bid 1^+, for player b to bid 1, for player p_0 to bid 0, for player p_1 to bid $\delta_1 - \epsilon$ and for player p_2 to bid δ_2. In this case the utility of the players from auctions I_1 and on will be: $u(a) = 2\epsilon$, $u(b) = \epsilon$, $u(p_0) = 0$, $u(p_1) = \epsilon$, $u(p_2) = 0$.

Using similar reasoning we deduce that player p_i can win I_i only if p_{i-1} has won I_{i-1}. If at any point some p_i does not win I_i then players a and b know that from that point onward no p_j can win auction I_j, and therefore they will get only utility ϵ from Z_1, Z_2. Thus there will be no reason for players a and b to allow unit-demand players to continue to win items, and thus the only equilibrium strategies from that point on will be for a to bid 1^+ on each of I_i and b to bid 1. This will lead to player a to get utility $O(\epsilon)$ from each auction for items $I_{i-1}, \ldots, I_2$, and player b to get no utility from these auctions. Thus, at any point in the auction, it is an equilibrium for players a and b to allow the unit demand player p_i to win auction I_i. This completes the proof of Theorem 1

Finally, as discussed throughout our analysis, the equilibrium described above survives IEWDS. The reason is that, for every item k and bidder i, the proposed equilibrium strategy for bidder i does not require that he bids more than his value for item k less his utility in the continuation game subject to not winning item k. As discussed in Paes Leme et al. [20], this property guarantees that no player is playing a weakly dominated strategy.

5   Extensions of the Lower Bound

We now provide extensions of our lower bound in Theorem 1, to show that linear inefficiency can occur under several important valuation classes.

Gross Substitutes and Capacitated. Since the classes GS and m-CAP include all additive and unit-demand valuations over m items, Theorem 1 immediately implies a linear price of anarchy for these classes. This also implies that, for any $\ell \leq m$, the price of anarchy for ℓ-capacitated valuations is at least linear in ℓ (e.g., by adding extra items of no value).

Budget-Additive. A valuation is budget additive if it can be written in the form $v(S) = \max\left\{B, \sum_{j \in S} v_j\right\}$. As it turns out, in the example in the previous section all valuations are budget additive. The additive players can be thought of as having infinite budget. Each of the unit-demand players p_i for $i \in [2, k]$ can be thought as budget-additive with a budget of δ_i and value δ_i for items I_i and I_{i+1} and 0 for everything else. Player p_1 has budget of 10 and additive value of δ_1 for I_1, 10 for Y and 0 for everything else. Player p_0 has budget $10 - \epsilon$ and additive value of $10 - \epsilon$ for each of Y, Z_1, Z_2 and 0 for everything else. Therefore the analysis in the previous section holds even for budget-additive valuations.

Additive valuations with budget constraints on payments. The same analysis applies when each player i has an additive valuation and a hard budget constraint B_i on his payment. Formally, if a player i receives a set S and pays a total price of p, then his utility $u_i(S, p)$ is $v_i(S) - p$ if $p \leq B_i$, or $-\infty$ otherwise.

We adapt Theorem 1 to the setting of budget constraints in a manner similar to budget-additive valuations. We set the budgets of the players as in the budget-additive case, interpreted as payment budgets rather than a cap on valuations.

There is an additional subtlety in the analysis, since it doesn't only matter whether a player won an item, but also at what price. As a result, the equilibrium in our example will change slightly. The additive bidders, in addition to letting bidder p_i win I_i, will also have to make him pay enough so that he has no remaining budget with which to win the subsequent item I_{i+1}. That is, the bids of the additive bidders on items I_i will be higher, in order to better exhaust the budgets of bidders p_i. The reason for this strategic budget exhaustion is that, if a player p_i has enough budget to purchase item I_{i+1}, the effect is similar to falling off the equilibrium path in the original example. The additive bidders are therefore incentivized to bid high enough to exhaust the budgets of the other players, and hence this is the only bidding behavior that survives IEWDS.

A more formal analysis of this setting appears in the full version of the paper. We conclude that the price of anarchy in this instance is $\Omega(k)$.

6   The Impact of Item Ordering

Our lower bound establishes that if items are sold sequentially, then arbitrarily inefficient outcomes can result at equilibrium even when all agents have gross substitutes valuations. The constructions depend on the items being sold in an arbitrary order. A natural question arises: does there always exist an order over the items such that the resulting outcome is efficient, or approximately efficient?

In this section we discuss this problem in the context of unit-demand bidders. Recall that, for unit-demand bidders, selling items in an arbitrary order always results in an outcome that achieves at least half of the optimal social welfare. Additionally, it is known by [20] that if any order is allowed, then the unique subgame-perfect equilibrium that survives IEWDS can be inefficient, achieving only a 3/2-approximation. This lower bound of 3/2 holds even for the special case of single-valued unit-demand bidders, where each player has a single value v_i for getting one item from some interest set S_i. We conjecture that, for the case of single-valued unit-demand bidders, if the auctioneer can choose the order in which the objects are sold, then it is possible to recover the optimal welfare at all natural equilibria. Indeed, we make a stronger conjecture: there exists an order in which the VCG outcome (allocation and payments) occurs at equilibrium.

Conjecture 1. For every instance of single-valued unit-demand bidders, there exists an order over the items such that the corresponding sequential auction admits a subgame perfect equilibrium that survives IEWDS and that replicates the VCG outcome.

Observe that such a result cannot hold for both additive and unit-demand bidders as is portrayed by our simple example in Section 3, where all items are identical and hence, under any ordering, the unique subgame-perfect equilibrium that survivies IEWDS is inefficient. Our conjecture also stems from the fact that for the case of single-valued unit-demand bidders the optimization problem is a matroid optimization problem. It is known by [20] that a form of sequential cut auction for matroids always leads to a VCG outcome. The difference is that sequential item-auctions do not correspond to auctions across cuts of the matroid. However, it is feasible that under some ordering the same behavior as in a sequential cut auction will be implemented.

As progress toward this conjecture, we will present a class of item orderings, the *augmenting path orderings*, which we believe always contains an ordering that satisfies Conjecture 1. In the full version we show that the 3/2 lower bound of [20] breaks if we only allow augmenting path orderings. We leave open the question of whether one of these orderings always yields a VCG outcome.

6.1   A Class of Orderings

Consider a profile of single-valued unit-demand valuations. Let x denote the VCG allocation (i.e., x_i is the item allocated to bidder i). We also write $x^{(-i)}$ to denote the VCG allocation when bidder i is excluded. For each i, the allocations x and $x^{(-i)}$ define a directed bipartite graph between players and objects, where there is an edge from item j to player k if $x_k^{(-i)} = j$ but $x_k \neq j$, and there is an edge from player k to item j if $x_k^{(-i)} \neq j$ but $x_k = j$. It is known that, for each player i, this graph is always a directed path from player i to some other player k; this is the *augmenting path for player i* and player k is the *price setter* of player i, i.e. the VCG price of player i is v_k. With no loss of generality we assume that every player has a price setter k.

Given a welfare-optimal matching π, that matches each player i to an item $\pi(i)$, consider the following forest construction. For each price setter k, in decreasing value order, we will construct a tree as follows. Consider all the items that are in the interest set of k, S_k, that are not yet in the forest. Add each such item to the tree as a child of player k. Next, from each such item j, consider its optimally matched player $\pi^{-1}(j)$ and add this player to the tree as a child of j. For each player i that was added, consider all items that are in the interest set of i, S_i, that are not yet in the forest, and add each of these items to the tree as a child of i. We continue this process, which is essentially a breadth-first traversal of the set of items, until there is no new item to be added.

The above process creates a forest that contains a node for each item, for each player that is allocated an item in the optimal allocation, and for each price setter. Additionally, each player belongs to the tree rooted at his price setter and his unique path in the tree to the price setter is an augmenting path in the initial bipartite graph. The reasoning is as follows: each tree contains all possible alternating paths ending at the price-setter, except alternating paths that contain items and players who have been included in the tree of a price setter with larger value. Since a player's price setter is the largest unallocated player with which he is connected, through an alternating path, the claim follows.

We will refer to the above forest as the *augmenting path graph* G. Given an augmenting path graph G, a *post-order item traversal of G* is a depth-first, post-order traversal of the nodes of G, restricted to the nodes corresponding to items and rooted at price setters. Note that this is an ordering over the items in the auction. We also assume that trees are traversed in decreasing order of price-setters. Also note that this order is not necessarily unique, as it does not specify the order in which the children of a given node should be traversed.

Definition 2. *The set of* augmenting path orderings *of the items is the set of orderings corresponding to post-order item traversals of G.*

Our (refined) conjecture is that, for every instance of single-valued unit-demand bidders, there exists an augmenting path ordering such that the corresponding sequential auction admits a subgame perfect equilibrium that replicates the VCG outcome. As an example, in the full version of the paper we show that this conjecture holds for the 3/2 lower bound example from [20]. We also show in the full version that it is not true that *all* augmenting path orderings lead to efficient outcomes at equilibrium: there are examples in which multiple augmenting path orderings exist, and some orderings lead to inefficient outcomes at equilibrium.

References

1. Ashenfelter, O.: How auctions work for wine and art. The Journal of Economic Perspectives 3(3), 23–36 (1989)
2. Bae, J., Beigman, E., Berry, R., Honig, M.L., Vohra, R.: Sequential Bandwidth and Power Auctions for Distributed Spectrum Sharing. IEEE Journal on Selected Areas in Communications 26(7), 1193–1203 (2008)

3. Bae, J., Beigman, E., Berry, R., Honig, M.L., Vohra, R.: On the efficiency of sequential auctions for spectrum sharing. In: 2009 International Conference on Game Theory for Networks, pp. 199–205 (May 2009)
4. Bhawalkar, K., Roughgarden, T.: Welfare guarantees for combinatorial auctions with item bidding. In: SODA (2011)
5. Bikhchandani, S.: Auctions of heterogeneous objects. Games and Economic Behavior 26(2), 193–220 (1999)
6. Blumrosen, L., Nisan, N.: Combinatorial Auctions. Camb. Univ. Press (2007)
7. Boutilier, C., Goldszmidt, M., Sabata, B.: Sequential Auctions for the Allocation of Resources with Complementarities. In: IJCAI 1999: Proceedings of the Sixteenth International Joint Conference on Artificial Intelligence, pp. 527–534 (1999)
8. Christodoulou, G., Kovács, A., Schapira, M.: Bayesian combinatorial auctions. In: Aceto, L., Damgård, I., Goldberg, L.A., Halldórsson, M.M., Ingólfsdóttir, A., Walukiewicz, I. (eds.) ICALP 2008, Part I. LNCS, vol. 5125, pp. 820–832. Springer, Heidelberg (2008)
9. Cohen, E., Feldman, M., Fiat, A., Kaplan, H., Olonetsky, S.: Truth, envy, and truthful market clearing bundle pricing. In: Chen, N., Elkind, E., Koutsoupias, E. (eds.) WINE 2011. LNCS, vol. 7090, pp. 97–108. Springer, Heidelberg (2011)
10. Rodriguez Gustavo, E.: Sequential auctions with multi-unit demands. The B.E. Journal of Theoretical Economics 9(1), 1–35 (2009)
11. Feldman, M., Fu, H., Gravin, N., Lucier, B.: Simultaneous auctions are (almost) efficient. In: STOC (2013)
12. Gale, I., Stegeman, M.: Sequential Auctions of Endogenously Valued Objects. Games and Economic Behavior 36(1), 74–103 (2001)
13. Gul, F., Stacchetti, E.: Walrasian equilibrium with gross substitutes. Journal of Economic Theory 87(1), 95–124 (1999)
14. Hassidim, A., Kaplan, H., Mansour, Y., Nisan, N.: Non-price equilibria in markets of discrete goods. In: EC 2011 (2011)
15. Huang, Z., Devanur, N.R., Malec, D.L.: Sequential auctions of identical items with budget-constrained bidders. CoRR, abs/1209.1698 (2012)
16. Lehmann, B., Lehmann, D., Nisan, N.: Combinatorial auctions with decreasing marginal utilities. In: EC 2001 (2001)
17. Preston McAfee, R.: Mechanism design by competing sellers. Econometrica 61(6), 1281–1312 (1993)
18. Milgrom, P.R., Weber, R.J.: A theory of auctions and competitive bidding II (1982)
19. Leme, R.P., Syrgkanis, V., Tardos, É.: The dining bidder problem: a la russe et a la francaise. SIGecom Exchanges 11(2) (2012)
20. Leme, R.P., Syrgkanis, V., Tardos, É.: Sequential auctions and externalities. In: SODA (2012)
21. Syrgkanis, V., Tardos, E.: Bayesian sequential auctions. In: EC (2012)
22. Syrgkanis, V., Tardos, E.: Composable and efficient mechanisms. In: STOC (2013)
23. Weber, R.J.: Multiple-object auctions. Discussion Paper 496, Kellog Graduate School of Management, Northwestern University (1981)

Competition in the Presence of Social Networks: How Many Service Providers Maximize Welfare?

Moran Feldman[1,*], Reshef Meir[2], and Moshe Tennenholtz[3]

[1] EPFL
moran.feldman@epfl.ch
[2] Harvard University
rmeir@seas.harvard.edu
[3] Technion-Israel Institute of Technology and Microsoft Research, Herzlia, Israel
moshet@microsoft.com

Abstract. Competition for clients among service providers is a classical situation discussed in the economics literature. While better service attracts more clients, in some cases clients may prefer to keep using a low quality service if their friends are also using the same service—a phenomenon largely encouraged by the Internet and online social networks. This is evident, for example, in competition between cloud storage service providers such as DropBox, Microsoft SkyDrive and Google Drive. In such settings, the utility of a client depends on both the proposed service level and the number of friends or colleagues using the same service.

We study how the welfare of the clients is affected by competition in the presence of social connections. Quite expectantly, competition among two firms can significantly increase the clients' social welfare in comparison with the monopoly case. However, we show that a further increase in competition triggered by the entry of additional firms may be hazardous for the society (*i.e.*, to the clients), which stands in contrast to the typical situation in competition. Indeed, we show via equilibrium analysis that the social benefit of additional firms beyond the duopoly is limited, whereas the potential loss from such an addition is unbounded.

1 Introduction

Competition between firms has received much attention in the economics literature [4,16,5], and recently also in the computer science literature [13,1,12,17]. In standard models of competition, firms compete over clients (or workers), by offering a certain level of service or payoff. The utility of the firm is derived from the set of clients that select it, and can be based either purely on their number, or on a more sophisticated combinatorial function. The utility of the clients on those models is assumed to be affected only by the service they receive and by their preferences over firms. In particular, it is assumed that clients are indifferent to the decisions of other clients.

---

* At the time of research, the first and second authors where interns at Microsoft Research, Herzlia, Israel.

Y. Chen and N. Immorlica (Eds.): WINE 2013, LNCS 8289, pp. 174–187, 2013.

This kind of competition leads to models where the introduction of additional firms always improves the total welfare of the clients, and the introduction of additional clients always improves the total welfare of the firms [10,16,5]. However, this kind of competition is an oversimplification of reality. In many real world scenarios, the decisions of clients have significant positive or negative effect on the utility of other clients, an effect known as *network externalities* (see the related work section). For example, people may prefer a restaurant that has fewer clients (*e.g.*, when they wish to maintain privacy), or alternatively choose one that is highly attended (*e.g.*, if they enjoy the crowd). In quite many cases, such preference may be based on social connections, as people prefer to spend time with their friends and benefit from their presence.

The above is particularly relevant for long-term selection of online services such as cloud storage services, social networks, cellular providers and, to some extent, e-mail providers, where the benefit one generates from a service highly depends on its adoption by colleagues and friends. For example, calls within a single cellular network may have fewer interruptions. Similarly, sharing files is easier between users who use the same cloud storage service provider.

A motivating example. As a concrete example, consider two competing cloud services, Grand-docs (G) offering 3GB of storage and Medium-drive (M) offering 5GB. Suppose a client of G called Alice considers moving to M. Alice wants to be able to share a workspace with her colleague Bob, which is another user of G. Alice values the convenience of sharing a platform with each single colleague as equivalent to 1GB of storage. Thus, Bob alone would not prevent her from switching to M. However if Alice has, say, five colleagues using G, and only one who uses M, then she will (perhaps reluctantly) keep using G. Eve, on the other hand, is a freelance who uses cloud services for storage only. Thus, she prefers M regardless of the actions of other clients.

In this work we consider a model where clients' utilities depend both on the price or quality of the offered service and on the number and identity of their friends who have chosen the same firm. Our model is a two-phase game. In the first phase firms commit to a particular level of service which can be measured, for example, by bandwidth, storage capacity or accessible content of interest.[1] In the second phase, each client independently selects a firm. Each firm gains a fixed value for every client it recruits, from which it subtracts the cost of the service. The utility of a client is composed of the service level offered by her chosen firm, and an additional utility the client gains for every friend selecting the same firm.

A monopoly of a single firm guarantees that all social links are used, but gives the firm no incentive to provide a decent service level. Given a set of service providers, with fixed offers, and assuming positive externalities, the society can still utilize all social connections, if all clients subscribe to a single provider (say, the one offering the best service). However, there may be idiosyncratic preferences over firms (see Remark 1), in which case such a partition would not be optimal. Further, even if clients do not discriminate among providers, there

---

[1] We consider services that are given for free, such as e-mail and some cloud services, but where service level varies.

may be many other partitions of clients that are also stable. Since some social connections are not exploited, these outcomes are less efficient, and we focus on the *worst case* partitions that are still stable.

Social connections may give rise to interesting dynamics in the market. For example, a new firm will have to provide a significantly higher level of service than an incumbent competitor, and even then it will only be able to attract clients who have low value for social connections. However, after this initial small wave, other clients may also move as now they might already have friends using the new service. A third and forth wave may occur, until every client achieves balance between her desired service level and social connections. While clients act myopically (as they may be unaware of the global structure of the network), firms try to predict the eventual partition of clients that will follow a given change in the service level. We, therefore, introduce an equilibrium concept that takes this behavior into consideration.

Our goal is to study the effect of competition on the *social welfare* of the clients, which is defined as the sum of their utilities, derived from both service and social connections. To that aim, we define the *clients' value of competition* of a given game as the ratio between the clients' social welfare under the *worst* equilibrium, and their social welfare under the monopoly outcome. We ask whether the value of competition increases or decreases with respect to the number of firms in the market. In particular, we are interested in how many firms should a market have to (approximately) maximize the social welfare.

Our contribution. On the conceptual side, our two-phase model of competition and the corresponding solution concept allows for a focused study of the interaction between the network structure, number of firms and clients' welfare, while isolating them from other factors which are kept simple.

Our first result is that while a monopoly may be infinitely worse for the clients than any equilibrium under two firms or more, further increasing the number of firms in the market is not always beneficial for the clients. Surprisingly, the entry of additional firms beyond two cannot increase the value of competition by more than a constant factor. In contrast, the value of competition can *decrease* linearly with the number of firms, *i.e.*, by an unbounded factor. This demonstrates that in markets where social connections play an important role too much competition may produce an adverse result.

We further study bounds on the value of competition under the special case of a complete social graph. In particular, we show that the value of competition may still decrease with the number of firms, but only when the number of clients is sufficiently large. A complete social graph is interesting because, in some sense, it represents the opposite case of models ignoring social effects. Finally, while the value of competition refers to the clients' welfare, one can also define a complementary concept with regard to the firms' revenue. We present a preliminary result in this direction.

Most proofs are omitted, and are available in the full version of the paper.[2]

---

[2] http://tinyurl.com/k9n36ha.

2   The Model

We consider a two-phase game with two types of players: firms and clients. The clients are the nodes of a graph that represents the social network. In the first phase, each firm declares a payoff or service level (*e.g.*, how much storage is allocated for clients joining it). In the second phase, the clients join firms. The objective of the firms is to get as many clients as possible, while supplying the least amount of service. The objective of the clients is to share a firm with as many of their neighbors as possible, while getting a high service level.

More formally, consider an undirected graph $G = \langle N, \Gamma \rangle$ whose nodes are the n clients. The edges of G have positive weights $w_{j,j'}$, reflecting the benefit to clients $c_j, c_{j'} \in N$ from sharing a firm with each other (see Remark 2 about asymmetric networks). We denote by $\Gamma(c_j) \subseteq N$ the set of neighbors of c_j, *i.e.*, all clients $c_{j'}$ for which $w_{j,j'} > 0$. In addition, each client $c_j \in N$ is associated with a constant a_j, which is the value c_j gets from each unit of good (*e.g.*, 1 Mb of disk space) it receives from the host firm. We denote by $\mathbf{a} \in \mathbb{R}_+^n$ the vector of clients' parameters, and assume that the information on $\mathbf{a}$ is implicitly contained in N (and thus in G). In other words, whenever we have a set N of clients, or a graph G, we also have a vector $\mathbf{a}$ associated with them.

In addition, the set F consists of m firms $f_1, f_2, \ldots, f_m$. Each firm $f_i \in F$ is associated with an integer constant r_i, representing the revenue of f_i from each client joining it. We similarly denote all firms' parameters by $\mathbf{r} \in \mathbb{R}_+^m$ (where information on $\mathbf{r}$ is implicitly contained in F). An instance of our game is, therefore, represented by a pair $I = \langle G, F \rangle$. We write $I_m = \langle G, m, r \rangle$ when F consists of m identical firms with $r_i = r$.

The strategies available to each firm are committing to a certain service level (payoff) x_i. An outcome of the game (also called *configuration*) can be written as $E = \langle \mathbf{x}, P \rangle$, where $\mathbf{x}$ is a payoff vector, and $P = (C_1, \ldots, C_m)$ is a partition of the clients to firms. We denote by $f(c, P) \in F$ the firm selected by client c, *i.e.*, $f(c, P) = f_i$ for all $c \in C_i$. Given an outcome $E = \langle \mathbf{x}, P \rangle$ in game $\langle G, F \rangle$, the utilities of the agents are as follows.

- The utility of each firm $f_i \in F$ is given by $v_i(E) = (r_i - x_i) \cdot |C_i|$.
- The utility of each client $c_j \in C_i$ is given by $u_j(E) = a_j \cdot x_i + \sum_{j' \in C_i \setminus \{c_j\}} w_{j,j'}$.

For any particular outcome E in game $\langle G, F \rangle$, we denote by $SW(G, F, E)$ (or $SW(I, E)$) the social welfare of the clients, *i.e.*, $SW(G, F, E) = \sum_{c_j \in N} u_j(E)$. In the rest of this paper, we sometimes omit the parameters I, G, F and E when they are clear from the context. The rest of this section is devoted for defining and explaining the solution concept we use.

Remark 1 [Preferences over firms]. In some models of competition each client is assumed to have preferences over the different firms, reflecting differences between the products not captured by the other parameters of the model [16,17,18]. To simplify the exposition, we do not explicitly consider preferences in the paper. However, most of our results (except in Section 5.1, where the complete graph is studied) hold even if we add preferences. See full version for details.

2.1 Client Dynamics

Suppose firms commit to some given service levels, $(x_1, \ldots, x_m)$. Clients can now choose which firms to join. Since the utility of each client is affected by the decision of her friends, there might not be dominant strategies. However, given any current partition of clients $P = (C_1, \ldots, C_m)$, every client c_j has a straight-forward best response, which is to join the firm f_i maximizing $a_j \cdot x_i + \sum_{j' \in C_i \setminus \{c_j\}} w_{j,j'}$. It is easy to see that a pure Nash equilibrium (PNE) for the clients exists (*e.g.*, when all clients join the firm offering the best service), however, there may be more than one such equilibrium.

Proposition 1. *If the strategies of the firms are fixed (i.e., firms are not players), and the clients switch strategies according to a best response dynamics, then the client strategies converge into a PNE.*

Remark 2 [Symmetry]. The proof of Proposition 1 is due to the fact that the clients' game admits a potential function. We note that there are other cases where Proposition 1 holds even without symmetry. For example, if c_j attributes the same positive value for every neighbor sharing a firm (say, some constant w_j). Interestingly, both symmetry and equal weight to neighbors turn out to be sufficient conditions for the existence of pure equilibrium in other models based on social connections [7,22]. We emphasize that all of our results in the rest of this paper hold even without the symmetry assumption, as long as the clients are guaranteed to converge to an equilibrium.

2.2 The Two-Phase Game and Equilibria Concepts

Knowing that for any profile of payoffs the clients must converge, we can define a game in extensive form. Our game proceeds in two phases as follows.

- In the first phase, each firm f_i declares a non-negative integer x_i, which is the payoff (or service level) f_i gives to each client joining it. Note that firms are not allowed to discriminate between clients. We denote the vector of firms' strategies by $\mathbf{x} \in \mathbb{N}^m$.
- In the second phase, each client c_j chooses a firm. Then, the clients follow a best response dynamics until they converge to a Nash equilibrium, *i.e.*, to a partition P.

Consider an outcome $E = (\mathbf{x}, P)$ obtained from this two-phase game. Clearly, the clients have no incentive to deviate in E, however, the firms might deviate. Once a firm deviates, the clients can reach a new equilibrium, in which case we get a new outcome E'. Given a firm f_i, a strategy x_i' and two outcomes E, E' which are PNEs for the clients, we say that a outcome E' is the *projected outcome* obtained from E via the deviation x_i' of firm f_i if:

- $x_i' > x_i$, *i.e.*, firm f_i offers a higher service level.
- There is a sequence of best-responses by clients starting from the state $((x_i', x_{-i}), P)$ that converges to E'.

The projected outcome is clearly unique if there are two firms, as clients will only join the deviating firm, until no more clients want to join. When there are three firms or more, the clients' best response dynamics might be able to reach multiple Nash equilibria. In such cases, the projected outcome is determined by one of these equilibria arbitrarily. Thus in what follows, we treat the projected outcome E' as unique given E, i and x'_i. Using the above definition of the projected outcomes, we are now ready to define our solution concept.

Definition 1. *An outcome $E = (\mathbf{x}, P)$ is a commitment equilibrium (CME) if: (a) E is a PNE for the clients; and (b) For any firm f_i and any $x'_i > x_i$, $v_i(E') \leq v_i(E)$, where E' is the projected outcome from E and x'_i.*

In other words, the last condition states that no firm is better off increasing its payment, assuming that following this deviation the game will reach the projected outcome associated with this deviation. The following section gives some theoretical and practical justifications for the definition of CME.

3   Properties of Commitment Equilibria

Suppose that firms announce some payoff vector $\mathbf{x}$, which results in a configuration $E = (\mathbf{x}, P)$. Moreover, suppose firm f_i deviates by announcing payoff $x'_i \neq x_i$ upon seeing the outcome E. In the unfolding of events, some clients may join f_i or desert it. This in turn may cause other clients to switch firms and so on. By Proposition 1, the clients will eventually reach a stable configuration $E' = ((x'_i, x_{-i}), P')$. The deviation is profitable for f_i if $v_i(E') > v_i(E)$.

Note that in theory, a firm may either gain by increasing its service level, thereby triggering a cascade of new clients joining it; or by decreasing payoff and thus reducing expenses However, deviations of the latter type are largely impractical in most situations. Often, the service level offered by the firm is considered by the clients as a *commitment*. This is why CME considers only deviations to a higher level of service. Decreasing the level of service is not considered an option.

Other justifications for the restrictions imposed by CMEs are given in the full version of the paper. The next two properties show that CMEs exist, and that they are closely related to other solution concepts.

Observation 2. *For every game instance, best response dynamics (of the firms) must converge into a CME.*

Proposition 3. *Given a commitment equilibrium E of the extensive form game, there exists a pure sub-game perfect equilibrium in which the utilities of all clients and firms are equal to their utilities in E.*

The last proposition allows us to think of CMEs as an equilibrium selection criterion, which favors sub-game perfect equilibria that are attained via a natural iterative process: if a firm deviates, the resulting configuration of the clients can be achieved from the original one by a sequence of best responses.

4   Benefits of Competition

In this section we discuss the possible benefit to the clients from competition between firms. To do that, we first define a way to measure the effect of competition on the social welfare of the clients. Given a network G, For every instance I we denote the social welfare in the *worst* CME by $SW^*(I)$. Under a monopoly, all clients select the single firm, and in the worst case get no service. We define the *monopoly welfare* $MW(G) = SW^*(G, 1, 0)$. Note that for $I = \langle G, F \rangle$, MW only depends on G.

The *clients' value of competition* of the instance $I = \langle G, F \rangle$ is now defined as the ratio $CVC(I) = SW^*(I)/MW(G)$. Thus, values of $CVC(I)$ greater than 1 indicate that the society (of clients) gains from the competition between firms, whereas lower values lower mean that the competition hurts the clients.

For reference, we also define $OPT(I)$ as the *maximal* social welfare of any outcome of I. It clearly holds that $OPT(I) \leq \sum_{j,j' \in N} w_{j,j'} + \max_{f_i \in F} r_i \cdot \sum_j a_j$, and when there are no preferences over firms, then this is an equality: in the best outcome all clients share the same firm, which provides maximal service.

There are several reasons for focusing on the worst equilibrium. First, firms may use non-binding agreements to settle on outcomes that are good for them and bad for the clients. Second, this is a worst case assumption allowing us to put a lower bound on the welfare in any other case. And finally, the *best* CME coincides with the optimal outcome described above, which is trivial.

In this section we focus on the network structure, and hence assume (unless explicitly mentioned otherwise), that $a_j = a$ for all $c_j \in N$ and $r_i = r$ for all $f_i \in F$. These simplifying assumptions will be relaxed in Section 5.

The clients' value of competition measures how much the clients gain from competition. It is natural to predict that competition will improve the outcome for the clients. Indeed, a duopoly (two firms) typically yields significantly higher welfare than a monopoly (although not always, see Prop. 8). The following proposition shows that the clients value of competition can be infinite. Informally, it implies that a second firm can significantly improve the total utility of the clients.

Proposition 4. *There is a game instance I where $CVC(I) = \infty$.*

Proof. Consider a game instance with two firms f_1 and f_2 having $r > 0$, and one client with $a = 1$. In every CME of this game, there must be a firm giving payoff of $x_i \geq r - 1$. Hence, $SW^*(I) \geq r - 1$. On the other hand, if there was a single firm, the utility the client would have been 0, as the firm would have paid 0 and the client has no neighbors. Thus $CVC(I) \geq (r - 1)/0 = \infty$.   □

In contrast to the potentially significant improvement in the social welfare produced by a second firm, the next theorem shows that additional firms can only have a limited positive effect on the clients. We prove this strong negative result by showing that a duopoly already extracts a constant fraction of the optimal social welfare (which in itself is a positive statement on duopolies).

Theorem 5. *Let $I_m = \langle G, m, r \rangle$ be an arbitrary game instance (with $a_j = a$ for all $c_j \in N$ and $r_i = r$ for all $f_i \in F$). Let B be any CME outcome in I_m for $m \geq 2$. Then, $SW(I_m, B) \leq \beta \cdot SW^*(I_2)$ for some constant $\beta < 7$.*

Proof sketch. Denote by A the worst CME of I_2, and let x_1 and x_2 denote the payoff levels of firms f_1 and f_2 in A, respectively. Also, denote by $C_1, C_2 \subseteq N$ the sets of clients of outcome A corresponding to firms f_1 and f_2, respectively, and let $n_1 = |C_1|, n_2 = |C_2|$. We assume, w.l.o.g., $n_1 \geq n_2$. For every client $c_j \in C_i$, we denote $\gamma_j = \sum_{c_{j'} \in C_i} w_{j,j'}$, and $\delta_j = \sum_{c_{j'} \in N \setminus C_i} w_{j,j'}$. We also use the average values $\gamma_i^* = \frac{1}{n_i} \sum_{c_j \in C_i} \gamma_j$ and $\delta_i^* = \frac{1}{n_i} \sum_{c_j \in C_i} \delta_j$.

Observe that for *every CME E*: $SW(I_m, E) \leq OPT(I_m) = OPT(I_2)$, and in particular this inequality holds for $E = B$. Therefore to prove a constant bound, it is sufficient to bound the *price of anarchy*[3] with 2 firms, *i.e.*, to show that $OPT(I_2) \leq O(SW(I_2, A))$. Let $u_j(A), u_j(OPT)$ be the utility of client j under the configurations A and OPT (in instance I_2). For $j \in C_i$, it holds that $u_j(A) = ax_i + \gamma_j$, whereas $u_j(OPT) = r + \gamma_j + \delta_j$.

The minimal increase in x_{-i} that can convince client $c_j \in C_i$ to switch firms is $\varepsilon_j = (\gamma_j/a + x_i) - (\delta_j/a + x_{-i} + 1) \geq 0$. Assume that clients in C_i are ordered by non-decreasing ε_j. By comparing firm's utility with and without the increase, we can show that f_i cannot gain by attracting clients $c_1, \ldots, c_j$:

$$(r - x_{-i}) \cdot n_{-i} \geq (r - (x_{-i} + \varepsilon_j))(n_{-i} + j) . \tag{1}$$

By rearranging, we now get: $r \leq \varepsilon_j \left(1 + \frac{n_{-i}}{j}\right) + x_{-i}$.

For any non-decreasing vector $\mathbf{z} = (z_1, \ldots, z_m)$ of non-negative numbers, denote its average by $z^* = \frac{1}{m} \sum_{j \leq m} z_j$. Let $\tau \in (0, 1)$ be some fraction, and let

$$\alpha_\tau = \max_{\mathbf{z} \geq 0} \{z_{\lceil \tau m \rceil}/z^*\}, \quad \Theta_\tau = \alpha_\tau(1 + 1/\tau) .$$

For example, if $\tau = 1/2$ (*i.e.*, $z_{\lceil \tau n_i \rceil}$ is the median of $\mathbf{z}$), then $\Theta_\tau = 6$.

Let $\varepsilon_i^* = \frac{1}{n_i} \sum_{c_j \in C_i} \varepsilon_j$. In what follows, we will take an arbitrary fraction τ, and prove our bound as a function of Θ_τ. We assume n_1, n_2 are sufficiently large so as to ignore rounding (*i.e.*, that $\lceil \tau n_i \rceil \cong \tau n_i$).

Applying the inequality above for $j = \tau n_i$ gives us

$$r \leq \left(1 + \frac{n_{-i}}{\tau n_i}\right) \varepsilon_{\tau n_i} + x_{-i} \leq \left(1 + \frac{n_{-i}}{\tau n_i}\right) \alpha_\tau(\gamma_i^*/a - \delta_i^*/a + x_i + 1). \tag{2}$$

In particular, for the larger firm f_1:

$$r \leq \left(1 + \frac{1}{\tau}\right) \alpha_\tau(\gamma_1^*/a - \delta_1^*/a + x_1 + 1) = \Theta_\tau(\gamma_1^*/a - \delta_1^*/a + x_1 + 1) , \quad \text{and}$$

$$\sum_{j \in C_1} u_j(OPT) \leq \sum_{c_j \in C_1} (ar + \delta_j + \gamma_j) = n_1(ar + \delta_1^* + \gamma_1^*)$$

$$\leq n_1(\Theta_\tau(\gamma_1^* + ax_1 + a) + \gamma_1^*) \leq (\Theta_\tau + 1) \sum_{j \in C_1} (\gamma_j + ax_1 + a) \cong (\Theta_\tau + 1) \sum_{j \in C_1} u_j(A) .$$

---

[3] The price of anarchy of a game is the ratio between the optimal social welfare achievable by any configuration, and the worst social welfare of any Nash equilibrium.

This means that at least the clients of the larger firm f_1 cannot gain on average more than a factor of $\Theta_\tau + 1$, plus some additive term $O(na)$ that does not depend on the welfare. Moreover, this term goes to zero when we use smaller minimal units of storage a. As for the smaller firm f_2 we consider two cases.

The first case is $\gamma_2^*/a - \delta_2^*/a + x_2 \geq \gamma_1^*/a - \delta_1^*/a + x_1$. In this case, for $i = 2$

$$r \leq \Theta_\tau(\gamma_1^*/a - \delta_1^*/a + x_1 + 1) \leq \Theta_\tau(\gamma_2^*/a - \delta_2^*/a + x_2 + 1) \ ,$$

and therefore, the same arguments used above can also be used to bound $\sum_{j \in C_2} u_j(OPT)$. This concludes the first case, as

$$OPT(I_2) \leq (\Theta_\tau + 1) \sum_{j \in C_1} u_j(A) + (\Theta_\tau + 1) \sum_{j \in C_2} u_j(A) + O(na) \cong (\Theta_\tau + 1) \cdot SW(A).$$

We now consider the second case, where $\gamma_2^*/a - \delta_2^*/a + x_2 < \gamma_1^*/a - \delta_1^*/a + x_1$. Denote $\gamma_i^{**} = \gamma_i^* + ax_i$. Using this notation, the above inequality becomes: $\gamma_2^{**} - \delta_2^* < \gamma_1^{**} - \delta_1^*$. Observe that:

$$SW(A) = \sum_{j \in C_1} u_j(A) + \sum_{j \in C_2} u_j(A) = n_1 \gamma_1^{**} + n_2 \gamma_2^{**} \ , \tag{3}$$

and therefore:

$$OPT(I_2) = \sum_{j \in C_1} u_j(OPT) + \sum_{j \in C_2} u_j(OPT) = n_1(ar + \gamma_1^* + \delta_1^*) + n_2(ar + \gamma_2^* + \delta_2^*)$$

$$\leq n_1(\alpha_\tau(\gamma_1^{**} - \delta_1^* + a)(1 + \frac{n_2}{\tau n_1}) + \gamma_1^* + \delta_1^*) + n_2(\alpha_\tau(\gamma_2^{**} - \delta_2^* + a)(1 + \frac{n_1}{\tau n_2}) + \gamma_2^* + \delta_2^*)$$

$$\cong (\alpha_\tau + 1)SW(A) + \alpha_\tau/\tau (n_2(\gamma_1^{**} - \delta_1^*) + n_1(\gamma_2^{**} - \delta_2^*)) \ , \qquad \text{(By Eq. (3))}$$

where the first inequality holds by applying Equation (2) once with $i = 1$, and once with $i = 2$. Finally, since $w \geq x, y \geq z$ implies $wz + xy \leq wy + xz$,

$$OPT(I_2) \lesssim (\alpha_\tau + 1)SW(A) + \frac{\alpha_\tau}{\tau}(n_1 \gamma_1^{**} + n_2 \gamma_2^{**}) = (\Theta_\tau + 1)SW(A) \ ,$$

where $w = (\gamma_1^{**} - \delta_1^*)$; $x = (\gamma_2^{**} - \delta_2^*)$; $y = n_1$; $z = n_2$.

This completes the proof of the inequality $SW(I_m, B) \leq (\Theta_\tau + 1)SW(I_2, A) + O(na)$. Since by selecting the median $\tau = 1/2$ we get $\Theta_\tau = 6$, the ratio is at most 7 (plus some additive term that diminishes with the resolution). $\qquad \square$

By optimizing the value of τ in the proof of Theorem 5, it can be shown that $\beta = \Theta_\tau \leq 6.828$ (for sufficiently large n). A possible extension of the theorem, which we leave open for future research, is how the ratio changes as a function of the vector $\mathbf{r}$ in the presence of heterogeneous firms (i.e., when not all the entries in r are identical). We conjecture that if the r_i values are close to one another, then the benefit of having more competing firms will still be limited.

The next theorem complements the upper bound with a lower bound of 2. It remains as an open question to close the gap between these two constants.

Proposition 6. For any $m > 2$, there exists an instance $I_m = \langle G, m, r \rangle$ s.t. $SW^*(I_m) \geq (2 - o(1))SW^*(I_2)$.

Notice that the proof of Theorem 5 in particular shows that the price of anarchy (for clients) is upper bounded by 6.828. Any better bound on the clients' value of competition that uses a similar proof technique must also translate into an upper bound on the price of anarchy. The following proposition shows that the price of anarchy is at least 4.26 in the worst case. Thus, matching the lower bound introduced by Proposition 6 will probably require different techniques.

Proposition 7. *There exists an instance* $I_2 = \langle G, 2, r \rangle$, *s.t.* $OPT(I_2) \geq (1 + \frac{1}{1 - \ln 2}) SW^*(I_2) \cong 4.26 SW^*(I_2)$.

5   The Cost of Competition

In this section we are interested in the question: "how low can the client value of competition be?". We start with a negative example, showing that the welfare of clients can linearly degrade with the number of firms, *i.e.*, that without further restrictions the cost of excessive competition is essentially unbounded.

While our construction uses a particular structure, we later show in Prop. 11 that a milder linear degradation may also occur under the complete graph.

Proposition 8. *For every* $m \geq 2, \varepsilon > 0$, *there exists a game instance* I_m *with* $CVC(I_m) \leq \frac{1}{m} + \varepsilon$.

Our next results show that Proposition 8 demonstrates the worst possible case. If the number of firms m is bounded then so is the value of competition. Interestingly, the proof of Theorem 9 provides a lower bound on the welfare not just in a CME, but in fact in any outcome where clients are stable (regardless of firms' strategies). The same is true for Theorem 10. In Theorem 9, the number of firms can also be replaced with the *maximum degree*.

Theorem 9. *For every game instance* $I = \langle G, F \rangle$ *with* m *firms,* $SW^*(I) \geq MW(G)/m$. *That is,* $CVC(I) \geq \frac{1}{m}$.

Proof. If there is only a single firm, all clients join it and get zero payoff. Hence, the monopoly welfare is $MW(G) = \sum_{j=1}^{n} \sum_{j' \neq j} w_{j,j'}$. Let us now focus on an arbitrary CME E of I. Consider a client c_j which joins firm f_i under E. Clearly, for every other firm $f_{i'}$ it must hold that:

$$a_j \cdot x_i + \sum_{j' \in C_i \setminus \{c_j\}} w_{j,j'} \geq a_j \cdot x_{i'} + \sum_{j' \in C_{i'} \setminus \{c_j\}} w_{j,j'} . \tag{4}$$

Observe that this inequality trivially holds also when $i = i'$. Hence, we can sum the inequalities for every $1 \leq i' \leq m$, and get:

$$m \cdot \left[a_j \cdot x_i + \sum_{j' \in C_i \setminus \{c_j\}} w_{j,j'} \right] \geq \sum_{i'=1}^{m} \left[a_j \cdot x_{i'} + \sum_{j' \in C_{i'} \setminus \{c_j\}} w_{j,j'} \right] \tag{5}$$

$$= \sum_{j' \neq j} w_{j,j'} + \sum_{i'=1}^{m} a_j \cdot x_{i'} \geq \sum_{j' \neq j} w_{j,j'} .$$

Rearranging, we get that the utility of c_j is at least $\sum_{j' \neq j} w_{j,j'}/m$. Hence, the total utility of all clients is at least: $\frac{1}{m} \cdot \sum_{j=1}^{n} \sum_{j' \neq j} w_{j,j'} = \frac{1}{m} \cdot MW(G)$.   □

5.1   The Complete Graph

The above results use examples of dense graphs to show that the clients' value of competition tend to be low. It thus makes sense to consider the complete graph, with equal edge weights (if different edge weights were allowed, non-complete graphs could be simulated by giving some edges very low weights, making them insignificant). For ease of notation, let us assume that all edge weights are 1. We note that the case of a complete graph models the situation where clients only care about the number of other clients sharing their firm, as in [15].

The main result of this section states that with complete graphs over a small set of clients, the loss due to competition (even with many firms) cannot be too high.

Theorem 10. *For any game instance $I = \langle G, F \rangle$ where G is a complete graph, it holds that $CVC(I) = \Omega(n^{-1/3})$.*

Moreover, the above bound is tight up to low order terms:

Proposition 11. *There is a family of instances $(I_m)_{m \geq 1}$, each with a complete social graph over $n(m)$ clients (where $n(m)$ is a bounded function of m), for which $CVC(I_m) = O(n^{-1/3})$.*

The instances constructed in the proof of the last proposition have the additional property that $CVC(I^n) = O(1/m)$. Hence, the proof also shows that the bound given by Theorem 9 is tight (up to lower order terms) even if we restrict ourselves to complete graphs (but allow r_i to vary, and allow $n = \Omega(m^3)$).

6   Firms' Revenue

While the main bulk of this paper is devoted to study the effect of increased competition on the welfare of clients, it is also important to understand how social connections change the revenue of the competing firms. In this section we provide a preliminary result in this direction. Given a game instance I, we define the *firms' revenue* as the sum of firms' utilities in the best CME of this game instance (best for the firms), *i.e.*, $FR(I) = \max\{\sum_{f_i \in F} v_i(E) \mid E \text{ is CME of } I\}$. The choice of the best CME in the definition of $FR(I)$ is justified by the observation that there always exists a CME with 0 utility for the firms (the one where $x_i = r_i$ for every firm f_i).

The example constructed in Proposition 6 can be analyzed to show that the addition of a third firm decreases the total revenue by half. The next theorem shows that similar examples exist for any value of m.

Theorem 12. *For any $m > 1$, there exists an instance $I_m = \langle G, m, r \rangle$ for which the total firms' revenue strictly decreases with the addition on an extra firm. Formally, $I_{m+1} = \langle G, m+1, r \rangle$ obeys $FR(I_{m+1}) \leq \frac{m-1}{m-2+\ln m} \cdot FR(I_m)$.*

The fact that increased competition can lead to lower revenue may sound trivial. However it should be noted that the marginal value of each client to a firm is *fixed*. Hence, without the network structure the number of firms does not affect the total revenue at all. It is the presence of social connections that makes competition potentially harmful for the firms. We leave for future research finding the maximal loss in revenue that may result by adding a firm. Another interesting question that we leave open is whether the firms' total revenue can also *increase* when a new firm is introduced into the game.

7 Discussion

We introduced a natural network-based model of competition with positive externalities between clients, and analyzed the effect of the number of firms on the welfare of clients.

7.1 Related Work

Katz and Shapiro [15] coined the term "network externalities" to denote situations where the decision of clients effect their neighbors in the network. They described a market where consumers' utility is partly derived from the *size* of the network they select, *i.e.*, the number of other clients selecting the same firm. Subsequently, Banerji and Dutta [3] studied an extension of the model that does take into account the structure of the network, by describing the interaction among groups of clients in the limited case of two firms.

Both of these papers, as well as our work, can be classified under the "macro approach" of Economides, which seeks to understand the effect of positive externalities on consumption patterns, rather than to explain their source [11].

The utility structure of the clients in our model is similar to the one in [15], but the clients are sensitive to the identity of their peers, rather than to their number only (as in [3]). Moreover, we assume a simple myopic behavior by clients whereas in the Katz and Shapiro model clients predict the *expected size* of the firms and act accordingly. Thus our equilibrium concepts are substantially different.

Beyond the technical differences between the models, our paper brings a novel perspective. In particular, the main focus of Banerji and Dutta is on the structure of the outcome partition in the case of two firms. Whereas we study the effect of the number of firms on the the clients' social welfare under network externalities. Similarly to Economides but due to different considerations, we arrive at the conclusion that excessive competition may compromise clients' welfare.

In particular, *compatibility* and *standardization* play a major role in some models of network externalities as in [15,11]. In our model the level of compatibility among competing services is assumed to be fixed (at least in short time scales). The strategic decisions of firms are therefore simplified to setting the price/service level, as in traditional Bertrand competition [4].

Other aspects of network externalities that have been studied focused on factors such as *price discrimination* [14], or particular diffusion dynamics [13].

A two-phase framework has been suggested as a model for competition in other domains, where firms first commit to strategies, and clients follow by playing a game induced by firms' actions. Examples of such games are available in the domain of *group buying*, where each vendor commits to a discount schedule based on quantity [8,9,18]. While the utility structure of both vendors and buyers in the group buying domain is substantially different, our work demonstrates how such a two-phase framework can be applied for modeling the effect of network externalities.

7.2   Conclusions

We showed that excessive competition can fragment the network and eliminate most of the clients' utility. On the other hand, two competing firms already guarantee at least a constant fraction of the clients' maximal welfare, and thus, the positive effect of adding more competitors is bounded, whereas the potential damage is much more significant. These results complement the findings by Economides and others on the potential damage in excessive competition, and provide some formal justification to statements that unregulated competition can be inefficient or even hazardous for society. For example, the necessity of regulation *against* competition has been discussed in domains such as labor markets, banking, and others [20,6,19]. A recent formal treatment of auctions with partial information reveals a similar effect to the one we found, showing that the entry of additional auctioneers incurs a loss on the bidders' welfare [21].

A loss of welfare is quite expected when there are *negative* externalities for firms' actions, such as pollution or waste of resources [2]. We emphasize, however, that the potential negative effect of competition in our model is not due to the typical race-to-the-bottom scenario, but rather due to (positive) externalities between the clients or workers themselves.

Future Work. In order to focus on the effect of network externalities, we simplified some factors that are necessary for a better understanding of real markets. Possible future work will consider non-linear utilities for the firms (reflecting, *e.g.*, decreasing marginal production costs that are typical to economies of scale), partial information of the network, and far-sighted strategies used by the clients.

This work outlined bounds on the value of competition assuming either arbitrary or complete networks. Networks in the real world tend to have certain characteristics and structure, that can possibly be exploited to get better bounds on the effect of competition under real world externalities.

Finally, we introduced only a preliminary result on the firms' revenue. Much more work is needed in order to gain an understanding of this quantity. For example, we showed that there is a class of instances where the introduction of a new firm decreases the firms' total revenue by at least a given factor. We conjecture that the converse does not hold, *i.e.*, that the introduction of additional firms can never increase the total revenue. This conjecture, if true, implies that both firms and clients are prone to the effects of excessive competition, and further emphasizes the importance of regulation.

References

1. Ashlagi, I., Tennenholtz, M., Zohar, A.: Competing schedulers. In: Proc. of 24th AAAI (2010)
2. Ayres, R.U., Kneese, A.V.: Production, consumption, and externalities. The American Economic Review 59(3), 282–297 (1969)
3. Banerji, A., Dutta, B.: Local network externalities and market segmentation. International Journal of Industrial Organization 27(5), 605–614 (2009)
4. Bertrand, J.: Book review of Theorie Mathematique de la Richesse Sociale. Journal des Savants, 499–508 (1883)
5. Bhaskar, V., Manning, A., To, T.: Oligopsony and monopsonistic competition in labor markets. The Journal of Economic Perspectives 16(2), 155–174 (2002)
6. Biggar, D., Heimler, A.: An increasing role for competition in the regulation of banks. Interntional competition network antitrust enforcement in regulated sectors subgroup 1 (2005), http://tinyurl.com/cgvu3u3
7. Biló, V., Celi, A., Flammini, M., Gallotti, V.: Social context congestion games. Theoretical Computer Science (to appear)
8. Chen, J., Chen, X., Song, X.: Comparison of the group-buying auction and the fixed pricing mechanism. Decision Support Systems 43(2), 445–459 (2007)
9. Chen, J., Kauffman, R.J., Liu, Y., Song, X.: Segmenting uncertain demand in group-buying auctions. Electronic Commerce Research and Applications 9(2), 126–147 (2010)
10. Deneckere, R., Davidson, C.: Incentives to form coalitions with bertrand competition. The RAND Journal of Economics 16(4), 473–486 (1985)
11. Economides, N.: The economics of networks. International Journal of Industrial Organization 14(6), 673–699 (1996)
12. Immorlica, N., Kalai, A.T., Lucier, B., Moitra, A., Postlewaite, A., Tennenholtz, M.: Dueling algorithms. In: Proc. of 43rd STOC, pp. 215–224 (2011)
13. Jin, Y., Sen, S., Guérin, R., Hosanagar, K., Zhang, Z.-L.: Dynamics of competition between incumbent and emerging network technologies. In: Proc. of 3rd NetEcon, pp. 49–54 (2008)
14. Jullien, B.: Competing in network industries: Divide and conquer. IDEI Working Paper (2000)
15. Katz, M.L., Shapiro, C.: Network externalities, competition, and compatibility. The American Economic Review 75(3), 424–440 (1985)
16. Kelso, A.S., Crawford, V.P.: Job matching, coalition formation, and gross substitutes. Econometrica 50(6), 1483–1504 (1982)
17. Lu, T., Boutilier, C.: Matching models for preference-sensitive group purchasing. In: Proc. of 13th ACM-EC, pp. 723–740 (2012)
18. Meir, R., Lu, T., Tennenholtz, M., Boutilier, C.: On the value of using group discounts under price competition. In: Proc. of 27th AAAI, pp. 683–689 (2013)
19. Government of NSW. Assessment against the competition test. Guide to Better Regulation, by the Department of Premier and Cabinet, Better Regulation Office, New South Wales (2008), http://tinyurl.com/cu6w8cp
20. O'Neill, R.: Protection racket: Big firms wanted the freedom to kill. they got it. Hazards 91 (August 2005), http://tinyurl.com/ca6g9x8
21. Polevoy, G., Smorodinsky, R., Tennenholtz, M.: Signalling competition and social welfare. arXiv:1203.6610 (2012) (manuscript)
22. Rahn, M., Schäfer, G.: Bounding the inefficiency of altruism through social contribution games. In: Chen, Y., Immorlica, N. (eds.) WINE 2013. LNCS, pp. 385–398. Springer, Heidelberg (2013)

Resolving Braess's Paradox in Random Networks*

Dimitris Fotakis[1], Alexis C. Kaporis[2], Thanasis Lianeas[1], and Paul G. Spirakis[3,4]

[1] Electrical and Computer Engineering, National Technical University of Athens, Greece
[2] Information and Communication Systems Dept., University of the Aegean, Samos, Greece
[3] Department of Computer Science , University of Liverpool, UK
[4] Computer Technology Institute and Press – Diophantus, Patras, Greece
fotakis@cs.ntua.gr, kaporisa@gmail.com, tlianeas@mail.ntua.gr,
P.Spirakis@liverpool.ac.uk, spirakis@cti.gr

Abstract. Braess's paradox states that removing a part of a network may improve the players' latency at equilibrium. In this work, we study the approximability of the best subnetwork problem for the class of random $\mathcal{G}_{n,p}$ instances proven prone to Braess's paradox by (Roughgarden and Valiant, RSA 2010) and (Chung and Young, WINE 2010). Our main contribution is a polynomial-time approximation-preserving reduction of the best subnetwork problem for such instances to the corresponding problem in a simplified network where all neighbors of s and t are directly connected by 0 latency edges. Building on this, we obtain an approximation scheme that for any constant $\varepsilon > 0$ and with high probability, computes a subnetwork and an ε-Nash flow with maximum latency at most $(1+\varepsilon)L^*+\varepsilon$, where L^* is the equilibrium latency of the best subnetwork. Our approximation scheme runs in polynomial time if the random network has average degree $O(\mathrm{poly}(\ln n))$ and the traffic rate is $O(\mathrm{poly}(\ln\ln n))$, and in quasipolynomial time for average degrees up to $o(n)$ and traffic rates of $O(\mathrm{poly}(\ln n))$.

1 Introduction

An instance of a (non-atomic) *selfish routing* game consists of a network with a source s and a sink t, and a traffic rate r divided among an infinite number of players. Every edge has a non-decreasing function that determines the edge's latency caused by its traffic. Each player routes a negligible amount of traffic through an $s - t$ path. Observing the traffic caused by others, every player selects an $s-t$ path that minimizes the sum of edge latencies. Thus, the players reach a *Nash equilibrium* (a.k.a., a *Wardrop equilibrium*), where all players use paths of equal minimum latency. Under some general assumptions on the latency functions, a Nash equilibrium flow (or simply a *Nash flow*) exists and the common players' latency in a Nash flow is essentially unique (see e.g., [14]).

Previous Work. It is well known that a Nash flow may not optimize the network performance, usually measured by the *total latency* incurred by all players. Thus, in the

---

* This research was supported by the project Algorithmic Game Theory, co-financed by the European Union (European Social Fund - ESF) and Greek national funds, through the Operational Program "Education and Lifelong Learning" of the National Strategic Reference Framework (NSRF) - Research Funding Program: THALES, investing in knowledge society through the European Social Fund, by the ERC project RIMACO, and by the EU FP7/2007-13 (DG INFSO G4-ICT for Transport) under Grant Agreement no. 288094 (Project eCompass).

Y. Chen and N. Immorlica (Eds.): WINE 2013, LNCS 8289, pp. 188–201, 2013.
© Springer-Verlag Berlin Heidelberg 2013

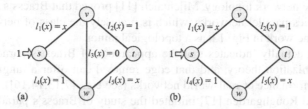

Fig. 1. (a) The optimal total latency is $3/2$, achieved by routing half of the flow on each of the paths (s, v, t) and (s, w, t). In the (unique) Nash flow, all traffic goes through the path (s, v, w, t) and has a latency of 2. (b) If we remove the edge (v, w), the Nash flow coincides with the optimal flow. Hence the network (b) is the *best subnetwork* of network (a).

last decade, there has been a significant interest in quantifying and understanding the performance degradation due to the players' selfish behavior, and in mitigating (or even eliminating) it using several approaches, such as introducing economic disincentives (tolls) for the use of congested edges, or exploiting the presence of centrally coordinated players (Stackelberg routing), see e.g., [14] and the references therein.

A simple way to improve the network performance at equilibrium is to exploit Braess's paradox [3], namely the fact that removing some edges may improve the latency of the Nash flow[1] (see e.g., Fig. 1 for an example). Thus, given an instance of selfish routing, one naturally seeks for the *best subnetwork*, i.e. the subnetwork minimizing the common players' latency at equilibrium. Compared against Stackelberg routing and tolls, edge removal is simpler and more appealing to both the network administrator and the players (see e.g., [6] for a discussion).

Unfortunately, Roughgarden [15] proved that it is NP-hard not only to find the best subnetwork, but also to compute any meaningful approximation to its equilibrium latency. Specifically, he proved that even for linear latencies, it is NP-hard to approximate the equilibrium latency of the best subnetwork within a factor of $4/3 - \varepsilon$, for any $\varepsilon > 0$, i.e., within any factor less than the worst-case Price of Anarchy for linear latencies. On the positive side, applying Althöfer's Sparsification Lemma [1], Fotakis, Kaporis, and Spirakis [6] presented an algorithm that approximates the equilibrium latency of the best subnetwork within an additive term of ε, for any constant $\varepsilon > 0$, in time that is subexponential if the total number of $s - t$ paths is polynomial, all paths are of polylogarithmic length, and the traffic rate is constant.

Interestingly, Braess's paradox can be dramatically more severe in networks with multiple sources and sinks. More specifically, Lin et al. [8] proved that for networks with a single source-sink pair and general latency functions, the removal of at most k edges cannot improve the equilibrium latency by a factor greater than $k + 1$. On the other hand, Lin et al. [8] presented a network with two source-sink pairs where the removal of a single edge improves the equilibrium latency by a factor of $2^{\Omega(n)}$. As for

---

[1] Due to space constraints, we have restricted the discussion of related work to the most relevant results on the existence and the elimination of Braess's paradox. There has been a large body of work on quantifying and mitigating the consequences of Braess's paradox on selfish traffic, especially in the areas of Transportation Science and Computer Networks. The interested reader may see e.g., [15,12] for more references.

the impact of the network topology, Milchtaich [11] proved that Braess's paradox does not occur in series-parallel networks, which is precisely the class of networks that do not contain the network in Fig. 1.a as a topological minor.

Recent work actually indicates that the appearance of Braess's paradox is not an artifact of optimization theory, and that edge removal can offer a tangible improvement on the performance of real-world networks (see e.g., [7,13,14,16]). In this direction, Valiant and Roughgarden [17] initiated the study of Braess's paradox in natural classes of random networks, and proved that the paradox occurs with high probability in dense random $\mathcal{G}_{n,p}$ networks, with $p = \omega(n^{-1/2})$, if each edge e has a linear latency $\ell_e(x) = a_e x + b_e$, with a_e, b_e drawn independently from some reasonable distribution. The subsequent work of Chung and Young [4] extended the result of [17] to sparse random networks, where $p = \Omega(\ln n/n)$, i.e., just greater than the connectivity threshold of $\mathcal{G}_{n,p}$, assuming that the network has a large number of edges e with small additive latency terms b_e. In fact, Chung and Young demonstrated that the crucial property for Braess's paradox to emerge is that the subnetwork consisting of the edges with small additive terms is a good expander (see also [5]). Nevertheless, the proof of [4,17] is merely existential; it provides no clue on how one can actually find (or even approximate) the best subnetwork and its equilibrium latency.

Motivation and Contribution. The motivating question for this work is whether in some interesting settings, where the paradox occurs, we can efficiently compute a set of edges whose removal significantly improves the equilibrium latency. From a more technical viewpoint, our work is motivated by the results of [4,17] about the prevalence of the paradox in random networks, and by the knowledge that in random instances some hard (in general) problems can actually be tractable.

Departing from [4,17], we adopt a purely algorithmic approach. We focus on the class of so-called *good* selfish routing instances, namely instances with the properties used by [4,17] to demonstrate the occurrence of Braess's paradox in random networks with high probability. In fact, one can easily verify that the random instances of [4,17] are good with high probability. Rather surprisingly, we prove that, in many interesting cases, we can efficiently approximate the best subnetwork and its equilibrium latency. What may be even more surprising is that our approximation algorithm is based on the expansion property of good instances, namely the very same property used by [4,17] to establish the prevalence of the paradox in good instances! To the best of our knowledge, our results are the first of theoretical nature which indicate that Braess's paradox can be efficiently eliminated in a large class of interesting instances.

Technically, we present essentially an approximation scheme. Given a good instance and any constant $\varepsilon > 0$, we compute a flow g that is an ε-Nash flow for the subnetwork consisting of the edges used by it, and has a latency of $L(g) \leq (1 + \varepsilon)L^* + \varepsilon$, where L^* is the equilibrium latency of the best subnetwork (Theorem 1). In fact, g has these properties with high probability. Our approximation scheme runs in polynomial time for the most interesting case that the network is relatively sparse and the traffic rate r is $O(\text{poly}(\ln \ln n))$, where n is the number of vertices. Specifically, the running time is polynomial if the good network has average degree $O(\text{poly}(\ln n))$, i.e., if $pn = O(\text{poly}(\ln n))$, for random $\mathcal{G}_{n,p}$ networks, and quasipolynomial for average degrees up to $o(n)$. As for the traffic rate, we emphasize that most work on selfish routing and selfish network design problems assumes that $r = 1$, or at least that r does not increase with the network's size (see e.g., [14] and the references therein). So, we can approximate, in polynomial-time, the best subnetwork for a large class of instances that, with high probability, include exponentially many $s - t$ paths and $s - t$ paths of length $\Theta(n)$. For such instances, a direct application of [6, Theorem 3] gives an exponential-time algorithm.

The main idea behind our approximation scheme, and our main technical contribution, is a polynomial-time approximation-preserving reduction of the best subnetwork problem for a good network G to a corresponding best subnetwork problem for a 0-latency simplified network G_0, which is a layered network obtained from G if we keep only s, t and their immediate neighbors, and connect all neighbors of s and t by direct edges of 0 latency. We first show that the equilibrium latency of the best subnetwork does not increase when we consider the 0-latency simplified network G_0 (Lemma 1). Although this may sound reasonable, we highlight that decreasing edge latencies to 0 may trigger Braess's paradox (e.g., starting from the network in Fig. 1.a with $l_3'(x) = 1$, and decreasing it to $l_3(x) = 0$ is just another way of triggering the paradox). Next, we employ Althöfer's Sparsification Lemma [1] (see also [9,10] and [6, Theorem 3]) and approximate the best subnetwork problem for the 0-latency simplified network.

The final (and crucial) step of our approximation preserving reduction is to start with the flow-solution to the best subnetwork problem for the 0-latency simplified network, and extend it to a flow-solution to the best subnetwork problem for the original (good) instance. To this end, we show how to "simulate" 0-latency edges by low latency paths in the original good network. Intuitively, this works because due to the expansion properties and the random latencies of the good network G, the intermediate subnetwork of G, connecting the neighbors of s to the neighbors of t, essentially behaves as a complete bipartite network with 0-latency edges. This is also the key step in the approach of [4,17], showing that Braess's paradox occurs in good networks with high probability (see [4, Section 2] for a detailed discussion). Hence, one could say that to some extent, the reason that Braess's paradox exists in good networks is the very same reason that the paradox can be efficiently resolved. Though conceptually simple, the full construction is technically involved and requires dealing with the amount of flow through the edges incident to s and t and their latencies. Our construction employs a careful grouping-and-matching argument, which works for good networks with high probability, see Lemmas 4 and 5.

We highlight that the reduction itself runs in polynomial time. The time consuming step is the application of [6, Theorem 3] to the 0-latency simplified network. Since such networks have only polynomially many (and very short) $s - t$ paths, they escape the hardness result of [15]. The approximability of the best subnetwork for 0-latency simplified networks is an intriguing open problem arising from our work.

Our result shows that a problem, that is NP-hard to approximate, can be very closely approximated in random (and random-like) networks. This resembles e.g., the problem of finding a Hamiltonian path in Erdös-Rényi graphs, where again, existence and construction both work just above the connectivity threshold, see e.g., [2]. However, not all hard problems are easy when one assumes random inputs (e.g., consider factoring or the hidden clique problem, for both of which no such results are known in full depth).

2 Model and Preliminaries

Notation. For an event E in a sample space, $\mathbb{P}[E]$ denotes the probability of E happening. We say that an event E occurs *with high probability*, if $\mathbb{P}[E] \geq 1 - n^{-\alpha}$, for some constant $\alpha \geq 1$, where n usually denotes the number of vertices of the network G to which E refers. We implicitly use the union bound to account for the occurrence of more than one low probability events.

Instances. A *selfish routing instance* is a tuple $\mathcal{G} = (G(V,E), (\ell_e)_{e \in E}, r)$, where $G(V, E)$ is an undirected network with a source s and a sink t, $\ell_e : \mathbb{R}_{\geq 0} \to \mathbb{R}_{\geq 0}$ is a non-decreasing latency function associated with each edge e, and $r > 0$ is the traffic rate. We let $\mathcal{P}$ (or $\mathcal{P}_G$, whenever the network G is not clear from the context) denote the (non-empty) set of simple $s - t$ paths in G. For brevity, we usually omit the latency functions, and refer to a selfish routing instance as (G, r).

We only consider linear latencies $\ell_e(x) = a_e x + b_e$, with $a_e, b_e \geq 0$. We restrict our attention to instances where the coefficients a_e and b_e are randomly selected from a pair of distributions $\mathcal{A}$ and $\mathcal{B}$. Following [4,17], we say that $\mathcal{A}$ and $\mathcal{B}$ are *reasonable* if:

- $\mathcal{A}$ has bounded range $[A_{\min}, A_{\max}]$ and $\mathcal{B}$ has bounded range $[0, B_{\max}]$, where $A_{\min} > 0$ and $A_{\max}, B_{\max}$ are constants, i.e., they do not depend on r and $|V|$.
- There is a closed interval $I_{\mathcal{A}}$ of positive length, such that for every non-trivial subinterval $I' \subseteq I_{\mathcal{A}}$, $\mathbb{P}_{a \sim \mathcal{A}}[a \in I'] > 0$.
- There is a closed interval $I_{\mathcal{B}}$, $0 \in I_{\mathcal{B}}$, of positive length, such that for every nontrivial subinterval $I' \subseteq I_{\mathcal{B}}$, $\mathbb{P}_{b \sim \mathcal{B}}[b \in I'] > 0$. Moreover, for any constant $\eta > 0$, there exists a constant $\delta_\eta > 0$, such that $\mathbb{P}_{b \sim \mathcal{B}}[b \leq \eta] \geq \delta_\eta$.

Subnetworks. Given a selfish routing instance $(G(V,E), r)$, any subgraph $H(V', E')$, $V' \subseteq V$, $E' \subseteq E$, $s, t \in V'$, obtained from G by edge and vertex removal, is a *subnetwork* of G. H has the same source s and sink t as G, and the edges of H have the same latencies as in G. Every instance $(H(V', E'), r)$, where $H(V', E')$ is a subnetwork of $G(V, E)$, is a *subinstance* of $(G(V, E), r)$.

Flows. Given an instance (G, r), a (feasible) *flow* f is a non-negative vector indexed by $\mathcal{P}$ such that $\sum_{q \in \mathcal{P}} f_q = r$. For a flow f, let $f_e = \sum_{q : e \in q} f_q$ be the amount of flow that f routes on edge e. Two flows f and g are *different* if there is an edge e with $f_e \neq g_e$. An edge e is used by flow f if $f_e > 0$, and a path q is used by f if $\min_{e \in q}\{f_e\} > 0$. We often write $f_q > 0$ to denote that a path q is used by f. Given a flow f, the latency of each edge e is $\ell_e(f_e)$, the latency of each path q is $\ell_q(f) = \sum_{e \in q} \ell_e(f_e)$, and the latency of f is $L(f) = \max_{q : f_q > 0} \ell_q(f)$. We sometimes write $L_G(f)$ when the network G is not clear from the context. For an instance $(G(V, E), r)$ and a flow f, we let $E_f = \{e \in E : f_e > 0\}$ be the set of edges used by f, and $G_f(V, E_f)$ be the corresponding subnetwork of G.

Nash Flow. A flow f is a *Nash (equilibrium) flow*, if it routes all traffic on minimum latency paths. Formally, f is a Nash flow if for every path q with $f_q > 0$, and every path q', $\ell_q(f) \leq \ell_{q'}(f)$. Therefore, in a Nash flow f, all players incur a common latency $L(f) = \min_q \ell_q(f) = \max_{q : f_q > 0} \ell_q(f)$ on their paths. A Nash flow f on a network $G(V, E)$ is a Nash flow on any subnetwork $G'(V', E')$ of G with $E_f \subseteq E'$.

Every instance (G, r) admits at least one Nash flow, and the players' latency is the same for all Nash flows (see e.g., [14]). Hence, we let $L(G, r)$ be the players' latency in some Nash flow of (G, r), and refer to it as the equilibrium latency of (G, r). For linear latency functions, a Nash flow can be computed efficiently, in strongly polynomial time, while for strictly increasing latencies, the Nash flow is essentially unique (see e.g., [14]).

ε-**Nash flow.** The definition of a Nash flow can be naturally generalized to that of an "almost Nash" flow. Formally, for some $\varepsilon > 0$, a flow f is an ε-Nash flow if for every path q with $f_q > 0$, and every path q', $\ell_q(f) \le \ell_{q'}(f) + \varepsilon$.

Best Subnetwork. Braess's paradox shows that there may be a subinstance (H, r) of an instance (G, r) with $L(H, r) < L(G, r)$ (see e.g., Fig. 1). The *best subnetwork H^** of (G, r) is a subnetwork of G with the minimum equilibrium latency, i.e., H^* has $L(H^*, r) \le L(H, r)$ for any subnetwork H of G. In this work, we study the approximability of the *Best Subnetwork Equilibrium Latency* problem, or BestSubEL in short. In BestSubEL, we are given an instance (G, r), and seek for the best subnetwork H^* of (G, r) and its equilibrium latency $L(H^*, r)$.

Good Networks. We restrict our attention to undirected $s - t$ networks $G(V, E)$. We let $n \equiv |V|$ and $m \equiv |E|$. For any vertex v, we let $\Gamma(v) = \{u \in V : \{u, v\} \in E\}$ denote the set of v's neighbors in G. Similarly, for any non-empty $S \subseteq V$, we let $\Gamma(S) = \bigcup_{v \in S} \Gamma(v)$ denote the set of neighbors of the vertices in S, and let $G[S]$ denote the subnetwork of G induced by S. For convenience, we let $V_s \equiv \Gamma(s)$, $E_s \equiv \{\{s, u\} : u \in V_s\}$, $V_t \equiv \Gamma(t)$, $E_t \equiv \{\{v, t\} : v \in V_t\}$, and $V_m \equiv V \setminus (\{s, t\} \cup V_s \cup V_t)$. We also let $n_s = |V_s|$, $n_t = |V_t|$, $n_+ = \max\{n_s, n_t\}$, $n_- = \min\{n_s, n_t\}$, and $n_m = |V_m|$. We sometimes write $V(G)$, $n(G)$, $V_s(G)$, $n_s(G)$, ..., if G is not clear from the context.

It is convenient to think that the network G has a layered structure consisting of s, the set of s's neighbors V_s, an "intermediate" subnetwork connecting the neighbors of s to the neighbors of t, the set of t's neighbors V_t, and t. Then, any $s - t$ path starts at s, visits some $u \in V_s$, proceeds either directly or through some vertices of V_m to some $v \in V_t$, and finally reaches t. Thus, we refer to $G_m \equiv G[V_s \cup V_m \cup V_t]$ as the *intermediate subnetwork* of G. Depending on the structure of G_m, we say that:

- G is a *random $\mathcal{G}_{n,p}$ network* if (i) n_s and n_t follow the binomial distribution with parameters n and p, and (ii) if any edge $\{u, v\}$, with $u \in V_m \cup V_s$ and $v \in V_m \cup V_t$, exists independently with probability p. Namely, the intermediate network G_m is an Erdös-Rényi random graph with $n - 2$ vertices and edge probability p, except for the fact that there are no edges in $G[V_s]$ and in $G[V_t]$.
- G is *internally bipartite* if the intermediate network G_m is a bipartite graph with independent sets V_s and V_t. G is *internally complete bipartite* if every neighbor of s is directly connected by an edge to every neighbor of t.
- G is *0-latency simplified* if it is internally complete bipartite and every edge e connecting a neighbor of s to a neighbor of t has latency function $\ell_e(x) = 0$.

The *0-latency simplification G_0* of a given network G is a 0-latency simplified network obtained from G by replacing $G[V_m]$ with a set of 0-latency edges directly connecting every neighbor of s to every neighbor of t. Moreover, we say that a 0-latency simplified network G is *balanced*, if $|n_s - n_t| \le 2n_-$.

Algorithm 1. Approximation Scheme for BestSubEL in Good Networks

---

Input: Good network $G(V, E)$, rate $r > 0$, approximation guarantee $\varepsilon > 0$
Output: Subnetwork H of G and ε-Nash flow g in H with $L(g) \leq (1 + \varepsilon)L(H^*, r) + \varepsilon$
1 if $L(G, r) < \varepsilon$, return G and a Nash flow of (G, r) ;
2 create the 0-latency simplification G_0 of G ;
3 if $r \geq (B_{\max} n_+)/(\varepsilon A_{\min})$, then let $H_0 = G_0$ and let f be a Nash flow of (G_0, r) ;
4 else, let H_0 be the subnetwork and f the $\varepsilon/6$-Nash flow of Thm. 2 applied with error $\varepsilon/6$;
5 let H be the subnetwork and let g be the ε-Nash flow of Lemma 5 starting from H_0 and f ;
6 return the subnetwork H and the ε-Nash flow g ;

We say that a network $G(V, E)$ is (n, p, k)-*good*, for some integer $n \leq |V|$, some probability $p \in (0, 1)$, with $pn = o(n)$, and some constant $k \geq 1$, if G satisfies that:

1. The maximum degree of G is at most $3np/2$, i.e., for any $v \in V$, $|\Gamma(v)| \leq 3np/2$.
2. G is an *expander graph*, namely, for any set $S \subseteq V$, $|\Gamma(S)| \geq \min\{np|S|, n\}/2$.
3. The edges of G have random reasonable latency functions distributed according to $\mathcal{A} \times \mathcal{B}$, and for any constant $\eta > 0$, $\mathbb{P}_{b \sim \mathcal{B}}[b \leq \eta/\ln n] = \omega(1/np)$.
4. If $k > 1$, we can compute in polynomial time a partitioning of V_m into k sets $V_m^1, \ldots, V_m^k$, each of cardinality $|V_m|/k$, such that all the induced subnetworks $G[\{s, t\} \cup V_s \cup V_m^i \cup V_t]$ are $(n/k, p, 1)$-good, with a possible violation of the maximum degree bound by s and t.

If G is a *random* $\mathcal{G}_{n,p}$ network, with n sufficiently large and $p \geq ck \ln n/n$, for some large enough constant $c > 1$, then G is an (n, p, k)-good network with high probability (see e.g., [2]), provided that the latency functions satisfy condition (3) above. As for condition (4), a random partitioning of V_m into k sets of cardinality $|V_m|/k$ satisfies (4) with high probability. Similarly, the random instances considered in [4] are good with high probability. Also note that the 0-latency simplification of a good network is balanced, due to (1) and (2).

3   The Approximation Scheme and Outline of the Analysis

In this section, we describe the main steps of the approximation scheme (see also Algorithm 1), and give an outline of its analysis. We let $\varepsilon > 0$ be the approximation guarantee, and assume that $L(G, r) \geq \varepsilon$. Otherwise, any Nash flow of (G, r) suffices.

Algorithm 1 is based on an approximation-preserving reduction of BestSubEL for a good network G to BestSubEL for the 0-latency simplification G_0 of G. The first step of our approximation-preserving reduction is to show that the equilibrium latency of the best subnetwork does not increase when we consider the 0-latency simplification G_0 of a network G instead of G itself. Since decreasing the edge latencies (e.g., decreasing $l_3'(x) = 1$ to $l_3(x) = 0$ in Fig. 1.a) may trigger Braess's paradox, we need Lemma 1, in Section 4, and its careful proof to make sure that zeroing out the latency of the intermediate subnetwork does not cause an abrupt increase in the equilibrium latency.

Next, we focus on the 0-latency simplification G_0 of G (step 2 in Alg. 1). We show that if the traffic rate is large enough, i.e., if $r = \Omega(n_+/\varepsilon)$, the paradox has a marginal influence on the equilibrium latency. Thus, any Nash flow of (G_0, r) is a $(1 + \varepsilon)$-approximation of BestSubEL (see Lemma 2 and step 4). If $r = O(n_+/\varepsilon)$, we use [6, Theorem 3] and obtain an $\varepsilon/6$-approximation of BestSubEL for (G_0, r) (see Theorem 2 and step 4).

We now have a subnetwork H_0 and an $\varepsilon/6$-Nash flow f that comprise a good approximate solution to BestSubEL for the simplified instance (G_0, r). The next step of our approximation-preserving reduction is to extend f to an approximate solution to BestSubEL for the original instance (G, r). The intuition is that due to the expansion and the reasonable latencies of G, any collection of 0-latency edges of H_0 used by f to route flow from V_s to V_t can be "simulated" by an appropriate collection of low-latency paths of the intermediate subnetwork G_m of G. In fact, this observation was the key step in the approach of [4,17] showing that Braess's paradox occurs in good networks with high probability. We first prove this claim for a small part of H_0 consisting only of neighbors of s and neighbors of t with approximately the same latency under f (see Lemma 4, the proof draws on ideas from [4, Lemma 5]). Then, using a careful latency-based grouping of the neighbors of s and of the neighbors of t in H_0, we extend this claim to the entire H_0 (see Lemma 5). Thus, we obtain a subnetwork H of G and an ε-Nash flow g in H such that $L(g) \leq (1 + \varepsilon)L(H^*, r) + \varepsilon$ (step 5).

We summarize our main result. The proof follows by combining Lemma 1, Theorem 2, and Lemma 5 in the way indicated by Algorithm 1 and the discussion above.

Theorem 1. *Let $G(V, E)$ be an (n, p, k)-good network, where $k \geq 1$ is a large enough constant, let $r > 0$ be any traffic rate, and let H^* be the best subnetwork of (G, r). Then, for any $\varepsilon > 0$, Algorithm 1 computes in time $n_+^{O(r^2 A_{max}^2 \ln(n_+)/\varepsilon^2)} \text{poly}(|V|)$, a flow g and a subnetwork H of G such that with high probability, wrt. the random choice of the latency functions, g is an ε-Nash flow of (H, r) and has $L(g) \leq (1+\varepsilon)L(H^*) + \varepsilon$.*

By the definition of reasonable latencies, A_{max} is a constant. Also, by Lemma 2, r affects the running time only if $r = O(n_+/\varepsilon)$. In fact, previous work on selfish network design assumes that $r = O(1)$, see e.g., [14]. Thus, if $r = O(1)$ (or more generally, if $r = O(\text{poly}(\ln \ln n))$) and $pn = O(\text{poly}(\ln n))$, in which case $n_+ = O(\text{poly}(\ln n))$, Theorem 1 gives a randomized polynomial-time approximation scheme for BestSubEL in good networks. Moreover, the running time is quasipolynomial for traffic rates up to $O(\text{poly}(\ln n))$ and average degrees up to $o(n)$, i.e., for the entire range of p in [4,17]. The next sections are devoted to the proofs of Lemmas 1 and 5, and of Theorem 2.

4 Network Simplification

We first show that the equilibrium latency of the best subnetwork does not increase when we consider the 0-latency simplification G_0 of a network G instead of G itself. We highlight that the following lemma holds not only for good networks, but also for any network with linear latencies and with the layered structure described in Section 2.

Lemma 1. *Let G be any network, let $r > 0$ be any traffic rate, and let H be the best subnetwork of (G, r). Then, there is a subnetwork H' of the 0-latency simplification of H (and thus, a subnetwork of G_0) with $L(H', r) \leq L(H, r)$.*

Proof sketch. We assume that all the edges of H are used by the equilibrium flow f of (H, r) (otherwise, we can remove all unused edges from H). The proof is constructive, and at the conceptual level, proceeds in two steps. For the first step, given the equilibrium flow f of the best subnetwork H of G, we construct a simplification H_1 of H that is internally bipartite and has constant latency edges connecting $\Gamma(s)$ to $\Gamma(t)$. H_1 also admits f as an equilibrium flow, and thus $L(H_1, r) = L(H, r)$. We can also show how to further simplify H_1 so that its intermediate bipartite subnetwork becomes acyclic.

The second part of the proof is to show that we can either remove some of the intermediate edges of H_1 or zero their latencies, and obtain a subnetwork H' of the 0-latency simplification of H with $L(H', r) \leq L(H, r)$. To this end, we describe a procedure where in each step, we either remove some intermediate edge of H_1 or zero its latency, without increasing the latency of the equilibrium flow.

Let us focus on an edge $e_{kl} = \{u_k, v_l\}$ connecting a neighbor u_k of s to a neighbor v_l of t. By the first part of the proof, the latency function of e_{kl} is a constant $b_{kl} > 0$. Next, we attempt to set the latency of e_{kl} to $b'_{kl} = 0$. We have also to change the equilibrium flow f to a new flow f' that is an equilibrium flow of latency at most L in the modified network with $b'_{kl} = 0$. We should be careful when changing f to f', since increasing the flow through $\{s, u_k\}$ and $\{v_l, t\}$ affects the latency of all $s - t$ paths going through u_k and v_l and may destroy the equilibrium property (or even increase the equilibrium latency). In what follows, we let r_q be the amount of flow moving from an $s - t$ path $q = (s, u_i, v_j, t)$ to the path $q_{kl} = (s, u_k, v_l, t)$ when we change f to f'. We note that r_q may be negative, in which case, $|r_q|$ units of flow actually move from q_{kl} to q. Thus, r_q's define a rerouting of f to a new flow f', with $f'_q = f_q - r_q$, for any $s - t$ path q other than q_{kl}, and $f'_{kl} = f_{kl} + \sum_q r_q$.

We next show how to compute r_q's so that f' is an equilibrium flow of cost at most L in the modified network (where we attempt to set $b'_{kl} = 0$). We let $\mathcal{P} = \mathcal{P}_{H_1} \setminus \{q_{kl}\}$ denote the set of all $s - t$ paths in H_1 other than q_{kl}. We let $\mathbf{F}$ be the $|\mathcal{P}| \times |\mathcal{P}|$ matrix, indexed by the paths $q \in \mathcal{P}$, where $\mathbf{F}[q_1, q_2] = \sum_{e \in q_1 \cap q_2} a_e - \sum_{e \in q_1 \cap q_{kl}} a_e$, and let $\mathbf{r}$ be the vector of r_q's. Then, the q-th component of $\mathbf{F}\mathbf{r}$ is equal to $\ell_q(f) - \ell_q(f')$. In the following, we consider two cases depending on whether $\mathbf{F}$ is singular or not.

If matrix $\mathbf{F}$ is non-singular, the linear system $\mathbf{F}\mathbf{r} = \varepsilon \mathbf{1}$ has a unique solution $\mathbf{r}_\varepsilon$, for any $\varepsilon > 0$. Moreover, due to linearity, for any $\alpha \geq 0$, the unique solution of the system $\mathbf{F}\mathbf{r} = \alpha \varepsilon \mathbf{1}$ is $\alpha \mathbf{r}_\varepsilon$. Therefore, for an appropriately small $\varepsilon > 0$, the linear system $Q_\varepsilon = \{\mathbf{F}\mathbf{r} = \varepsilon \mathbf{1}, f_q - r_q \geq 0 \ \forall q \in \mathcal{P}, f_{kl} + \sum_q r_q \geq 0, \ell_{q_{kl}}(f') \leq L + b_{kl} - \varepsilon\}$ admits a unique solution $\mathbf{r}$. We keep increasing ε until one of the inequalities of Q_ε becomes tight. If it first becomes $r_q = f_q$ for some path $q = (s, u_i, v_j, t) \in \mathcal{P}$, we remove the edge $\{u_i, v_j\}$ from H_1 and adjust the constant latency of e_{kl} so that $\ell_{q_{kl}}(f') = L - \varepsilon$. Then, the flow f' is an equilibrium flow of cost $L - \varepsilon$ for the resulting network, which has one edge less than the original network H_1. If $\sum_q r_q < 0$ and it first becomes $\sum_q r_q = -f_{kl}$, we remove the edge e_{kl} from H_1. Then, f' is an equilibrium flow of cost $L - \varepsilon$ for the resulting network, which again has one edge less than H_1. If $\sum_q r_q > 0$ and it first becomes $\ell_{q_{kl}}(f') = L + b_{kl} - \varepsilon$, we set the constant latency of the edge e_{kl} to $b'_{kl} = 0$. In this case, f' is an equilibrium flow of cost $L - \varepsilon$ for the resulting network that has one edge of 0 latency more than the initial network H_1. Moreover, we can show that if q_{kl} is disjoint from the paths $q \in \mathcal{P}$, the fact that the intermediate network H_1 is acyclic implies that the matrix F is positive definite, and thus non-singular. Therefore, if q_{kl} is disjoint from the paths in $\mathcal{P}$, the procedure above leads to a decrease in the equilibrium latency, and eventually to setting $b'_{kl} = 0$.

If F is singular, we can compute r_q's so that f' is an equilibrium flow of cost L in a modified network that includes one edge less than the original network H_1. If F is singular, the homogeneous linear system $\boldsymbol{Fr} = \boldsymbol{0}$ admits a nontrivial solution $\boldsymbol{r} \neq \boldsymbol{0}$. Moreover, due to linearity, for any $\alpha \in \mathbb{R}$, $\alpha\boldsymbol{r}$ is also a solution to $\boldsymbol{Fr} = \boldsymbol{0}$. Therefore, the linear system $Q_0 = \{\boldsymbol{Fr} = \boldsymbol{0}, f_q - r_q \geq 0 \ \forall q \in \mathcal{P}, f_{kl} + \sum_q r_q \geq 0\}$ admits a solution $\boldsymbol{r} \neq \boldsymbol{0}$ that makes at least one of the inequalities tight. We recall that the q-th component of $\boldsymbol{Fr}$ is equal to $\ell_q(f) - \ell_q(f')$. Therefore, for the flow f' obtained from the particular solution $\boldsymbol{r}$ of Q_0, the latency of any path $q \in \mathcal{P}$ is equal to L. If $\boldsymbol{r}$ is such that $r_q = f_q$ for some path $q = (s, u_i, v_j, t) \in \mathcal{P}$, we remove the edge $\{u_i, v_j\}$ from H_1 and adjust the constant latency of e_{kl} so that $\ell_{q_{kl}}(f') = L$. Then, the flow f' is an equilibrium flow of cost L for the resulting network, which has one edge less than the original network H_1. If $\boldsymbol{r}$ is such that $\sum_q r_q = -f_{kl}$, we remove the edge e_{kl} from H_1. Then, f' is an equilibrium flow of cost L for the resulting network, which again has one edge less than H_1.

Each time we apply the procedure above either we decrease the number of edges of the intermediate network by one or we increase the number of 0-latency edges of the intermediate network by one, without increasing the latency of the equilibrium flow. So, by repeatedly applying these steps, we end up with a subnetwork H' of the 0-latency simplification of H with $L(H', r) \leq L(H; r)$.   □

5   Approximating the Best Subnetwork of Simplified Networks

We proceed to show how to approximate the BestSubEL problem in a balanced 0-latency simplified network G_0 with reasonable latencies. We may always regard G_0 as the 0-latency simplification of a good network G. We first state two useful lemmas about the maximum traffic rate r up to which BestSubEL remains interesting, and about the maximum amount of flow routed on any edge / path in the best subnetwork.

Lemma 2. *Let G_0 be any 0-latency simplified network, let $r > 0$, and let H_0^* be the best subnetwork of (G_0, r). For any $\varepsilon > 0$, if $r > \frac{B_{\max} n_+}{A_{\min} \varepsilon}$, then $L(G_0, r) \leq (1+\varepsilon)L(H_0^*, r)$.*

Proof. We assume that $r > \frac{B_{\max} n_+}{A_{\min} \varepsilon}$, let f be a Nash flow of (G_0, r), and consider how f allocates r units of flow to the edges of $E_s \equiv E_s(G_0)$ and to the edges $E_t \equiv E_t(G_0)$. For simplicity, we let $L \equiv L(G_0, r)$ denote the equilibrium latency of G_0, and let $A_s = \sum_{e \in E_s} 1/a_e$ and $A_t = \sum_{e \in E_t} 1/a_e$.

Since G_0 is a 0-latency simplified network and f is a Nash flow of (G_0, r), there are $L_1, L_2 > 0$, with $L_1 + L_2 = L$, such that all used edges incident to s (resp. to t) have latency L_1 (resp. L_2) in the Nash flow f. Since $r > \frac{B_{\max} n_+}{A_{\min}}$, $L_1, L_2 > B_{\max}$ and all edges in $E_s \cup E_t$ are used by f. Moreover, by an averaging argument, we have that there is an edge $e \in E_s$ with $a_e f_e \leq r/A_s$, and that there is an edge $e \in E_t$ with $a_e f_e \leq r/A_t$. Therefore, $L_1 \leq (r/A_s) + B_{\max}$ and $L_2 \leq (r/A_t) + B_{\max}$, and thus, $L \leq \frac{r}{A_s} + \frac{r}{A_t} + 2B_{\max}$.

On the other hand, if we ignore the additive terms b_e of the latency functions, the optimal average latency of the players is $r/A_s + r/A_t$, which implies that $L(H_0^*, r) \geq r/A_s + r/A_t$. Therefore, $L \leq L(H_0^*, r) + 2B_{\max}$. Moreover, since $r > \frac{B_{\max} n_+}{A_{\min} \varepsilon}$, $A_s \leq n_s/A_{\min}$, and $A_t \leq n_t/A_{\min}$, we have that:

$$L(H_0^*, r) \geq \frac{r}{A_s} + \frac{r}{A_t} \geq \frac{B_{\max} n_s}{A_{\min} \varepsilon} \frac{A_{\min}}{n_s} + \frac{B_{\max} n_t}{A_{\min} \varepsilon} \frac{A_{\min}}{n_t} \geq 2B_{\max}/\varepsilon$$

Therefore, $2B_{\max} \leq \varepsilon L(H_0^*, r)$, and $L \leq (1 + \varepsilon)L(H_0^*, r)$. $\square$

Lemma 3. *Let G_0 be a balanced 0-latency simplified network with reasonable latencies, let $r > 0$, and let f be a Nash flow of the best subnetwork of (G_0, r). For any $\varepsilon > 0$, if $\mathbb{P}_{b \sim \mathcal{B}}[b \leq \varepsilon/4] \geq \delta$, for some constant $\delta > 0$, there exists a constant $\rho = \frac{24 A_{\max} B_{\max}}{\delta \varepsilon A_{\min}^2}$ such that with probability at least $1 - e^{-\delta n_-/8}$, $f_e \leq \rho$, for all edges e.*

Approximating the Best Subnetwork of Simplified Networks. We proceed to derive an approximation scheme for the best subnetwork of any simplified instance (G_0, r).

Theorem 2. *Let G_0 be a balanced 0-latency simplified network with reasonable latencies, let $r > 0$, and let H_0^* be the best subnetwork of (G_0, r). Then, for any $\varepsilon > 0$, we can compute, in time $n_+^{O(A_{\max}^2 r^2 \ln(n_+)/\varepsilon^2)}$, a flow f and a subnetwork H_0 consisting of the edges used by f, such that (i) f is an ε-Nash flow of (H_0, r), (ii) $L(f) \leq L(H_0^*, r) + \varepsilon/2$, and (iii) there exists a constant $\rho > 0$, such that $f_e \leq \rho + \varepsilon$, for all e.*

Theorem 2 is a corollary of [6, Theorem 3], since in our case the number of different $s - t$ paths is at most n_+^2 and each path consists of 3 edges. So, in [6, Theorem 3], we have $d_1 = 2$, $d_2 = 0$, $\alpha = A_{\max}$, and the error is ε/r. Moreover, we know that any Nash flow g of (H_0^*, r) routes $g_e \leq \rho$ units of flow on any edge e, and that in the exhaustive search step, in the proof of [6, Theorem 3], one of the acceptable flows f has $|g_e - f_e| \leq \varepsilon$, for all edges e (see also [6, Lemma 3]). Thus, there is an acceptable flow f with $f_e \leq \rho + \varepsilon$, for all edges e. In fact, if among all acceptable flows enumerated in the proof of [6, Theorem 3], we keep the acceptable flow f that minimizes the maximum amount flow routed on any edge, we have that $f_e \leq \rho + \varepsilon$, for all edges e.

6   Extending the Solution to the Good Network

Given a good instance (G, r), we create the 0-latency simplification G_0 of G, and using Theorem 2, we compute a subnetwork H_0 and an $\varepsilon/6$-Nash flow f that comprise an approximate solution to BestSubEL for (G_0, r). Next, we show how to extend f to an approximate solution to BestSubEL for the original instance (G, r). The intuition is that the 0-latency edges of H_0 used by f to route flow from V_s to V_t can be "simulated" by low-latency paths of G_m. We first formalize this intuition for the subnetwork of G induced by the neighbors of s with (almost) the same latency B_s and the neighbors of t with (almost) the same latency B_t, for some B_s, B_t with $B_s + B_t \approx L(f)$. We may think of the networks G and H_0 in the lemma below as some small parts of the original network G and of the actual subnetwork H_0 of G_0. Thus, we obtain the following lemma, which serves as a building block in the proof of Lemma 5.

Lemma 4. *We assume that $G(V, E)$ is an $(n, p, 1)$-good network, with a possible violation of the maximum degree bound by s and t, but with $|V_s|, |V_t| \leq 3knp/2$, for some constant $k > 0$. Also the latencies of the edges in $E_s \cup E_t$ are not random, but there exist constants $B_s, B_t \geq 0$, such that for all $e \in E_s$, $\ell_e(x) = B_s$, and for all $e \in E_t$, $\ell_e(x) = B_t$. We let $r > 0$ be any traffic rate, let H_0 be any subnetwork of the 0-latency simplification G_0 of G, and let f be any flow of (H_0, r). We assume that there exists a constant $\rho' > 0$, such that for all $e \in E(H_0)$, $0 < f_e \leq \rho'$. Then, for any $\epsilon_1 > 0$, with high probability, wrt. the random choice of the latency functions of G, we can compute in $\text{poly}(|V|)$ time a subnetwork G' of G, with $E_s(G') = E_s(H_0)$ and $E_t(G') = E_t(H_0)$, and a flow g of (G', r) such that (i) $g_e = f_e$ for all $e \in E_s(G') \cup E_t(G')$, (ii) g is a $7\epsilon_1$-Nash flow in G', and (iii) $L_{G'}(g) \leq B_s + B_t + 7\epsilon_1$.*

Proof sketch. For convenience and wlog., we assume that $E_s(G) = E_s(H_0)$ and that $E_t(G) = E_t(H_0)$, so that we simply write V_s, V_t, E_s, and E_t from now on. For each $e \in E_s \cup E_t$, we let $g_e = f_e$. So, the flow g satisfies (i), by construction.

We compute the extension of g through G_m as an "almost" Nash flow in a modified version of G, where each edge $e \in E_s \cup E_t$ has a capacity $g_e = f_e$ and a constant latency $\ell_e(x) = B_s$, if $e \in E_s$, and $\ell_e(x) = B_t$, if $e \in E_t$. All other edges e of G have an infinite capacity and a (randomly chosen) reasonable latency function $\ell_e(x)$.

We let g be the flow of rate r that respects the capacities of the edges in $E_s \cup E_t$, and minimizes $\text{Pot}(g) = \sum_{e \in E} \int_0^{g_e} \ell_e(x) dx$. Such a flow g can be computed in strongly polynomial time (see e.g., [18]). The subnetwork G' of G is simply G_g, namely, the subnetwork that includes only the edges used by g. It could have been that g is not a Nash flow of (G, r), due to the capacity constraints on the edges of $E_s \cup E_t$. However, since g is a minimizer of $\text{Pot}(g)$, for any $u \in V_s$ and $v \in V_t$, and any pair of $s - t$ paths q, q' going through u and v, if $g_q > 0$, then $\ell_q(g) \leq \ell_{q'}(g)$. Thus, g can be regarded as a Nash flow for any pair $u \in V_s$ and $v \in V_t$ connected by g-used paths.

To conclude the proof, we adjust the proof of [4, Lemma 5], and show that for any $s - t$ path q used by g, $\ell_q(g) \leq B_s + B_t + 7\epsilon_1$. To prove this, we let $q = (s, u, \ldots, v, t)$ be the $s - t$ path used by g that maximizes $\ell_q(g)$. We show the existence of a path $q' = (s, u, \ldots, v, t)$ in G of latency $\ell_{q'}(g) \leq B_s + B_t + 7\epsilon_1$. Therefore, since g is a minimizer of $\text{Pot}(g)$, the latency of the maximum latency g-used path q, and thus the latency of any other g-used $s - t$ path, is at most $B_s + B_t + 7\epsilon_1$, i.e., g satisfies (iii). Moreover, since for any $s - t$ path q, $\ell_q(g) \geq B_s + B_t$, g is an $7\epsilon_1$-Nash flow in G'. $\quad\square$

Grouping the Neighbors of s and t. Let us now consider the entire network G and the entire subnetwork H_0 of G_0. Lemma 4 can be applied only to subsets of edges in $E_s(H_0)$ and in $E_t(H_0)$ that have (almost) the same latency under f. Since H_0 does not need to be internally complete bipartite, there may be neighbors of s (resp. t) connected to disjoint subsets of V_t (resp. of V_s) in H_0, and thus have quite different latency. Hence, to apply Lemma 4, we partition the neighbors of s and the neighbors of t into classes V_s^i and V_t^j according to their latency. For convenience, we let $\epsilon_2 = \varepsilon/6$, i.e., f is an ϵ_2-Nash flow, and $L \equiv L_{H_0}(f)$. By Theorem 2, applied with error $\epsilon_2 = \varepsilon/6$, there exists a ρ such that for all $e \in E(H_0)$, $0 < f_e \leq \rho + \epsilon_2$. Therefore, $L \leq 2A_{\max}(\rho + \epsilon_2) + 2B_{\max}$ is bounded by a constant.

We partition the interval $[0, L]$ into $\kappa = \lceil L/\epsilon_2 \rceil$ subintervals, where the i-th subinterval is $I^i = (i\epsilon_2, (i+1)\epsilon_2]$, $i = 0, \ldots, \kappa - 1$. We partition the vertices of V_s (resp. of V_t)

that receive positive flow by f into κ classes V_s^i (resp. V_t^i), $i = 0, \ldots, \kappa - 1$. Precisely, a vertex $x \in V_s$ (resp. $x \in V_t$), connected to s (resp. to t) by the edge $e_x = \{s, x\}$ (resp. $e_x = \{x, t\}$), is in the class V_s^i (resp. in the class V_t^i), if $\ell_{e_x}(f_{e_x}) \in I_i$. If a vertex $x \in V_s$ (resp. $x \in V_t$) does not receive any flow from f, x is removed from G and does not belong to any class. Hence, from now on, we assume that all neighbors of s and t receive positive flow from f, and that $V_s^0, \ldots V_s^{\kappa-1}$ (resp. $V_t^0, \ldots, V_t^{\kappa-1}$) is a partitioning of V_s (resp. V_t). In exactly the same way, we partition the edges of E_s (resp. of E_t) used by f into k classes E_s^i (resp. E_t^i), $i = 0, \ldots, \kappa - 1$.

To find out which parts of H_0 will be connected through the intermediate subnetwork of G, using the construction of Lemma 4, we further classify the vertices of V_s^i and V_t^j based on the neighbors of t and on the neighbors of s, respectively, to which they are connected by f-used edges in the subnetwork H_0. In particular, a vertex $u \in V_s^i$ belongs to the classes $V_s^{(i,j)}$, for all j, $0 \leq j \leq \kappa - 1$, such that there is a vertex $v \in V_t^j$ with $f_{\{u,v\}} > 0$. Similarly, a vertex $v \in V_t^j$ belongs to the classes $V_t^{(i,j)}$, for all i, $0 \leq i \leq \kappa - 1$, such that there is a vertex $u \in V_s^i$ with $f_{\{u,v\}} > 0$. A vertex $u \in V_s^i$ (resp. $v \in V_t^j$) may belong to many different classes $V_s^{(i,j)}$ (resp. to $V_t^{(i,j)}$), and that the class $V_s^{(i,j)}$ is non-empty iff the class $V_t^{(i,j)}$ is non-empty. We let $k \leq \kappa^2$ be the number of pairs (i, j) for which $V_s^{(i,j)}$ and $V_t^{(i,j)}$ are non-empty. We note that k is a constant, i.e., does not depend on $|V|$ and r. We let $E_s^{(i,j)}$ be the set of edges connecting s to the vertices in $V_s^{(i,j)}$ and $E_t^{(i,j)}$ be the set of edges connecting t to the vertices in $V_t^{(i,j)}$.

Building the Intermediate Subnetworks of G. The last step is to replace the 0-latency simplified parts connecting the vertices of each pair of classes $V_s^{(i,j)}$ and $V_t^{(i,j)}$ in H_0 with a subnetwork of G_m. We partition, as in condition (4) in the definition of good networks, the set V_m of intermediate vertices of G into k subsets, each of cardinality $|V_m|/k$, and associate a different such subset $V_m^{(i,j)}$ with any pair of non-empty classes $V_s^{(i,j)}$ and $V_t^{(i,j)}$. For each pair (i, j) for which the classes $V_s^{(i,j)}$ and $V_t^{(i,j)}$ are non-empty, we consider the induced subnetwork $G^{(i,j)} \equiv G[\{s, t\} \cup V_s^{(i,j)} \cup V_m^{(i,j)} \cup V_t^{(i,j)}]$, which is an $(n/k, p, 1)$-good network, since G is an (n, p, k)-good network. Therefore, we can apply Lemma 4 to $G^{(i,j)}$, with $H_0^{(i,j)} \equiv H_0[\{s, t\} \cup V_s^{(i,j)} \cup V_t^{(i,j)}]$ in the role of H_0, the restriction $f^{(i,j)}$ of f to $H_0^{(i,j)}$ in the role of the flow f, and $\rho' = \rho + \epsilon_2$. Moreover, we let $B_s^{(i,j)} = \max_{e \in E_s^{(i,j)}} \ell_e(f_e)$ and $B_t^{(i,j)} = \max_{e \in E_t^{(i,j)}} \ell_e(f_e)$ correspond to B_s and B_t, and introduce constant latencies $\ell'_e(x) = B_s^{(i,j)}$ for all $e \in E_s^{(i,j)}$ and $\ell'_e(x) = B_t^{(i,j)}$ for all $e \in E_t^{(i,j)}$, as required by Lemma 4. Thus, we obtain, with high probability, a subnetwork $H^{(i,j)}$ of $G^{(i,j)}$ and a flow $g^{(i,j)}$ that routes as much flow as $f^{(i,j)}$ on all edges of $E_s^{(i,j)} \cup E_t^{(i,j)}$, and satisfies the conclusion of Lemma 4, if we keep in $H^{(i,j)}$ the constant latencies $\ell'_e(x)$ for all $e \in E_s^{(i,j)} \cup E_t^{(i,j)}$.

The final outcome is the union of the subnetworks $H^{(i,j)}$, denoted H (H has the latency functions of the original instance G), and the union of the flows $g^{(i,j)}$, denoted g, where the union is taken over all k pairs (i, j) for which the classes $V_s^{(i,j)}$ and $V_t^{(i,j)}$ are non-empty. By construction, all edges of H are used by g. Using the properties of the construction above, we can show that if $\epsilon_1 = \varepsilon/42$ and $\epsilon_2 = \varepsilon/6$, the flow g is an ε-Nash flow of (H, r), and satisfies $L_H(g) \leq L_{H_0}(f) + \varepsilon/2$. Thus, we obtain:

Lemma 5. *Let any $\varepsilon > 0$, let $k = \lceil 12(A_{\max}(\rho + \varepsilon) + B_{\max})/\varepsilon \rceil^2$, let $G(V, E)$ be an (n, p, k)-good network, let $r > 0$, let H_0 be any subnetwork of the 0-latency simplification of G, and let f be an $(\varepsilon/6)$-Nash flow of (H_0, r) for which there exists a constant $\rho' > 0$, such that for all $e \in E(H_0)$, $0 < f_e \leq \rho'$. Then, with high probability, wrt. the random choice of the latency functions of G, we can compute in $\mathrm{poly}(|V|)$ time a subnetwork H of G and an ε-Nash flow g of (H, r) with $L_H(g) \leq L_{H_0}(f) + \varepsilon/2$.*

References

1. Althöfer, I.: On Sparse Approximations to Randomized Strategies and Convex Combinations. Linear Algebra and Applications 99, 339–355 (1994)
2. Bollobás, B.: Random Graphs, 2nd edn. Cambridge Studies in Advanced Mathematics, vol. 73. Cambridge University Press (2001)
3. Braess, D.: Über ein paradox aus der Verkehrsplanung. Unternehmensforschung 12, 258–268 (1968)
4. Chung, F., Young, S.J.: Braess's paradox in large sparse graphs. In: Saberi, A. (ed.) WINE 2010. LNCS, vol. 6484, pp. 194–208. Springer, Heidelberg (2010)
5. Chung, F., Young, S.J., Zhao, W.: Braess's paradox in expanders. Random Structures and Algorithms 41(4), 451–468 (2012)
6. Fotakis, D., Kaporis, A.C., Spirakis, P.G.: Efficient methods for selfish network design. Theoretical Computer Science 448, 9–20 (2012)
7. Kelly, F.: The mathematics of traffic in networks. In: Gowers, T., Green, J., Leader, I. (eds.) The Princeton Companion to Mathematics. Princeton University Press (2008)
8. Lin, H.C., Roughgarden, T., Tardos, É., Walkover, A.: Stronger bounds on Braess's paradox and the maximum latency of selfish routing. SIAM Journal on Discrete Mathematics 25(4), 1667–1686 (2011)
9. Lipton, R.J., Markakis, E., Mehta, A.: Playing Large Games Using Simple Strategies. In: Proc. of the 4th ACM Conference on Electronic Commerce (EC 2003), pp. 36–41 (2003)
10. Lipton, R.J., Young, N.E.: Simple Strategies for Large Zero-Sum Games with Applications to Complexity Theory. In: Proc. of the 26th ACM Symposium on Theory of Computing (STOC 1994), pp. 734–740 (1994)
11. Milchtaich, I.: Network Topology and the Efficiency of Equilibrium. Games and Economic Behavior 57, 321–346 (2006)
12. Nagurney, A., Boyce, D.: Preface to "On a Paradox of Traffic Planning". Transportation Science 39(4), 443–445 (2005)
13. Pas, E.I., Principio, S.L.: Braess's paradox: Some new insights. Transportation Research Part B 31(3), 265–276 (1997)
14. Roughgarden, T.: Selfish Routing and the Price of Anarchy. MIT Press (2005)
15. Roughgarden, T.: On the Severity of Braess's Paradox: Designing Networks for Selfish Users is Hard. Journal of Computer and System Sciences 72(5), 922–953 (2006)
16. Steinberg, R., Zangwill, W.I.: The prevalence of Braess' paradox. Transportation Science 17(3), 301–318 (1983)
17. Valiant, G., Roughgarden, T.: Braess's paradox in large random graphs. Random Structures and Algorithms 37(4), 495–515 (2010)
18. Végh, L.A.: Strongly polynomial algorithm for a class of minimum-cost flow problems with separable convex objectives. In: Proc. of the 44th ACM Symposium on Theory of Computing (STOC 2012), pp. 27–40 (2012)

Truthfulness Flooded Domains
and the Power of Verification for Mechanism Design*

Dimitris Fotakis and Emmanouil Zampetakis

School of Electrical and Computer Engineering
National Technical University of Athens, 15780 Athens, Greece
fotakis@cs.ntua.gr, mzampet@corelab.ntua.gr

Abstract. We investigate the reasons that make symmetric partial verification essentially useless in virtually all domains. Departing from previous work, we consider any possible (finite or infinite) domain and general symmetric verification. We identify a natural property, namely that the correspondence graph of a symmetric verification M is strongly connected by finite paths along which the preferences are consistent with the preferences at the endpoints, and prove that this property is sufficient for the equivalence of truthfulness and M-truthfulness. In fact, defining appropriate versions of this property, we obtain this result for deterministic and randomized mechanisms with and without money. Moreover, we show that a slightly relaxed version of this property is also necessary for the equivalence of truthfulness and M-truthfulness. Our conditions provide a generic and convenient way of checking whether truthful implementation can take advantage of any symmetric verification scheme in any domain. Since the simplest case of symmetric verification is local verification, our results imply, as a special case, the equivalence of local truthfulness and global truthfulness in the setting without money. To complete the picture, we consider asymmetric verification, and prove that a social choice function is M-truthfully implementable by some asymmetric verification M if and only if f does not admit a cycle of profitable deviations.

1 Introduction

In mechanism design, a principal seeks to implement a social choice function that maps the private preferences of some strategic agents to a set of possible outcomes. Exploiting their power over the outcome, the agents may lie about their preferences if they find it profitable. Trying to incentivize truthfulness, the principal may offer payments to (or collect payments from) the agents or find ways of partially verifying their statements, thus restricting the false statements available to them. A social choice function is *truthfully implementable* (or implementable, in short) if there is a payment scheme under which truthtelling becomes a dominant strategy of the agents. Since many social choice functions are not implementable, a central research direction in mechanism design is

---

* This research was supported by the project AlgoNow, co-financed by the European Union (European Social Fund - ESF) and Greek national funds, through the Operational Program "Education and Lifelong Learning" of the National Strategic Reference Framework (NSRF) - Research Funding Program: THALES, investing in knowledge society through the European Social Fund.

Y. Chen and N. Immorlica (Eds.): WINE 2013, LNCS 8289, pp. 202–215, 2013.

to identify sufficient and necessary conditions under which large classes of functions are truthfully implementable. In this direction, we seek a deeper understanding of the power of partial verification in mechanism design, as far as truthful implementation is concerned, a question going back to the work of Green and Laffont [9].

The Model. For the purposes of this work, it is without loss of generality to consider mechanism design with a single agent, also known as the *principal-agent* setting (see e.g., [2,3] for an explanation). In this setting, the principal wants to implement a *social choice function* $f : D \to O$, where O is the set of possible *outcomes* and D is the *domain* of agent's preferences. Formally, D consists of the agent's *types*, where each type $x : O \to \mathbb{R}$ gives the utility of the agent for each outcome. The agent's type is private information. So, based on the agent's declared type x, the principal computes the outcome $o = f(x)$. A function f is (truthfully) *implementable* if for each type x, with $o = f(x)$, and any other type y, with $o' = f(y)$, $x(o) \geq x(o')$. Then, declaring her real type x is a dominant strategy of the agent. Otherwise, the agent may misreport a type y that results in a utility of $x(o') > x(o)$ under her true type x. This undesirable situation is usually corrected with a payment scheme $p : O \to \mathbb{R}$, that compensates the agent for telling the truth. Then, a function f is (truthfully) *implementable with payments* p (or, in general, implementable with money) if for each type x, with $o = f(x)$, and any other type y, with $o' = f(y)$, $x(o) + p(o) \geq x(o') + p(o')$.

Gui, Müller, and Vohra [10] cast this setting in terms of a (possibly infinite) directed graph G on vertex set D. For each ordered pair of types x and y, G has a directed edge (x, y). Given the social choice function f, we obtain an edge-weighted version of G, denoted G_f, where the weight of each edge (x, y) is $x(o) - x(o')$, with $o = f(x)$ and $o' = f(y)$. This corresponds to the gain of the agent if instead of misreporting y, she reports her true type x. Then, a social choice function f is truthfully implementable if and only if G_f does not contain any negative edges. Moreover, Rochet's theorem [14] implies that a function f is truthfully implementable with money if and only if G_f does not contain any directed negative cycles (see also [17]).

There are many classical impossibility results stating that natural social choice functions (or large classes of them) are not implementable, even with the use of money (see e.g., [12]). Virtually all such proofs seem to crucially exploit that the agent can declare any type in the domain. Hence, Nisan and Ronen [13] suggested that the class of implementable functions could be enriched if we assume *partial verification* [9], which restricts the types that the agent can misreport. Formally, we assume a *correspondence function* (or simply, a *verification*) $M : D \to 2^D$ such that if the agent's true type is x, she can only misreport a type in $M(x) \subseteq D$. As before, we can cast M as a (possibly infinite) directed *correspondence graph* G_M on D. For each ordered pair of types x and y, G_M has a directed edge (x, y) if $y \in M(x)$. Given the social choice function f, we obtain the edge-weighted version $G_{M,f}$ of G_M by letting the edge weights be as in G_f. A social choice function f is M-truthfully implementable (resp. with money) if and only if $G_{M,f}$ does not contain any negative edges (resp. directed negative cycles).

Previous Work. Every function f can be implemented by an appropriately strong verification scheme combined with payments (see also Section 5). So, the problem now is to come up with a meaningful verification M, which is either inherent in or naturally enforceable for some interesting domains and allows for a few non-implementable functions to be M-truthfully implementable. To this end, previous work has considered two kinds of verification, namely *symmetric* and *asymmetric* verification.

Symmetric verification naturally applies to convex domains (e.g., Combinatorial Auctions) and to domains with an inherent notion of distance (e.g., Facility Location, Voting). The idea is that every type x can only declare some type y not far from x. A typical example is M^ε verification where each type x can declare any type y in a ball of radius ε around x. Another typical example is M^{swap} verification, naturally applicable to Voting and to ordinal preference domains. In M^{swap} verification, each type x is as a linear order on O and can declare any type y obtained from x by swapping two adjacent outcomes. Rather surprisingly, previous work provides strong evidence that symmetric verification does not give any benefit to the principal, as far as truthful implementation is concerned. In particular, the strong and elegant result of Archer and Kleinberg [2] and its extension by Berger, Müller, and Naeemi [5] imply that M^ε verification does not help in convex domains. Formally, the results of [2,5,6] imply that for any convex domain, truthfulness with money is equivalent with M^ε-truthfulness with money. Similarly, Caragiannis, Elkind, Szegedy, and Yu [6] proved that M^{swap} verification does not help in the domain of Voting.

As far as implementation without money is concerned, the research on the power of symmetric verification is closely related to the research about sufficient and necessary conditions under which weaker properties are equivalent to global truthfulness. Even though the motivation for studying weaker properties may be more general (see e.g., [15,2,7,16]), in the absence of money, local truthfulness is essentially a special case of symmetric verification. In this research agenda, Sato [16] considered M^{swap} verification (under the name of adjacent manipulation truthfulness) for ordinal preference domains, and proved that if $G_{M^{\text{swap}}}$ is strongly connected by paths satisfying the no-restoration property, then truthful implementation and M^{swap}-truthful implementation are equivalent. He also proved that the universal domain, that includes all linear orders on O, and single-peaked domains have the no-restoration property, and thus, for these domains, truthful implementation is equivalent to M^{swap}-truthful implementation. Independently, Carroll [7] obtained similar results for convex domains, for the universal domain, and for single-peaked and single-crossing domains, which also extend to randomized mechanisms. Carroll also gave a necessary condition for the equivalence of local and global truthfulness in a specific domain with cardinal preferences.

On the other hand, asymmetric verification is "one-sided". Given a social choice function f, a typical example of asymmetric verification is when the agent can only lie either by overstating or by understating her utility. E.g., for Scheduling on related machines, the machine can only lie by overstating its speed [4], for Combinatorial Auctions, the agent can only underbid on her preferred sets [11], and for Facility Location, the agent can only understate her distance to the nearest facility [8]. The use of asymmetric verification has led to strong positive results about the truthful implementation of natural social choice functions in several important domains (see e.g., [4,11,8] and the references there in). The intuition is that the mechanism discourages one direction of lying, while the other direction of lying is forbidden by the verification.

Motivation and Contribution. Our work is motivated by the general observation, stated explicitly in [6], that even very strict symmetric verification schemes do not help in truthful implementation, while strong positive results are possible with simple asymmetric verification. So, we seek a deeper understanding of the reasons that make symmetric verification essentially useless in virtually all domains, and some formal justification behind the success of asymmetric verification.

Departing from previous work, we consider any possible (finite or infinite) domain D and very general classes of partial verification. To formalize the notions of symmetric and asymmetric verification, we say that a verification M is symmetric if the presence of a directed edge (x, y) in G_M implies the presence of the reverse edge (y, x), and asymmetric if G_M is an acyclic tournament.

Our main result is a general and unified explanation about the weakness of symmetric verification. In Section 3, we identify a natural property, namely that the correspondence graph G_M is strongly connected by finite paths along which the preferences are consistent with the preferences at the endpoints. In fact, we define three versions of this property depending on whether we consider implementation by deterministic truthful mechanisms (strict order-preserving property), by deterministic mechanisms that use payments (strict difference-preserving property), and by randomized truthful-in-expectation mechanisms (difference-convex property). Despite the slightly different definitions, the essence of the property is the same, but stronger versions of it are required as the mechanisms become more powerful. We show that for any (finite or infinite) domain D and any symmetric verification M that satisfies the corresponding version of the property, deterministic / randomized truthful implementation (resp. with money) is equivalent to deterministic / randomized M-truthful implementation (resp. with money). In all cases, the proof is simple and elegant, and only exploits an elementary combinatorial argument on the paths of G_M. With this general sufficient condition for the equivalence of truthfulness and M-truthfulness, we simplify, unify, and strengthen several known results about symmetric verification and local truthfulness without money. E.g., we obtain, as simple corollaries, the equivalence of truthful and M^ε-truthful implementation for any convex domain (even with money) and for Facility Location, and the equivalence of truthfulness and M^{swap}-truthfulness for Voting.

In Section 4, we identify necessary conditions for the equivalence of truthfulness and M-truthfulness, for any symmetric verification M. These are relaxed versions of the sufficient conditions, and require that the correspondence graph G_M is strongly connected by finite preference preserving paths. Otherwise, we show how to find a separator of G_M, which in turn, leads to the definition of a function that is M-truthfully implementable, but not implementable. We also observe that the necessary condition is violated by the domain of 2-Facility Location. To conclude the discussion about symmetric verification, we close the small gap between the sufficient and necessary properties, and present the first known condition that is both sufficient and necessary for the equivalence of truthful and M-truthful implementation. Overall, our conditions provide a generic and convenient way of checking whether truthful implementation can take advantage of any symmetric verification scheme in any domain.

Finally, in Section 5, we consider asymmetric verification, and prove that a social choice function f is M-truthfully implementable by some asymmetric verification M if and only if the subgraph of G_f consisting of negative edges is acyclic (Theorem 8). This result provides strong formal evidence about the power of asymmetric verification, since, as we discuss in Section 5, any reasonable social choice function f should not have a cycle in G_f that entirely consists of negative edges. Moreover, we prove that given any function f truthfully implementable by payments p, an asymmetric verification that truthfully implements f can be directly obtained by p (Proposition 1).

Comparison to Previous Work. The strict order-reserving property, which we employ as a sufficient condition for deterministic truthful implementation without money, is similar to the no-restoration property of [16]. However, the results of [16] are restricted to finite domains with ordinal preferences and to M^{swap} verification. Our results are far more general, since we manage, in Theorem 1, to extend the equivalence of truthful and M-truthful implementation, under the strict order-preserving property, to any (even infinite) domain and to any symmetric verification. Moreover, our necessary property generalizes and unifies the necessary conditions of both [7,16].

We also note that our results in case of deterministic implementation with money are not directly comparable to the strong and elegant results about local truthfulness with money in convex domains (see e.g., [2,1]). For instance, if we restrict Theorem 3 to convex domains and compare it to [2, Theorem 3.8], our result is significantly weaker, since it starts from a much stronger hypothesis (see also the discussion in Section 3.2). On the other hand, Theorem 3 is more general, in the sense that it applies to any symmetric strict difference-preserving verification and to arbitrary (even non-convex) domains.

2   Notation and Preliminaries

The basic model and most of the notation are introduced in Section 1. Next, we discuss some conventions, give some definitions, and state some useful facts.

Ordinal Preferences. We always assume that each type x is a function from O to $\mathbb{R}$. However, in case of deterministic mechanisms without money, when the preferences are ordinal, we only care about the relative order of the outcomes in each type.

Truthful Implementation. A social choice function $f : D \to O$ is *M-truthfully implementable* if for every type x and any $y \in M(x)$, $x(f(x)) \geq x(f(y))$. A social choice function f is *M-truthfully implementable with money* if there is a payment scheme $p : O \to \mathbb{R}$ such that for every type x and any $y \in M(x)$, $x(f(x)) + p(f(x)) \geq x(f(y)) + p(f(y))$. If there is no verification, i.e., if for all types x, $M(x) = D$, we say that f is *truthfully implementable* and *truthfully implementable with money*, respectively. We say that truthfulness (resp. with money) is equivalent to M-truthfulness (resp. with money) if for every function f, f is truthfully implementable (resp. with money) iff it is M-truthfully implementable (resp. with money). In what follows, we use the terms *mechanism* and *social choice function* interchangeably.

Randomized Mechanisms. A randomized mechanism $f : D \to \Delta(O)$ maps each type x to a probability distribution over O. A randomized mechanism is (resp. *M-*)*universally truthful* if it is a probability distribution over deterministic (resp. M-)truthful mechanisms (even with money). For truthfulness-in-expectation, we assume, for simplicity, that O is finite, and let $f_o(x)$ be the probability of the outcome o if the agent reports x. Then, a randomized mechanism f is (resp. *M-*)*truthful-in-expectation* if for every type x and any $y \in D$ (resp. $y \in M(x)$), $\sum_{o \in O} f_o(x)x(o) \geq \sum_{o \in O} f_o(y)x(o)$.

A randomized mechanism f is (resp. M-)*truthful-in-expectation with money* if there are payments $p : O \to \mathbb{R}$ such that for every $x \in D$ and any $y \in D$ (resp. $y \in M(x)$), $\sum_{o \in O} f_o(x)(x(o) + p(o)) \geq \sum_{o \in O} f_o(y)(x(o) + p(o))$.

Correspondence Graph. A verification M can be represented by the directed *correspondence graph* $G_M = (D, \{(x, y) : y \in M(x)\})$. Given a social choice function f, we let the edge-weighted graph

$$G_{M,f} = (D, \{(x, y) : y \in M(x)\}, w), \quad \text{where } w(x, y) = x(f(x)) - x(f(y))$$

A k-cycle (resp. k-path) in G_M is a directed cycle (resp. path) consisting of k edges. We say that an edge (x, y) of $G_{M,f}$ is negative if $w(x, y) < 0$. We say that a cycle in $G_{M,f}$ is negative if the total weight of its edges is negative. We let $G^-_{M,f}$ denote the subgraph of $G_{M,f}$ that consists of all its negative edges. If there is no verification, we refer to $G_{D,f}$, $G^-_{D,f}$ as G_f, G^-_f. Also, given a graph G, we let $V(G)$ be its vertex set and $E(G)$ be its edge set.

A social choice function f is M-truthfully implementable iff $G_{M,f}$ does not contain any negative edges. Furthermore, Rochet [14] proved that a social choice function f is M-truthfully implementable with money if and only if the correspondence graph $G_{M,f}$ does not have any finite negative cycles.

Symmetric and Asymmetric Verification. We say that a verification M is symmetric if G_M is *symmetric*, i.e., for each directed edge $(x, y) \in E(G_M)$, $(y, x) \in E(G_M)$. We say that a verification M is *asymmetric* if G_M is an acyclic tournament.

Weak Monotonicity and Cycle Monotonicity. A social choice function f satisfies M-*weak-monotonicity* if for every $x \in D$ and any $y \in M(x)$, $x(f(x)) + y(f(y)) \geq x(f(y)) + y(f(x))$. Equivalently, f is M-weakly-monotone iff $G_{M,f}$ does not contain any negative 2-cycles. A function f satisfies M-*cycle-monotonicity* if for all $k \geq 1$, and all $x_1, \ldots, x_k \in D$, such that $x_{i+1} \in M(x_i)$, $\sum_{i=1}^k x_i(f(x_i)) \geq \sum_{i=1}^k x_{i-1}(f(x_i))$, where the subscripts are modulo k. Equivalently, f is M-cyclic-monotone iff $G_{M,f}$ does not contain any finite negative cycles. If there is no verification, we simply say that f is weakly-monotone and cyclic-monotone, respectively.

Convex Domains. A domain D is *convex* if for every $x, y \in D$ and any $\lambda \in [0, 1]$, the function $z : O \to \mathbb{R}$, with $z(a) = \lambda x(a) + (1 - \lambda)y(a)$, for each $a \in O$, is also in D.

Strategic Voting. We have k candidates and select one of them based on the preferences of n agents. Hence, $O = \{o_1, \ldots, o_k\}$ is the set of candidates, $V = \{v_1, \ldots, v_n\}$ is the set of voters, and the type of each voter is a linear order over O.

k-**Facility Location.** In k-Facility Location, we place $k \geq 1$ facilities on the real line based on the preferences of n agents. The type of each agent i is determined by $x_i \in \mathbb{R}$, and the set of outcomes is $O = \mathbb{R}^k$. The utility of agent i from an outcome $(y_1, \ldots, y_k) \in O$ is $- \min_j |x_i - y_j|$. If $k = 1$, we simply refer to Facility Location.

M^ε **and** M^{swap} **Verification.** In case of a convex domain or Facility Location, given an $\varepsilon > 0$, we let $M^\varepsilon(x) = \{y \in D : ||x - y|| \leq \varepsilon\}$, for all x, where $|| \cdot ||$ is the l_2 distance in $\mathbb{R}^O$ for convex domains and $|x - y|$ for Facility Location. If we have a domain D where the agent's types are linear orders on O, for any type $x \in D$, $M^{\mathrm{swap}}(x)$ is the set of all linear orders on O obtained from x by swapping two adjacent outcomes in x.

3 Sufficient Conditions for Truthful Implementation

Without any assumptions on the domain, symmetric verification is not sufficient for the equivalence of truthfulness and M-truthfulness. Next, we assume that the correspondence graph G_M is symmetric and strongly connected by finite paths along which the preferences are consistent with the preferences at the endpoints. We prove that this suffices for the equivalence of truthfulness and M-truthfulness, even for infinite domains.

3.1 Deterministic Mechanisms

We start with a sufficient condition for a symmetric verification M (and its correspondence graph) under which any deterministic M-truthful mechanism is also truthful.

Definition 1 (Order-Preserving Path). *Given a verification M, an $x-y$ path p in G_M is order-preserving if for all outcomes $a, b \in O$, with $x(a) > x(b)$ and $y(a) \geq y(b)$, and for any intermediate type w in p, $w(a) > w(b)$. A $x - y$ path p in G_M is strict order-preserving if for every type w in p, the subpath of p from x to w is order-preserving.*

Intuitively, if the endpoints x and y of an order-preserving path p agree that outcome a is preferable to outcome b, any intermediate type w in p should also agree on this. Following Definition 1, we say that a verification M is *symmetric* (resp. *strict*) *order-preserving* if M is symmetric and for any types $x, y \in D$, there is a *finite* (resp. strict) order-preserving $x - y$ path in the correspondence graph G_M. Next, we show that:

Theorem 1. *Let M be a symmetric strict order-preserving verification. Then, truthfulness is equivalent to M-truthfulness.*

Proof. If a social function is truthfully implementable, it is also M-truthfully implementable. For the converse, we use induction on the length of the strict order-preserving paths in G_M. Technically, for sake of contradiction, we assume that there is a function f that is M-truthfully implementable, but not implementable. Therefore, all edges in $G_{M,f}$ are non-negative, but there is a negative edge $(x, z) \in E(G_f)$.

Since M is symmetric strict order-preserving, there is a finite strict order-preserving $x - z$ path p in $G_{M,f}$. In particular, we let $p = (x = v_0, v_1, v_2, \ldots, v_k = z)$, and let i, $2 \leq i \leq k$, be the smallest index such that the edge $(x, v_i) \in E(G_f)$ is negative. For convenience, we let $y = v_i$ and $w = v_{i-1}$. We note that by the definition of i, the edge $(x, w) \in E(G_f)$ is non-negative, and also since f is M-truthfully implementable, the edges $(w, y), (y, w) \in E(G_{M,f})$ are non-negative (see also Fig. 1.i).

For convenience, we let $a = f(x)$, $b = f(w)$, $c = f(y)$ denote the outcome of f at x, y, and w, respectively. Since the edge (x, y) is negative, $a \neq c$. Moreover, by the definition of i (and of y), $b \neq c$. By the discussion above, we have that $x(c) > x(a) \geq x(b)$ and $y(c) \geq y(b)$. Therefore, since the $x - z$ path is strict order-preserving, and thus its $x - y$ subpath is order-preserving, we obtain that $w(c) > w(b)$, a contradiction to the hypothesis that the edge $(w, y) \in E(G_{M,f})$ is non-negative. Therefore there is no negative edge in G_f, which implies that f is truthfully implementable.   □

If D is finite, we can show that for a symmetric verification, the strict order-preserving property is equivalent to the order-preserving property. Thus, we obtain that:

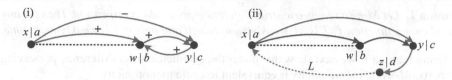

Fig. 1. (i) The part of G_f considered in the proof of Theorem 1. (ii) The part of G_f considered in the proof of Theorem 3. The label of each node consists of the type and the outcome of f.

Theorem 2. *Let M be a symmetric order-preserving verification in a finite domain D. Then, truthfulness is equivalent to M-truthfulness.*

Applications. Theorems 1 and 2 provide a generic and convenient way of checking whether truthful implementation can take any advantage of symmetric verification. E.g., one can verify that for any convex domain D, M^ε verification is strict order-preserving, and that for Strategic Voting, M^{swap} verification is order-preserving. Thus, we obtain alternative (and very simple) proofs of [6, Theorems 3.1 and 3.3]. Moreover, our corollary about M^{swap} verification implies the main result of [16]. Similarly, we can show that for the Facility Location domain, which is non-convex, M^ε verification is strict order-preserving. Thus, for Facility Location, a mechanism is truthful iff it is M^ε-truthful.

3.2 Deterministic Mechanisms with Money

Next, we extend the notion of order-preserving paths to mechanisms with money. Since utilities are not ordinal anymore, we use the notion of difference-preserving paths, which takes into account the difference between the utility of different outcomes. Formally, given a verification M, an $x - y$ path p in G_M is *difference-preserving* if for any intermediate type w in p and for all outcomes $a, b \in O$, if $x(a) - x(b) \neq y(a) - y(b)$,

- $w(a) - w(b) \in (\min\{x(a) - x(b), y(a) - y(b)\}, \max\{x(a) - x(b), y(a) - y(b)\})$
- $w(a) - w(b) = x(a) - x(b)$, if $x(a) - x(b) = y(a) - y(b)$.

As for order-preserving paths, if both endpoints x and y of a difference-preserving path p prefer a to b, any type w in p should also prefer a to b. Moreover, the strength of w's reference for a, i.e., $w(a) - w(b)$, should lie between the strength of x's and of y's preference for a. In fact, the difference-preserving property is a stronger version of the increasing difference property in [5, Definition 5]. Similarly, an $x - y$ path p in G_M is *strict difference-preserving* if for every type w in p, the subpath of p from x to w is also difference-preserving. A verification M is *symmetric* (resp. *strict*) *difference-preserving* if M is symmetric and for any $x, y \in D$, there is a *finite* (resp. strict) difference-preserving $x - y$ path in G_M.

We proceed to show that the symmetric strict difference-preserving property is sufficient for the equivalence of M-truthfulness with money and truthfulness with money. The proof is based on the equivalence of cycle monotonicity and truthful implementation with money. As a first step, we employ a proof similar to that of Theorem 1, and show that under the symmetric strict difference-preserving property, for any function f, $G_{M,f}$ does not have any negative 2-cycles iff G_f does not have any negative 2-cycles.

Lemma 1. *Let M be a symmetric strict difference-preserving verification. Then for any social choice function f, f is M-weakly monotone if and only if f is weakly-monotone.*

Using Lemma 1, we next show that under the symmetric strict difference-preserving property, M-cycle monotonicity is equivalent to cycle monotonicity.

Theorem 3. *Let M be a symmetric strict difference-preserving verification. Then for any social choice function f, f is M-truthfully implementable with money if and only if f is truthfully implementable with money.*

Proof. If f is truthfully implementable with money, it is also M-truthfully implementable with money. For the converse, we show that if $G_{M,f}$ does not have any negative cycles, then G_f does not have any negative cycles as well. In what follows, we assume that G_f does not have any negative 2-cycles, since otherwise, by Lemma 1, f is not M-weakly monotone, and thus, not truthfully implementable with money.

For sake of contradiction, we assume that G_f includes some negative cycle with more than 2 (and a finite number of) edges. In particular, we let $C = (x, y, z, \ldots, x)$ be any such cycle. The existence of such a cycle C is guaranteed by Rochet's theorem. Moreover, C contains at least one edge $(x, y) \in E(G_f) \setminus E(G_{M,f})$, because C is not present in $G_{M,f}$. Since M is a symmetric strict difference-preserving verification, there is a finite strict difference-preserving $x - y$ path $p = (v_0 = x, v_1, \ldots, v_k = y)$. For convenience, we let $w = v_{k-1}$ be the last node before y in p, let $a = f(x)$, $b = f(w)$, $c = f(y)$, and $d = f(z)$ be the outcome of f at x, w, y, and z, respectively, and let L be the total length of the $z - x$ path used by C (see also Fig. 1.ii).

Since the cycle C is negative, $x(a) - x(c) + y(c) - y(d) + L < 0$. Moreover, since G_f does not contain any negative 2-cycles, $x(c) - x(b) \le y(c) - y(b)$. Otherwise, since w belongs to a difference-preserving $x - y$ path, we would have that $y(c) - y(b) < w(c) - w(b)$, which implies that the 2-cycle (w, y, w) is negative. Hence, since w belongs to a difference-preserving $x - y$ path, $x(c) - x(b) \le w(c) - w(b)$. Therefore,

$$x(a) - x(b) + w(b) - w(c) + y(c) - y(d) + L \le x(a) - x(c) + y(c) - y(d) + L < 0$$

So, we have that the cycle $C_1 = (x, w = v_{k-1}, y, \ldots, z)$ is also negative.

Since p is strict difference-preserving, the path $p' = (x = v_0, v_1, \ldots, v_{k-1} = w)$ is also difference-preserving. Therefore, using the same argument, we can prove that the cycle $C_2 = (x, v_{k-2}, v_{k-1}, y, \ldots, z)$ is also negative. Repeating the same process $k - 1$ times, we obtain that the cycle $C_{k-1} = (x = v_0, v_1, \ldots, v_{k-1}, y, \ldots, z)$ is also negative. However, all the edges (v_i, v_{i+1}), $i = 0, \ldots, k - 1$, of the strict difference-preserving $x - y$ path p belong to G_M. Hence, the edge $(x, y) \in E(G_f) \setminus E(G_{M,f})$ in C is replaced by k edges of $E(G_{M,f})$ in C_{k-1}. Therefore, the negative cycle C_{k-1} has one edge not in $E(G_{M,f})$ less than the original negative cycle C. Repeating the same process for every edge of C not in $E(G_{M,f})$, we obtain a negative cycle C' with all edges in $E(G_{M,f})$. This is a contradiction, since it implies that f is not M-truthfully implementable with money. $\square$

Since M^ε verification is symmetric and strict difference-preserving for any convex domain, Theorem 3 implies that for convex domains, M^ε-truthful implementation with money is equivalent to truthful implementation with money. This result is also a corollary of [2, Theorem 3.8], but here we obtain it through a completely different approach. In particular, Archer and Kleinberg [2] proved that if there is no "local" negative cycle C in G_f, where "local" means that C can fit in a small area of the convex domain D, then G_f does not contain any negative cycles, and thus, f is truthfully implementable with money. On the other hand, we prove here that if G_f does not contain any negative cycles consisting of "local" edges, then G_f does not contain any negative cycles. So, in our case, the hypothesis is much stronger, since it excludes the existence of negative cycles that consist of "local" edges, but may cover an arbitrarily large area of the convex domain D. In this sense, if we restrict Theorem 3 to convex domains, our result is different in nature and weaker than [2, Theorem 3.8]. Nevertheless, Theorem 3 is quite more general, in the sense that it applies to any symmetric strict difference-preserving verification and to arbitrary (even non-convex) domains.

3.3 Randomized Truthful-in-Expectation Mechanisms

A general condition is sufficient and/or necessary for the equivalence between universal truthfulness and M-universal truthfulness in randomized mechanisms, iff it is sufficient and/or necessary for the equivalence between truthfulness and M-truthfulness in deterministic mechanisms. Hence, all the results of Sections 3.1, 3.2, and 4 directly apply to randomized universally-truthful mechanisms (also with money).

A similar, but more interesting, correspondence holds for the case of randomized truthful-in-expectation mechanisms. For simplicity, we assume here that the set of outcomes $O = \{o_1, \dots, o_m\}$ is finite. With each type $x : O \mapsto \mathbb{R}$, we associate a new type $X : \Delta(O) \mapsto \mathbb{R}$, such that for each probability distribution q over outcomes, the utility $X(q)$ is the expected utility of x wrt. q. Formally, $X(q) = \sum_{i=1}^{m} q_i x(o_i)$. We let D' be the set of these new types. By definition, there is an one-to-one correspondence between types in D and types in D'. Hence, a social choice function $f : D \to \Delta(O)$ corresponds to a (deterministic) social choice function $f' : D' \to \Delta(O)$. Moreover, (resp. given a verification M) f is (resp. M-)truthful-in-expectation iff f' is (resp. M-)truthful.

As before, we seek a general condition under which truthfulness-in-expectation is equivalent to M-truthfulness-in-expectation. For each type $X \in D'$, corresponding to type $x \in D$, we define $M'(X) = \{Y \in D' : y \in M(x)\}$. Now, the results of Sections 3.1, 3.2, and 4 directly apply to the new domain D' with verification M'. We note that if M is symmetric, then M' is symmetric as well. Hence, for a result that directly applies to the original verification M and domain D, we need a property of the paths in G_M that guarantees that the corresponding paths in $G_{M'}$ are order-preserving.

An $x-y$ path p in G_M is *difference-convex* if for any type w in p, there is a $\lambda \in (0,1)$, such that for all $a, b \in O$, $w(a) - w(b) = \lambda(x(a) - x(b)) + (1 - \lambda)(y(a) - y(b))$. Similarly, an $x - y$ path p in G_M is *strict difference-convex* if for every type w in p, the subpath of p from x to w is also difference-convex. A verification M is called *symmetric* (resp. *strict*) *difference-convex* if M is symmetric and for any $x, y \in D$, there is a *finite* (resp. strict) difference-convex $x - y$ path in G_M. For truthfulness-in-expectation, we quantify the utility of each type x for each outcome. Hence, the difference-convex property is a stronger version of the difference-preserving property, which in turn, is a stronger version of the order-preserving property.

Lemma 2. *If an $x - y$ path p in G_M is (resp. strict) difference-convex, then the corresponding $X - Y$ path p' in $G_{M'}$ is (resp. strict) difference-preserving, and thus, (resp. strict) order-preserving.*

Although the difference-convex property seems quite strong, a slight deviation from it results in paths in $G_{M'}$ that are not difference-preserving. In this sense, the difference-convex property and Lemma 2 are tight.

By the discussion above, Lemma 2, Theorem 1, and Theorem 3 imply that:

Theorem 4. *Let M be a symmetric strict difference-convex verification. Then, truthfulness-in-expectation (resp. with money) is equivalent to M-truthfulness-in-expectation (resp. with money).*

4   Necessary Conditions for Truthful Implementation

Next, we study relaxed versions of the sufficient conditions in Section 3, and show that they are necessary conditions for the equivalence of truthfulness and M-truthfulness.

Deterministic Mechanisms. Given an outcome $a \in O$, we say that an $x - y$ path p in G_M is *a-preserving* if for all outcomes $b \in O$, with $x(a) > x(b)$ and $y(a) \geq y(b)$, and for any intermediate type w in p, $w(a) > w(b)$. Namely, if the endpoints x and y of p agree that a is preferable to b, any intermediate type w in p should also prefer a to b. A verification M is called *symmetric outcome-preserving* if M is symmetric and for all types $x, y \in D$ and all outcomes $a \in O$, there is a *finite* a-preserving $x - y$ path p in G_M. Though quite close to each other, the order-preserving property implies the outcome-preserving property, but not vice versa. Specifically, an a-preserving path p may not be order-preserving, because the relative preference order of some outcomes, other than a, may change in the intermediate nodes of p.

Theorem 5. *Let M be a symmetric verification that is not outcome-preserving. Then, there exists a function g which is M-truthfully implementable, but not implementable.*

Proof. Since M is not outcome-preserving, there exists a pair of types $x, y \in D$ and an outcome $a \in O$, such that any finite $x - y$ path in G_M violates the a-preserving property. Thus, all $x - y$ paths in G_M consist of at least 2 edges (a single edge is trivially order-preserving). Then, we construct a certificate that M is not outcome-preserving, which is a separator of x and y in G_M, and based on this, we define a function g that is M-truthfully implementable, but not truthfully implementable.

For every finite $x - y$ path p in G_M, we let t_p denote the first intermediate type in p and o_p denote an outcome, such that $x(a) > x(o_p) \wedge y(a) \geq y(o_p) \wedge t_p(o_p) \geq t_p(a)$. Namely, for every finite $x - y$ path p, t_p and o_p provide a certificate that p violates the a-preserving property. We let $O_{xy} = \{o_p \in O : p$ is a finite $x - y$ path$\}$ be the set of outcomes in these certificates, and let $C_{xy} = \{z \in D \setminus \{y\} : \exists b \in O_{xy}$ with $z(b) \geq z(a)\}$ be a set of types that can be used as certificates along with the outcomes in O_{xy}. For convenience, we simply use C instead of C_{xy}. The crucial observation is that for every finite $x - y$ path p in G_M, $t_p \in C$, and thus, C is a separator of x and y in G_M.

Let A be the set of types in the connected component[1] that contains x, obtained from G_M after we remove C, and let $B = D \setminus (A \cup C)$. Since $y \notin C$, by definition, and for every finite $x - y$ path p, $t_p \in C$, y is in B. We consider the following function:

$$g(z) = \begin{cases} \arg\max_{b \in O_{xy}} \{z(b)\} & z \in A \cup C \\ a & z \in B \end{cases}$$

By the definition of C, every type in $A \cup B$ prefers a to any outcome in O_{xy}. However, by the definition of A and B, no type $z \in A$ has a neighbor in B, since otherwise, we could find a finite path from x to G_M^y. Therefore, for any $z \in A$, all z's neighbors G_M are in $A \cup C$, and thus $g(z)$ is z's best outcome in its G_M neighborhood. Similarly, every type $z \in C$ prefers any type in O_{xy} to a, and every type $z \in B$ prefers a to any outcome in O_{xy}, by the definition of C. Hence, g is M-truthfully implementable. On the other hand, g is not truthfully implementable, because x prefers a to any outcome in O_{xy}, and thus has an incentive to misreport y, if we do not have any verification. $\square$

Theorem 5 provides a convenient way of checking whether truthful implementation cannot take any advantage of symmetric verification. E.g., we can show that for the domain of 2-Facility Location, M^ε verification is not outcome-preserving, and thus, there are such social choice functions that become truthful with M^ε verification.

Deterministic Mechanisms with Money. We obtain here a necessary condition for the equivalence of weak and M-weak monotonicity. Given a verification M and $a, b \in O$, an $x - y$ path p in G_M, with $x(a) - x(b) \neq y(a) - y(b)$, is *difference (a, b)-preserving* if for any type w in p, $w(a) - w(b) \in (\min\{x(a) - x(b), y(a) - y(b)\}, \max\{x(a) - x(b), y(a) - y(b)\})$. A verification M is *symmetric difference outcome-preserving* if M is symmetric and for any types $x, y \in D$ and all outcomes $a, b \in O$, there is a *finite difference (a, b)-preserving* $x - y$ path p in G_M. As before, the difference-preserving property implies the difference outcome-preserving property, but not vice versa. By a proof similar to that of Theorem 5, we can show that:

Theorem 6. *Let M be a symmetric verification which is not difference outcome-preserving. Then, there is a social choice function g which is M-weakly monotone, but not weakly monotone.*

Sufficient and Necessary Condition. Closing the small gap between the order-preserving and outcome-preserving properties, we present a condition that is both sufficient and necessary for the equivalence of truthful and M-truthful implementation. Given a social choice function f, a $x - y$ path $p = (x = v_0, v_1, \ldots, v_k, v_{k+1} = y)$ in G_M is *f-preserving* if for any type v_i, $1 \leq i \leq k + 1$ in p, and for all outcomes $a \in O$, with $x(f(v_i)) > x(a)$ and $v_i(f(v_i)) \geq v_i(a)$, $v_{i-1}(f(v_i)) > v_{i-1}(a)$. A verification M is *symmetric function-preserving* if M is symmetric and for any M-truthfully implementable function f and all types $x, y \in D$, there is a *finite f-preserving* $x - y$ path in G_M. Using the techniques in the proofs of Theorems 1 and 5, we can show that:

Theorem 7. *Let M be a symmetric verification. Then, truthful implementation is equivalent to M-truthful implementation if and only if M is function-preserving.*

---

[1] If D is finite, we use the standard graph-theoretic definition of connected components. If D is infinite, A includes x and all types $w \in D$ reachable from x through a finite path.

5 On the Power of Asymmetric Verification

Intuitively, one should expect that asymmetric verification is powerful due to requirement that the correspondence graph should be acyclic. In fact, if we consider any asymmetric verification M, since G_M does not have any negative cycles, Rochet's theorem implies that any social choice function f is M-truthfully implementable with money. We next show a natural characterization of the social choice functions that can be M-truthfully implemented (without money), for some asymmetric verification M.

Theorem 8. *Let f be any social choice function. There is an asymmetric verification M such that f is M-truthfully implementable iff G_f^- is a directed acyclic graph.*

Proof. Let M be an asymmetric verification that truthfully implements f. Hence, G_M is an acyclic tournament and $G_{M,f}$ does not contain any any edges of G_f^-. Therefore, if we arrange the vertices of G_f on the line according to the (unique) topological ordering of $G_{M,f}$, all edges of G_f not included in $G_{M,f}$ are directed from right to left. Therefore, the edges of G_f^- cannot form a cycle. For the converse, let f be a social choice function with an acyclic G_f^-. We consider a topological ordering of G_f^- and remove any edge of G_f directed from left to right. This removes all edges of G_f^- and leaves an acyclic subgraph G_f', since all its edges are directed from right to left. Moreover, for every pair of types x, y, we remove one of the edges (x, y) and (y, x). Hence, G_f' is an acyclic tournament without any negative edges. Therefore, f is M-truthfully implementable for the asymmetric verification M corresponding to G_f'. $\qquad\square$

Reasonable social choice functions should have an acyclic G_f^-. This is true for all functions maximizing the social welfare and all functions truthfully implementable with money. Although one may construct examples of functions f where G_f^- contains cycles, such functions (and such cycles) are hardly natural. For instance, a 2-cycle (x, y, x) in G_f^- indicates that type x prefers outcome $f(y)$ to $f(x)$, while type y prefers outcome $f(x)$ to $f(y)$. But then, one may change f to f', with $f'(x) = f(y)$, $f'(y) = f(x)$, and $f'(z) = f(z)$ for any other type z. Thus, one eliminates the cycle (x, y, x) and the social welfare is strictly greater using f' allocation.

We can extend the construction in the proof of Theorem 8 to a *universal asymmetric verification*, which can truthfully implement any social choice function with acyclic G_f^-. Applying this, we can show that in the Facility Location domain, the function $F_{\max}(x) = (\min x + \max x)/2$, that minimizes the maximum distance of the agents to the facility, can be truthfully implemented with verification $M_{\max}(x_i) = \{y : |y - F_{\max}(x_{-i})| \le |x_i - F_{\max}(x_{-i})|\}$. Similarly, we can show that in the domain of Strategic Voting, Plurality can be truthfully implemented by an asymmetric verification where the voters are not allowed to misreport a higher preference for the winner of the election. Moreover, we can show that Borda Count can be truthfully implemented by an asymmetric verification where the voters are not allowed to misreport either a higher preference for the winner of the election or a lower preference for some of the remaining candidates.

Asymmetric Verification and Payments. The absence of negative cycles in G_f implies the absence of cycles in G_f^-. Thus, Theorem 8, combined with Rochet's theorem, shows that for any function f truthfully implementable with money, there is an asymmetric verification M that truthfully implements f. Extending the proof of Theorem 8, can can show that such an asymmetric verification M can be directly obtained from any payment scheme that implements f.

Proposition 1. *Let f be a social choice function truthfully implementable by payments $p : D \mapsto \mathbb{R}$. Then, removing all edges $(x, y) \in E(G_f)$ with $p(f(x)) > p(f(y))$ results in an asymmetric verification M that truthfully implements f (without money).*

References

1. Archer, A., Kleinberg, R.: Characterizing truthful mechanisms with convex type spaces. ACM SIGecom Exchanges 7(3) (2008)
2. Archer, A., Kleinberg, R.: Truthful germs are contagious: A local-to-global characterization of truthfulness. In: Proc. of the 9th ACM Conference on Electronic Commerce (EC 2008), pp. 21–30 (2008)
3. Auletta, V., Penna, P., Persiano, G., Ventre, C.: Alternatives to truthfulness are hard to recognize. Autonomous Agents and Multi-Agent Systems 22(1), 200–216 (2011)
4. Auletta, V., De Prisco, R., Penna, P., Persiano, G.: The power of verification for one-parameter agents. Journal of Computer and System Sciences 75, 190–211 (2009)
5. Berger, A., Müller, R., Naeemi, S.H.: Characterizing incentive compatibility for convex valuations. In: Mavronicolas, M., Papadopoulou, V.G. (eds.) SAGT 2009. LNCS, vol. 5814, pp. 24–35. Springer, Heidelberg (2009); Updated version as Research Memoranda 035, Maastricht Research School of Economics of Technology and Organization
6. Caragiannis, I., Elkind, E., Szegedy, M., Yu, L.: Mechanism design: from partial to probabilistic verification. In: Proc. of the 13th ACM Conference on Electronic Commerce (EC 2012), pp. 266–283 (2012)
7. Carroll, G.: When are local incentive constraints sufficient? Econometrica 80(2), 661–686 (2012)
8. Fotakis, D., Tzamos, C.: Winner-imposing strategyproof mechanisms for multiple Facility Location games. Theoretical Computer Science 472, 90–103 (2013)
9. Green, J., Laffont, J.: Partially verifiable information and mechanism design. Review of Economic Studies 53(3), 447–456 (1986)
10. Gui, H., Müller, R., Vohra, R.V.: Dominant strategy mechanisms with multi-dimensional types. Discussion Paper 1392, Northwestern University (2004)
11. Krysta, P., Ventre, C.: Combinatorial auctions with verification are tractable. In: de Berg, M., Meyer, U. (eds.) ESA 2010, Part II. LNCS, vol. 6347, pp. 39–50. Springer, Heidelberg (2010)
12. Nisan, N.: Introduction to Mechanism Design (for Computer Scientists). In: Algorithmic Game Theory, ch. 9, pp. 209–241 (2007)
13. Nisan, N., Ronen, A.: Algorithmic mechanism design. Games and Economic Behavior 35, 166–196 (2001)
14. Rochet, J.C.: A necessary and sufficient condition for rationalizability in a quasi-linear context. Journal of Mathematical Economics 16(2), 191–200 (1987)
15. Saks, M.E., Yu, L.: Weak monotonicity suffices for truthfulness on convex domains. In: Proc. of the 6th ACM Conference on Electronic Commerce (EC 2005), pp. 286–293 (2005)
16. Sato, S.: A sufficient condition for the equivalence of strategy-proofness and nonmanipulability by preferences adjacent to the sincere one. J. Economic Theory 148, 259–278 (2013)
17. Vohra, R.V.: Mechanism Design: A Linear Programming Approach. Cambridge University Press (2011)

A Protocol for Cutting Matroids Like Cakes*

Laurent Gourvès[1,2], Jérôme Monnot[1,2], and Lydia Tlilane[2,1]

[1] CNRS, LAMSADE UMR 7243
[2] PSL*, Université Paris-Dauphine, France
`firstname.lastname@dauphine.fr`

Abstract. We study a problem that generalizes the fair allocation of indivisible goods. The input is a matroid and a set of agents. Each agent has his own utility for every element of the matroid. Our goal is to build a base of the matroid and provide worst case guarantees on the additive utilities of the agents. These utilities are private, an assumption that is commonly made for the fair division of *divisible* resources, Since the use of an algorithm is not appropriate in this context, we resort to protocols, like in cake cutting problems. Our contribution is a protocol where the agents can interact and build a base of the matroid. If there are up to 8 agents, we show how everyone can ensure that his worst case utility for the resulting base is the same as those given by Markakis and Psomas [18] for the fair allocation of indivisible goods, based on the guarantees of Demko and Hill [8].

1 Introduction

We study a problem defined on a matroid $\mathcal{M}$ and a set N of n agents. The agents have non-negative and additive utilities for the subsets of elements of $\mathcal{M}$. The aim is to find a single base B of $\mathcal{M}$ and provide some guarantees on the agents' utility for B. Our problem is a generalization of the allocation of indivisible goods. Dealing with matroids – a classical structure in combinatorial optimization – allows to cover applications whose feasibility constraints are more complex. Let us give a concrete example.

Example 1. A department of computer science is composed of n teams which attend a common seminar. The seminar consists of m fixed dates and one has to select m speakers out of a pool of candidates. The candidates have preliminarily given their availabilities for the m dates. Each team has its own interest for the candidates but the order by which the talks are given does not matter. A solution is a subset of speakers, under the constraint that a feasible assignment (one available speaker per date) exists. The head of the department, who is in charge of the program of the seminar, needs to find a feasible solution. To be fair, he also has to take the interest of all teams into account.

---

* This work is supported by French National Agency (ANR), project COCA ANR-09-JCJC-0066-01.

Y. Chen and N. Immorlica (Eds.): WINE 2013, LNCS 8289, pp. 216–229, 2013.

Matroids are defined in Section 3 and the matroid structure of Example 1 is clarified in Section 4. Beforehand, let us situate our work. The allocation of scarce resources (e.g. water, bandwidth, grants) is a recurrent problem. A challenge is to propose methods which lead to fair and efficient solutions. The distinction between divisible and indivisible goods is typically made. Another important information is about agents' utilities: are they publicly known or private?

Though a large body of literature is devoted to the case of divisible goods [23,6,7,19,25,20], the computer science community is paying more attention to the allocation of indivisible resources [16,3,9,2,14,1,18,5,13], especially when there is no monetary compensation. In case of public utilities, an algorithm can determine the allocation. In case of private utilities, the agents may take part of the determination of the final solution via a *protocol*. Our work is related to the indivisible goods. Let us review its connections with previous results.

In [8], Demko and Hill consider the problem of allocating indivisible goods with additive utilities. They show the existence of an allocation with an explicit guarantee on the utility of the poorest agent. This guarantee is $V_n(\alpha)$ where n is the number of agents and α is defined as the largest utility for a single good, over the whole set of agents in a normalized instance. V_n is a nonincreasing function of α. Markakis and Psomas revisit the work of Demko and Hill with a constructive approach [18]. They propose a polynomial time algorithm called ALLOCATE. It outputs an allocation such that agent i's utility is at least $V_n(\alpha_i)$ where α_i is agent i's maximum utility for a single good. Since V_n is nonincreasing and $\alpha \geq \alpha_i$, it follows that $V_n(\alpha_i) \geq V_n(\alpha)$.

The allocation of indivisible goods can be extended to the determination of a single base of a matroid. This is done in [10] where a new guarantee $W_n(\alpha_i)$ is provided for every agent i such that α_i is agent i's maximum utility for an element of the matroid in a normalized instance. W_n is a non-monotonic function satisfying $W_n(x) \geq V_n(x)$ for all x and n. This guarantee is obtained via a polynomial time algorithm called THRESHOLD which is an extension of ALLOCATE.

Both ALLOCATE and THRESHOLD work with publicly known utilities. In this article, we study *private* utilities in the generalized context of matroids. So, we borrow ideas from cake cutting problems. We propose a deterministic protocol for finding a base of a matroid such that every agent's utility is at least $V_n(\alpha_i)$. In this context, α_i is agent i's largest utility for a single element of the matroid. Up to our knowledge, there is no previous work on protocols for matroids. Moreover, protocols for indivisible goods under private utilities are not numerous (see chapter 2 of [7] for Knaster's *procedure of sealed bids* and Lucas' *method of markers*). The result that is closest to our work is a randomized protocol for two agents inspired by the famous *Divide-and-Choose* procedure [3].

The paper is organized as follows. Related works are presented in Section 2. A formal presentation of our model is given in Section 3. Section 4 contains some general properties on matroids. Our first contribution consists of two protocols for 2 and 3 agents presented in Sections 5 and 6, respectively. They introduce a general protocol given in Section 7 which works for up to 8 agents. We conclude this article with a discussion. Due to space limitations, some proofs are omitted.

2   Related Work

The problem of fairly allocating a given set of goods to a given set $N = \{1, \ldots, n\}$ of agents has received a lot of attention in economic theory (social science in particular) and, more recently, in theoretical computer science. The problem admits numerous variants and one can list some of them by answering to the following questions. Are the goods divisible or indivisible? Are the agents' utilities (for portions or subsets of the goods) public or private? Which notion of fairness is cast? The last question is itself a vast topic of research [7,19,25].

The representation and the manipulation of agents' utility functions for every possible portion of the goods can be a barrier. To avoid this, the community has mainly concentrated on additive utilities.

Let us introduce some notations for the rest of this section: S denotes the entire set of goods, S_i refers to the share of an agent $i \in N$, $(S_i)_{i \in N}$ is an allocation (profile of disjoint shares) and $u_i(S_i)$ is agent i's (non-negative) utility for S_i. We also assume that $u_i(S)$ is normalized to 1, and $u_i(S_i') \leq u_i(S_i)$ whenever S_i contains S_i'.

The allocation of divisible resources, under private utilities of the agents, is commonly known as the *cake cutting* problem. *Divide-and-Choose* (a.k.a *Cut-and-Choose*), a long known protocol, achieves *envy-freeness* for the case of two agents. Envy-freeness is reached once $u_i(S_i) \geq u_i(S_j)$ for every pair of agents i, j. *Proportionality* is less demanding than envy-freeness since it requires that $u_i(S_i) \geq u_i(S)/n = 1/n$. Historically, proportional protocols for $n \geq 3$ agents exist since the 1940's [23]. However, the first envy-free protocol for any number of agents dates back to 1995 [6].

The interest for the problem when S consists of $m \geq n$ indivisible items, and no monetary compensation is possible, is more recent. Let us first briefly review the case of public and *additive* utilities. By additive it is meant that $u_i(S_i) = \sum_{j \in S_i} u_i(j)$ where $u_i(j)$ is a convenient abbreviation of $u_i(\{j\})$. Thus, an agent's utility for any bundle of items can be derived from his utilities for the single items. In this context, one can mention the contribution of Lipton *et al.*, [16]. Since no envy-free allocation is guaranteed to exist in the indivisible setting (proportional allocations are also not guaranteed), they seek for *minimum-envy* allocations. Unfortunately, for any constant c, there can be no 2^{m^c}-approximation algorithm unless P=NP. However, the problem is tractable for the minimization of the *envy-ratio* $\max_{i,i' \in N \times N}\{1, u_i(S_i)/u_i(S_{i'})\}$.

An important body of research deals with the design of polynomial approximation algorithms for the maximization of the poorest agent's utility. This optimization challenge is known as the *Max-Min allocation* or the *Santa Claus* problem [3,9,2,14,1,24].

Bezáková and Dani give a $1/(m - n + 1)$-approximation algorithm and show that no ρ-approximation algorithm with $\rho > 0.5$ is likely to exist [3]. Golovin [9] provides an algorithm whose solution guarantees a utility of OPT/k to at least $\lceil (1 - 1/k)n \rceil$ agents, for any given $k \in \mathbb{Z}_+$, where OPT is the optimal value of the problem. A $\Omega(1/\sqrt{n})$-approximation is also given for a subcase called "Big goods/Small goods": small goods have utility in $\{0, 1\}$ and big goods have utility in $\{0, x\}$ for some $x > 1$. Bansal and Sviridenko [2] study a restricted case in which $u_i(\{j\}) \in \{0, p_j\}$ and provide an algorithm with an approximation factor of $\Omega(\log \log \log n / \log \log n)$. Khot and Ponnuswami [14] and Asadpour and Saberi [1] provide approximation algorithms for the general *Santa Claus* problem with ratios $(2n - 1)^{-1}$ and $\Omega\left(1/\left(\sqrt{n} \log^3 n\right)\right)$, respectively. If the agents have homogeneous utilities for the items then Woeginger's polynomial-time approximation scheme can be used [24].

Though focused on the utility of the poorest agent, the approach of Demko and Hill [8] differs from the Santa Claus problem. The goal is to give an absolute value $t_n \in [0, 1]$ such that an allocation $(S_i)_{i \in N}$ satisfying $\forall i \in N$, $u_i(S_i) \geq t_n$ exists in any case. An immediate answer would be $t_n = 0$ if one thinks of the case of two agents both having a utility of 1 for an item i^* and a null utility for any other item $i \neq i^*$: there must be one of these agents who does not receive i^* and thus $t_n = 0$. Besides n, Demko and Hill's approach comprises a parameter α in its input, that is the maximum utility an agent can have for a single item (the instance is normalized, so $u_i(S) = 1$ for every agent i). We have $\alpha = \max_{i,j \in N \times S} u_i(j)$ in general and $\alpha = 1$ in the previous example with two single-minded agents. Demko and Hill give a nonincreasing function $V_n : [0, 1] \to [0, 1/n]$ (see Definition 1) and prove that $t_n \geq V_n(\alpha)$.

Definition 1. *[12,18] Given any integer $n \geq 2$, let $V_n : [0, 1] \to [0, n^{-1}]$ be the unique nonincreasing function satisfying $V_n(x) = 1/n$ for $x = 0$, whereas for*

$$x \in \left(0, \tfrac{1}{n-1}\right]; \; V_n(x) = \begin{cases} 1 - p(n - 1)x, & x \in I(n, p) \\ \dfrac{p}{(p+1)n-1}, & x \in NI(n, p) \end{cases} \; \textit{for some integer } p \geq 1,$$

where $I(n, p) = \left[\frac{p+1}{p((p+1)n-1)}, \frac{1}{pn-1}\right]$, $NI(n, p) = \left(\frac{1}{(p+1)n-1}, \frac{p+1}{p((p+1)n-1)}\right)$ *and* $V_n(x) = 0$ *for* $x \in \left(\frac{1}{n-1}, 1\right]$. *We add to the definition* $V_1(x) = 1$ *for all* $x \in [0, 1]$.

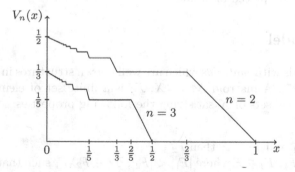

Fig. 1. V_n for $n = 2$ and $n = 3$

It is noteworthy that, within the class of nonincreasing functions, V_n is shown to be the best lower bound for t_n [8]. Thus, t_n is equal to V_n in $I(n, p)$ for every integer $p \geq 1$, i.e. within every interval where V_n is decreasing.

Since only n and α are retained, Demko and Hill's approach is partially oblivious of the instance and t_n may be much lower than the optimal value of the Santa Claus problem. This explains the apparent paradox of having a 0.5-inapproximability result for the Santa Claus problem and the possibility to build, in polynomial time, an allocation $(S_i)_{i \in N}$ satisfying $u_i(S_i) \geq V_n(\alpha)$ for all $i \in N$. Indeed, Markakis and Psomas [18] have recently proposed a polynomial algorithm, called ALLOCATE, which returns a solution $(S_i)_{i \in N}$ satisfying $\forall i \in N$, $u_i(S_i) \geq V_n(\alpha_i)$. Here, α_i is agent i's maximum utility for a single item (in a normalized instance) and so $\alpha = \max_{i \in N} \alpha_i$. Since V_n is nonincreasing, $V_n(\alpha_i) \geq V_n(\alpha)$ for all $i \in N$. In other words, ALLOCATE guarantees $V_n(\alpha)$ for everyone and possibly more if not the poorest.

Very recently, a non-monotone function W_n satisfying $W_n(x) \geq V_n(x)$ for all $x \in [0,1]$ and $\min_{i \in N} W_n(\alpha_i) \leq t_n$ has been proposed [10]. The new function W_n applies on a matroid problem which generalizes the problem of allocating indivisible goods. A guarantee of $W_n(\alpha_i)$ is obtained via a deterministic algorithm called THRESHOLD which is an extension of ALLOCATE.

ALLOCATE and THRESHOLD deal with public utilities because their input comprises the agents' utilities. The problem of allocating m indivisible items under *private* and *additive* utilities has been investigated by Bezáková and Dani for two agents [3]. They revisit *Divide-and-Choose* and propose a randomized version in which the utility of the poorest agent is at least $1/2$ in expectation. In fact, Bezáková and Dani resort to randomization because, as previously mentioned, the worst case utility of an agent is 0 if α is put aside ($V_n(x) = 0$ when $x \geq \frac{1}{n-1}$). The problem of extending the protocol to a larger number of agents is posed by Bezáková and Dani as a future work. Up to our knowledge, existing protocols are not directly comparable to our work. Knaster's *procedure of sealed bids* makes monetary compensations. Lucas' *method of markers* relies on a strong linearity assumption saying that every player can equally divide the items in contiguous bundles if they are placed on a line.

3   The Model

This article deals with *matroids* which are well known structures in combinatorial optimization [15]. A matroid $\mathcal{M} = (X, \mathcal{F})$ is a finite set of elements X and a collection $\mathcal{F}$ of subsets of X satisfying the following properties:

(i) $\emptyset \in \mathcal{F}$,
(ii) if $F_2 \subseteq F_1$ and $F_1 \in \mathcal{F}$ then $F_2 \in \mathcal{F}$,
(iii) for every $F_1, F_2 \in \mathcal{F}$ where $|F_1| < |F_2|$, $\exists e \in F_2 \backslash F_1$ such that $F_1 \cup \{e\} \in \mathcal{F}$.

Every element of $\mathcal{F}$ that is inclusion-wise maximal is called a *base*. Moreover, without loss of generality, we assume that $\forall e \in X$, $\{e\} \in \mathcal{F}$. More details on matroids are given in Section 4.

The input of our problem is a matroid $\mathcal{M} = (X, \mathcal{F})$ and a set of agents $N = \{1, ..., n\}$. The output is a single base B of $\mathcal{F}$ that is shared by the agents. Each agent $i \in N$ has a non-negative utility $u_i(e)$ for every element $e \in X$.

The utility functions of the agents are additive, i.e. $u_i(X') = \sum_{e \in X'} u_i(e)$ for all $X' \subseteq X$ and $i \in N$. These utility functions are also private, so the resulting base is not constructed directly. Instead, B is built by the agents via a protocol.

The resulting base B can be decomposed in n disjoint subsets $B_1, \ldots, B_n$ such that for all $i \in N$, B_i is the *contribution* of agent i. A worst case analysis being conducted, we can suppose that the utility of any agent for B is reduced to his utility for his contribution. Indeed $u_i(B) = u_i(B_i) + u_i(B \setminus B_i) \geq u_i(B_i)$.

Besides its correctness, the protocol should be *fair*, i.e. offering a guarantee on the utility of every agent. Following the approach of Demko and Hill [8] and Markakis and Psomas [18], the guarantee of agent i depends on α_i, which is defined as agent i's maximum utility for a single element.

Let $OPT_i(\mathcal{M})$ be the value of a base that maximizes u_i for $i \in N$. We assume, without loss of generality, that after a possible rescaling, the instance is normalized so that $OPT_i(\mathcal{M}) = 1$ for all $i \in N$. Thus, α_i is defined as $\alpha_i = \max_{e \in X} \frac{u_i(e)}{OPT_i(\mathcal{M})} = \max_{e \in X} u_i(e)$ for every agent $i \in N$.

Our contribution is a deterministic protocol for up to 8 agents. We prove that the agents can enjoy the guarantee given in [18], that is $V_n(\alpha_i), \forall i \in N$. We also elaborate a strategy that agents can adopt if they want to meet the aforementioned guarantees. These strategies are based on polynomial time algorithms.

4   Matroid Properties

In a matroid $\mathcal{M} = (X, \mathcal{F})$, the elements of $\mathcal{F}$ and $2^X \setminus \mathcal{F}$ are called *independent* sets and *dependent* sets, respectively. The *bases* of a matroid are its inclusion-wise maximal independent sets. All bases of a matroid $\mathcal{M}$ have the same cardinality $r(\mathcal{M})$, defined as the *rank* of $\mathcal{M}$. The set of bases of $\mathcal{M}$ is denoted by $\mathcal{B}$.

Matroid theory has significantly contributed to the understanding of some important combinatorial structures. The forests of a multigraph is a typical example of a matroid, called the *graphic matroid*. The bases are the spanning trees if the graph is connected. Another example is the *partition matroid* where the set of elements X is partitioned into k disjoint sets $X_1, \ldots, X_k$ for some integer $k \geq 1$. Given non-negative integers b_i ($i = 1, \ldots, k$), the sets $F \subseteq X$ satisfying $|F \cap X_i| \leq b_i$ form a matroid. Notably, allocating a set of m indivisible items to n agents can be seen as a partition matroid. Build m sets $X_i = \{i^1, i^2, \ldots, i^n\}$ and let $b_i = 1$ for $i \in [m]$. Taking i^k means allocating item i to agent k. We also cite the *uniform matroid* which is formed by all subsets of length at most k elements for some integer $k \geq 1$. The uniform matroid can model some multi-winner election problems [21,17,22]. Last example is the *transversal matroid* defined by m (not necessarily disjoint) sets $X_1, \ldots, X_m$, subsets of a ground set X and $\mathcal{F} = \{T \in 2^X : T$ is a partial transversal of $X\}$ where a *partial transversal* is a set $T \subseteq X$ such that an injective map $\Phi : T \to [m]$ satisfying $t \in X_{\Phi(t)}$ exists. Example 1 given in Introduction can be modeled as a transversal matroid such that X is the set of candidates and X_i is the set of available candidates during date i for all $i = 1, \ldots, m$.

Since we work on a general matroid, our results apply to all these problems.

222   L. Gourvès, J. Monnot, and L. Tlilane

In the presence of a non-negative weight function $w : X \to \mathbb{R}^+$, we use the convenient shorthand notation $w(X') = \sum_{x \in X'} w(x)$ for all $X' \subseteq X$. A *weighted matroid* is a matroid where each element e has a weight $w(e) \geq 0$ and we denote it by $\mathcal{M} = (X, \mathcal{F}, w)$.

Given a weighted matroid $\mathcal{M} = (X, \mathcal{F}, w)$, a classical optimization problem consists in computing a base $B \in \mathcal{B}$ that maximizes $w(B)$. This problem is solved by the famous polynomial time GREEDY algorithm described for example in [15]. The maximum weight of a base and the subset of bases which are maximal for w are denoted by $OPT_w(\mathcal{M})$ and $\mathcal{B}_w^*$, respectively. We assume, without loss of generality, that after a possible rescaling, the instance is normalized so that $OPT_w(\mathcal{M}) = 1$.

The time complexity of matroid algorithms depends on the difficulty of testing if a set $F \in \mathcal{F}$. We always assume that this test is made in $O(1)$ time.

Given a matroid $\mathcal{M} = (X, \mathcal{F})$ and a subset $X' \subset X$, if $X' \in \mathcal{F}$ then the *contraction* of $\mathcal{M}$ by X', denoted by $\mathcal{M}/X'$, is the structure $(X \setminus X', \mathcal{F}')$ where $\mathcal{F}' = \{F \subseteq X \setminus X' : F \cup X' \in \mathcal{F}\}$. It is well known that $\mathcal{M}/X'$ is a matroid.

Every matroid satisfies the *multiple exchange* property [11]: Let A and B be bases of a matroid $\mathcal{M}$, and let $\{A_1, \ldots, A_n\}$ be a partition of A. Then there exists a partition $\{B_1, \ldots, B_n\}$ of B such that $A \setminus A_i \cup B_i$, $1 \leq i \leq n$ are all bases of $\mathcal{M}$. This result is existential but the construction of $\{B_1, \ldots, B_n\}$ can be done in polynomial time [4].

Let us give some general lemmas that are used later.

Lemma 1. *Let $\mathcal{M} = (X, \mathcal{F}, w)$ be a weighted matroid. Given a maximum weight base $A^* \in \mathcal{B}_w^*$ and a partition $\{A_1, \ldots, A_n\}$ of another base A, there exists a partition $\{A_1^*, \ldots, A_n^*\}$ of A^* satisfying $\min_{i \in [n]} w(A_i^*) \geq \min_{i \in [n]} w(A_i)$.*

Lemma 2. *Let A be a base of a weighted matroid $\mathcal{M} = (X, \mathcal{F}, w)$ partitioned into $\{A_1, \ldots, A_n\}$ such that $n \geq 1$. There exists a permutation σ of $\{1, \ldots, n\}$ such that for all $i = 1, \ldots, n$, $OPT_w \left(\mathcal{M} / \left(\cup_{j \leq i} A_{\sigma(j)} \right) \right) \geq \frac{n-i}{n} OPT_w(\mathcal{M})$. Moreover, the permutation σ can be built in $O(n^2 |X| \ln |X|)$ time.*

Lemma 3. *Let S be an independent set of a weighted matroid $\mathcal{M} = (X, \mathcal{F}, w)$ such that $OPT_w(\mathcal{M}) \geq \rho_0$ and $OPT_w(\mathcal{M}/S) < \rho_1 \leq \rho_0$. Then for every base T of $\mathcal{M}/S$, $OPT_w(\mathcal{M}/T) \geq \rho_0 - \rho_1$.*

Lemma 4. *Given two integers n, k such that $n \geq 2$ and $n > k \geq 1$, and a real $x \in I(n, p) \cup NI(n, p)$ where $p \geq 1$ is an integer, we have that*

1. $\frac{k}{n} V_k \left(\frac{n}{k} x \right) \geq V_n(x)$,
2. $\frac{k}{n} V_k \left(\frac{n}{k} x \right) \geq \frac{k+1}{n} V_{k+1}(\frac{n}{k+1} x)$.

5   A Protocol for Two Agents

Divide-and-Choose (a.k.a. *Cut-and-Choose*) is a well known envy-free protocol for two agents on a *divisible* resource. One agent divides the resource into what

---

Protocol 1. DIVIDE-AND-CHOOSE

Data: $\mathcal{M} = (X, \mathcal{F})$, Agents 1 and 2.

1 Agent 1 computes a base $A \in \mathcal{B}$ and a partition $\{A_1, A_2\}$ of A.

2 Agent 2 chooses one part $A_i \in \{A_1, A_2\}$ and gives it to Agent 1. Then, Agent 2 completes A_i by adding his own part B_{3-i} such that $A_i \cup B_{3-i}$ is a base of $\mathcal{M}$.

---

he believes are equal halves, and the other agent chooses the half he prefers. We propose in Protocol 1 a similar protocol which deals with matroids.

Agent 1 is called the *divider* and let us see how he can guarantee to himself a utility of $V_2(\alpha_1)$ with the use of polynomial algorithms. Agent 1 computes a base A that he partitions in two parts A_1 and A_2 and his contribution is one of them. Thus, his utility is at least $\min\{u_1(A_1), u_1(A_2)\}$. Let $A^\star$ be a base that maximizes u_1, obtained by applying GREEDY [15]. We have $u_1(A^\star) = 1$ by the normalization hypothesis. Using Lemma 1 with $n = 2$ and $w = u_1$, there exists a partition $\{A_1^\star, A_2^\star\}$ of $A^\star$ such that $\min\{u_1(A_1^\star), u_1(A_2^\star)\} \geq \min\{u_1(A_1), u_1(A_2)\}$ for every bases A. So, the divider never loses by partitioning $A^\star$ in $\{A_1^\star, A_2^\star\}$. Once $A^\star$ is computed, one can use ALLOCATE [18] on the following input: the elements of $A^\star$ and two fictitious agents, both having u_1 as utility function for the items. The result is a bi-partition of $A^\star$ into $\{A_1^\star, A_2^\star\}$ such that $\min\{u_1(A_1^\star), u_1(A_2^\star)\} \geq V_2(\alpha_1)$ because Agent 1's largest utility for an element of $A^\star$ is α_1. Since ALLOCATE and GREEDY are polynomial, the divider can guarantee to himself a utility of $V_2(\alpha_1)$ in polynomial time.

Agent 2 is called the *chooser* and he has in hand the partition $\{A_1, A_2\}$ communicated by the divider. Let us see how he can guarantee to himself a utility of $\frac{1}{2} \geq V_2(\alpha_2)$ with the use of polynomial algorithms. Let $B^\star$ be a base that maximizes u_2. By the multiple exchange property, there exists a partition of $B^\star$ into $B_1^\star$ and $B_2^\star$ such that $A_1 \cup B_2^\star$ and $A_2 \cup B_1^\star$ are two bases of $\mathcal{M}$. We have that $u_2(B^\star) = u_2(B_1^\star) + u_2(B_2^\star) = 1$ by the normalization hypothesis. Use GREEDY to complete A_1 and A_2 into the bases $A_1 \cup B_2$ and $A_2 \cup B_1$ of $\mathcal{M}$, respectively. B_2 and $B_2^\star$ are both bases of $\mathcal{M}/A_1$ but B_2 is optimal for the matroid $\mathcal{M}/A_1$. We get that $u_2(B_2) \geq u_2(B_2^\star)$ and also $u_2(B_1) \geq u_2(B_1^\star)$ with similar arguments. Thus, $\max\{u_2(B_1), u_2(B_2)\} \geq \max\{u_2(B_1^\star), u_2(B_2^\star)\} \geq \frac{1}{2}(u_2(B_1^\star) + u_2(B_2^\star)) = \frac{1}{2}$.

Proposition 1. *Using Protocol 1, Agents 1 and 2 can guarantee to themselves $V_2(\alpha_1)$ and $\frac{1}{2} = \frac{1}{2}V_1(2\alpha_2) \geq V_2(\alpha_2)$, respectively.*

6 A Protocol for Three Agents

When there is a third agent, the protocol described in Protocol 2 is more involved: the chooser should ask the third agent before realizing any action.

Again, Agent 1 is called the *divider* and he can guarantee to himself a utility of $V_3(\alpha_1)$ with polynomial algorithms. At Step 1 of the protocol, it suffices to build a base A that maximizes u_1 with GREEDY and cut it in 3 parts with ALLOCATE. Three disjoint sets $\{A_1, A_2, A_3\}$ are obtained, such that

---

Protocol 2. DIVIDE-ASK-AND-CHOOSE

Data: $\mathcal{M} = (X, \mathcal{F})$, Agents 1, 2 and 3.

1 Agent 1 computes a base A that he partitions into $\{A_1, A_2, A_3\}$.

2 Agent 2 chooses A_i for some $i \in \{1, 2, 3\}$ and proposes to Agent 3 to give this part to Agent 1.

3 Agent 3 agrees: A_i is the contribution of Agent 1. Apply DIVIDE-AND-CHOOSE on $\mathcal{M}/A_i$ such that Agent 2 is the divider and Agent 3 the chooser.

4 Agent 3 refuses: Apply DIVIDE-AND-CHOOSE on $\mathcal{M}/A_i$ such that Agent 1 is the divider and Agent 2 the chooser. Let F be the contribution of Agents 1 and 2. Agent 3 completes F by adding his own part.

$\min\{u_1(A_1), u_1(A_2), u_1(A_3)\} \geq V_3(\alpha_1)$ and $A_1 \cup A_2 \cup A_3$ is a base of $\mathcal{M}$. At Step 4 of Protocol 2, Agent 1 is the divider on the contracted matroid $\mathcal{M}/A_i$ and he can submit $\{A_1, A_2, A_3\} \setminus \{A_i\}$. By construction, Agent 1's utility is still $V_3(\alpha_1)$.

Let us focus on Agent 2. At Step 2 of the protocol, we suppose that Agent 2 chooses $i \in \{1, 2, 3\}$ such that $OPT_2(\mathcal{M}/A_i) \geq \frac{2}{3}OPT_2(\mathcal{M}) = \frac{2}{3}$. Using Lemma 2, i exists and can be obtained in polynomial time. At Step 3 of the protocol, Agent 2 has to apply DIVIDE-AND-CHOOSE on $\mathcal{M}/A_i$ as a divider. The revised parameter in this contracted matroid is denoted by $\tilde{\alpha}_2$ and $\tilde{\alpha}_2 \leq \alpha_2/OPT_2(\mathcal{M}/A_i) \leq 3\alpha_2/2$ by hypothesis. Agent 2 can compute a base B of $\mathcal{M}/A_i$ satisfying $u_2(B) = OPT_2(\mathcal{M}/A_i)$ with GREEDY. He can use ALLO-CATE in order to partition B into $\{B_1, B_2\}$ such that $\min\{u_2(B_1), u_2(B_2)\} \geq V_2(\tilde{\alpha}_2)OPT_2(\mathcal{M}/A_i) \geq \frac{2}{3}V_2(\tilde{\alpha}_2)$. Since either B_1 or B_2 belongs to the final base, Agent 2's utility is at least $\frac{2}{3}V_2(\tilde{\alpha}_2)$. Using the facts that V_2 is non-increasing, $\tilde{\alpha}_2 \leq 3\alpha_2/2$ and item 1 of Lemma 4 with $n = 3$ and $k = 2$, we get that $\frac{2}{3}V_2(\tilde{\alpha}_2) \geq \frac{2}{3}V_2\left(\frac{3}{2}\alpha_2\right) \geq V_3(\alpha_2)$; so, Agent 2's utility is at least $\frac{2}{3}V_2\left(\frac{3}{2}\alpha_2\right) \geq V_3(\alpha_2)$. At the fourth step of the protocol, Agent 2 has to apply DIVIDE-AND-CHOOSE (Protocol 1) on $\mathcal{M}/A_i$ as a chooser. We know from the analysis of DIVIDE-AND-CHOOSE that the chooser's utility is at least the utility of a maximum base, divided by 2. Hence, Agent 2's utility is at least $\frac{1}{2}OPT_2(\mathcal{M}/A_i) \geq \frac{1}{2}\frac{2}{3} = \frac{1}{3}$ by hypothesis. In all, Agent 2 can guarantee to himself a utility of $\min\left\{\frac{2}{3}V_2\left(\frac{3}{2}\alpha_2\right), \frac{1}{3}\right\} = \frac{2}{3}V_2\left(\frac{3}{2}\alpha_2\right)$ by item 2 of Lemma 4.

For Agent 3, suppose that he agrees to give A_i to Agent 1 if and only if $OPT_3(\mathcal{M}/A_i) \geq \frac{2}{3}$ (we explain why in Assumption 1). At Step 3 of the protocol, Agent 3 applies DIVIDE-AND-CHOOSE on $\mathcal{M}/A_i$ as a chooser. Again, the chooser's utility is at least the utility of a maximum base, divided by 2. So in this case, the utility of Agent 3 is at least $\frac{1}{2}OPT_3(\mathcal{M}/A_i) \geq \frac{1}{2}\frac{2}{3} = \frac{1}{3} = \frac{1}{3}V_1(3\alpha_3)$ by hypothesis. Note that $\frac{1}{3}V_1(3\alpha_3) \geq V_3(\alpha_3)$ by item 1 of Lemma 4. At Step 4 of the protocol, Agent 3 disagreed to give A_i to Agent 1, so $OPT_3(\mathcal{M}/A_i) < \frac{2}{3}$ by hypothesis. Agent 3 has to complete F, a base of $\mathcal{M}/A_i$ communicated by the other agents, with F_3 such that $F \cup F_3$ is a base of $\mathcal{M}$. $F \cup A_i$ is a base of $\mathcal{M}$ and using Lemma 3, we know that $OPT_3(\mathcal{M}/F) \geq OPT_3(\mathcal{M}) - OPT_3(\mathcal{M}/A_i) \geq 1 - \frac{2}{3} = \frac{1}{3}$. Thus, the completion of F with GREEDY leads to a set F_3 valued at least $\frac{1}{3}$ by Agent 3.

Proposition 2. *Using Protocol 2, Agents 1, 2 and 3 can guarantee to themselves* $V_3(\alpha_1)$, $\frac{2}{3}V_2\left(\frac{3}{2}\alpha_2\right) \geq V_3(\alpha_2)$ *and* $\frac{1}{3} = \frac{1}{3}V_1(3\alpha_3) \geq V_3(\alpha_3)$, *respectively.*

7   A Protocol for $n \leq 8$ Agents

We present a protocol for $n \leq 8$ agents in Protocol 3, which is a generalization of Protocols 1 and 2. The high level idea is the following: A first agent, called the *divider* and denoted by i_1, finds a base and partitions it into n parts. A second agent, called the *pivot* and denoted by i_2, determines an ordering (permutation) σ of the n parts. Next, the protocol is to find a $k \in [n]$ such that the k first parts $\sigma(1), ..., \sigma(k)$ or the $n - k$ last parts $\sigma(k + 1), ..., \sigma(n)$ are shared by k or $n - k$ agents, respectively. This gives a partial solution which is completed into a base. A recursive call of the protocol arranges a sharing of the complement by $n - k$ or k agents, respectively.

We note that if we make a recursive call on a subset of agents $S \subseteq N$ such that the previous divider i_1 is in S, then there is no need to choose a different agent as the divider. We can also keep the shares $A_1, ..., A_n$ that i_1 has previously built. Moreover, making a recursive call on the protocol for one agent (Steps 4 and 26 of Protocol 3) is similar to use Algorithm 4 (DIVIDE) which describes the task of the divider because this unique agent has just to find a base of a matroid which is directly given by DIVIDE.

Let us give an example on the application of Protocol 3. Consider a matroid $\mathcal{M} = (X, \mathcal{F})$, a set of agents $N = \{1, ..., 8\}$ such that $i_1 = 1$ is the divider and $i_2 = 2$ is the pivot or the agent who builds σ (see Level 1 on Figure 2). Suppose that at Step 6, we find $N_0 = N$, $N_1 = \{3, 4, 5, 6\}$, $N_2 = \{3, 7\}$, $N_3 = \{3, 4\}$ and $N_4 = \{5, 6, 7, 8\}$. We stop at N_4 because $|N_k| < n - k - 2$ for $k = 1, 2, 3$ and $|N_4| \geq n - 4 - 2 = 2$, so $k_0 = 4$. We move on Step 12 since $|N_4| = 4 > n - 4 - 1 = 3$ and we conclude that $J_1 = \{2, 7, 5, 6\}$. Then, in Step 14, the agents of J_1 have to apply Protocol 3 to share the matroid $\mathcal{M}/(\cup_{j\leq 4}A_{\sigma(j)})$ where agent 2 is the divider and they obtain B_{J_1} (see Level 2 on Figure 2). We do not impose that Agent 2 is the divider, we just need to choose an agent from J_1. Now, we move on Step 18 because $N_3 \not\subseteq N_4$. We get $p = 2$ and we move on Step 24 because $|N_2\backslash N_4| \neq |N_3\backslash N_4|$. Then, $i_3 = 3$ and the part of Agent 3 is a base B_{i_3} of the matroid $\mathcal{M}/(\cup_{j\leq k_0-1}A_{\sigma(j)} \cup B_{J_1})$ as given in Step 26 (see Level 3). $A_{\sigma(3)}$ is the part of Agent 1 and the agents of $N\backslash(J_1 \cup \{i_1, i_3\}) = \{4, 8\}$ have to share the matroid $\mathcal{M}/(B_{J_1} \cup B_{i_3} \cup A_{\sigma(3)})$ (see Level 4).

We first prove that Protocol 3 is well defined, i.e. a **return** instruction is reached in any case and the algorithm terminates.

Lemma 5. *If* $n = |N| \leq 8$ *then Protocol 3 is well defined.*

7.1   The Divider's Point of View

We assume that Agent $i_1 \in N$ is the divider and let us see how he can guarantee a utility of $V_n(\alpha_{i_1})$ to himself.

Protocol 3. Protocol

Data: a matroid $\mathcal{M} = (X, \mathcal{F})$, a set $N = \{1, \ldots, n\}$ of $n \geq 1$ agents, a divider $i_1 \in N$ who is (if possible) the same as in the previous call, n parts $\{A_1, \cdots, A_n\}$ partitioned by the divider i_1.

Result: a base of $\mathcal{M}$.

1   **if** $n = 1$ **then** assign part A_1 to agent i_1 and **return** A_1.

2   pick any $i_2 \in N \setminus \{i_1\}$ (the pivot), then i_2 renames the n parts $\{A_1, \ldots, A_n\}$ (he produces a permutation σ of N) such that $\forall k \in \{1, \ldots, n-2\}$, agent i_2 agrees to share the matroid $\mathcal{M}/(\cup_{j \leq k} A_{\sigma(j)})$ with $n-k-1$ other agents.

3   **if** $n = 2$ **then**

4 | assign part $A_{\sigma(1)}$ to agent i_1 and **return** $A_{\sigma(1)} \cup \mathrm{DIVIDE}(\mathcal{M}/A_{\sigma(1)}, i_2, 1)$.

5   **else**

6 | $N_0 \leftarrow N$ and for $k = 1, \ldots, n-2$, let $N_k \subseteq N \setminus \{i_1, i_2\}$ be the set of agents who agree to share the matroid $\mathcal{M}/(\cup_{j \leq k} A_{\sigma(j)})$ with $n-k-1$ other agents.

7 | $k_0 \leftarrow \min\{k \geq 1 : |N_k| \geq n - k - 2\}$.

8 | **if** $|N_{k_0}| \leq n - k_0 - 1$ **then**

9 | **if** $|N_{k_0}| = n - k_0 - 2$ **then** $J_0 \leftarrow N_{k_0} \cup \{i_1, i_2\}$ **else** $J_0 \leftarrow N_{k_0} \cup \{i_2\}$.

10 | $B_{J_0} \leftarrow \mathtt{Protocol}(\mathcal{M}/(\cup_{j \leq k_0} A_{\sigma(j)}), J_0, i_1$ if $i_1 \in J_0$ and i_2 otherwise, $\{A_{\sigma(j)}, k_0 + 1 \leq j \leq n\}$ if $i_1 \in J_0$ and $\mathrm{DIVIDE}(\mathcal{M}/(\cup_{j \leq k_0} A_{\sigma(j)}), i_2, n - k_0)$ otherwise).

11 | **return** $B_{J_0} \cup \mathtt{Protocol}(\mathcal{M}/B_{J_0}, N \setminus J_0, i_1$ if $i_1 \in N \setminus J_0$ and any $i_1' \in N \setminus J_0$ otherwise, $\{A_{\sigma(j)}, j \leq k_0\}$ if $i_1 \in N \setminus J_0$ and $\mathrm{DIVIDE}(\mathcal{M}/B_{J_0}, i_1', k_0)$ otherwise).

12 | **else**

13 | Choose a set J_1 of $n - k_0$ agents as follows: $J_1 = \{i_2\}$ at the beginning, next add new agents from $N_{k_0} \cap N_{k_0 - 1}$, next add new agents from $N_{k_0} \cap N_{k_0 - 2}$, and so on until $N_{k_0} \cap N_0$.

14 | $B_{J_1} \leftarrow \mathtt{Protocol}(\mathcal{M}/(\cup_{j \leq k_0} A_{\sigma(j)}), J_1, i_2, \mathrm{DIVIDE}(\mathcal{M}/(\cup_{j \leq k_0} A_{\sigma(j)}), i_2, n - k_0))$.

15 | **if** $N_{k_0 - 1} \subseteq N_{k_0}$ **then**

16 | assign part $A_{\sigma(k_0)}$ to agent i_1.

17 | **return** $B_{J_1} \cup A_{\sigma(k_0)} \cup \mathtt{Protocol}(\mathcal{M}/(B_{J_1} \cup A_{\sigma(k_0)}), N \setminus (J_1 \cup \{i_1\}), i_1' \in N \setminus (J_1 \cup \{i_1\}), \mathrm{DIVIDE}(\mathcal{M}/(B_{J_1} \cup A_{\sigma(k_0)}), i_1', |N \setminus (J_1 \cup \{i_1\})|))$.

18 | **else**

19 | $p \leftarrow |N_{k_0 - 1} \setminus N_{k_0}|$.

20 | **if** $|N_{k_0 - p} \setminus N_{k_0}| = |N_{k_0 - 1} \setminus N_{k_0}|$ **then**

21 | $J_2 \leftarrow N_{k_0 - p} \setminus N_{k_0}$.

22 | $B_{J_2} \leftarrow \mathtt{Protocol}(\mathcal{M}/(\cup_{j \leq k_0 - p} A_{\sigma(j)} \cup B_{J_1}), J_2, \text{any } i_1' \in J_2, \mathrm{DIVIDE}(\mathcal{M}/(\cup_{j \leq k_0 - p} A_{\sigma(j)} \cup B_{J_1}), i_1', |J_2|))$.

23 | **return** $B_{J_1} \cup B_{J_2} \cup \mathtt{Protocol}(\mathcal{M}/(B_{J_1} \cup B_{J_2}), N \setminus (J_1 \cup J_2), i_1, \{A_{\sigma(j)}, j \leq k_0 - p\})$.

24 | **else**

25 | let $i_3 \in N_{k_0 - 1}$ and if possible $i_3 \in N_{k_0 - 2}$.

26 | $B_{i_3} \leftarrow \mathrm{DIVIDE}(\mathcal{M}/(\cup_{j \leq k_0 - 1} A_{\sigma(j)} \cup B_{J_1}), i_3, 1)$.

27 | assign part $A_{\sigma(k_0 - 1)}$ to agent i_1.

28 | **return** $B_{J_1} \cup B_{i_3} \cup A_{\sigma(k_0 - 1)} \cup \mathtt{Protocol}(\mathcal{M}/(B_{J_1} \cup B_{i_3} \cup A_{\sigma(k_0 - 1)}), N \setminus (J_1 \cup \{i_3, i_1\}), \text{any } i_1' \in N \setminus (J_1 \cup \{i_3, i_1\}), \mathrm{DIVIDE}(\mathcal{M}/(B_{J_1} \cup B_{i_3} \cup A_{\sigma(k_0 - 1)}), i_1', |N \setminus (J_1 \cup \{i_3, i_1\})|))$.

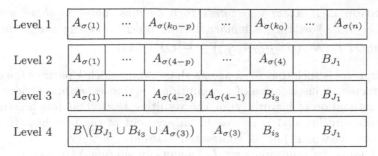

Fig. 2. Example of an execution of Protocol 3

The divider is asked to produce a base A and a partition of it into n parts $A_1, \ldots, A_n$. The protocol is such that the contribution of i_1 is one of these parts. Therefore, it is in his interest to maximize $\min_{i \in N} u_{i_1}(A_i)$. Lemma 1 states that Agent i_1 may choose a base A^* maximum for u_{i_1} because there exists a partition $A_1^*, \ldots, A_n^*$ of A^* satisfying $\min_{i \in N} u_{i_1}(A_i^*) \geq \min_{i \in N} u_{i_1}(A_i)$. The base A^* can be constructed in polynomial time with GREEDY and we know that $u_{i_1}(A^*) = 1$ by the normalization assumption. Let β be agent i_1's maximum utility for an element of A^*, i.e. $\beta = \max_{e \in A^*} u_{i_1}(e)$. By construction of A^*, $\beta = \alpha_{i_1}$ since an element of maximum weight is always taken at first by GREEDY. Using ALLOCATE, one can find a partition of A^* satisfying $\min_{i \in N} u_{i_1}(A_i^*) \geq V_n(\alpha_{i_1}) u_{i_1}(A^*) = V_n(\alpha_{i_1})$.

Proposition 3. *Using Protocol 3, Agent i_1 (the divider) can guarantee to himself $V_n(\alpha_{i_1})$ in polynomial time.*

To summarize, the way the divider can initially guarantee to himself $V_n(\alpha_{i_1})$ is described in Algorithm 4.

Algorithm 4. DIVIDE

   Data: A weighted matroid $\mathcal{M} = (X, \mathcal{F}, u_i)$, Agent $i \in N$, an integer $n \geq 1$.
1 Agent i computes an optimal base A of the matroid $\mathcal{M}$ by applying GREEDY.
2 Normalize the weights of A i.e. $u_i(e) \leftarrow u_i(e)/u_i(A)$ for all $e \in A$.
3 Agent i partitions A into n parts $\{A_1, ..., A_n\}$ by applying ALLOCATE such that the elements are those of A and there are n fictitious agents, all of them have the same utility u_i.
4 **return** $\{A_1, ..., A_n\}$

7.2   The Non-dividers' Points of View for $n \geq 3$ Agents

In the protocol, each non-divider $i \in N \backslash \{i_1\}$ is asked if he agrees to share $\mathcal{M}$ contracted on some parts given by the divider. To show how a non-divider can obtain a certain guarantee on his utility, we make the following assumption.

Assumption 1. *Given $S \subset N$, Agent i agrees to share a base of $\mathcal{M}/(\cup_{j \in S} A_{\sigma(j)})$ with $n - |S| - 1$ other agents if $OPT_i\left(\mathcal{M}/(\cup_{j \in S} A_{\sigma(j)})\right) \geq \frac{n - |S|}{n} OPT_i(\mathcal{M})$.*

Observation 1. *If Agent i disagrees then he must agree to share a base of $\mathcal{M}/(\cup_{j \in N \setminus S} A_{\sigma(j)})$ with $|S| - 1$ other agents. Indeed, one can use Lemma 3 to show that $OPT_i\left(\mathcal{M}/(\cup_{j \in N \setminus S} A_{\sigma(j)})\right) \geq \frac{|S|}{n} OPT_i(\mathcal{M})$.*

The protocol is such that $n - k$ agents share $\mathcal{M}$ in which k parts of the divider are contracted. A justification of Assumption 1 is that every non-divider makes a rough estimation of his utility for the resulting base. This rough estimation is to ensure a utility of $1/n$ in case of an even cut of the best base (from the non-divider's viewpoint) of the contracted matroid.

After finding the appropriate set J of agents for sharing $\mathcal{M}/(\cup_{j \in S} A_{\sigma(j)})$, next lemma shows that every agent of J satisfies Assumption 1 in Protocol 3.

Lemma 6. *If $n \leq 8$ then at each recursive call of Protocol 3, a set J of agents and an independent set $B_J \in \mathcal{F}$ such that $OPT_j(\mathcal{M}/B_J) \geq \frac{|J|}{n} OPT_j(\mathcal{M})$ for all $j \in J \setminus \{i_1\}$ are found.*

We suppose that $B = \{B_1, \ldots, B_n\}$ is the final base returned by Protocol 3 where B_i is the contribution of agent $i \in N$. When $n \leq 8$, Protocol 3 gives a guarantee at least as good as the guarantee of ALLOCATE [18] but Protocol 3 works with matroids (a generalization of the allocation of indivisible goods) and it copes with private utilities (as opposed, ALLOCATE deals with public utilities).

Theorem 1. *If $n \leq 8$, then Protocol 3 applied to $\mathcal{M} = (X, \mathcal{F})$, $N = \{1, \ldots, n\}$, a divider $i_1 \in N$ and $\{A_1, ..., A_n\} = Divide(\mathcal{M}, i_1, n)$, returns a base $B = \{B_1, ..., B_n\}$ such that $\forall i \in N$, $u_i(B_i) \geq V_n(\alpha_i)$.*

8   Discussion

The protocol presented in this paper is valid for $n \leq 8$ agents. We managed to produce a protocol for $n = 9$ and $n = 10$ agents but the number of cases to check has grown rapidly and we were unable to write a concise version of the protocol and its analysis. However, we conjecture that a protocol exists for any number of agents n.

We note that we work with Hill's function V_n whereas a slightly better guarantee, through a new function W_n, was recently proposed, especially as W_n is valid for matroids [10]. This is due to the non-monotonicity of W_n and we need a monotonicity argument in the proof of Theorem 1.

References

1. Asadpour, A., Saberi, A.: An approximation algorithm for max-min fair allocation of indivisible goods. SIAM J. Comput. 39(7), 2970–2989 (2010)
2. Bansal, N., Sviridenko, M.: The santa claus problem. In: Kleinberg, J.M. (ed.) STOC, pp. 31–40. ACM (2006)
3. Bezáková, I., Dani, V.: Allocating indivisible goods. SIGecom Exchanges 5(3), 11–18 (2005)

4. Bixby, R.E., Cunningham, W.H.: Matroid Optimization and Algorithms. In: Handbook of Combinatorics. North Holland (1995)
5. Bouveret, S., Lemaître, M., Fargier, H., Lang, J.: Allocation of indivisible goods: a general model and some complexity results. In: AAMAS, pp. 1309–1310 (2005)
6. Brams, S.J., Taylor, A.D.: An envy-free cake division protocol. The American Mathematical Monthly 102(1), 9–18 (1995)
7. Brams, S.J., Taylor, A.D.: Fair division: From cake cutting to dispute resolution. Cambridge University Press (1996)
8. Demko, S., Hill, T.P.: Equitable distribution of indivisible objects. Mathematical Social Sciences 16(2), 145–158 (1988)
9. Golovin, D. Max-min fair allocation of indivisible goods. Tech. Rep. 2348, Computer Science Department, Carnegie Mellon University (2005)
10. Gourvès, L., Monnot, J., Tlilane, L.: A matroid approach to the worst case allocation of indivisible goods. In: IJCAI, pp. 136–142 (2013)
11. Greene, C., Magnanti, T.L.: Some abstract pivot algorithms. SIAM Journal on Applied Mathematics 29(3), 530–539 (1975)
12. Hill, T.P.: Partitioning general probability measures. The Annals of Probability 15(2), 804–813 (1987)
13. Kalinowski, T., Nardoytska, N., Walsh, T.: A social welfare optimal sequential allocation procedure. IJCAI, 227–233 (2013)
14. Khot, S., Ponnuswami, A.K.: Approximation algorithms for the max-min allocation problem. In: Charikar, M., Jansen, K., Reingold, O., Rolim, J.D.P. (eds.) APPROX and RANDOM 2007. LNCS, vol. 4627, pp. 204–217. Springer, Heidelberg (2007)
15. Korte, B., Vygen, J.: Combinatorial Optimization: Theory and Algorithms. Springer (2007)
16. Lipton, R., Markakis, E., Mossel, E., Saberi, A.: On approximately fair allocations of indivisible goods. In: Breese, J., Feigenbaum, J., Seltzer, M. (eds.) ACM Conference on Electronic Commerce, pp. 125–131. ACM (2004)
17. Lu, T., Boutilier, C.: Budgeted social choice: From consensus to personalized decision making. In: IJCAI, pp. 280–286. AAAI Press (2011)
18. Markakis, E., Psomas, C.-A.: On worst-case allocations in the presence of indivisible goods. In: Chen, N., Elkind, E., Koutsoupias, E. (eds.) WINE 2011. LNCS, vol. 7090, pp. 278–289. Springer, Heidelberg (2011)
19. Moulin, H.: Fair division and collective welfare. MIT Press (2003)
20. Procaccia, A.D.: Cake cutting: not just child's play. Commun. ACM 56(7), 78–87 (2013)
21. Procaccia, A.D., Rosenschein, J.S., Zohar, A.: On the complexity of achieving proportional representation. Social Choice and Welfare 30(3), 353–362 (2008)
22. Skowron, P., Faliszewski, P., Slinko, A.: Fully proportional representation as resource allocation: Approximability results. In: IJCAI, pp. 353–359 (2013)
23. Steinhaus, H.: The problem of fair division. Econometrica 16, 101–104 (1948)
24. Woeginger, G.: A polynomial-time approximation scheme for maximizing the minimum machine completion time. Oper. Res. Lett. 20, 149–154 (1997)
25. Young, H.P.: Equity: In theory and practice. Princeton University Press (1994)

Quantitative Comparative Statics
for a Multimarket Paradox

Tobias Harks[1] and Philipp von Falkenhausen[2],*

[1] Maastricht University
t.harks@maastrichtuniversity.nl
[2] Technical University Berlin
falkenhausen@math.tu-berlin.de

1 Introduction

Comparative statics is a well established research field where one analyzes how changes in parameters of a strategic game affect the resulting equilibria. Examples of such parameter changes include tax/subsidy changes or production cost shifts in oligopoly models.

While classic comparative statics is mainly concerned with qualitative approaches (e.g., deciding whether a marginal parameter change improves or hurts equilibrium profits or welfare), our approach is to capture the *maximum possible* effect that the change of a parameter – shift of the inverse demand function in our case – can have. This worst-case approach exhibits both

1. significance: are changes in a given parameter worth considering?
2. robustness: how sensitive is the game to changes of a parameter?

Significance is a crucial motivation of both the analysis of an effect and discussion of whether it can be put to use (à la 'should a new tax be introduced?'). Robustness on the other hand is important when there is uncertainty about the values of parameters and when parameters change over time. To address these issues, we propose a quantitative approach.

We apply our quantitative approach to the multimarket oligopoly model introduced by Bulow, Geanakoplos and Klemperer [1]. They investigated how "changes in one market have ramifications in a second market" and discovered that a positive price shock in a firm's monopoly market can have a negative effect on the firm's profit by influencing competitors' strategies in a different market, cf. Section VII in [1]. Motivated by a counterintuitive example with two firms where a positive price shock reduced the monopolist's profit by 0.76%, they introduced the classification of markets in terms of strategic substitutes and strategic complements.[1] Our paper is about rigorously quantifying profit effects induced by price shocks in multimarket Cournot oligopolies.

---

* This research was supported by the Deutsche Forschungsgemeinschaft within the research training group 'Methods for Discrete Structures' (GRK 1408).
[1] In a market with *strategic substitutes*, more aggressive play by a firm leads to less aggressive play of the competitors on that market; with *strategic complements*, more aggressive play results in more aggressive play of the competitors.

Y. Chen and N. Immorlica (Eds.): WINE 2013, LNCS 8289, pp. 230–231, 2013.
© Springer-Verlag Berlin Heidelberg 2013

2 Our Results

We exactly quantify the worst-case profit reduction for the case of two markets with affine price functions and firms with convex cost technologies. We show that the worst case loss of the monopoly firm is at most 25% no matter how many firms compete on the second market. In particular we show for the setting of the example in [1] involving only one additional firm on the second market that the worst case loss in profit is bounded by 6.25%.

We prove this result by first establishing some basic characteristics of equilibria before and after a price shock, e.g. uniqueness of equilibria and monotonicity of quantities. Given these characteristics, we subsequently obtain a set of worst case instances by iteratively restricting the cost and price functions. In particular, we show that the monopolist's profit loss is maximized by two factors: competitors behave most aggressively when they have linear cost functions (leading to maximal strategic substitution), while in contrast for the monopolist a non-linear cost function simulating hard production capacities is worst case. As a byproduct of our analysis, we exactly quantify the magnitude of strategic substitution for any market model where competitors have linear cost functions and prices are affine.

A question dual to the above question is: How much can a firm gain from a negative price shock in its monopoly market? Our results imply that this gain is at most 33%.

We complement our bounds by concrete examples of markets where these bounds are attained.

Example Application. Profit gains from negative price shocks can occur in international trade, as noted by Bulow et al. [1, Sec. VI (C)]. Consider two markets located in separate countries with convex cost technologies, one of which is a monopoly market for firm a. A tax change in the country of the monopolist can be considered a price shock. A government may decide to increase domestic taxes in order to increase firm a's profitability in the foreign market. Our results imply that this positive effect can be significant as it may increase the profitability by up to 33% of current profits.

A full version of this paper is available at http://arxiv.org/abs/1307.5617.

References

1. Bulow, J., Geanakoplos, J., Klemperer, P.: Multimarket oligopoly: Strategic substitutes and complements. J. Polit. Econ. 93(3), 488–511 (1985)

Price of Anarchy for the N-Player Competitive Cascade Game with Submodular Activation Functions

Xinran He* and David Kempe**

Computer Science Department, University of Southern California,
941 Bloom Walk, Los Angeles, CA, USA

Abstract. We study the Price of Anarchy (PoA) of the competitive cascade game following the framework proposed by Goyal and Kearns in [11]. Our main insight is that a reduction to a Linear Threshold Model in a time-expanded graph establishes the submodularity of the social utility function. From this observation, we deduce that the game is a valid utility game, which in turn implies an upper bound of 2 on the (coarse) PoA. This cleaner understanding of the model yields a simpler proof of a much more general result than that established by Goyal and Kearns: for the N-player competitive cascade game, the (coarse) PoA is upper-bounded by 2 under any graph structure. We also show that this bound is tight.

Keywords: Competitive cascade game, Price of Anarchy, Submodularity, Valid utility game, Influence maximization.

1 Introduction

The processes and dynamics by which information and behaviors spread through social networks have long interested scientists within many areas [18]. Understanding such processes has the potential to shed light on human social structure, and to impact the strategies used to promote behaviors or products. While the interest in the subject is long-standing, the recent increased availability of social network and information diffusion data (through sites such as Facebook and LinkedIn) has put into relief algorithmic questions within the area, and led to widespread interest in the topic within the computer science community.

One particular application that has been receiving interest in enterprises is to use word-of-mouth effects as a tool for viral marketing. Motivated by the marketing goal, mathematical formalizations of influence maximization have been proposed and extensively studied by many researchers [9,14,17,23,24,8,7,16]. Influence maximization is the problem of selecting a small set of seed nodes in a social network, such that their overall influence on other nodes in the network — defined according to particular models of diffusion — is maximized.

When considering the word-of-mouth marketing application, it is natural to realize that multiple companies, political movements, or other organizations may use diffusion in a social network to promote their products simultaneously. For example, Samsung

---

* Supported in part by NSF Grant 0545855.

** Supported in part by NSF Grant 0545855, an ONR Young Investigator Award, and a Sloan Rseearch Fellowship.

may try to promote their new Galaxy phone, while Apple tries to advertise their new iPhone. Companies will necessarily end up in competition with each other, so it becomes essential to understand the outcome of competitive diffusion phenomena in the network.

Motivated by the above scenarios, several models for competitive diffusion have been proposed and studied [2,4,7,12,20,11,1,21]. Past work tends to follow one of two assumptions about the timing of players' moves. The first approach is to assume that all but one of the competitors have already chosen their strategies, and to study the algorithmic problem of finding the best response [2,4,7,12,5,6]. The goal may be maximizing one's own influence [2,4,7] or minimizing the influence of the competitors [12,5]. The other approach is to model the competition as a simultaneous game, in which all companies pick their strategies at the same time [1,11,20,21]. The final influence is determined by the initial seed set of every company and the underlying diffusion process.

In this paper, we follow the second approach. In the game, the players are companies (or other organizations) who try to promote their competing products in the social network through word-of-mouth marketing. The players simultaneously allocate resources to individuals in the social network in order to seed them as initial adopters of their products. These resources could be free samples, time spent explaining the advantages of the product, or monetary rewards. Based on the allocated resources, the nodes choose which of the products to adopt initially. Subsequently, the diffusion of the adoption of products proceeds according to the local adoption dynamics. The goal for each player is to maximize the coverage of his[1] own product.

The local adoption dynamics play a vital role in determining properties of the game. In this paper, we follow the framework proposed recently by Goyal and Kearns [11]. Their model decomposes the local adoption decisions into two stages: *switching* and *selection*. In the switching stage, the user decides whether to adopt any product or company at all. This decision is based on the set of neighbors who have already adopted one of the products. If the user decides to adopt a product, in the following selection stage, she decides which company's product to adopt based on the fraction of neighbors who have adopted the product from each company.

For example, assume that iPhone and Galaxy are the only two smartphones available. In the switching stage, a user decides whether to adopt a smartphone or not, based on the fraction of her neighbors who have already bought a smartphone. If she has decided to adopt a smartphone, in the selection stage, she decides whether to choose an iPhone or Galaxy based on the fraction of iPhone users and Galaxy users among her friends. The two stages are modeled using a *switching function* $f_v(\alpha_1 + \alpha_2)$, which gives the probability that the user adopts one of the products, and the *selection function* $g_v(\alpha_1, \alpha_2)$, which determines the probability that the user chooses the product of a specific company. Here α_1 and α_2 are the fractions of the user's friends who have already adopted the product from the two competitors. The details of the model are presented in Section 2.

Under this framework, Goyal and Kearns have studied the *Price of Anarchy* (PoA) of the two-player competitive cascade game. Informally, the PoA is a measure of the

---

[1] Throughout the paper, to simplify the distinction of roles, we consistently use "she" to denote individuals in the social network and "he" to denote the players, i.e., the companies.

maximum potential inefficiency created by non-cooperative activity. (The precise definition of the PoA is given in Section 2.3.) Goyal and Kearns have shown that the PoA under the switching-selection model with concave switching functions and linear selection functions is upper-bounded by 4.[2]

In this paper, we show that a stronger PoA bound for the Goyal-Kearns model follows from several well-understood and general phenomena. The key observation is that by considering a time-expanded graph, the Goyal-Kearns model can be considered an instance of a general threshold model. Then, the result of Mossel and Roch [17] guarantees that the social utility function is submodular, and a simple coupling argument establishes that players' utility functions are competitive. With a submodular social utility function, the game is a valid utility game. (This type of proof was used previously by Bharathi et al. [2].) Finally, for valid utility games, the results of Vetta [22] and Blum et al. and Roughgarden [3,19] establish a (coarse) Price of Anarchy of at most 2.

Thanks to the above understanding, we obtain a much more general result with a much simpler proof. We show that the PoA is upper-bounded by 2 for the competitive cascade game with an arbitrary number of players and any graph structure with submodular activation functions $f_v(\cdot)$. We formally state this result in Theorem 1. Moreover, by utilizing the result of Roughgarden in [19], we show that our bound not only holds for the PoA under pure or mixed Nash equilibria but also for the coarse PoA. We also show that the proposed PoA bound is tight.

Theorem 1. *The coarse PoA is upper-bounded by 2 under the switching-selection model with concave switching functions and linear selection functions.*[3]

Our result on the PoA bound holds under a generalized version of the framework used in [11]. First, and most importantly, our model allows for an arbitrary number of players. Second, we allow multiple players to target the same individual and allow each player to put multiple units of budget on the same individual.[4] This generalization enlarges the strategy space from sets to multisets and somewhat complicates the analysis of our model. Third, we associate each individual in the network with a weight measuring the importance of the node. Fourth, we generalize the adoption functions defined on the fraction of already adopting neighbors to arbitrary set functions defined on the individuals who have previously adopted the product.

1.1   Related Work

Our work is mainly motivated by [11], lying at the intersection of influence analysis in social networks and traditional game theory research. The model in [11] and the differences compared to our work are discussed in detail above and in Section 2.

---

[2] In fact, they proved that the PoA upper bound holds in a more general model, which we will discuss in Section 2.

[3] Similar to the result by Goyal and Kearns, our PoA upper bound extends to a more general model. We define this more general model in Section 2, and state and prove the more general result in Section 3.

[4] The model proposed by Goyal and Kearns [11] allows for multiple units of budget on the same individual, but the proof does not explicitly cover this extension.

Submodularity has been a recurring topic in the study of diffusion phenomena [17,14,12,2,4,5]. [14,17] have shown that influence coverage is submodular under local dynamics with submodularity. The submodularity of global influence coverage can be utilized to design efficient algorithm for either maximizing the influence [14] or minimizing the influence of the competitors [12]. Submodularity has also been applied in the analysis of a competitive influence game by Bharathi et al. [2]. Bharathi et al. use a similar approach as we do in this paper; they also bound the PoA bound by showing that the game is a valid utility game. However, they analyze the competitive cascade game under a simpler diffusion model. Under their model, a node adopts the product from the neighbor who first succeeds in activating her; a continuous timing component ensures that this node is unique with probability 1.

In the proof for the PoA bound of the competitive cascade game, we are drawing heavily on previous research on the PoA for valid utility games [22,19,3]. Vetta first showed that for a valid utility game, the PoA for pure Nash equilibria is upper-bounded by 2 in [22]. Blum et al. and Roughgarden later generalized Vetta's result to the coarse PoA in [19,3].

Several other game-theoretic approaches have been considered for competitive diffusion in social networks [21,1,20,10,6]. [20] mainly focuses on the efficient computation of the Nash strategy instead of the theoretical bound of the PoA. [6] focuses on studying the algorithmic problem of finding the best response. Though [1,21,10] studied the competitive cascade game from a game-theoretic perspective, they mainly focused on the existence of pure Nash equilibria. [1] mainly focuses on the existence of pure Nash equilibria under a deterministic threshold model. [10] also tries to characterize the structure of the pure Nash equilibria in the game. The PoA is studied in [21]; however, they studied the PoA bound of pure Nash equilibria and used a deterministic diffusion model instead of the stochastic dynamics we use in our work. In their model, the PoA is unbounded as in the Goyal-Kearns model with non-concave switching functions. As noted by [4,5], small differences in the diffusion model can lead to dramatically different behaviors of the model.

2 Models and Preliminaries

In this section, we define basic notation, present the different models of diffusion and the N-player game, and include other definitions of concepts used in our proof. In the game, the players allocate resources to the nodes in the graph $G = (V, E)$ to win them as initial adopters of their products. Then, the adoption of products propagates according to the local dynamics, described in detail in Section 2.1. The formal definition of the game is presented in Section 2.3.

Throughout, we use the following conventions for notation. Players are typically denoted by i, j, k, while nodes are u, v, w. For sets, functions, etc., the identity of a player is applied as a superscript, while that of a node (and time step) is applied as a subscript. Vectors are written in boldface, including vectors of sets; in particular, we frequently write $S = (S^1, \ldots, S^N)$ for the vector of sets of nodes belonging to the different players.

2.1   General Adoption Model

The *general adoption model* is a generalization of the switching-selection model described in Section 1. Each node in G is in one of the $N + 1$ states $\{0, 1, \ldots, N\}$. A node v in state $i > 0$ means that individual v has adopted the product of player i, while state 0 means that she has not adopted the product of any player. In this case, we also say that v is *inactive*. Conversely, we say that node v is *activated* if she is in one of the states $i > 0$. Initially, all nodes are inactive. The diffusion of the adoption of products is a process described by nodes' state changes. We assume that the process is *progressive*, meaning that a node can change her state at most once, from 0 to some $i > 0$, and must remain in that state subsequently.

The diffusion process works in two stages. We call the first stage *Seeding* and the second stage *Diffusion*. In the first stage, the initial seeds of all players are decided based on the budgets that each player allocates to the nodes. The initial seeds are used as starting points for the diffusion stage. In the second stage, the adoption propagates according to certain local dynamics based on the nodes who have previously adopted the products.

Seeding stage: The strategy M^i of player i is a multiset of nodes. We define α_v^i as the number of times that v appears in player i's multiset. For each node $v \in V$, if $\sum_{i=1}^{N} \alpha_v^i = 0$, the initial state of node v is 0; otherwise, the initial state of node v is one of $\{1, 2, \ldots, N\}$ with probabilities $(\frac{\alpha_v^1}{Z_v}, \ldots, \frac{\alpha_v^N}{Z_v})$, where $Z_v = \sum_{i=1}^{N} \alpha_v^i$ is simply the normalizing constant. The decisions for different nodes are independent. Thus, if no player selects a node, the node remains inactive. Otherwise, the players win the node as an initial adopter with probability proportional to the number of times they select the node.

Diffusion stage: The important part of diffusion is the local dynamics deciding when a node gets influenced, i.e., changes her state from 0 to i. Let S^i be the set of nodes in state i. A node v who is still in state 0 changes into state $1, \ldots, N, 0$ according to the probabilities

$$(h_v^1(\boldsymbol{S}), \ldots, h_v^N(\boldsymbol{S}), 1 - \sum_{i=1}^{N} h_v^i(\boldsymbol{S})).$$

We call $h_v^i(S^1, \ldots, S^N)$ the *adoption function* of node v for product i. It gives the probability that a still inactive node v adopts product i given that S^j is the current set of nodes in state j. The adoption functions must satisfy the following two conditions:

$$0 \le h_v^i(\boldsymbol{S}) \le 1, \quad \forall v \in V, i = 1, \ldots, N$$
$$\textstyle\sum_{i=1}^{N} h_v^i(\boldsymbol{S}) \le 1, \forall v \in V.$$

We call $H_v(\boldsymbol{S}) = \sum_{i=1}^{N} h_v^i(\boldsymbol{S})$ the *activation probability*; it gives the probability that v adopts any product and changes from state 0 to any state $i > 0$.

Equipped with the local dynamics of adoption, we still need to define in what order nodes' states are updated. In the general adoption model, we assume that an update schedule is given in advance to determine the order of updates. The update schedule is a finite sequence Q of nodes $\langle v_1, \ldots, v_\ell \rangle$, of length ℓ. A node could occur multiple times in the sequence.

Nodes' states are updated according to the order prescribed by the sequence. Let S_t^i be the set of nodes in state i after the first t updates; S_0^i is the seed set of player i resulting from the seeding stage. In each round t, the state of node v_t is updated according to the local dynamics of adoption and previously activated nodes, namely $S_{t-1} = (S_{t-1}^1, \ldots, S_{t-1}^N)$. If node v_t is already in state $i > 0$, she remains in state i. Otherwise, she changes into state $1, \ldots, N, 0$ according to the probabilities

$$(h_v^1(S_{t-1}), \ldots, h_v^N(S_{t-1}), 1 - \sum_{i=1}^{N} h_v^i(S_{t-1})).$$

The states of all other nodes remain the same. The updates in different rounds are independent. The diffusion stage ends after the ℓ update steps. The prescribed update sequence makes this model different from the previously studied Independent Cascade and Threshold Models. We discuss the difference and some implications in more detail after defining the Threshold Model in Section 2.4.

2.2 Useful Properties

We next identify three important properties that make the model more tractable analytically: (1) additivity of the activation probability H_v, (2) competitiveness of the adoption function h_v and (3) submodularity of the activation function f_v.

Definition 1. *The total activation probability $H_v(S) = \sum_{i=1}^{N} h_v(S)$ is additive if and only if H_v can be written as $H_v(S) = f_v(\bigcup_{i=1}^{N} S^i)$ for some monotone set function defined on V. We call $f_v(S)$ the* activation function *for v when H_v is additive.*

Additivity implies that the probability for a node to adopt the product and change from inactive to active only depends on the set of already activated nodes and not on which specific products they have adopted. For example, the probability that one adopts a smartphone only depends on who has already adopted one, independent of who is using iPhone and who is using Galaxy.

To simplify notation, we define $S^{-i} = \bigcup_{k \neq i} S^k$, and $\boldsymbol{S}^{-i} = (S^k)_{k \neq i}$.

Definition 2. *The adoption function $h_v^i(\boldsymbol{S})$ for player i is* competitive *if and only if $h_v^i(\boldsymbol{S}) \geq h_v^i(\hat{\boldsymbol{S}})$ whenever $\hat{S}^i \subseteq S^i$ and $S^{-i} \subseteq \hat{S}^{-i}$.*

Competitiveness means that the adoption function for player i is monotone increasing in the set of nodes that have adopted product i and monotone decreasing in the set of nodes that have adopted some competitor's products.[5]

---

[5] This assumption is reasonable when the reputation of the product is already well-established. However, when a new product comes out, the presence of competitors may help popularize the product, by increasing its overall exposure or perceived importance or relevance. These effects could lead to more purchases even for one particular company i. This subtle distinction is discussed more in Section 2.4.

Definition 3. *The activation function f_v is submodular if and only if for any two set $S \subseteq T \subseteq V$ and any node $u \in V$,*

$$f_v(S \cup \{u\}) - f_v(S) \geq f_v(T \cup \{u\}) - f_v(T).$$

Submodularity of activation functions implies that the overall activation probability has diminishing returns. It intuitively means that the first friend to buy and recommend a smartphone has more influence than a friend who recommends it after many others.

Goyal and Kearns have shown in [11] that the switching-selection model with concave switching functions and linear selection functions is a special case of the general adoption model with competitive adoption functions and additive activation probabilities. In addition, due to the concavity of the switching function f_v, the activation functions in the general adoption model are also submodular. Therefore, we have the following lemma:

Lemma 1. *Every instance of the switching-selection model with concave switching functions and linear selection functions is an instance of the general adoption model satisfying all three of the above properties.*

Lemma 1 allows us to prove our PoA bound only for the general adoption model; it then implies Theorem 1.

2.3   The Game

The competitive cascade game is an N-player game on a given graph $G = (V, E)$. The structure of the graph as well as all adoption functions are known to all the players. Each player i is a company. The strategy for each player i is a multiset M^i of nodes; we use $M = (M^1, \ldots, M^N)$ to denote the strategy vector for all players and α_v^i for the number of times that node v appears in M^i.

Players' strategies are constrained by their budgets B^i, in that they must satisfy $|M^i| \leq B^i$. We further allow node-specific constraints requiring that $\alpha_v^i \leq K_v^i$ for given node-specific budgets $K_v^i \leq B^i$. These may constrain players from investing a lot of resources into particularly hard-to-reach nodes; however, the node-specific constraints mostly serve to simplify notation in some later proofs. We say that a strategy M^i is *feasible* if all of the above conditions are satisfied.

All players simultaneously allocate their budgets to the nodes of G. Given the choices that the players make, the payoffs are determined by the general adoption model as the coverage of the player's product among the individuals in G. Each node v in the graph is associated with a weight $\omega_v \geq 0$, measuring the importance of the node. The payoff function of player i is $\sigma^i(M) = \mathbb{E}[\sum_{v \in S_\ell^i} \omega_v]$, the expected sum of weights from nodes having adopted i's product after all ℓ update steps.

The social utility $\gamma(S_0) = \sum_i \sigma^i(M)$ is the sum of weights from nodes adopting any of the products.[6] Notice that when the activation probabilities H_v are additive

---

[6] This definition implicitly assumes that the product carries a value for those who adopt it; thus, society is better off when more people adopt at least one product.

(Definition 1), $\gamma(\cdot)$ only depends on $S_0 = \bigcup_{i=1}^{N} S_0^i$, the set of nodes activated after the seeding stage (but not on which company they chose).

To simplify notation, we define $(M^{-k}, \tilde{M}^k) = (M^1, \ldots, M^{k-1}, \tilde{M}^k, M^{k+1}, \ldots, M^N)$, and in particular $(M^{-k}, \emptyset^k) = (M^1, \ldots, M^{k-1}, \emptyset, M^{k+1}, \ldots, M^N)$.

We say that a strategy profile M is a *pure Nash equilibrium* if no player has an incentive to change his strategy. Namely, for any player i,

$$\sigma^i(M) \geq \sigma^i(M^{-i}, \tilde{M}^i) \quad \text{for all feasible } \tilde{M}^i.$$

Let **OPT** be a strategy profile maximizing the social utility function, and $\mathrm{EQ}_{\mathrm{pure}}$ the set of all pure Nash equilibria. The price of anarchy of pure Nash equilibria is defined as follows:

$$\text{Pure Price of Anarchy} = \max_{M \in \mathrm{EQ}_{\mathrm{pure}}} \frac{\gamma(\mathbf{OPT})}{\gamma(M)}.$$

However, the competitive cascade game could have no pure Nash equilibrium [21]. Thus, we extend our analysis to more general equilibrium concepts. A *coarse (correlated) equilibrium* of a game is a joint probability distribution $\mathbf{P}$ with the following property [19]: if M is a random variable with distribution $\mathbf{P}$, then for each player i, and all feasible $\hat{M}^i$:

$$\mathbb{E}_{M \sim \mathbf{P}}[\sigma^i(M)] \geq \mathbb{E}_{M^{-i} \sim \mathbf{P}^{-i}}[\sigma^i(M^{-i}, \hat{M}^i)].$$

Similar to the PoA for pure Nash equilibria, the *coarse price of anarchy* is defined as

$$\text{Coarse Price of Anarchy} = \max_{\mathbf{P} \in \mathrm{EQ}_{\mathrm{coarse}}} \frac{\gamma(\mathbf{OPT})}{\mathbb{E}_{M \sim \mathbf{P}} \gamma(M)},$$

where $\mathrm{EQ}_{\mathrm{coarse}}$ is the set of all coarse equilibria.

2.4   The Threshold Model

Our analysis will be based on a careful reduction of the general adoption model to the general threshold model defined in [14,15]. In the general threshold model (with $N = 1$), every node v in the network has an associated activation function $\hat{f}_v(\cdot)$. At the beginning of the process, each node draws a threshold θ_v independently and uniformly from $[0, 1]$. Starting from an initially active set S_0, a node becomes active at time t (i.e., is a member of S_t) if and only if $\hat{f}_v(S_{t-1}) \geq \theta_v$. The process ends when for one round, no new node has become active (which is guaranteed to happen in at most $|V|$ steps). If t is the time when this happens, the influence of the initial set S_0 is defined as $\sigma_\omega(S_0) = \mathbb{E}[\sum_{v \in S_t} \omega_v]$. In a beautiful piece of work, Mossel and Roch established the following theorem about the function σ_ω:

Theorem 2 (Mossel-Roch [17]). *If f_v is monotone and submodular for every node v in the graph, then σ_ω is monotone and submodular under the general threshold model.*

Given the apparent similarity between the general adoption model and the general threshold model (say, for $N = 1$), it is illuminating to consider the ways in which the models differ, and the implications for the competitive cascade game. In the general adoption model, a sequence of nodes to update is given, and nodes only consider changing their state when they appear in the sequence. By contrast, in the general threshold model, nodes consider changing their state in each round.

So at first, it appears as though a sequence repeating $|V|$ times a permutation of all $|V|$ nodes would allow a reduction from the threshold model to the adoption model. However, note that in the adoption model, each node makes an *independent* random choice whether to change her state in each round, whereas in the threshold model, the random choices are coupled via the threshold θ_v, which stays constant throughout the process. In particular, if $f_v(S_0) > 0$, then a node appearing often enough in the update sequence will eventually be activated with probability converging to 1, whereas this need not be the case in the general threshold model.

The increase in activation probability caused by multiple occurrences in the update sequence has powerful implications for the competitive game. It allows us to establish rather straightforwardly the competitiveness (Definition 2) of each player's objective function, and the submodularity (Definition 3) of the social utility. By contrast, Borodin et al. [4] show that both properties fail to hold for most natural definitions of competitive threshold games. At the heart of the counter-examples in [4] lies the following kind of dynamic: At time 1, a node u recommends to v the use of a Galaxy phone, but fails to convince v. At time 2, another node w recommends to v the use of an iPhone. If v decides to adopt a smartphone at time 2, most natural versions of a threshold model (as well as under the general adoption model) allow for an adoption of a Galaxy phone as well. This "extra chance" results in synergistic effects between competitors, and thus breaks competitiveness. Under the model of [11], this problem is side-stepped. v will only consider adopting a smartphone in step 2 when she appears in the sequence at time 2; in that case, adoption of a Galaxy phone in step 2 will be considered independently of whether w has adopted an iPhone. This observation fleshes out the discussion alluded to in Footnote 5.

2.5   Valid Utility Games

A valid utility game [22] is defined on a ground set V with social utility function γ defined on subsets of V. The strategies of the game are sets $S_0^i \subseteq V$ (it is possible that not all sets are allowed as strategies for some or all players), and the payoff functions are σ^i for each player i. The social utility is defined on the union of all players' sets, $\gamma(\bigcup_{i=1}^N S_0^i)$. The definition requires that three conditions hold: (1) The social utility function $\gamma(\cdot)$ is submodular; (2) For each player i, $\sigma^i(S_0) \geq \gamma(S_0) - \gamma(S_0^{-i}, \emptyset^i)$; (3) $\sum_{i=1}^N \sigma^i(S_0) \leq \gamma(S_0)$.

3   Upper Bound on the Coarse Price of Anarchy

In this section, we present our main result: the upper bound on the coarse PoA with submodular activation functions. We prove the upper bound on the PoA by showing that the competitive cascade game is a valid utility game. We note, however, that the strategy space of our competitive cascade game consists of multisets, whereas the standard definition of utility games has only sets as strategies. In order to deal with this subtle technical issue, we use the following lemma, whose proof is given in the appendix.

Lemma 2. *Let* $\mathcal{G} = (G, \{h_v^i\}, \{K_v^i\}, \{B^i\}, \{\omega_v\}, Q)$ *be an arbitrary instance of the competitive cascade game. Then, there exists an instance* $\hat{\mathcal{G}} = (\hat{G}, \{\hat{h}_v^i\}, \{\hat{K}_{\hat{v}}^i\}, \{\hat{B}^i\}, \{\hat{\omega}_v\}, \hat{Q})$ *with the same set of players, and the following properties:*

1. $\sum_{i=1}^N \hat{K}_{\hat{v}}^i \le 1$ *for all* $v \in \hat{V}$. *(At most one player is allowed to target a node, and with at most one resource.)*
2. *For every player* i, *there are mappings* $\mu^i, \hat{\mu}^i$ *mapping* i's *strategies in* $\mathcal{G}$ *to his strategies in* $\hat{\mathcal{G}}$ *and vice versa, respectively, satisfying the following property: If for all* i, *either* $\mu^i(M^i) = \hat{M}^i$ *or* $\hat{\mu}^i(\hat{M}^i) = M^i$, *then for all* i, $\sigma^i(\boldsymbol{M}) = \hat{\sigma}^i(\hat{\boldsymbol{M}})$.

In particular, Lemma 2 implies that the social utility is also preserved between the two games, and strategies M^i are best responses to $M^j, j \ne i$ if and only if the $\hat{M}^i$ are best response to $\hat{M}^j, j \ne i$. (Otherwise, a player could improve his payoff in the other game by switching to $\mu^i(M^i)$ or $\hat{\mu}^i(\hat{M}^i)$.) In this sense, Lemma 2 establishes that for every competitive cascade game instance, there is an "equivalent" instance in which each node can be targeted by at most one player, and with at most one resource. Therefore, we will henceforth assume without loss of generality that the strategy space for each player consists only of sets.

In fitting the competitive cascade game into the valid utility game framework, the ground set of the game is V, and the payoff function of player i is $\sigma^i(\boldsymbol{S}) = \mathbb{E}[\sum_{v \in S_t^i} \omega_v]$: the sum of weights from the nodes $v \in V$ in state i at the end of the updating sequence. Because of additivity, the social utility function depends only on S_0. That is, the following is well-defined: $\gamma(\boldsymbol{S_0}) = \gamma(S_0) = \sum_i \sigma^i(\boldsymbol{S_0})$. Therefore, the third condition of a valid utility game (sum boundedness) is satisfied trivially. Below, we will prove the following two lemmas:

Lemma 3. *Assume that for every node* v, *the total activation probability* $H_v(\boldsymbol{S})$ *is additive, and the activation function* $f_v(S)$ *is submodular. Then, the social utility function* $\gamma(S_0)$ *is submodular and monotone.*

Lemma 4. *If for every node* v *and ever player* k, $H_v(\cdot)$ *is additive and* $h_v^k(\cdot)$ *is competitive, then for each player* i, *we have* $\sigma^i(\boldsymbol{S_0}) \ge \gamma(\boldsymbol{S_0}) - \gamma(\boldsymbol{S_0^{-i}}, \emptyset^i)$.

Lemmas 3 and 4 together establish that the competitive cascade game is a valid utility game. Example 1.4 in [19] shows that the coarse PoA of valid utility games is at most 2 (Vetta [22] establishes the same for the PoA), proving the following main result of our paper:

Theorem 3. *Assume that the following conditions hold:*

1. *For every node* v, *the total activation probability* $H_v(\boldsymbol{S})$ *is additive.*
2. *For every node* v, *the activation function* $f_v(S)$ *is submodular.*
3. *For every player* i *and node* v *in the graph, the adoption function* $h_v^i(\boldsymbol{S})$ *is competitive.*

Then, the upper bound on the PoA (and coarse PoA) is 2 in the competitive cascade game.

By Lemma 1, the switching-selection model with concave switching functions and linear selection functions is a special case of the general adoption model with competitive adoption functions, additive activation probabilities and submodular activation functions. Therefore, Theorem 1 follows naturally as a corollary of Theorem 3.

Proof of Lemma 3. We build an instance of the general threshold model whose influence coverage function $\sigma_w(S_0)$ is exactly the same as $\gamma(S_0)$. The idea is that for additive functions, the social utility does not depend on which node chooses which company, so the game is reduced to the case of just a single influence. The update sequence can be emulated with a time-expanded layered graph.

The time-expanded graph G_ℓ is defined as follows.[7] For each node v of the original graph, we have $\ell + 1$ nodes $\hat{v}_0, \hat{v}_1, \ldots, \hat{v}_\ell$ in G_ℓ. We use $L_t = \{\hat{v}_t \mid v \in V\}$ to denote the set of nodes in layer t. The activation functions are defined as follows:

1. In layer 0, $\hat{f}_{\hat{v}_0} \equiv 0$ for every node $v \in V$.
2. In layer t, $1 \leq t \leq \ell$, consider a node v with switching function f_v. If v is the tth element of the updating sequence ($v = v_t$), we set

$$\hat{f}_{\hat{v}_t}(S) = \begin{cases} 1 & \text{if } \hat{v}_{t-1} \in S \\ f_v(\{u \mid \hat{u}_{t-1} \in S\}) & \text{otherwise}; \end{cases}$$

otherwise we set

$$\hat{f}_{\hat{v}_t}(S) = \begin{cases} 1 & \text{if } \hat{v}_{t-1} \in S \\ 0 & \text{otherwise}. \end{cases}$$

Finally, the total influence is defined as $\sigma_w(S_0) = \mathbb{E}[\sum_{\hat{v}_\ell \in \hat{S}} w_v]$, where $\hat{S}$ is the set of nodes activated in the threshold model once no more activations occur. In the instance, each layer L_t emulates one update in the original update sequence of the general adoption model.

For each node $\hat{v}_t$ in the layered graph, $f_{\hat{v}_t}(S)$ is submodular and additive. The submodularity and monotonicity for the 0-1 activation functions are trivially satisfied. For the nodes in the update sequence, submodularity holds because we assumed the $f_v(S)$ to be submodular, and monotonicity follows because the $H_v(S)$ are additive.

Next, we show that $\gamma(S_0) = \sigma_w(S_0)$, by using a straightforward coupling between the general threshold model and the general adoption model. According to the construction of G_ℓ, the state changes for all nodes except $\hat{v}_t$ (where v_t is the tth element of the updating sequence) are deterministic. Therefore, we only need to draw the thresholds $\Theta = \langle \theta_1, \theta_2, \ldots, \theta_\ell \rangle$ for the ℓ nodes in the update sequence: they are drawn independently and uniformly from $[0, 1]$. In the general adoption model, when updating the tth node v_t in the sequence, if node v_t is still inactive, she becomes active if and only if $f_v(S_{t-1}) \geq \theta_t$. If the node is already activated, she remains activated in the same state. By induction on t, $v \in S_t$ if and only if $\hat{v}_t \in \hat{S} \cap L_t$. Thus, the outcomes of the two

---

[7] All activation information is encoded in the activation functions. Therefore, there is no need to explicitly specify the edges of G_ℓ.

processes are the same pointwise over threshold vectors Θ: $\gamma(S_0|\Theta) = \sigma_\omega(S_0|\Theta)$. In particular, their expectations are thus the same.

Finally, Theorem 2 establishes the monotonicity and submodularity of $\sigma_\omega(S_0)$, and thus also $\gamma(S_0)$. ∎

Proof of Lemma 4. We begin by showing that under the assumptions of the lemma, $\sigma^i(S_0) \leq \sigma^i(S_0^{-k}, \emptyset^k)$ for all players $k, i, k \neq i$. To do so, we exhibit a simple coupling of the general adoption processes for the two initial states S_0 and $(S_0^{-k}, \emptyset^k)$, essentially identical to one used in the proof of Lemma 1 in [11]. Notice that the activation functions are additive; therefore, we can combine all states $k \neq i$ into one state, which we denote by $-i$.

The activation process is defined by the way in which nodes decide whether to update their state, and if so, to which new state. An equivalent way of describing the choice is as follows: for each step t of the update sequence, we draw an independent uniformly random number $z_t \in [0, 1]$. In step t, assuming that node v_t is still in state 0, she changes her state to:

- state i if $z_t \in [0, h_{v_t}^i(S_{t-1}))$.
- state $-i$ if $z_t \in [h_{v_t}^i(S_{t-1}), f_{v_t}(\bigcup_{j=1}^N S_{t-1}^j))$.
- state 0 otherwise.

To couple the two random processes with starting conditions $(S_0^{-k}, \emptyset^k)$ and S_0, we simply choose the same values z_t for both. Let X_t^j denote the set of nodes in state j, $j \in \{i, -i, 0\}$ after t updates with starting condition $(S_0^{-k}, \emptyset^k)$. Y_t^j is defined analogously, with starting condition S_0.

Conditioned on any choice of $(z_1, \ldots, z_\ell)$, a simple induction proof using competitiveness of the h_v^i and monotonicity of the f_v shows that for each time t, $X_t^i \supseteq Y_t^i$, $X_t^0 \supseteq Y_t^0$, and thus also $X_t^{-i} \subseteq Y_t^{-i}$. Therefore, at the end of the update sequence, the desired inequality holds pointwise over $(z_1, \ldots, z_\ell)$, and in particular in expectation, as claimed. Finally, having established that $\sigma^i(S_0) \leq \sigma^i(S_0^{-k}, \emptyset^k)$, we use it in the following calculations:

$$\gamma(S_0) - \gamma(S_0^{-i}, \emptyset^i) = \sum_k (\sigma^k(S_0) - \sigma^k(S_0^{-i}, \emptyset^i))$$

$$= \sigma^i(S_0) + \sum_{k \neq i} (\sigma^k(S_0) - \sigma^k(S_0^{-i}, \emptyset^i))$$

$$\leq \sigma^i(S_0).$$

∎

4 Tightness of the PoA Upper Bound

We give an instance of the competitive cascade game in the (more restrictive) switching-selection model to show that our upper bound of 2 for the PoA in Theorem 3 is tight.

Let N be the number of players. The graph consists of a star with one center and N leaves, as well as N isolated nodes. Each player has only one unit of budget, and the update sequence is any permutation of the nodes in the star graph. The switching functions are the constant 1 for all nodes in the graph, which implies that if a node has any neighbor who has adopted the product, the node also adopts the product. The selection functions are simply the fraction of neighbors who have adopted the product previously. Under this instance, the unique Nash equilibrium has every player allocating his unit of budget to the center node of the star graph. By placing the budget at the center node, the expected payoff for each player is $\frac{N+1}{N}$, while placing it on any other node at most leads to a payoff of 1. However, the strategy that optimizes the social utility is to place one unit of budget at the center node of the star graph while placing all others at the isolated nodes. Thus, the PoA (and also Price of Stability) is $\frac{2N}{N+1}$. As N goes to infinity, the lower bound on the PoA tends to 2. Therefore, we have proved the following proposition:

Proposition 1. *The upper bound of 2 on the PoA (and thus also coarse PoA) is tight for the competitive cascade game even for the simpler switching-selection model.*

5   Conclusion and Future Work

We have studied the efficiency of resource allocation at equilibria of the competitive cascade game in terms of the *Price of Anarchy* (PoA). We have shown that an improved bound compared to [11] follows from several well-understood and general phenomena. This cleaner approach has led to a simpler proof of a more general result: for the N-player competitive cascade game, the coarse PoA is upper-bounded by 2 under any graph structure. We have also shown that this bound is tight.

It is open whether the same (or a slightly weaker) bound can be guaranteed without the assumption of submodularity of the activation functions (but assuming competitiveness and additivity). The techniques from [11] can be generalized to give an upper bound of $2N$ in this case, but do not directly yield any better bounds.

At a more fundamental level, it would be desirable to broaden the models considered for competitive cascades. Most positive results on either algorithmic questions or the PoA — the present one included — rely on submodularity properties of the particular modeling choices. (That such properties are also at the heart of the model of Goyal and Kearns is our main insight here.) It would be desirable to find models for which positive results — algorithmic or game-theoretic — can be obtained without requiring submodularity. Furthermore, most work on cascade models so far has assumed that nodes only adopt a single product. In many cases, products may be *partly* in competition, but not fully so. One of the few papers to consider a model with partial compatibility between products is [13]; an exploration of the game-theoretic implications of such a model would be of interest.

References

1. Alon, N., Feldman, M., Procaccia, A.D., Tennenholtz, M.: A note on competitive diffusion through social networks. Information Processing Letters 110, 221–225 (2010)
2. Bharathi, S., Kempe, D., Salek, M.: Competitive influence maximization in social networks. In: Deng, X., Graham, F.C. (eds.) WINE 2007. LNCS, vol. 4858, pp. 306–311. Springer, Heidelberg (2007)

3. Blum, A., Hajiaghayi, M., Ligett, K., Roth, A.: Regret minimization and the price of total anarchy. In: Proc. 39th ACM Symp. on Theory of Computing, pp. 373–382 (2008)
4. Borodin, A., Filmus, Y., Oren, J.: Threshold models for competitive influence in social networks. In: Saberi, A. (ed.) WINE 2010. LNCS, vol. 6484, pp. 539–550. Springer, Heidelberg (2010)
5. Budak, C., Agrawal, D., Abbadi, A.E.: Limiting the spread of misinformation in social networks. In: 20th Intl. World Wide Web Conference, pp. 665–674 (2011)
6. Carnes, T., Nagarajan, C., Wild, S.M., van Zuylen, A.: Maximizing influence in a competitive social network: A follower's perspective. In: Proc. Intl. Conf. on Electronic Commerce, ICEC (2007)
7. Chen, W., Collins, A., Cummings, R., Ke, T., Liu, Z., Rincon, D., Sun, X., Wang, Y., Wei, W., Yuan, Y.: Influence maximization in social networks when negative opinions emerge and propagate. In: Proc. 11th SIAM Intl. Conf. on Data Mining, pp. 379–390 (2011)
8. Chen, W., Wang, Y., Yang, S.: Efficient influence maximization in social networks. In: Proc. 15th Intl. Conf. on Knowledge Discovery and Data Mining, pp. 199–208 (2009)
9. Domingos, P., Richardson, M.: Mining the network value of customers. In: Proc. 7th Intl. Conf. on Knowledge Discovery and Data Mining, pp. 57–66 (2001)
10. Dubey, P., Garg, R., De Meyer, B.: Competing for customers in a social network: The quasi-linear case. In: Spirakis, P.G., Mavronicolas, M., Kontogiannis, S.C. (eds.) WINE 2006. LNCS, vol. 4286, pp. 162–173. Springer, Heidelberg (2006)
11. Goyal, S., Kearns, M.: Competitive contagion in networks. In: Proc. 43rd ACM Symp. on Theory of Computing, pp. 759–774 (2012)
12. He, X., Song, G., Chen, W., Jiang, Q.: Influence blocking maximization in social networks under the competitive linear threshold model. In: Proc. 12th SIAM Intl. Conf. on Data Mining (2012)
13. Immorlica, N., Kleinberg, J., Mahdian, M., Wexler, T.: The role of compatibility in the diffusion of technologies through social networks. In: Proc. 9th ACM Conf. on Electronic Commerce, pp. 75–83 (2007)
14. Kempe, D., Kleinberg, J., Tardos, E.: Maximizing the spread of influence in a social network. In: Proc. 9th Intl. Conf. on Knowledge Discovery and Data Mining, pp. 137–146 (2003)
15. Kempe, D., Kleinberg, J.M., Tardos, É.: Influential nodes in a diffusion model for social networks. In: Caires, L., Italiano, G.F., Monteiro, L., Palamidessi, C., Yung, M. (eds.) ICALP 2005. LNCS, vol. 3580, pp. 1127–1138. Springer, Heidelberg (2005)
16. Kimura, M., Saito, K.: Tractable models for information diffusion in social networks. In: Fürnkranz, J., Scheffer, T., Spiliopoulou, M. (eds.) PKDD 2006. LNCS (LNAI), vol. 4213, pp. 259–271. Springer, Heidelberg (2006)
17. Mossel, E., Roch, S.: Submodularity of influence in social networks: From local to global. SIAM J. Comput. 39(6), 2176–2188 (2010)
18. Rogers, E.: Diffusion of innovations, 5th edn. Free Press (2003)
19. Roughgarden, T.: Intrinsic robustness of the price of anarchy. In: Proc. 40th ACM Symp. on Theory of Computing, pp. 513–522 (2009)
20. Tsai, J., Nguyen, T.H., Tambe, M.: Security games for controlling contagion. In: Proc. 27th AAAI Conf. on Artificial Intelligence (2012)
21. Tzoumas, V., Amanatidis, C., Markakis, E.: A game-theoretic analysis of a competitive diffusion process over social networks. In: Goldberg, P.W. (ed.) WINE 2012. LNCS, vol. 7695, pp. 1–14. Springer, Heidelberg (2012)
22. Vetta, A.: Nash equlibria in competitive societies with applications to facility location, traffic routing and auctions. In: Proc. 43rd IEEE Symp. on Foundations of Computer Science, pp. 416–425 (2002)

23. Wang, C., Chen, W., Wang, Y.: Scalable influence maximization for independent cascade model in large-scale social networks. Data Mining and Knowledge Discovery Journal 25(3), 545–576 (2012)
24. Wang, Y., Cong, G., Song, G., Xie, K.: Community-based greedy algorithm for mining top-k influential nodes in mobile social networks. In: Proc. 16th Intl. Conf. on Knowledge Discovery and Data Mining, pp. 1039–1048 (2010)

A   Proof of Lemma 2

We restate Lemma 2 for convenience.

Lemma 2. *Let $\mathcal{G} = (G, \{h_v^i\}, \{K_v^i\}, \{B^i\}, \{\omega_v\}, Q)$ be an arbitrary instance of the competitive cascade game. Then, there exists an instance $\hat{\mathcal{G}} = (\hat{G}, \{\hat{h}_v^i\}, \{\hat{K}_{\hat{v}}^i\}, \{\hat{B}^i\}, \{\hat{\omega}_v\}, \hat{Q})$ with the same set of players, and the following properties:*

1. *$\sum_{i=1}^{N} \hat{K}_{\hat{v}}^i \le 1$ for all $v \in \hat{V}$. (At most one player is allowed to target a node, and with at most one resource.)*
2. *For every player i, there are mappings $\mu^i, \hat{\mu}^i$ mapping i's strategies in $\mathcal{G}$ to his strategies in $\hat{\mathcal{G}}$ and vice versa, respectively, satisfying the following property: If for all i, either $\mu^i(M^i) = \hat{M}^i$ or $\hat{\mu}^i(\hat{M}^i) = M^i$, then for all i, $\sigma^i(M) = \hat{\sigma}^i(\hat{M})$.*

Proof.   Given $\mathcal{G}$, we construct $\hat{\mathcal{G}}$ as a game on a graph with three layers.

Nodes: The first layer contains, for each node $v \in V$ and player i, a set of K_v^i new nodes $V_v^i = \{v_1, \ldots, v_{K_v^i}\}$. The second layer of $\hat{G}$ contains, for each node $v \in V$, one node v' connected to all nodes in V_v^i. The third layer is a copy of the original graph G.

Node Budgets: For player i and any node $v \in G$, we set $\hat{K}_{\hat{v}}^i = 1$ for all nodes $\hat{v} \in V_v^i$, and $\hat{K}_{\hat{v}}^i = 0$ for all other nodes (including all nodes in layers 2 and 3). In other words, player i may only target nodes that are in V_v^i for some $v \in G$.

Budgets: We set $\hat{B}^i = B^i$, for all players i.

Weights: We set $\hat{\omega}_v \equiv 0$ for all nodes in the first layer. If a node v appears (at least once) in Q, then we set $\hat{\omega}_v = \omega_v$ in the third layer and $\hat{\omega}_{v'} = 0$ in the second layer. If v does not appear in Q, then we set $\hat{\omega}_v = 0$ in the third layer and $\hat{\omega}_{v'} = \omega_v$ in the second layer. Thus, players are interested in influencing nodes in the second or third layer, depending on whether the node can be influenced via the update sequence, or must be influenced by direct targeting.

Adoption Functions: Different adoption functions are used for the nodes in different layers:

1. In layer 1, $\hat{h}_{\hat{v}}^i(\cdot) \equiv 0$ for any player i and node $\hat{v}$.
2. In layer 2, for a player i and node v,

$$\hat{h}_{v'}^i(\boldsymbol{S}) = \begin{cases} 0 & \text{if } \bigcup_{k=1}^{N}(S^k \cap V_v^k) = \emptyset \\ \frac{|S^i \cap V_v^i|}{\sum_k |S^k \cap V_v^k|} & \text{otherwise.} \end{cases}$$

3. To simplify notation, we define $A^i = \{v \in V | v \in S^i \text{ or } v' \in S^i\}$. Then in layer 3, for player i and node v,

$$\hat{h}_v^i(S) = \begin{cases} 0 & \text{if } v' \in S^j \text{ for some } j \neq i \\ 1 & \text{if } v' \in S^i \\ h_v^i(A^1, \ldots, A^N) & \text{otherwise.} \end{cases}$$

Notice that the game $\hat{G}$ satisfies competitiveness, additivity and submodularity whenever the game G satisfies all these three properties.

Update Sequence: The update sequence is $\hat{Q} = \langle v'_1, v'_2, \ldots, v'_{|V|}, v_1, \ldots, v_\ell \rangle$, where $Q = \langle v_1, \ldots, v_\ell \rangle$ is the update sequence of the original instance and $v'_1, v'_2, \ldots, v'_{|V|}$ are all the nodes in the second layer, in some arbitrary order. The first $|V|$ updates in $\hat{G}$ emulate the seeding stage in G, and the remaining ℓ updates emulate the update sequence Q.

Payoffs and Social Utility: The players' payoff functions $\hat{\sigma}^i(M)$ and the social utility $\hat{\gamma}(M)$ are defined as usual in terms of the other modeling parameters.

The mappings μ^i are defined as follows. Let M^i be i's strategy in G, characterized by the budgets α_v^i that i puts on nodes v. For each node $v \in V$, we choose an arbitrary (but fixed) set $\hat{M}_v^i$ of α_v^i nodes in V_v^i. Player i's strategy is $\hat{M}^i = \hat{\mu}^i(M^i) = \bigcup_v \hat{M}_v^i$. Conversely, we define $\hat{\mu}^i$ as follows: For any strategy profile $\hat{M}$ of $\hat{G}$, and for each node $v \in V$, we set $\alpha_v^i = |\hat{M}^i \cap V_v^i|$. $\hat{\mu}^i(\hat{M}^i)$ is the strategy in which player i puts α_v^i resources on node v, for all v.

The first claim of the lemma holds by definition. For the second claim, consider two strategy profiles M and $\hat{M}$, such that for all players i, either $\hat{M}^i = \mu^i(M^i)$ or $M^i = \hat{\mu}^i(\hat{M}^i)$. We show that $\hat{\sigma}^i(\hat{M}) = \sigma^i(M)$ for each player i. To do so, we exhibit a coupling of the random choices between the two games G and $\hat{G}$. The coupling is quite similar to the one used in the proof of Lemma 4.

For the seeding stage of G, an equivalent way of describing the choice is as follows: for each node $v \in V$, we draw an independent uniformly random number $z_v \in [0, 1]$. The state of node v is

- 0 if $Z_v = \sum_{i=1}^N \alpha_v^i = 0$,
- $i > 0$ if $z_v \in \left(\frac{\sum_{j=1}^{i-1} \alpha_v^j}{Z_v}, \frac{\sum_{j=1}^i \alpha_v^j}{Z_v} \right]$.

Similarly, for the updates in the diffusion stage for both G and $\hat{G}$, an equivalent way of describing the update in step t is the following. Draw an independent uniformly random number $z_t \in [0, 1]$. If node v_t is in a state $i > 0$, she retains her current state. Otherwise, she changes her state to

- $i > 0$ if $z_t \in [\sum_{j=1}^{i-1} h_{v_t}^j(S_{t-1}), \sum_{j=1}^i h_{v_t}^j(S_{t-1}))$,
- 0 otherwise.

To couple the two random processes, we simply choose the same values $z_v = \hat{z}_{t_{v'}}$ and $z_t = \hat{z}_{t+|V|}$, where $t_{v'}$ is the order of node v' in the update sequence $\hat{Q}$.

Since the strategy $\hat{M}$ consists of sets instead of multisets, the seeding stage of $\hat{\mathcal{G}}$ is deterministic. In $\hat{\mathcal{G}}$, a node v is initially in state $i > 0$ if and only if player i selects her as a seed. Let $\hat{S}_0^i$ be the set of activated nodes after the seeding stage with strategy profile $\hat{M}$. We have $|\hat{S}_0^i \cap V_v^i| = \alpha_v^i$, for all players i and nodes v.

Conditioned on any fixed choice of the z_v (and thus $\hat{z}_1, \ldots, \hat{z}_{|V|}$), we have $\boldsymbol{S_0} = \hat{\boldsymbol{S}}_{|V|}$, where $\boldsymbol{S_0}$ is the vector of sets of nodes in state i after the seeding stage with strategy profile M, and $\hat{\boldsymbol{S}}_{|V|}$ is the vector of sets of nodes in layer 2 of $\hat{\mathcal{G}}$ in state i after the first $|V|$ update steps with strategy profile M.

Finally, a simple induction proof over the ℓ steps of the update sequence Q shows that for each time t, we have the following property: (1) if v appears in Q at least once before time step t, then $v \in S_t^i$ if and only if $v \in \hat{S}_{t+|V|}^i$. (2) if v does not appear in Q before time step t, then $v \in S_t^i$ if and only if $v' \in \hat{S}_{t+|V|}^i$. Applying this result after all ℓ steps, we obtain that each node v appearing in Q has $v \in S_\ell^i$ if and only if $v \in \hat{S}_{\ell+|V|}^i$, and each node v not appearing in Q has $v \in S_\ell^i$ if and only if $v' \in \hat{S}_{\ell+|V|}^i$. Notice that the corresponding nodes v or v' in $\hat{\mathcal{G}}$ are exactly the ones inheriting the weight of node v in G, implying that the payoff of each player i is the same pointwise in $\mathcal{G}$ and $\hat{\mathcal{G}}$. Thus, each player's expected payoff is also the same in the two games, completing the proof. ∎

Designing Profit Shares in Matching and Coalition Formation Games[*]

Martin Hoefer[1] and Lisa Wagner[2]

[1] Max-Planck-Institut für Informatik and Saarland University, Germany
mhoefer@mpi-inf.mpg.de
[2] Dept. of Computer Science, RWTH Aachen University, Germany
lwagner@cs.rwth-aachen.de

Abstract. Matching and coalition formation are fundamental problems in a variety of scenarios where agents join efforts to perform tasks, such as, e.g., in scientific publishing. To allocate credit or profit stemming from a joint project, different communities use different crediting schemes in practice. A natural and widely used approach to profit distribution is equal sharing, where every member receives the same credit for a joint work. This scheme captures a natural egalitarian fairness condition when each member of a coalition is critical for success. Unfortunately, when coalitions are formed by rational agents, equal sharing can lead to high inefficiency of the resulting stable states.

In this paper, we study the impact of changing profit sharing schemes in order to obtain good stable states in matching and coalition formation games. We generalize equal sharing to sharing schemes where for each coalition each player is guaranteed to receive at least an α-share. This way the coalition formation can stabilize on more efficient outcomes. In particular, we show a direct trade-off between efficiency and equal treatment. If k denotes the size of the largest possible coalition, we prove an asymptotically tight bound of $k^2\alpha$ on prices of anarchy and stability. This result extends to polynomial-time algorithms to compute good sharing schemes. Further, we show improved results for a novel class of matching problems that covers many well-studied cases, including two-sided matching and instances with integrality gap 1.

1 Introduction

Matching problems are central to a variety of research at the intersection of computer science and economics. The standard model of matching with preferences is stable matching, in which a set of agents strives to group into pairs, and each agent has an ordinal preference list over all possible partners. In this case, a matching is stable if it has no blocking pair, i.e., no pair of players could both improve by pairing up and dropping their current partners. Applications of this model include, e.g., matching in job markets, hospitals, colleges, social

---

[*] Supported by DFG Cluster of Excellence MMCI and grant Ho 3831/3-1.

Y. Chen and N. Immorlica (Eds.): WINE 2013, LNCS 8289, pp. 249–262, 2013.
© Springer-Verlag Berlin Heidelberg 2013

networks, or distributed systems [2, 8, 15, 17, 18]. Numerous extensions of this standard model have been treated in the past [15, 22].

While the basic stable matching model uses ordinal preferences, many applications allow cardinal preferences to express incentives in terms of profit or reward. Perhaps the most prominent case of cardinal preferences studied in the literature are correlated preferences, in which each matched pair generates a profit that is shared equally among the involved agents. This model has favorable properties, e.g., existence of a stable matching is guaranteed by a potential function argument, and convergence time of improvement dynamics is polynomial [1]. These conditions extend even to hedonic coalition formation games, when instead of matching pairs the agents construct a partition into coalitions of $k > 2$ players. However, these properties come at a cost – the total reward of every stable coalition structure can be up to k times smaller than in optimum.

For agents working on a joint project, equal sharing implements a natural egalitarian fairness condition. For example, in mathematics and theoretical computer science it is common practice to list authors in alphabetical order, which gives equal credit to every author involved in a paper. This is justified by the argument that a ranking of ideas that led to the results in a paper is often impossible. On the other hand, in many other sciences the author sequence gives different credit to the different authors involved in the project. In some cases, these approaches are overruled by the community which gives most credit for a paper to its most prominent author (or to authors that are PhD students). Naturally, such different profit sharing schemes generate different incentives for the agents to form coalitions. In this paper, we study the impact of profit distribution on fairness and efficiency in the resulting coalition formation games. It is known that completely arbitrary sharing can lead to non-existence of stable states or arbitrarily high price of anarchy [3]. Similar to recent work [20], our focus is to design good profit or credit distribution schemes such that the stable states implement good outcomes. In this direction, it is not difficult to observe that using arbitrarily low profit shares we can stabilize every optimal partition. However, such sharing schemes are clearly undesirable when we want to maintain egalitarian fairness conditions. In our analysis, we provide asymptotically tight bounds on the inherent tension between efficiency and equal treatment. Further, we give efficient algorithms to compute good sharing schemes and show complementing hardness results. Before we state our results, we start with a formal description of the model.

1.1   Stable Matchings and Coalition Structures

We assume that there is a simple, undirected graph $G = (V, E)$, where V is the set of *agents* and E the set of possible *projects* or *edges*. In the matching case, we assume each edge is a pair $e = \{u, v\} \in E$ and yields some *profit* $w(e) > 0$ that is to be shared among u and v. Our goal is to design a profit distribution scheme d with $d_u(e), d_v(e) \in [0, 1]$ and $d_u(e) + d_v(e) = 1$ for all $e \in E$. This implies that u gets individual profit $d_u(e)w(e)$ when being matched in e. The profits yield an instance of stable matching with cardinal preferences. The *stable matchings*

$M \subset E$ are matchings that allow no *blocking pair* – no pair $\{u,v\} \in E \setminus M$ of agents that can both strictly increase their individual profit by destroying their incident edge in M (if any) and creating $\{u,v\}$. The social welfare of a matching M is $w(M) = \sum_{e \in M} w(e)$. We denote by M^* a (possibly non-stable) optimum matching that maximizes social welfare. The *price of anarchy/stability* (denoted PoA/PoS) is the ratio of $w(M^*)/w(M)$, where M is the worst/best stable matching, respectively. While the set of stable matchings depends on d, the social welfare of a particular matching is independent of d, and so is the set of optimum matchings.

We generalize this scenario to hedonic coalition formation with arbitrary coalition size in a straightforward way. Instead of edges e we are given a set of possible *hyperedges* or *coalitions* $C \subseteq 2^V$. Each $S \in C$ fulfills $|S| \geq 2$ and yields profit $w(S) > 0$. The distribution scheme has $d_u(S) \in [0,1]$ and $\sum_{u \in S} d_u(S) = 1$. Then d again specifies the fraction of $w(S)$ allocated to $u \in S$ when coalition S forms. A *coalition structure* $\mathcal{S} \subseteq C$ is a collection of sets from C that is mutually disjoint. $\mathcal{S}$ is *core-stable* if there is no *blocking coalition* – no $S \in C \setminus \mathcal{S}$ of agents that each and all strictly improve their individual profit by destroying their incident coalition in $\mathcal{S}$ (if any) and creating S. Observe that the usual definition of core-stability involves all possible coalitions $S \subseteq 2^V$. We can easily allow this by assuming that $w(\{v\}) = 0$ for all $v \in V$ and $w(S) = -1$ for all $S \in 2^V \setminus C$ with $|S| > 1$. Definitions of social welfare and prices of anarchy and stability extend in the obvious way. An instance is called *inclusion monotone* if for $S, S' \in C$ and $S' \subset S$, we have $w(S)/|S| \geq w(S')/|S'|$. We denote by $k = \max_{S \in C} |S|$. Stable matchings are exactly core-stable coalition structures when $|S| = 2$ for all $S \in C$. By definition, every such instance is inclusion monotone.

Our aim is to design d in order to obtain good core-stable coalition structures. To characterize the tension between stability and equal treatment, a distribution scheme is termed α-*bounded* if $d_u(S) \geq \alpha$ for all $S \in C$ and all $u \in S$. If d is α-bounded, the resulting instance of hedonic coalition formation is termed α-*egalitarian*.

Throughout the paper, we assume a profit of 0 for singleton coalitions, which is in some sense without loss of generality. Suppose we have $\{v\} \in C$ with $w(\{v\}) > 0$ for a node $v \in V$. Such a player will participate in a coalition only if he receives profit at least $w(\{v\})$. Thus, we can reduce the profit of every coalition $S \in C$ with $v \in S$ by this amount. By executing this step for every player and every coalition, we obtain an instance with the desired properties. Coalitions that arrive at zero profit in this way can be disregarded, as they can be assumed to be neither part of any equilibrium nor in the optimum solution. Applying our algorithms to the remaining instance, we strive to equally distribute the surplus that the coalition generates over individually required profits. This objective is closely related to Nash bargaining solutions [21]. Note that in the remaining instance with all $w(\{v\}) = 0$, we get larger prices of anarchy and stability. As our bounds apply to all instances of this sort, they continue to hold accordingly for all instances with arbitrary positive $w(\{v\})$.

1.2   Results and Overview

In Section 2 we characterize the effect of α-boundedness on the resulting prices of anarchy and stability. We provide asymptotically tight bounds on prices of anarchy and stability depending on α. Given an optimum coalition structure $\mathcal{S}^*$, we show how to design a distribution scheme d that guarantees (1) existence of a core-stable coalition structure and (2) a price of stability of $\max\{1, k^2\alpha\}$, for any $\alpha \in [0, 1/k]$. This result shows, in particular, that for every $\alpha \leq 1/k^2$, we can construct α-bounded schemes with an optimal core-stable coalition structure. This is asymptotically tight – using α-bounded schemes we cannot achieve a price of stability of less than $(k^2 - k)\alpha$, i.e., the price of stability for α-bounded schemes is in $\Theta(k^2)\alpha$. Conversely, this bound translates into a bound on $\alpha \leq \delta/(k^2 - k)$ to guarantee price of stability at most δ. For inclusion monotone instances, we can also provide the same upper bound of $\max\{1, k^2\alpha\}$ on the price of anarchy, i.e., in such instances and $\alpha \leq 1/k^2$, every core-stable structure is optimal. In contrast, there exist instances that are not inclusion monotone, in which α-bounded schemes cannot guarantee a price of anarchy of 1, even for arbitrarily small $\alpha > 0$.

While computing $\mathcal{S}^*$ is NP-hard, we can also combine our algorithms with efficient approximation algorithms for the set packing problem of optimizing social welfare. If $\mathcal{S}^*$ is an arbitrary ρ-approximation to the optimum solution, our algorithms can be used to construct in polynomial time an α-bounded distribution scheme that guarantees price of stability of $\rho \cdot \max\{1, k^2\alpha\}$. The same result can be achieved for the price of anarchy in inclusion monotone instances.

In addition, we study a problem inspired by computing core imputations in coalitional games. For a given coalition structure with profits we aim to determine a distribution scheme with largest α that stabilizes a given optimum solution $\mathcal{S}^*$. This problem is shown NP-hard whenever we have coalitions of size $k \geq 3$. The problem remains hard for $k = 2$ if instead of a solution, we have a given bound W, and the goal is to maximize α such that at least one solution of social welfare at least W is stable.

In Section 3 we study stable matching games. As the general results from the previous section carry over, we concentrate on a subclass of instances that we term acyclic alternating. This includes many standard cases such as bipartite matching or instances for which the standard matching LP is integral. In this case, we can show that even $1/3$-bounded distribution schemes yield a PoS of 1. In addition, given an instance and any solution M, an α-bounded distribution scheme stabilizing M with maximal α can be found efficiently.

1.3   Related Work

We study profit distribution in cardinal stable matching and more general games. Stable matching has been extensively studied [22] and the literature on the problem is too vast to survey here. Directly related to our work are [4,5] which address the price of anarchy in stable matching and related models. Very recently, we have studied the price of anarchy under different edge-based profit sharing schemes [3]. In contrast, this paper concentrates on designing profit shares to guarantee good stable matchings.

Profit sharing in more general coalition formation games has been studied recently [6] in a related model, where coalitions are represented by resources. Agents can join and leave a resource/coalition unilaterally. The authors focus on submodular profit functions and three particular sharing schemes. For the resulting games, they derive results on existence of pure Nash equilibrium, price of anarchy, and convergence of improvement dynamics. In contrast to this model, we do not restrict the number of coalitions that can be formed simultaneously and assume coalitional deviations and core-stability.

In cooperative game theory, profit sharing has been a major focus over the last decades. For example, core stability in the classic transferable-utility cooperative matching game assumes that the total profit of a global maximum matching is distributed to all agents such that every subset S of agents receives in sum at least the value of a maximum weight matching for S. Computing such imputations is closely related to LP duality [13]. Computing different solution concepts in this game has also been of interest [7,12,19]. In contrast, we assume utility transfer only within coalitions and evaluate the quality of a scheme based on the price of anarchy for coalitional stability concepts in the resulting coalition formation game. Additionally, we focus on trade-offs between efficiency and equality.

Computing stability concepts in hedonic coalition formation games is a recent line of research in computational social choice [10,16]. Many stability concepts are NP- or PLS-hard to compute. This holds even in the case of additive-separable coalition profits, which can be interpreted by an underlying graph structure with weighted edges, and the profit of a coalition is measured by the total edge weights covered by the coalition [14,23]. In addition, some price of anarchy results recently appeared in [9]. While our main focus are structures inspired by matching problems, designing profit shares in the additive-separable case can be formulated in our model, and it represents an interesting avenue for future work.

Designing good cost sharing schemes to minimize prices of anarchy and stability [11,24] in resulting strategic games is a topic of recent interest in algorithmic game theory.

2   Coalition Formation

We start by analyzing the relation between α-boundedness and the PoS/PoA. At first we will see that we can give non-trivial upper bounds on the PoS and PoA subject to α. In addition, given an optimum solution we can compute a distribution scheme that obtains these bounds.

Theorem 1. *For any $\alpha \in \left[0, \frac{1}{k}\right]$, there is a distribution scheme $d(\alpha)$ that is α-bounded and results in a PoS of at most $\max\{1, k^2\alpha\}$. If further the instance is inclusion monotone, the distribution scheme ensures a PoA of at most $\max\{1, k^2\alpha\}$. Given any social optimum $\mathcal{S}^*$, we can compute the distribution scheme in polynomial time.*

Algorithm 1. Ensuring PoS

Data: Instance $(N, \mathcal{C}, w)$, social optimum $\mathcal{S}^*$, bound α

1 set $i = 0$, $\mathcal{C}_0 = \mathcal{C}$ and $\mathcal{S} = \emptyset$;
2 **while** $\mathcal{C}_i \neq \emptyset$ **do**
3   choose S with $w(S) = max\{w(S) \mid S \in \mathcal{C}_i\}$;
4   set $i = i + 1$;
5   **if** $S \in \mathcal{S}^*$ **then**
6          $\mid$ set $S_i^* = S$;
7   **else if** $S \notin \mathcal{S}^*$ and $\alpha w(S) < \frac{1}{k} w(S')$ for some $S' \in \mathcal{S}^*$ **then**
8   choose $S' \in \mathcal{S}^*$ with $\alpha w(S) < \frac{1}{k} w(S')$;
9   set $S_i^* = S'$;
10   **else**
11         $\mid$ set $S_i^* = S$;
12   set $\mathcal{C}_i = \mathcal{C}_{i-1} \setminus \{S_i^*\}$ and $\mathcal{S} = \mathcal{S} \cup \{S_i^*\}$;
13   **foreach** $u \in S_i^*$ **do**
14         $\mid$ set $d_u(S_i^*) = \frac{1}{|S_i^*|}$;
15   **foreach** $S' \in \mathcal{C}_i$ with $S' \cap S_i^* \neq \emptyset$ **do**
16   choose $u \in S' \cap S_i^*$ and set $d_u(S') = \alpha$;
17   **foreach** $u' \in S' \setminus \{u\}$ **do**
18             $\mid$ set $d_{u'}(S') = \frac{1-\alpha}{|S'|}$;
19   set $\mathcal{C}_i = \mathcal{C}_i \setminus \{S'\}$;

Proof. We provide algorithms that compute the suitable distribution schemes. For the PoS see Algorithm 1 and for the PoA see Algorithm 2. The idea of both algorithms to always consider the worthiest remaining coalition S and use it to decide which coalition S_i^* to stabilize next. If S is part of the optimal coalition structure $\mathcal{S}^*$ we make it S_i^*. Otherwise, if S is overlapping with some worthy enough coalition S' of $\mathcal{S}^*$, we pick S' as S_i^*. Thus in both cases we stabilize an edge of the optimal coalition structure. If the coalition S is not in $\mathcal{S}^*$ but too worthy to be outbid by a $\frac{1}{k}$-share of some overlapping coalition of $\mathcal{S}^*$, we set $S = S_i^*$ instead. As that only happens when the value difference is quite big and the number of affected optimal coalitions per occurrence is limited, this way we get a good bound on how much we loose against the optimum. In $\mathcal{S}$ we keep track of the stable solution. To ensure that S_i^* is stable, $w(S_i^*)$ is shared equally and all overlapping coalitions such that the players joint with S_i^* only receive an α-share.

We start with proving that in both cases $\mathcal{S}$ is core-stable. Obviously $\mathcal{S}$ is a coalition structure. The crucial point for the algorithms to work is as follows. All coalitions S' distributed in round i are of value at most $w(S)$ for the initially chosen S of round i. Hence, for every S' of round i at least one of the players they share with S_i^* wants to stay at S_i^* as by choice of S_i^* $\frac{1}{|S_i^*|} w(S_i^*) \geq \frac{1}{k} w(S_i^*) > \alpha w(S')$. Furthermore, in Algorithm 2 for all coalitions S' of round i the players they share with S_i^* actually prefer S_i^* for the same reason. Then obviously $\mathcal{S}$ is stable as for every $S^+ \in \mathcal{C} \setminus \mathcal{S}$ we have $S^+ \in \mathcal{C}_{i-1} \setminus \mathcal{C}_i$ for some i. Hence there is some agent in $S^+ \cap S_i^*$ (namely u of Line 16 in Algorithm 1 respectively Line 17 in Algorithm 2) which refuses to deviate from S_i^* to S^+.

Algorithm 2. Ensuring PoA

Data: Instance $(N, \mathcal{C}, w)$, $\mathcal{C}$ inclusion monotone, social optimum S^*, bound α

1   set $i = 0$, $\mathcal{C}_0 = \mathcal{C}$ and $\mathcal{S} = \emptyset$;
2   **while** $\mathcal{C}_i \neq \emptyset$ **do**
3   choose S with $w(S) = max\{w(S) \mid S \in \mathcal{C}_i\}$;
4   set $i = i + 1$;
5   **if** $S \in S^*$ **then**
6   set $S_i^* = S$;
7   **else if** $S \notin S^*$ and $\alpha w(S) < \frac{1}{k} w(S')$ *for some* $S' \in S^*$ **then**
8   choose $S' \in S^*$ with $\alpha w(S) < \frac{1}{k} w(S')$;
9   set $S_i^* = S'$;
10   **else**
11   set $S_i^* = S$;
12   set $i = i + 1$, $\mathcal{C}_i = \mathcal{C}_{i-1} \setminus \{S_i^*\}$ and $\mathcal{S} = \mathcal{S} \cup \{S_i^*\}$;
13   **foreach** $u \in S_i^*$ **do**
14   set $d_u(S_i^*) = \frac{1}{|S_i^*|}$;
15   **foreach** $S' \in \mathcal{C}_i$ with $S' \cap S_i^* \neq \emptyset$ **do**
16   **foreach** $u \in S'$ **do**
17   **if** $u \in S_i^*$ **then**
18   set $d_u(S') = \alpha$;
19   **else**
20   set $d_u(S') = \frac{1 - \alpha|S_i^* \cap S'|}{|S' \setminus S_i^*|}$;
21   set $\mathcal{C}_i = \mathcal{C}_i \setminus \{S'\}$;

For Algorithm 2 we show that further there is no other core-stable state under $d(\alpha)$. To see this assume some other coalition structure S' and consider some coalition S^+ of $\mathcal{S} \setminus S'$ with i minimal such that $S^+ \in \mathcal{C}_{i-1} \setminus \mathcal{C}_i$. Now $S^+ = S_i^*$ and all coalitions of S' which intersect with S^+ where distributed in the same round i – as otherwise there would have been a coalition of an earlier round in $\mathcal{S} \setminus S'$. Thus all involved players want to deviate to S_i^* because they are either unmatched or get worse profit from their coalition in S' than from S_i^*.

Hence, we can use $\mathcal{S}$ to give an upper bound on the PoS respectively the PoA. We compare $\mathcal{S}$ to the optimal outcome S^* we used for the algorithm. For each coalition in $\mathcal{S} \cap S^*$ both structures give the same value. Next we assign each coalition $S \in S^* \setminus \mathcal{S}$ to the coalition S_i^* for i such that $S \in \mathcal{C}_{i-1} \setminus \mathcal{C}_i$. Now each S_i^* has at most k coalitions S assigned to it as the size of S_i^* limits the number of mutually disjoint coalitions intersecting with S_i^*. Further, by the choice of S_i^* each of the Ss fulfills $\alpha w(S) < \frac{1}{k} w(S_i^*)$. That is, $\mathcal{S}$ looses at most $k^2 \alpha w(S_i^*)$ compared to S^* for every coalition S_i^* in $\mathcal{S} \setminus S^*$. This gives us a PoS and a PoA at most $k^2 \alpha$. $\square$

The previous proof can be applied directly even if S^* is not an optimum solution. Optimality of S^* only served to establish a relation to the optimum value for social welfare. Hence, if we run Algorithms 1 and 2 on a ρ-approximate solution $\mathcal{S}$, we obtain core-stable states for which social welfare is at most $k^2 \alpha$ worse than

$w(\mathcal{S})$. This allows to obtain α-bounded distribution schemes with bounded PoS and PoA in polynomial time.

Corollary 1. *Given any coalition structure $\mathcal{S}$ that is a ρ-approximation to the optimum, Algorithm 1 computes an α-bounded distribution scheme such that the PoS is at most $\rho \cdot \max\{1, k^2\alpha\}$. The same result holds for Algorithm 2 and PoA in inclusion monotone instances.*

Proposition 1. *There exist instances with n agents and $\mathcal{C}$ not inclusion monotone, in which the PoA is at least $2 - \frac{4}{n+2}$ for every distribution scheme.*

Note that for the extreme of equal sharing of the biggest coalitions $\alpha = \frac{1}{k}$, we get an upper bound of k for the PoS and the PoA while for $\alpha = \frac{1}{k^2}$ we reach PoS = PoA = 1. In particular for every $\alpha \leq \frac{1}{k^2}$ we can always assure optimality of core-stable coalition structures.

Conversely, for all k and $\alpha = \frac{1}{k}$ we can show tightness through the example $N = \{1, \ldots, k^2\}$ with $w(\{1, \ldots, k\}) = 1 + \epsilon$, $w(\{i + jk \mid j = 0 \ldots k - 1\}) = 1$, $i = 1 \ldots k$ and $w(S) = 0$ for every other coalition S. For $\epsilon \to 0$ this leads to a PoS of k.

Furthermore there are instances of α-egalitarian games where PoS $\in \Theta(k^2)\alpha$.

Proposition 2. *For every $k > 2$, there is an instance in which every α-bounded distribution scheme yields a PoS of at least $\max\{1, (k^2 - k)\alpha\}$.*

Remark. *In reverse, there are instances where for PoS at most δ the required α lies between $\frac{\delta}{k^2-k}$ and $\frac{\delta}{k^2}$. Our algorithms compute a distribution scheme for $\alpha = \frac{\delta}{k^2}$.*

We next consider deciding if a given α is small enough to allow for a distribution scheme with a core-stable coalition structure that obtains guaranteed total profit. Equivalently, we consider finding the smallest α such that an α-bounded distribution scheme yields a stable structure with a certain social welfare.

Theorem 2. *It is NP-hard to decide whether for a given $\alpha > 0$ and a given value $W > 0$ there is an α-bounded distribution scheme that admits a core-stable coalition structure $\mathcal{S}$ such that $\sum_{S \in \mathcal{S}} w(S) \geq W$. This holds even for instances with $k = 2$.*

Corollary 2. *Given $\alpha > 0$, it is NP-hard to decide the largest reachable social welfare value by a core-stable solution under a α-bounded distribution scheme.*

Corollary 3. *Given $W > 0$, it is NP-hard to decide the value of the largest α such that some α-bounded distribution scheme can stabilize at least one coalition structure of value at least W.*

Intuitively, finding the largest α gets easier when the coalition structure to be stabilized is some social optimum given in advance. Sadly, for $k > 2$ we again show NP-hardness of this problem. Conversely for $k = 2$ we will in Section 3 below provide an algorithm implementing this task in polynomial time under mild additional constrains.

Theorem 3. *Let $k \geq 3$. Given an optimal coalition structure S^*, it is NP-hard to decide whether for a given α there is an α-bounded distribution scheme such that S^* becomes core-stable. This even holds for instances with all coalitions of size exactly k.*

3 Stable Matchings and Acyclic Alternating Paths

At first we note that if some matching M is not inclusion maximal, that is, can be enlarged to a bigger matching by adding some edge, it cannot be stabilized for any $\alpha > 0$. In contrast, for $\alpha = 0$ it is easy to see that every coalition structure can be stabilized. Inclusion maximality can easily be tested, so we will only deal with inclusion maximal matchings from now on. The matching case is a subclass of coalition formation with $k = 2$. Hence, some properties from Section 2 translate directly:

- The PoS and the PoA are bounded by 4α, and we can compute a suitable distribution scheme in polynomial time. In particular, for $\alpha = \frac{1}{4}$ we can ensure a PoS and a PoA of 1, while for $\alpha = \frac{1}{2}$ the PoS can go up to 2.
- For a given α and a given value W, it is NP-hard to decide whether there is an α-bounded distribution scheme that admits a stable matching M with $\sum_{e \in M} w(e) \geq W$.
- For a given α, it is NP-hard to decide the value of the best matching which can be stabilized by some α-bounded distribution scheme.
- For a given W, it is NP-hard to decide the value of the largest α such that some α-bounded distribution scheme can stabilize at least one matching of value at least W.

However, the lower bounds on the PoS in terms of α given in Proposition 2 do not extend, as they only hold for $k > 2$. However, the simple example of 4 players, a path $e_1 = \{1, 2\}, e_2 = \{2, 3\}, e_3 = \{3, 4\}$ and profits $w(e_1) = w(e_2) = 1$ and $w(e_2) = \frac{1 - \alpha + \epsilon}{\alpha}$ for some small enough $\epsilon > 0$ already gives a lower bound of $\frac{2\alpha}{1 - \alpha}$ which coincides with the upper bound of 2 at the extreme point of $\alpha = \frac{1}{2}$. Obviously e_1 and e_3 both can offer $1 - \alpha$ to the vertices of the inner edge, but need $\alpha \frac{1 - \alpha + \epsilon}{\alpha} = 1 - \alpha + \epsilon > 1 - \alpha$. This leads to a PoS of $\frac{2\alpha}{1 - \alpha + \epsilon} \xrightarrow{\epsilon \to 0} \frac{2\alpha}{1 - \alpha}$.

For the remainder of this section, we will show improved results by restricting our attention to a subclass of matching instances, which we term *acyclic alternating*. For defining this subclass we make the following observations.

Suppose we want to stabilize some matching M. Consider an edge $e \notin M$ which has a common endpoint v with some $e' \in M$ such that $w(e) \leq w(e')$. Such edge never is a blocking pair for M if we assign only $\alpha w(e)$ of e to v, as at least $\alpha w(e') \geq \alpha w(e)$ is offered to v by e'. Hence, for all following analyses we will assume that the distribution schemes assign $\alpha w(e)$ of e to v for all edges $e \in E \setminus M$ with $w(e) \leq w(e')$ for some adjacent $e' \in M$ with $e \cap e' = \{v\}$ and not handle them explicitly anymore. Instead, we only focus on the edges $e \in E \setminus M$ for which all adjacent $e' \in M$ have $w(e') < w(e)$. We call such edges *dominating* and their adjacent matching edges *dominated*. We denote the subgraph of G

consisting of the dominating and dominated edges by $G_d(M)$ and the set of these edges by $E_d(M)$. If every path in $G_d(M)$ which alternates between M and $E \setminus M$ is acyclic, we call $G_d(M)$ *acyclic alternating*. We note that for optimal M, $G_d(M)$ cannot contain any even cycles alternating between M and $E \setminus M$ as this would contradict the optimality of M. An acyclic alternating $G_d(M)$ resulting from some optimal matching M allows us to show improved bounds.

Let us first note that the restriction to graphs with $G_d(M)$ acyclic alternating for some or even every optimal matching M is not a drastic cutback, as it covers interesting subclasses of well-studied matching problems.

Proposition 3. *Let $LP(G, w)$ the LP-relaxation for the problem of finding a maximum weight matching in a graph G with edge weights w and M^* be a maximum weight matching. Then we have*

$$\{(G, w) \mid G \text{ bipartite}\}$$
$$\subsetneq \{(G, w) \mid LP(G, w) \text{ has integrality gap } 1\}$$
$$\subsetneq \{(G, w) \mid \forall M^* : G_d(M^*) \text{ is acyclic alternating}\}$$
$$\subsetneq \{(G, w) \mid \exists M^* : G_d(M^*) \text{ is acyclic alternating}\}.$$

Now we will see that the acyclic alternating property can actually help improving the lower bound on α needed for a PoS of 1:

Theorem 4. *For any optimal matching M^* such that $G_d(M^*)$ is acyclic alternating, there is an α-bounded distribution scheme that stabilizes M^* with $\alpha = \frac{1}{3}$, and this bound is tight. Given such an M^*, the distribution scheme can be computed in polynomial time.*

Note that this bound gives a real improvement compared to graphs without acyclic alternating structure:

Proposition 4. *There are matching games where a PoS of 1 requires $\alpha = 1 - \sqrt{\frac{1}{2}} \approx 0.2929 < \frac{1}{3}$.*

In Section 2 we observed NP-hardness of deciding whether a given α suffices to stabilize some optimal matching. Using the acyclic alternating property, we can optimize α to stabilize arbitrary matchings. Hence, this property helps not only in stabilizing optimal but arbitrary matchings.

Theorem 5. *Given a matching M such that $G_d(M)$ is acyclic alternating and some $\alpha \in [0, \frac{1}{2}]$, we can decide in polynomial time if there is an α-bounded distribution scheme stabilizing M.*

Proof. We have already seen how to find a distribution scheme for $\alpha \leq \frac{1}{3}$ in Theorem 4. Here we will treat the slightly more general approach as shown in Algorithm 3. We first describe the intuitive idea behind the algorithm. The main idea of the algorithm for $G_d(M)$ is using that in every round there is an edge for which the profit of one agent is already determined by the algorithm. In particular, for $e \in M$ there are only edges $e' \notin M$ on one side that we have to

---

Algorithm 3. Computing a Distribution Scheme for M

    Data: Instance (G, w), matching M, bound α

1   set $S = E$;
2   **foreach** $e \in E \setminus M$ **do**
3   **if** $\exists e' \in M : e \cap e' = \{u\}$ *and* $w(e) \leq w(e')$ **then**
4   share e such that $\alpha w(e)$ is offered to u;
5   set $S = S \setminus \{e\}$;
6   **foreach** $e = \{u, v\} \in M$ **do**
7   set $s_u = \alpha w(e)$, $s_v = \alpha w(e)$ and $rest_e = (1 - 2\alpha)w(e)$;
8   **while** $S \neq \emptyset$ **do**
9   **if** $\exists e = \{u, v\} \in M : u \notin e' \forall e' \in S \setminus \{e\}$ **then**
10   set $s_v = s_v + rest_e$ and $rest_e = 0$;
11   share e such that s_u is offered to u and s_v is offered to v;
12   set $S = S \setminus \{e\}$;
13   **foreach** $e' \in S$ *with* $e' \cap e = \{v\}$ **do**
14   **if** $s_v \geq \alpha w(e')$ **then**
15   share e' such that $\alpha w(e')$ is offered to v;
16   set $S = S \setminus \{e'\}$;
17   **if** $\exists e = \{u, v\} \in E \setminus M : u \notin e' \forall e' \in S \cap M$ **then**
18   **if** $\exists e' \in S \cap M : v \in e'$ *and* $s_v + rest_{e'} \geq \alpha w(e)$ **then**
19   **if** $s_v < \alpha w(e)$ **then**
20   set $s_v = \alpha w(e)$ and $rest_{e'} = rest_{e'} - (\alpha w(e) - s_v)$;
21   share e such that $\alpha w(e)$ is offered to v;
22   set $S = S \setminus \{e\}$;
23   **else**
24   **return** 'M cannot be stabilized with lower bound α';

---

make non-blocking and for $e \in E_d(M) \setminus M$ there is only one $e' \in M$ left on one side which can be used to make e non-blocking. This property is due to the fact that no alternating path in $G_d(M)$ contains a cycle. Deciding the distribution for $e \in E_d(M) \setminus M$ is easy. We give the smallest possible value $\alpha w(e)$ to the side where an edge $e' \in M$ is supposed to ensure non-blocking status of e. In contrast, for matching edges $e \in M$ we have to be more careful. We start by giving each side only the minimal portion of $\alpha w(e)$ and keep the rest as buffer. Now every time we encounter an $e' \in E_d(M) \setminus M$ which can only be stabilized by e, we raise the share on that side just as much as needed using up some of the buffer. If such e' are left only on one side, we can push all the remaining buffer to that side and check whether (some of) the e' become non-blocking by this assignment.

More formally consider the execution of Algorithm 3. At first we show that we will not get stuck in the while-loop, that is, in every execution of the loop at least one edge is removed from S (or we find out that M cannot be stabilized and stop). Assume conversely that although $S \neq \emptyset$, there neither is a matching edge with no adjacent edges at one side nor a non matching edge with no adjacent matching edge at one side. Then S contains a cycle which alternates between matching and non matching edges in contradiction to the properties of $G_d(M)$.

Next we will see that, if the algorithm does not terminate early, the final output is an α-bounded distribution scheme which stabilizes M. Every edge e of $E \setminus M$ is shared $(\alpha w(e), (1-\alpha)w(e))$ and every edge $e = \{u, v\}$ in M is shared according to (s_u, s_v) which is a valid distribution as throughout the algorithm $s_u + s_v + rest_e = w(e)$, where the buffer $rest_e = 0$ in the end. Obviously we never exceed nor undercut the value of an edge. Further, s_u and s_v start at $\alpha w(e)$ and increase monotonically. Thus, our distribution respects the lower bound of α. We claim that every $e \in E_d(M) \setminus M$ that is dropped from S is already stabilized by an offer according to the current value s_u of some incident agent u. Then, with S empty in the end and all matching edges shared according to the s_u-values (that increase over the run of the algorithm) M is indeed stable. If e has been removed at Line 5, the agent (denoted u) to which $\alpha w(e)$ is offered will get at least as much from its incident matching edge e'. If e has been removed at Line 16, it is stabilized by the matching edge which is removed along with it. Further, each edge removed at Line 22 is stabilized as well, as the s_v-value is adjusted to be large enough if it has not been before. Hence M is stable under the given distribution.

Now assume that the algorithm terminates early declaring that M cannot be stabilized. Obviously, at the point where this decision is made, the currently examined edge $e \in E_d(M) \setminus M$ cannot be made non-blocking anymore, because either the incident matching edges where fully shared and removed already without the share being large enough to make e non-blocking and delete it from S (Line 16) or the current offer combined with the remaining buffer of the incident matching edge is smaller than $\alpha w(e)$. Thus, we have to show that we did not offer a bigger share than needed to the other side earlier. In the beginning each s_v is set to $\alpha w(e)$ where e is the matching edge containing v, so nothing is wasted. Now there are only two points where s_v is enlarged. If s_v is changed in Line 10, we already know there are only edges on the side of v which remain to be stabilized, as we have seen above that every dropped edge is already stabilized. Thus by giving the rest of the buffer to v we waste nothing for the other side. The other time s_v is changed is at Line 20 where it is enlarged to meet the minimal possible offer of the currently examined non matching edge $e = \{u, v\}$. Now e has been picked because there is no matching edge left on one side of it, that is, either u is not matched in M, then e has to be stabilized at v and thus it is necessary to rise s_v, or the matching edge of u is already deleted from S. But when deleting a matching edge we always ensure to delete all non matching edges which get stabilized by the matching edge as well. Hence, again it is necessary to stabilize e at v, and the algorithm only terminates early, if it is not possible to stabilize M while respecting the α-bound. Together with the fact that the algorithm provides an α-bounded distribution scheme under which M is stable, if it terminates with $S = \emptyset$, this proves the theorem. $\square$

Proposition 5. *Suppose we are given a matching M and an α-bounded distribution scheme with maximal α stabilizing M. There are at most $|E|^3$ many relevant values for such a maximal α, which can be computed in polynomial time. This holds even if $G_d(M)$ is not acyclic alternating.*

Proof. Consider some edge $e = \{u,v\} \in M$ and let $L_e = \{e'|u \in e' \in E \setminus M, w(e') > w(e)\} \cup \{e_a\}$ and $R_e = \{e'|v \in e' \in E \setminus M, w(e') > w(e)\} \cup \{e_a\}$ where we assume e_a to be some auxiliary edge of value $w(e_a) = w(e)$. Now let $e_1 \in L_e$ be the edge of highest value in L_e which has to be put to non-blocking status by e and, similarly, $e_2 \in R_e$ be the edge of highest value in R_e which has to be put to non-blocking status by e. If on some side there are no edges which have to be handled, we choose e_a to ensure an offer of α. Now the largest α which allows e putting both e_1 and e_2 (and all smaller edges on the respective sides) to non-blocking fulfills exactly $w(e) = \alpha w(e_1) + \alpha w(e_2)$. As $|M|$, $|L_e|$ and $|R_e|$ are all of size at most $|E|$, the number of different such α-values arising from M is limited by $|E|^3$. We claim that the maximum α for an α-bounded distribution scheme stabilizing M must be among these candidate values. Assume conversely the optimal α^* does not fulfill the equation $w(e) = \alpha^* w(e_1) + \alpha^* w(e_2)$ for any $e \in M$, $e_1 \in L_e$, $e_2 \in R_e$. Now consider some α^*-bounded distribution scheme d which stabilizes M. For each $e \in M$ let e_1^* be the worthiest edge of L_e which is non-blocking and e_2^* the worthiest edge of R_e which is non-blocking because of e under d. We know that $\alpha^* w(e_1^*) + \alpha^* w(e_2^*) \lesssim w(e)$ for every $e \in M$. Let $\alpha^+ = \min\{\alpha \mid \alpha w(e_1^*) + \alpha w(e_2^*) = w(e) \text{ for some } e \in M\}$. Then $\alpha^+ > \alpha^*$ and we can stabilize M with an α^+-bounded distribution scheme in the following way. We share each edge e' in $E \setminus M$ such that α^+ is offered to (one of) the matching edge which ensures non-blocking status for e' in d, and we share each $e \in M$ such that $d_u(e)w(e) \geq \alpha^+ w(e_1^*)$ and $d_v(e)w(e) \geq \alpha^+ w(e_2^*)$ for its respective e_1^* and e_2^*. This contradicts maximality of α^* and completes the proof.   □

Corollary 4. *Given a matching M such that $G_d(M)$ is acyclic alternating, we can in polynomial time find the maximal bound α for which M can be stabilized.*

Observe that for general matching games, the relevant α-values can be bounded and computed in the same way, even if $G_d(M)$ is not acyclic alternating. However, in general it is not clear how to use this information to construct an optimal distribution scheme, as it remains to decide which matching edges have to stabilize which non-matching edges within cycles.

The characterization for the number of candidate values for optimal α can be directly generalized to larger coalitions using the same arguments. However, we have already seen in Theorem 3 that even the knowledge of the optimal value for α does not help in finding a stabilizing distribution scheme efficiently.

Acknowledgement. The authors thank Elliot Anshelevich for insightful discussions about the results in this paper.

References

1. Ackermann, H., Goldberg, P., Mirrokni, V., Röglin, H., Vöcking, B.: Uncoordinated two-sided matching markets. SIAM J. Comput. 40(1), 92–106 (2011)
2. Akkaya, K., Guneydas, I., Bicak, A.: Autonomous actor positioning in wireless sensor and actor networks using stable matching. Intl. J. Parallel, Emergent and Distrib. Syst. 25(6), 439–464 (2010)

3. Anshelevich, E., Bhardwaj, O., Hoefer, M.: Friendship and stable matching. In: Bodlaender, H.L., Italiano, G.F. (eds.) ESA 2013. LNCS, vol. 8125, pp. 49–60. Springer, Heidelberg (2013)
4. Anshelevich, E., Das, S., Naamad, Y.: Anarchy, stability, and utopia: Creating better matchings. Auton. Agents Multi-Agent Syst. 26(1), 120–140 (2013)
5. Anshelevich, E., Hoefer, M.: Contribution games in networks. Algorithmica 63(1-2), 51–90 (2012)
6. Augustine, J., Chen, N., Elkind, E., Fanelli, A., Gravin, N., Shiryaev, D.: Dynamics of profit-sharing games. In: Proc. 22nd Intl. Joint Conf. Artif. Intell. (IJCAI), pp. 37–42 (2011)
7. Biró, P., Kern, W., Paulusma, D.: Computing solutions for matching games. Int. J. Game Theory 41(1), 75–90 (2011)
8. Blum, Y., Roth, A., Rothblum, U.: Vacancy chains and equilibration in senior-level labor markets. J. Econom. Theory 76, 362–411 (1997)
9. Branzei, S., Larson, K.: Coalitional affinity games and the stability gap. In: Proc. 21st Intl. Joint Conf. Artif. Intell. (IJCAI), pp. 79–84 (2009)
10. Cechlárova, K.: Stable partition problem. In: Encyclopedia of Algorithms (2008)
11. Chen, H.-L., Roughgarden, T., Valiant, G.: Designing network protocols for good equilibria. SIAM J. Comput. 39(5), 1799–1832 (2010)
12. Chen, N., Lu, P., Zhang, H.: Computing the nucleolus of matching, cover and clique games. In: Proc. 26th Conf. Artificial Intelligence, AAAI (2012)
13. Deng, X., Ibaraki, T., Nagamochi, H.: Algorithmic aspects of the core of combinatorial optimization games. Math. Oper. Res. 24(3), 751–766 (1999)
14. Gairing, M., Savani, R.: Computing stable outcomes in hedonic games. In: Kontogiannis, S., Koutsoupias, E., Spirakis, P.G. (eds.) SAGT 2010. LNCS, vol. 6386, pp. 174–185. Springer, Heidelberg (2010)
15. Gusfield, D., Irving, R.: The Stable Marriage Problem: Structure and Algorithms. MIT Press (1989)
16. Hajduková, J.: Coalition formation games: A survey. Intl. Game Theory Rev. 8(4), 613–641 (2006)
17. Hoefer, M.: Local matching dynamics in social networks. Inf. Comput. 222, 20–35 (2013)
18. Hoefer, M., Wagner, L.: Locally stable marriage with strict preferences. In: Fomin, F.V., Freivalds, R., Kwiatkowska, M., Peleg, D. (eds.) ICALP 2013, Part II. LNCS, vol. 7966, pp. 620–631. Springer, Heidelberg (2013)
19. Kern, W., Paulusma, D.: Matching games: The least core and the nucleolus. Math. Oper. Res. 28(2), 294–308 (2003)
20. Kleinberg, J., Oren, S.: Mechanisms for (mis)allocating scientific credit. In: Proc. 43rd Symp. Theory of Computing (STOC), pp. 529–538 (2011)
21. Kleinberg, J., Tardos, É.: Balanced outcomes in social exchange networks. In: Proc. 40th Symp. Theory of Computing (STOC), pp. 295–304 (2008)
22. Manlove, D.: Algorithmics of Matching Under Preferences. World Scientific (2013)
23. Sung, S.-C., Dimitrov, D.: Computational complexity in additive hedonic games. Europ. J. Oper. Res. 203(3), 635–639 (2010)
24. von Falkenhausen, P., Harks, T.: Optimal cost sharing for resource selection games. Math. Oper. Res. 38(1), 184–208 (2013)

Jealousy Graphs: Structure and Complexity of Decentralized Stable Matching

Moshe Hoffman, Daniel Moeller, and Ramamohan Paturi

University of California, San Diego
http://www.ucsd.edu

Abstract. The stable matching problem has many applications to real world markets and efficient centralized algorithms are known. However, little is known about the decentralized case. Several natural randomized algorithmic models for this setting have been proposed but they have worst case exponential time in expectation. We present a novel structure associated with a stable matching on a matching market. Using this structure, we are able to provide a finer analysis of the complexity of a subclass of decentralized matching markets.

Keywords: decentralized stable matching, market algorithms.

1 Introduction

The stable matching problem and its variants have been widely studied due to real world market applications, such as assigning residents to hospitals, women to sororities, and students to public schools [1–3]. In a seminal paper, Gale and Shapley first proposed an algorithm to find a stable matching in the basic two-sided (bipartite) version [4]. Others have subsequently investigated the structure of the set of stable matchings [1]. However, most prior work involves centralized algorithms to find stable matchings where the entire set of preferences is known to some central authority. In some cases the algorithms are not totally centralized, but the participants are subject to strict protocols where only one side of the market can make proposals. Nevertheless, many applications of stable matchings have no central authority or enforcement of protocols, such as college admissions and the computer scientist job market. Therefore we investigate this problem in a decentralized setting, where members of both sides of the market can make proposals.

One major open question in decentralized stable matching concerns whether natural and efficient algorithms exist. To this end, Yariv argues that natural processes will find a stable matching and provides experimental support [5]. Roth and Vande Vate propose a class of randomized algorithms to model the decentralized setting and show that algorithms in this class converge to a stable matching with probability one [6]. At each step these algorithms match two participants who form a blocking pair (who prefer to be matched with each other over its partner) of the current matching. However, they present no expected time

Y. Chen and N. Immorlica (Eds.): WINE 2013, LNCS 8289, pp. 263–276, 2013.

complexity. Ackerman et al. investigate one particular algorithm in this class, the better response algorithm (or random better response dynamics). In each step of this algorithm, one blocking pair is chosen uniformly at random. For this algorithm, they show worst case instances that take exponential time to reach a stable matching in expectation [7].

Since the better response algorithm is natural but takes exponential time in the worst case, can we find a natural subclass of matching markets which do not require exponential time? Ackermann et al. show that the better response algorithm only requires polynomial time for one class of problem instances, those with correlated preferences [7]. However, correlated preferences require that a participant obtains the same benefit from a partnership as its partner. This significantly limits the preference structures allowed in the matching market. Therefore, we investigate other structural properties of stable matching markets which facilitate faster convergence.

In this paper we make progress toward answering the previous question by expanding the subclass of markets with polynomial time convergence guarantees. For this purpose we associate a directed graph, called the *jealousy graph*, with each stable matching. It turns out that this structure is a key factor in determining the convergence time of the better response algorithm. The jealousy graph is a directed graph where a vertex v corresponds to a pair in the stable matching and an edge (u, v) is present if one member of the pair v prefers a member of the pair u to its partner in the stable matching. The strongly connected component graph of this jealousy graph provides a *decomposition* for that stable matching. Our intent is to formalize a notion of structure using jealousy graphs and the corresponding decompositions. In particular, we find that the strongly connected components of this graph give insight into the complexity of that market. Gusfield and Irving provide a structural property of stable matchings which describes the set of stable matchings and the relation between them, whereas our structures relate to individual stable matchings and the distributed process by which these stable matchings are achieved [1].

With a decomposition, we associate a size and depth. Our main result, Theorem 24, states that for a matching market of size n with a decomposition of size c and depth d, the convergence time is $O(c^{O(cd)} n^{O(c+d)})$. Therefore, for constant size and depth decompositions, we demonstrate that the better response algorithm requires only polynomial time in expectation to converge for an expanded class of matching markets. This indicates that the jealousy graph and decomposition structures partially answer the convergence questions of the decentralized stable matching problem. As an application of our work, we demonstrate how Theorem 24 provides theoretical justification for the simulated results of Boudreau [8]. We also conjecture that these structures will provide a means of predicting which stable matchings are likely to be achieved when there are multiple stable matchings, a question that others in the literature have investigated [5, 9, 10].

In the remainder of this section we formalize our model and present the basic concepts. In section 2, we present some useful structural properties of the jealousy graph and decomposition. In section 3, we have our convergence result. Section 4 contains the application of our work to [8], and section 5 is our conclusion.

1.1 Basic Stable Matching Concepts

We start with the basic definitions of matching markets and stable matchings. Those familiar with the matching literature will notice that we restrict preferences to be complete and strict.

Definition 1. (S, P) *is a matching market if* $S = M \bigcup W$ *for some disjoint sets* M, W, $|M| = |W|$ *and* $P = \{\succ_s\}_{s \in S}$ *where, for* $s \in M$, $\succ_s$ *is a total order over* W, *and for* $s \in W$, $\succ_s$ *is a total order over* M.

We say a matching market has size n if $|M| = |W| = n$.

Definition 2. *A matching on the set* S *is a function* $\mu : S \to S$ *such that* $\forall s \in S$, $\mu(\mu(s)) = s$, $s \in M \Rightarrow \mu(s) \in W \bigcup \{s\}$ *and* $s \in W \Rightarrow \mu(s) \in M \bigcup \{s\}$.

We say that a participant $s \in S$ is unmatched by a matching μ if $\mu(s) = s$. We also assume that all participants prefer to be matched to anyone than to be unmatched. Observe that μ can be thought of as a collection of pairs (m, w) if we allow self loops (s, s) for unmatched participants.

Definition 3. *A matching on the set* S *is a perfect matching if* $\mu(s) \neq s$ *for all* $s \in S$.

Given a matching, if there were a man and a woman who each preferred the other to their partner, this causes the matching to be unstable. Therefore any stable matching must have no such pairs. We call such a pair a blocking pair, defined formally here:

Definition 4. *Let* (S, P) *be a matching market and* μ *be any matching on* S. *A blocking pair for* μ *in* (S, P) *is a pair* (m, w) *such that* $m \in M$, $w \in W$, $\mu(m) \neq w$, $w \succ_m \mu(m)$, *and* $m \succ_w \mu(w)$.

Definition 5. *Let* (S, P) *be a matching market. A matching* μ *on* S *is a stable matching for* (S, P) *if it has no blocking pairs in* (S, P).

The following three concepts will be useful since we will deal with subsets of the matching market.

Definition 6. *A balanced subset of a matching market* (S, P), $S = M \bigcup W$, *is a subset* $S' \subseteq S$ *such that* $|S' \bigcap M| = |S' \bigcap W|$.

Definition 7. *A matching* μ *is locally perfect on a balanced subset* $S' \subseteq S$ *if* $\mu(S') \subseteq (S')$ *and* $\mu \restriction_{S'}$ *is a perfect matching on* S'.

Definition 8. *Let* μ *be a stable matching on a matching market* (S, P). *A matching* μ' *is* μ-*stable on a balanced subset* $S' \subseteq S$ *if* $\mu' \restriction_{S'} = \mu \restriction_{S'}$.

1.2   Better Response Algorithm

The class of algorithms introduced by Roth and Vande Vate [6] involve randomly choosing a blocking pair of the current matching and creating a new matching by matching the participants in the blocking pair with each other. This resolves the chosen blocking pair.

Definition 9. *A blocking pair* (x, y) *in a matching* μ *is resolved by forming a new matching* μ' *where* $\mu'(x) = y$, $\mu'(\mu(x)) = \mu(x)$ *if* $\mu(x) \neq x$, $\mu'(\mu(y)) = \mu(y)$ *if* $\mu(y) \neq y$, *and* $\mu'(s) = \mu(s)$ *for* $s \notin \{x, y, \mu(x), \mu(y)\}$.

This process is repeated until a stable matching is reached. The *better response algorithm* defined in [7], is the algorithm in this class where the blocking pair is chosen uniformly at random from all blocking pairs of the current matching. Note that this algorithm results in a sequence of matchings. A valid sequence of matchings is any sequence where each matching is formed by resolving one blocking pair in the previous matching.

We focus on the better response algorithm since the uniform distribution on blocking pairs facilitates our analysis and we believe it provides insight into the more general class of algorithms. This algorithm also serves as a model of a distributed stable matching market.

1.3   Jealousy Graph and Related Definitions

In order to analyze matching markets, we represent the preference structure as a directed graph. While we lose some of the preference information, we retain critical relationships relative to the stable partners. In section 3 we will provide bounds on convergence based on this simpler structure.

Definition 10. *The jealousy graph of a stable matching* μ *on a matching market* (S, P) *is defined as the graph* $J_\mu = (V, E)$ *where, for each pair* $\{x, \mu(x)\}$, $x \in S$, *there is a vertex* $v_{\{x,\mu(x)\}} \in V$ *and* $E = \{(u_{\{x,y\}}, v_{\{x',y'\}}) | u_{\{x,y\}}, v_{\{x',y'\}} \in V$, *and either* $x \succ_{y'} x'$ *or* $y \succ_{x'} y'\}$.

The jealousy graph can provide insight into the complexity of stabilization. For example, suppose the jealousy graph for a stable matching μ is one large clique. Even when all but one pair of the participants are matched with their partner in μ, there are still many blocking pairs. Therefore, the better response algorithm would be unlikely to choose the blocking pair that would result in a stable matching. This greatly hinders convergence to the stable matching.

On the other hand, suppose the jealousy graph for μ is a DAG. Then there is at least one vertex with no incoming edges. This means each partner in the corresponding pair is the other's first preference. Consequently, this will remain a blocking pair until it is resolved, so we would expect such a pair to be resolved in $O(n^2)$ time under the better response dynamics. Moreover, once resolved, the match will remain unbroken since neither partner will ever be involved in any blocking pairs. Ignoring this pair will result in at least one other source vertex of the graph. Inductively, these pairs will be resolved in $O(n^2)$ expected time.

This results in an expected convergence time of $O(n^3)$ for the matching market. It should be noted that the class of correlated markets, for which Ackermann et al. prove the better response algorithm requires only polynomial time, falls into this special case.

When it is a DAG, the jealousy graph provides an order in which the pairs will likely be resolved to reach μ, namely, a topological sorted order. However, a matching market might not fall into this extreme case as there could be cycles in the jealousy graph. Therefore, we define a decomposition which is a DAG obtained from the jealousy graph.

Definition 11. *Let J_μ be the jealousy graph of a stable matching μ for a matching market (S, P). A μ-decomposition, ρ_μ is a graph of components of J_μ such that if u, v are in the same strongly connected component of J_μ then they are in the same component in ρ_μ and if edge (A, B) is in ρ_μ then there is a path from a vertex in A to a vertex in B in J_μ.*

We call the strongly connected components of J_μ stable components. Observe that ρ_μ is a directed acyclic graph. Therefore it induces a partial order on the stable components. Sometimes it will be simpler to refer to the decomposition as $\rho_\mu = (\Pi, \preceq)$ where Π is a partition of S into sets corresponding to the stable components of ρ_μ and $\preceq$ is the induced partial order on those components. As a slight abuse of notation, we will use the term stable component to refer to both the connected component in the decomposition and the set of participants corresponding to this component.

In dealing with partial orders we will use the concept of a downset. A *downset* of a partially ordered set Π with partial order $\preceq$ is any set such that for $A, B \in \Pi$, if A is in the set and $B \preceq A$, then B is in the set. The downset of an element $A \in \Pi$ is $Down(A) = \{B | B \preceq A\}$. When the elements of Π are sets themselves, as in the case of decompositions, we will denote union of sets in $Down(A)$ as $\mathbf{D}(A) = \bigcup_{B \in Down(A)} B$.

For our complexity results we need the following two notions:

Definition 12. *The depth of a stable component A of a μ-decomposition, ρ_μ, is the length of the longest path in ρ_μ from any source vertex to v_A. The depth of ρ_μ is defined as $\max_{A \in \rho_\mu} depth(A)$.*

We will say that a stable component A is on level j if $depth(A) = j$. Minimal stable components are on level 0. Intuitively, we would expect components on lower levels to converge to the stable matching sooner than those on higher levels.

Definition 13. *The size of a μ-decomposition, ρ_μ, is defined as $\max_{A \in \rho_\mu} size(A)$.*

Intuitively, components with smaller sizes can have less internal thrashing so they will converge to the stable matching more quickly than larger components.

2   Structural Results

2.1   Any Digraph Can Be a Jealousy Graph

These structural notions would not be very enlightening if all matching markets had the similar jealousy graphs and decompositions. However, the following result shows that any directed graph is the jealousy graph associated with a stable matching for some matching market. The proof can be found in the full paper.

Theorem 14. *Given any directed graph G with n vertices, there is a set $S = \{m_i, w_i | i = 1, 2, \ldots, n\}$ and preferences $P = \{\succ_{m_i}, \succ_{w_i} | 1 \leq i \leq n\}$ such that (S, P) is a matching market with a stable matching μ where $\mu(m_i) = w_i$ and $J_\mu = G$.*

2.2   Properties of Decompositions

In this section we prove several structural properties of the jealousy graphs and decompositions essential to our main convergence result. The proofs are in the full paper. The first property says that if there is a path from one vertex to another in the jealousy graph, then the first vertex must be in the downset of any component containing the second vertex.

Lemma 15. *Given a matching market (S, P) with a stable matching μ, let J_μ be the jealousy graph associated with μ. Let $v_{\{m,w\}}$ and $v_{\{m',w'\}}$ be vertices in J_μ. Suppose $v_{\{m',w'\}} \in A$ for a stable component A of a μ-decomposition $\rho_\mu = (\Pi, \preceq)$. If there is a path from $v_{\{m,w\}}$ to $v_{\{m',w'\}}$, then $m, w \in \mathbf{D}(A)$.*

Using this lemma, we prove that no member of a stable component can prefer anyone outside of the downset of that component to his stable partner.

Lemma 16. *Given a matching market (S, P) with a stable matching μ, let $\rho_\mu = (\Pi, \preceq)$ be a μ-decomposition. For $A \in \Pi$, $a \in A$, $s \in S - \mathbf{D}(A)$, $\mu(a) \succ_a s$.*

A further property is that if there are two stable matchings with distinct decompositions, the intersection of the downsets of stable components must be mapped to itself in both stable matchings.

Lemma 17. *Given a matching market (S, P) with stable matchings μ, μ', let ρ_μ and $\rho_{\mu'}$ be respective decompositions. Let A be $\mathbf{D}(X)$ for some stable component X of ρ_μ and B be $\mathbf{D}Y$ for some stable component Y of $\rho_{\mu'}$. Then $\mu(A \cap B) = \mu'(A \cap B) = A \cap B$.*

Our final result shows that forming a stable matching on the downset of a stable component cannot increase the size or depth of the decomposition of another stable matching.

Lemma 18. *Given a matching market (S, P) with stable matchings μ, μ', let ρ_μ and $\rho_{\mu'}$ be respective decompositions. Suppose the size of ρ_μ is c and the depth is d. Let A be a stable component of $\rho_{\mu'}$. Then there is a stable matching μ'' such that $\mu'' \upharpoonright_{\mathbf{D}_{\mu'}(A)} = \mu' \upharpoonright_{\mathbf{D}_{\mu'}(A)}$ and $\mu'' \upharpoonright_{S - \mathbf{D}_{\mu'}(A)} = \mu \upharpoonright_{S - \mathbf{D}_{\mu'}(A)}$. There is also a μ''-decomposition on $S - \mathbf{D}_{\mu'}(A)$ of size at most c and depth at most d.*

3   Convergence

In this section we prove our convergence result. The proof uses two main ideas. First, in the following sequence of lemmas, we show that a stable component will converge to a locally perfect matching in time that is only polynomially dependent on the size of the entire market. Then the proof of Theorem 24 uses this to bound the time it takes for all components of the decomposition to reach a stable matching.

For this section we will assume (S, P) is a matching market of size n, μ is a stable matching on S, and $(\Pi, \preceq)$ be a μ-decomposition.

The following lemma says that if a matching is not locally perfect on a stable component of a μ-decomposition, then there is a blocking pair which is in μ between two members of that component.

Lemma 19. *Let $A \in \Pi$ and $X = \mathbf{D}(A) - A$. Let μ' be the current matching. If μ' has no matches between members of X and members of A and μ' is not locally perfect on A, then there is a blocking pair (x,y) for μ' such that $x, y \in A$ and $\mu(x) = y$.*

Proof. Since μ' is not a locally perfect matching on A there must be some $x_0 \in A$ such that $\mu'(x_0) = x_0$ or $\mu'(x_0) \in S - X - A$. Let $y_0 = \mu(x_0)$. Now since μ is a stable matching, $y_0 \succ_{x_0} \mu'(x_0)$. If $x_0 \succ_{y_0} \mu'(y_0)$ then (x_0, y_0) is a blocking pair of μ' and $\mu(x_0) = y_0$.

Otherwise $\mu'(y_0) \succ_{y_0} x_0$, so $\mu'(y_0) \in \mathbf{D}(A)$. In fact, $\mu' \in A$ since μ' has no matches between members of A and X. Let $x_1 = \mu'(y_0)$ and $y_1 = \mu(x_1)$. Since μ is a stable matching, $y_1 \succ_{x_1} y_0$ or else (x_1, y_0) would form a blocking pair for μ. Now if $x_1 \succ_{y_1} \mu'(y_1)$, (x_1, y_1) is a blocking pair of μ' and $\mu(x_1) = y_1$ so we have our result. Otherwise we repeat in the same manner to form a sequence of pairs $\{(x_i, y_i)\}$ such that $x_i, y_i \in A$, $\mu(x_i) = y_i$, $\mu'(y_i) = x_{i+1}$, $y_i \succ_{x_i} \mu'(x_i)$, and $x_{i+1} \succ_{y_i} x_i$ for all i. But this cannot cycle since no participant is repeated. This is because at each step we add a new pair x_i, y_i where $\mu(x_i) = y_i$ and either $\mu'(x_0) = x_0$ or $\mu'(x_0) \notin A$, so x_0 cannot be repeated. Furthermore, it cannot go forever since A is finite. Therefore the sequence must terminate at some index k and (x_k, y_k) is a blocking pair for μ'.   □

Next we place a lower bound on the probability that we make some progress toward the μ-stable matching when a stable component of the decomposition is not in a locally perfect matching.

Lemma 20. *Let $A \in \Pi$ be a stable component of size at most c and $X = \mathbf{D}(A) - A$. Let μ' be any matching on S that is not a locally perfect matching on A. Then starting from μ', if no matches are formed between a member of A and a member of X, the probability that the first blocking pair resolved between two members of A is a pair in μ is at least $\frac{1}{c^2}$.*

Proof. Lemma 19 shows there will be one blocking pair which is in μ until the matching becomes locally perfect on A. In order for the matching to become locally perfect on A, a blocking pair must be resolved between two members of A. Therefore since there will be at most c^2 blocking pairs involving two members of A and at least one of them is in μ, there is a $\frac{1}{c^2}$ probability that the first blocking pair resolved between members of A is in μ.

Using this lemma, we bound the probability that a component of the decomposition will make some progress toward the μ-stable matching each time the matching is not locally perfect on it.

Lemma 21. *Let $A \in \Pi$ be a stable component of size at most c and $X = \mathbf{D}(A) - A$. Let μ_0 be any matching on S such that $\mu_0 \restriction_A$ contains m of the pairs in μ where $0 \le m < c$. Let $\mu_0, \mu_1, \ldots, \mu_t$ be any valid sequence of matchings under the better response dynamics starting from μ_0 such that*

1. *μ_t is locally perfect on A*
2. *μ_i is not locally perfect on A for some $i, 0 \le i < t$*
3. *μ_k does not have any matches between a member of A and a member of X for some $k, 0 \le k \le t$*

Then the probability that $\exists j, 0 < j \le t, \mu_j \restriction_A$ contains at least $m+1$ of the pairs in μ is at least $\frac{1}{c^4}$.

Proof. Assume $\mu_0, \mu_1, \ldots, \mu_t$ is such a sequence, and i is the first index such that μ_i is not locally perfect. Without loss of generality assume $k = t$ is the first index $k > i$ such that μ_k is locally perfect on A. This assumption is valid because, if there is at least a probability p of some event occurring in a subsequence, then there is clearly at least a probability p of that event occurring in the entire sequence.

There are two cases: either μ_0 is locally perfect on A or not.

case i: Assume μ_0 is not locally perfect, so $i = 0$. Then in order to reach μ_t there must be at least one match formed between two members of A. Let $j > 0$ be the first index in the sequence such that μ_j was formed by resolving a blocking pair between two members of A. Since no one in A prefers anyone in $S - \mathbf{D}(A)$ to his partner in μ, $\mu_{j-1} \restriction_A$ has m pairs in μ. By lemma 20 there is at least $\frac{1}{c^2}$ probability that the first blocking pair resolved between two members of A is in μ. This will result in $\mu_j \restriction_A$ having $m + 1$ pairs in μ.

case ii: If μ_0 is locally perfect, so $i > 0$. There are two ways to transition from μ_{i-1} to μ_i. One is for a blocking pair of μ_{i-1} between a member of A and a member of $S - X - A$ to be resolved. Since this cannot involve a member of A who is with his partner in μ according to μ', $\mu_i \restriction_A$ has m pairs that are in μ. Therefore this case reduces to the first case where the initial matching is not perfect.

The other way to transition from μ_{i-1} to μ_i is for a blocking pair between two members of A to be resolved, leaving two unmatched members of A, say x, y. The blocking pair cannot involve two pairs of μ or else it would be a blocking pair for μ. If it involves no pairs of μ then again this case reduces to the first case.

In the last case, $\mu_i \restriction_A$ has $m - 1$ pairs that are in μ. We cannot reach μ_t without resolving a blocking pair between two members of A. Let $l > i$ be the first index after i in the sequence such that μ_l was formed by resolving a blocking pair between two members of A. Then $\mu_{l-1} \restriction_A$ must have $m - 1$ pairs that are in μ. By lemma 20 there is at least $\frac{1}{c^2}$ probability that $\mu_l \restriction_A$ has m pairs that are in μ. If this occurs, the blocking pair resolved to transition to μ_l cannot involve both x and y because they are not partners in μ. Thus, at least one of x or y is still not matched to someone in A. Therefore, μ_l is not a locally perfect matching on A. Then by the first case, we have at least $\frac{1}{c^2}$ probability that for some j, $l < j \le t$, $\mu_j \restriction_A$ has $m + 1$ pairs that are in μ. This gives us a total probability of at least $\frac{1}{c^4}$ that $\mu_j \restriction_A$ has $m + 1$ pairs that are in μ for some j, $0 < j \le t$.

We now bound the expected number of times each stable component will have to become not locally perfect before it becomes μ-stable. The proof can be found in the full paper.

Lemma 22. *Let $A \in \Pi$ be a stable component of size at most c and $X = \mathbf{D}(A) - A$. Let μ' be any matching on S. Then starting from μ', if no matches are formed between a member of A and a member of X, the expected number of distinct times the matching needs to transition from a locally perfect matching on A to a matching that is not locally perfect on A before it reaches a μ-stable matching on A is at most $c^{4(c+1)}$.*

The final lemma we need shows that when the matching is not locally perfect on a stable component of the decomposition, it will reach a perfect matching in time that depends only linearly in n in expectation, provided there is no interference from members of lower stable components.

Lemma 23. *Let $A \in \Pi$ be a stable component of size at most c and $X = \mathbf{D}(A) - A$. Let μ' be any matching on S which is not locally perfect on A. Then starting from μ', if no matches are formed between a member of A and a member of X, the expected time reach a matching which is locally perfect on A is at most cn^{2c}.*

Proof. Lemma 19 implies that for any given matching, either the matching is locally perfect on A or there is a blocking pair between two members of A which is a pair in μ. Since the size of A is at most c, there are at most c such pairs. Therefore if all of them are resolved in c consecutive steps, the resulting matching will be locally perfect on A. Alternatively if after fewer than c steps of resolving blocking pairs that are in μ we reach a matching with no such blocking pairs, then the matching must already be locally perfect on A. For any given matching there are at most n^2 total blocking pairs so the probability of resolving a blocking pair between two members of A that is a pair in μ is at least $\frac{1}{n^2}$. But then the probability of resolving up to c of them and reaching a locally perfect matching in c or fewer steps is at least $\frac{1}{n^{2c}}$.

Therefore, in expectation we will have to repeat the process of making c steps at most n^{2c} times before reaching a locally perfect matching on A. This leads to at most cn^{2c} steps in expectation.

Finally we will show that the expected convergence time for the better response dynamics is linear in the total number of participants but possibly exponential in the size of the largest stable component and depth of the decomposition. The special case where the size of the decomposition is 1 includes the correlated preferences of Ackermann et al.

Theorem 24 (Convergence). *Suppose μ is a stable matching. Suppose the depth of $(\Pi, \preceq)$ is d and the size of the largest stable component of Π is no more than c. Then the expected time to converge to a stable matching is $O(c^{O(cd)}n^{O(c+d)})$. If $c = 1$, then the expected time is $O(n^3)$.*

Proof. Suppose μ' is another stable matching. First, suppose that for any stable component A' of a μ'-decomposition, a μ'-stable matching is never reached on $\mathbf{D}_{\mu'}(A')$.

Consider the μ-decomposition graph for $(\Pi, \preceq)$. Recall that a stable component A is on level j if $depth(A) = j$. For convenience, let level $d + 1$ be an empty dummy level at the top. Since the depth is d, there are exactly $d + 1$ levels. We proceed by bounding the expected time for one level to reach a μ stable matching, and then recurse on the higher levels.

Let $T(l)$ denote the expected time for the participants in stable components on levels l and above to reach a stable matching without resolving blocking pairs involving any members of stable components on lower levels. Let n_l be the number of stable components on level l. Note that since there are at most n stable components of $\mathcal{D}$, $n_1 + \ldots + n_d \leq n$. We will show that $T(0) = O(c^{O(cd)}n^{O(c+d)})$.

First observe that $T(d + 1) = 0$ since there are no stable components at level $d + 1$.

Now consider $T(l)$ for $l < d + 1$.

When one of the n_l stable components A on level l is not in a locally perfect matching. Then by Lemma 23, we know it will take cn^{2c} steps in expectation to reach a locally perfect matching on A. Also, by lemma 21 we know it has at least $\frac{1}{c^2}$ probability of reaching a matching whose restriction to A has a greater number of pairs that are in μ than the current matching, before it reaches a locally perfect matching.

On the other hand, when all n_l stable components are in locally perfect matchings, then there are two cases:

If there is a blocking pair between two members of stable components on level l it will remain there until the matching becomes not locally perfect on at least one stable component on level l. Since there are at most n^2 blocking pairs, it will take at most n^2 steps in expectation for the matching to become not locally perfect on at least one stable component on level l.

If there are no such blocking pairs, it might be required for the higher levels to reach a stable matching before exposing a blocking pair involving a participant on level l. If no matches are formed involving any members of components on level l or lower, the expected time for the remaining stable components to reach a stable matching is given by $T(l + 1)$. Once the higher levels have reached a stable matching, the only blocking pairs not involving members of levels below l are between a member of a stable component on level l and a member of a stable component on a higher level. Unless all stable components on level l and above are in a stable matching, at least one such blocking pair must exist. Therefore it will only take 1 more step to reach a matching which is not locally perfect on one stable component on level l.

Consequently, it will take at most $n^2 + T(l+1) + 1$ steps to reach a matching that is not locally perfect on one stable component on level l. Again, by Lemma 23, we know it will take cn^{2c} steps in expectation to reach a locally perfect matching on A. By Lemma 22, we know in expectation, for each stable component on level l, it will take at most $c^{4(c+1)}$ transitions from a locally perfect matching to a matching which is not locally perfect on that stable component it reaches a μ-stable matching. This means that in expectation it will take at most $n_l c^{4(c+1)}$ of these transitions total before all stable components on level l reach a μ-stable matching.

Therefore, in the worst case, it will take $(n^2 + T(l+1) + 1)$ steps to transition from a locally perfect matching to a matching that is not locally perfect on one of the stable components on level l. Then it will take at most cn^{2c} steps to reach a matching which is locally perfect on that stable component. Furthermore, this process needs to be repeated no more than $n_l c^{4(c+1)}$ times in expectation in order for all stable components on level l to reach a μ-stable matching.

Once all stable components on level l have reached a μ-stable matching, all that remains is for the higher levels to reach a stable matching, which takes $T(l+1)$ time in expectation.

This yields the following formula:

$$T(l) \le n_l c^{4(c+1)}(cn^{2c} + n^2 + T(l+1) + 1) + T(l+1) \le 2n_l c^{4(c+1)}(cn^{2c} + T(l+1))$$

Solving this recursion for $T(0)$, we obtain

$$T(0) \le 2n_0 c^{4(c+1)}(cn^{2c} + T(1))$$

$$T(0) \le (cn^{2c}) \sum_{i=1}^{d+1} (2c^{4(c+1)})^i \prod_{j=0}^{i-1} n_j$$

so since $n_i + 1 \le O(n)$ for all i, $T(0) = O(c^{O(cd)} n^{O(c+d)})$.

This is the expected time to reach the stable matching μ. Now suppose for some stable component A' of a μ'-decomposition for some other stable matching μ', a μ'-stable matching is reached on $\mathbf{D}_{\mu'}(A')$. By Lemma 18, this will not increase the size or depth of the remaining decomposition. Therefore, if this happens before μ is reached, it will only decrease the convergence time.

Finally, as a special case assume $c = 1$. In this case a locally perfect matching on a stable component is a μ-stable matching. By lemma 23 it will take at most n^2 steps for a stable component on level l to reach a μ-stable matching. Since there are n_l components on level l, $T(l) \le n_l n^2 + T(l+1) \le \sum_{i=1}^{d-l} n_i n^2$ so $T(0) = \le \sum_{i=1}^{d} n_i n^2 = n^3$.

4 Correlated and Intercorrelated Preferences

We have shown bounds on convergence time but this is only relevant if there is variation in the jealousy graph structures of real markets. While randomly generated preferences tend to have decompositions that are close to the trivial decomposition, which is the entire set, real-world markets tend to have some structure. Here we show that two classes of preferences found in real world markets, correlated and intercorrelated preferences, exhibit decompositions with small size components. Partially correlated preferences are often used by modelers [8, 11] and are natural in many matching markets (e.g. mate selection) where preferences are based on a mixture of universally desirable features (e.g. intelligence) and idiosyncratic tastes (e.g. shared hobbies). Note that the correlated preferences discussed here differ from the correlated preferences of Ackemann et al. Intercorrelation exists when the preferences of the men relate to the preferences of the women. See [12] for examples of markets with intercorrelation. Boudreau showed that more correlation and intercorrelation lead to faster convergence of the better response algorithm [8]. We provide similar plots in Figures 1(b) and 1(d). Theorem 24 provides theoretical justification for these simulated results.

As described in [11, 13], correlated preferences are generated using scores of the form:

$$S_{mw} = \eta_{mw} + U I_w$$

where S_{mw} is the score man m gives woman w composed of his individual score η_{mw} and a correlation factor $U \in [0, \infty)$ multiplied by the consensus score of w, I_w. η_{mw} and I_w are chosen uniformly at random from $[0, 1]$. The men then rank the women in order from lowest score to highest. Women's preferences are generated analogously. For various values of U we generate 100 preferences with correlation factor U. For each set of preferences we find the decomposition with smallest size and report the average of these sizes. We also compute the average minimal depth in the same manner. The results are shown in figure 1(a). At $U = 0$, the average size is close to n and the depth is close to 1. As U goes to ∞, the average size approaches 1 and the depth approaches n. These are the parameters of perfectly correlated preferences. This shows that as the amount of correlation varies, so do the size and depth of the decompositions. Figure 1(b) shows the log of the average convergence time over 100 trials for each of the 100 correlated preferences generated.

As in [12], intercorrelated preferences can be generated using scores of the form:

$$S_{m_i w_j} = \eta_{m_i w_j} + V * |i - j|_n$$

where $S_{m_i w_j}$ is the score man m_i gives woman w_j. As with correlated preferences, $\eta_{m_i w_j}$ is his individual score. Here V is the intercorrelation factor and $|i - j|_n = \min(|i - j|, n - |i - j|)/(\frac{n}{2})$ represents the "distance" man m_i is from woman w_j. 1(c) and 1(d) are generated in the same manner as 1(a) and 1(b), respectively. These plots show that as preferences become more intercorrelated,

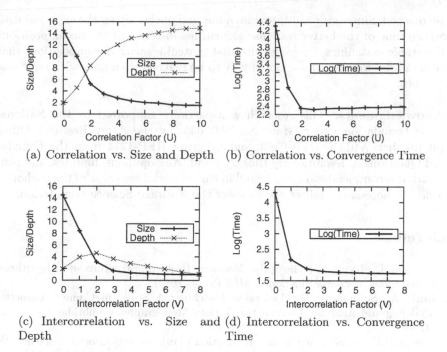

(a) Correlation vs. Size and Depth (b) Correlation vs. Convergence Time

(c) Intercorrelation vs. Size and (d) Intercorrelation vs. Convergence
Depth   Time

Fig. 1. Jealousy Graphs vs. Correlation and Intercorrelation. (a) The jealousy graph
parameters change as preferences become more correlated. (b) Convergence time decreases as preferences become more correlated. (c) The jealousy graph parameters
change as preferences become more intercorrelated. (d) Convergence time decreases as
preferences become more intercorrelated.

the size and depth of the decompositions decrease. As Theorem 24 explains, this
decreases the convergence time of the better response algorithm as intercorrelation increases.

5   Conclusion

We have introduced a new way of viewing stable matching problems in terms of
their jealousy graphs and μ-decompositions. We demonstrate that these concepts
are useful in analyzing the convergence time of the better response algorithm
and guarantee polynomial convergence on a subclass of matching markets. Furthermore, these theoretical results apply to a broad range of markets since they
provide a notion of structure which extends beyond the well-studied notions of
correlation and intercorrelation.

One open question involves the exponential dependency on the depth of the
decomposition. While we know that the exponential dependency on size cannot
be removed, it remains an open question whether we can improve this bound in
terms of the depth. Another open problem concerns which matching is most likely

to be reached. Since our result provides a method of classifying the expected convergence time of the better response algorithms in terms of the decompositions of the stable matchings, we conjecture that matchings with decompositions that have small size and depth are more likely to be reached than ones with large size and depth.

Acknowledgments. This research was partially supported by the National Science Foundation under Grant No. 0905645, by the Army Research Office grant number W911NF-11-1-0363, and by grant RFP-1211 from the Foundational Questions in Evolutionary Biology Fund. Any opinions, findings, and conclusions or recommendations expressed in this material are those of the author(s) and do not necessarily reflect the views of the National Science Foundation.

References

1. Gusfield, D., Irving, R.: The Stable Marriage Problem: Structure and Algorithms. Foundations of Computing Series. MIT Press (1989)
2. Roth, A., Sotomayor, M.: Two-sided Matching: A Study in Game - Theoretic Modeling and Analysis. Econometric Society Monographs. Cambridge University (1990)
3. Knuth, D.E.: Stable marriage and its relation to other combinatorial problems: An introduction to the mathematical analysis of algorithms, vol. 10. Amer. Mathematical Society (1997)
4. Gale, D., Shapley, L.S.: College admissions and the stability of marriage. The American Mathematical Monthly 69(1), 9–15 (1962)
5. Echenique, F., Yariv, L.: An experimental study of decentralized matching (2011) (working paper, Caltech)
6. Roth, A.E., Vande Vate, J.H.: Random paths to stability in two-sided matching. Econometrica 58(6), 1475–1480 (1990)
7. Ackermann, H., Goldberg, P.W., Mirrokni, V.S., Röglin, H., Vöcking, B.: Uncoordinated two-sided matching markets. SIAM J. Comput. 40(1), 92–106 (2011)
8. Boudreau, J.W.: Preference structure and random paths to stability in matching markets. Economics Bulletin 3(67), 1–12 (2008)
9. Boudreau, J.W.: An exploration into why some matchings are more likely than others. In: MATCH-UP 2012: the Second International Workshop on Matching Under Preferences, p. 39 (2012)
10. Pais, J., Pinter, A., Veszteg, R.F.: Decentralized matching markets: a laboratory experiment (2012)
11. Celik, O.B., Knoblauch, V.: Marriage matching with correlated preferences. Economics Working Papers, 200716 (2007)
12. Boudreau, J.W., Knoblauch, V.: Marriage matching and intercorrelation of preferences. Journal of Public Economic Theory 12(3), 587–602 (2010)
13. Caldarelli, G., Capocci, A.: Beauty and distance in the stable marriage problem. Physica A: Statistical Mechanics and its Applications 300(1), 325–331 (2001)

Linear Regression as a Non-cooperative Game

Stratis Ioannidis[1] and Patrick Loiseau[2]

[1] Technicolor Palo Alto, CA, USA
stratis.ioannidis@technicolor.com
[2] EURECOM, France
patrick.loiseau@eurecom.fr

Abstract. Linear regression amounts to estimating a linear model that maps features (e.g., age or gender) to corresponding data (e.g., the answer to a survey or the outcome of a medical exam). It is a ubiquitous tool in experimental sciences. We study a setting in which features are public but the data is private information. While the estimation of the linear model may be useful to participating individuals, (if, e.g., it leads to the discovery of a treatment to a disease), individuals may be reluctant to disclose their data due to privacy concerns. In this paper, we propose a generic game-theoretic model to express this trade-off. Users add noise to their data before releasing it. In particular, they choose the variance of this noise to minimize a cost comprising two components: (a) a privacy cost, representing the loss of privacy incurred by the release; and (b) an estimation cost, representing the inaccuracy in the linear model estimate. We study the Nash equilibria of this game, establishing the existence of a unique non-trivial equilibrium. We determine its efficiency for several classes of privacy and estimation costs, using the concept of the price of stability. Finally, we prove that, for a specific estimation cost, the generalized least-square estimator is optimal among all linear unbiased estimators in our non-cooperative setting: this result extends the famous Aitken/Gauss-Markov theorem in statistics, establishing that its conclusion persists even in the presence of strategic individuals.

Keywords: Linear regression, Gauss-Markov theorem, Aitken theorem, privacy, potential game, price of stability.

1 Introduction

The statistical analysis of personal data is a cornerstone of several experimental sciences, such as medicine and sociology. Studies in these areas typically rely on experiments, drug trials, or surveys involving human subjects. Data collection has also become recently a commonplace—yet controversial—aspect of the Internet economy: companies such as Google, Amazon and Netflix maintain and mine large databases of behavioral information (such as, e.g., search queries or past purchases) to profile their users and personalize their services. In turn, this has raised privacy concerns from consumer advocacy groups, regulatory bodies, as well as the general public.

Y. Chen and N. Immorlica (Eds.): WINE 2013, LNCS 8289, pp. 277–290, 2013.

The desire for privacy incentivizes individuals to lie about their private information–or, in the extreme, altogether refrain from any disclosure. For example, an individual may be reluctant to participate in a medical study collecting biometric information, concerned that it may be used in the future to increase her insurance premiums. Similarly, an online user may not wish to disclose her ratings to movies if this information is used to infer, e.g., her political affiliation. On the other hand, a successful data analysis may also provide a utility to the individuals from which the data is collected. This is evident in medical studies: an experiment may lead to the discovery of a treatment for a disease, from which an experiment subject may clearly benefit. In the cases of commercial data mining, users may benefit both from overall service improvements, as well as from personalization. If such benefits outweigh privacy considerations, users may consent to the collection and analysis of their data, e.g., by participating in a clinical trial, completing a survey, or using an online service.

In this paper, we approach the above issues through a non-cooperative game, focusing on a statistical analysis task called *linear regression*. We consider the following formal setting. A set of individuals $i \in \{1, \ldots, n\}$ participate in an experiment, in which they are about to disclose to a data analyst a private variable $y_i \in \mathbb{R}$–e.g., the answer to a survey or the outcome of a medical test. Each individual i is associated with a feature vector $\boldsymbol{x}_i \in \mathbb{R}^d$, capturing public information such as, e.g., age, gender, etc. The analyst wishes to perform *linear regression* over the data, i.e., compute a vector $\boldsymbol{\beta} \in \mathbb{R}^d$ such that:

$$y_i \approx \boldsymbol{\beta}^T \boldsymbol{x}_i, \qquad \text{for all } i \in \{1, \ldots, n\}.$$

However, individuals *do not* reveal their true private variables to the analyst in the clear: instead, before reporting these values, they *first add noise*. In our examples above, such noise addition aims to protect against, e.g., future use of the individual's biometric data by an insurance company, or inference of her political affiliation from her movie ratings. The higher the variance of the noise an individual adds, the better the privacy that she attains, as her true private value is obscured. On the other hand, high noise variance may also hurt the accuracy of the analyst's estimate of $\boldsymbol{\beta}$, the linear model computed in aggregate across multiple individuals. As such, the individuals need to strike a balance between the privacy cost they incur through disclosure and the utility they accrue from accurate model prediction.

Our contributions can be summarized as follows.

(i) We model interactions between individuals as a non-cooperative game, in which each individual selects the variance level of the noise to add to her private variable strategically. An individual's decision minimizes a cost function comprising two components: (a) a *privacy cost*, that is an increasing function of the added noise variance, and (b) an *estimation cost*, that decreases as the accuracy of the analyst's estimation of $\boldsymbol{\beta}$ increases. Formally, the estimation cost increases with the covariance matrix of the estimate of $\boldsymbol{\beta}$, when this estimate is computed through a least-squares minimization.

(ii) We characterize the Nash equilibria of the above game. In particular, we show that the above setting forms a potential game. Moreover, under appropriate assumptions on the privacy and estimation costs, there exists a unique pure Nash equilibrium at which individual costs are bounded.

(iii) Armed with this result, we determine the game's efficiency, providing bounds for the price of stability for several cases of privacy and estimation costs.

(iv) Finally, we turn our attention to the analyst's estimation algorithm. We show that, among the class of unbiased, linear estimators, generalized least squares is the estimator that yields the most accurate estimate, at equilibrium. In a formal sense, this extends the Aitken theorem in statistics, which states that generalized least squares estimation yields minimal variance among linear unbiased estimators. Our result implies that this optimality persists even if individuals strategically choose the variance of their data.

The remainder of this paper is organized as follows. We present related work in Section 2. Section 3 contains a review of linear regression and the definition of our non-cooperative game. We characterize Nash equilibria in Section 4 and discuss their efficiency in Section 5. Our Aitken-type theorem is in Section 6, and our conclusions in Section 7. Due to space constraints, long proofs are relegated to our technical report [1].

2   Related Work

Perturbing a dataset before submitting it as input to a data mining algorithm has a long history in privacy-preserving data-mining (see, e.g., [2, 3]). Independent of an algorithm, early research focused on perturbing a dataset prior to its public release [4, 5]. Perturbations tailored to specific data mining tasks have also been studied in the context of reconstructing the original distribution of the underlying data [6], building decision trees [6], clustering [7], and association rule mining [8]. We are not aware of any study of such perturbation techniques in a non-cooperative setting, where individuals add noise strategically.

The above setting differs from the more recent framework of ϵ-differential privacy [9, 10], which has also been studied from the prespective of mechanism design [11–13]. In differential privacy, noise is added to the *output* of a computation, which is subsequently publicly released. Differential privacy offers a strong guarantee: changing an individual's input alters the distribution of the perturbed output at most by an $\exp \epsilon \approx 1 + \epsilon$ factor. The analyst performing the computation is a priori trusted; as such, individuals submit unadulterated inputs. In contrast, the classic privacy-preserving data-mining setting we study here assumes an untrusted analyst, which motivates input perturbation.

In experimental design [14, 15], an analyst observes the public features of a set of experiment subjects, and determines which experiments to conduct with the objective of learning a linear model. The quality of an estimated model is quantified through a scalarization of its variance [16]. Though many such scalarizations exist, we focus here on non-negative scalarizations, to ensure meaningful notions of efficiency (as determined by the price of stability, *c.f.* Section 5).

Several papers study problems of statistical inference from the perspective of mechanism design. Horel *et al.* [17] study a version of the experimental design problem in which subjects report their private values truthfully, but may lie about the costs they require for their participation. Closer to our setting, Dekel *et al.* [18] consider a broad class of regression problems in which participants may misreport their private values, and determine loss functions under which empirical risk minimization is group strategy-proof–the special case of linear regression is also treated, albeit in a more restricted setting, in [19]. Our work differs in considering noise addition as a non-cooperative game, and studying the efficiency of its Nash equilibria, rather than mechanism design issues.

Our model has analogies to models used in *public good* provision problems (see, e.g., [20] and references therein). Indeed, the estimate variance reduction can be seen as a public good in that, when an individual contributes her data, all other individuals in the game benefit. Moreover, the perturbation technique used in our proof of Theorem 6 is similar to techniques used in public good models introduced in the context of traffic congestion [21, 22].

3   Model Description

In this section, we give a detailed description of our linear regression game and the players involved. Before discussing strategic considerations, we give a brief technical review of linear models, as well as key properties of least squares estimators; all related results presented here are classic (see, e.g., [23]).

Notational conventions. We use boldface type (e.g., x, y, β) to denote vectors (all vectors are column vectors), and capital letters (e.g., A, B, V) to denote matrices. As usual, we denote by S_+^d, $S_{++}^d \subset \mathbb{R}^{d \times d}$ the sets of (symmetric) positive semidefinite (PSD) and positive definite matrices of size $d \times d$, respectively. For two positive semidefinite matrices $A, B \in S_+^d$, we write that $A \succeq B$ if $A - B \in S_+^d$; recall that $\succeq$ defines a partial order over S_+^d. We say that $F : S_+^d \to \mathbb{R}$ is non-decreasing in the positive semidefinite order if $F(A) \geq F(A')$ for any two $A, A' \in S_+^d$ such that $A \succeq A'$. Moreover, we say that a matrix-valued function $F : \mathbb{R}^n \to S_+^d$ is *matrix convex* if $\alpha F(\lambda) + (1 - \alpha)F(\lambda') \succeq F(\alpha\lambda + (1 - \alpha)\lambda')$ for all $\alpha \in [0, 1]$ and $\lambda, \lambda' \in \mathbb{R}^n$. Given a square matrix $A = [a_{ij}]_{1 \leq i,j \leq d} \in \mathbb{R}^{d \times d}$, we denote by trace$(A)$ its *trace* (i.e., the sum of its diagonal elements), and by $\|A\|_F$ its *Frobenious norm* (i.e., the ℓ_2-norm of its d^2 elements).

3.1   Linear Models

Consider a set of n individuals, denoted by $N \equiv \{1, \cdots, n\}$. Each individual $i \in N$ is associated with a vector $x_i \in \mathbb{R}^d$, the *feature vector*, which is public; for example, this vector may correspond to publicly available demographic information about the individual, such as age, gender, etc. Each $i \in N$ is also associated with a private variable $y_i \in \mathbb{R}$; for example, this may express the likelihood that this individual contracts a disease, the concentration of a substance in her blood or an answer that she gives to a survey.

Throughout our analysis, we assume that the individual's private variable y_i is a linear function of her public features x_i. In particular, there exists a vector $\beta \in \mathbb{R}^d$, the *model*, such that the private variables are given by

$$y_i = \beta^T x_i + \epsilon_i, \quad \text{for all } i \in N, \tag{1}$$

where the "inherent noise" variables $\{\epsilon_i\}_{i \in N}$ are i.i.d. zero-mean random variables in $\mathbb{R}$ with finite variance σ^2. We stress that we make no further assumptions on the noise; in particular, we *do not* assume it is Gaussian.

An analyst wishes to observe the y_i's and infer the model $\beta \in \mathbb{R}^d$. This type of inference is ubiquitous in experimental sciences, and has a variety of applications. For example, the magnitude of β's coordinates captures the effect that features (e.g., age or weight) have on y_i (e.g., the propensity to get a disease), while the sign of a coordinate captures positive or negative correlation. Knowing β can also aid in prediction: an estimate of private variable $y \in \mathbb{R}$ of a new individual with features $x \in \mathbb{R}^d$ is given by the inner product $\beta^T x$.

We note that the linear relationship between y_i and x_i expressed in (1) is in fact quite general. For example, the case where $y_i = f(x) + \epsilon_i$, where f is a polynomial function of degree 2, reduces to a linear model by considering the transformed feature space whose features comprise the monomials $x_{ik} x_{ik'}$, for $1 \le k, k' \le d$. More generally, the same principle can be applied to reduce to (1) any function class spanned by a finite set of basis functions over $\mathbb{R}^d$ [23].

3.2 Generalized Least Squares Estimation

We consider a setup in which the individuals intentionally *perturb* or *distort* their private variable by adding excess noise. In particular, each $i \in N$ computes $\tilde{y}_i = y_i + z_i$ where z_i is a zero-mean random variable with variance σ_i^2; we assume that $\{z_i\}_{i \in N}$ are independent, and are also independent of the inherent noise variables $\{\epsilon_i\}_{i \in N}$. Subsequently, each individual reveals to the analyst (a) the perturbed variable $\tilde{y}_i$ and (b) the variance σ_i^2. As a result, the aggregate variance of the reported value is $\sigma^2 + \sigma_i^2$.

In turn, having access to the perturbed variables $\tilde{y}_i$, $i \in N$, and the corresponding variances, the analyst estimates β through *generalized least squares* (GLS) estimation. For $i \in N$, let $\lambda_i \equiv \frac{1}{\sigma^2 + \sigma_i^2}$ be the inverse of the aggregate variance. Denote by $\lambda = [\lambda_i]_{i \in N}$ the vector of inverses and by $\Lambda = \text{diag}(\lambda)$ the diagonal matrix whose diagonal is given by vector λ. Then, the generalized least squares estimator is given by:

$$\hat{\beta}_{\text{GLS}} = \underset{\beta \in \mathbb{R}^d}{\arg\min} \left(\sum_{i \in N} \lambda_i (\tilde{y}_i - \beta^T x_i)^2 \right) = (X^T \Lambda X)^{-1} X^T \Lambda \tilde{y} \tag{2}$$

where $\tilde{y} = [\tilde{y}_i]_{i \in N}$ is the n-dimensional vector of perturbed variables, and $X = [x_i^T]_{i \in N} \in \mathbb{R}^{n \times d}$ the $n \times d$ matrix whose rows comprise the transposed feature vectors. Throughout our analysis, we assume that $n \ge d$ and that X has rank d.

Note that $\tilde{y} \in \mathbb{R}^n$ is a random variable and as such, by (2), so is $\hat{\beta}_{\text{GLS}}$. It can be shown that $\mathbb{E}(\hat{\beta}_{\text{GLS}}) = \beta$ (i.e., $\hat{\beta}_{\text{GLS}}$ is unbiased), and

$$V(\lambda) \equiv Cov(\hat{\beta}_{\text{GLS}}) = \mathbb{E}\left[(\hat{\beta}_{\text{GLS}} - \beta)^T (\hat{\beta}_{\text{GLS}} - \beta)\right] = (X^T \Lambda X)^{-1}.$$

The covariance V captures the uncertainty of the estimation of β. The matrix

$$A(\lambda) \equiv X^T \Lambda X = \sum_{i \in N} \lambda_i x_i x_i^T$$

is known as the *precision* matrix. It is positive semidefinite, i.e., $A(\lambda) \in S_+^d$, but it may not be invertible: this is the case when $\text{rank}(X^T \Lambda) < d$, i.e., the vectors x_i, $i \in N$, for which $\lambda_i > 0$, do not span $\mathbb{R}^d$. Put differently, if the set of individuals providing useful information does not include d linearly independent vectors, there exists a direction $x \in \mathbb{R}^d$ that is a "blind spot" to the analyst: the analyst has no way of predicting the value $\beta^T x$. In this degenerate case the number of solutions to the least squares estimation problem (2) is infinite, and the covariance is not well-defined (it is infinite in all such directions x). Note however that, since X has rank d (and hence $X^T X$ is invertible), the set of λ for which the precision matrix is invertible is non-empty. In particular, it contains $(0, 1/\sigma^2]^n$ since $A(\lambda) \in S_{++}^d$ if $\lambda_i > 0$ for all $i \in N$.

3.3   User Costs and a Non-cooperative Game

The perturbations z_i are motivated by privacy concerns: an individual may be reluctant to grant unfettered access to her private variable or release it in the clear. On the other hand, it may be to the individual's advantage that the analyst learns the model β. In our running medical example, learning that, e.g., a disease is correlated to an individual's weight or her cholesterol level may lead to a cure, which in turn may be beneficial to the individual.

We model the above considerations through cost functions. Recall that the action of each individual $i \in N$ amounts to choosing the noise level of the perturbation, captured by the variance $\sigma_i^2 \in [0, \infty]$. For notational convenience, we use the equivalent representation $\lambda_i = 1/(\sigma^2 + \sigma_i^2) \in [0, 1/\sigma^2]$ for the action of an individual. Note that $\lambda_i = 0$ (or, equivalently, infinite variance σ_i) corresponds to no participation: in terms of estimation through (2), it is as if this perturbed value is not reported.

Each individual $i \in N$ chooses her action $\lambda_i \in [0, 1/\sigma^2]$ to minimize her cost

$$J_i(\lambda_i, \lambda_{-i}) = c_i(\lambda_i) + f(\lambda), \tag{3}$$

where we use the standard notation λ_{-i} to denote the collection of actions of all players but i. The cost function $J_i : \mathbb{R}_+^n \to \mathbb{R}_+$ of player $i \in N$ comprises two non-negative components. We refer to the first component $c_i : \mathbb{R}_+ \to \mathbb{R}_+$ as the *privacy cost*: it is the cost that the individual incurs on account of the privacy violation sustained by revealing the perturbed variable. The second component is the *estimation cost*, and we assume that it takes the form $f(\lambda) = F(V(\lambda))$, if $A(\lambda) \in S_{++}^d$, and $f(\lambda) = \infty$ otherwise. The mapping $F : S_{++}^d \to \mathbb{R}_+$ is known as a *scalarization* [16]. It maps the covariance matrix $V(\boldsymbol{\lambda})$ to a scalar value $F(V(\boldsymbol{\lambda}))$, and captures how well the analyst can estimate the model $\boldsymbol{\beta}$. The estimation cost $f : \mathbb{R}^n_+ \to \bar{\mathbb{R}}_+ = \mathbb{R}_+ \cup \{\infty\}$ is the so-called *extended-value extension* of $F(V(\boldsymbol{\lambda}))$: it equals $F(V(\boldsymbol{\lambda}))$ in its domain, and $+\infty$ outside its domain. Throughout our analysis, we make the following two assumptions:

Assumption 1. *The privacy costs $c_i : \mathbb{R}_+ \to \mathbb{R}_+$, $i \in N$, are twice continuously differentiable, non-negative, non-decreasing and strictly convex.*

Assumption 2. *The scalarization $F : S^d_{++} \to \mathbb{R}_+$ is twice continuously differentiable, non-negative, non-constant, non-decreasing in the positive semidefinite order, and convex.*

The monotonicity and convexity assumptions in Assumptions 1 and 2 are standard and natural. Increasing λ_i (i.e., decreasing the noise added by the individual) leads to a higher privacy cost. In contrast, increasing λ_i can only decrease the estimation cost: this is because decreasing the noise of an individual also decreases the variance in the positive semidefinite sense (as the matrix inverse is a PSD-decreasing function). Note that it is possible to relax Assumptions 1 and 2 (in particular, amend the twice-continuous differentiability assumption) without affecting most of our results, at the expense of an increased technical complexity in our proofs. We therefore focus on the above two assumptions for the sake of simplicity.

As a consequence of Assumption 2, the extended-value extension f is convex. The convexity of $F(V(\cdot))$ follows from the fact that it is the composition of the non-decreasing convex function $F(\cdot)$ with the matrix convex function $V(\cdot)$; the latter is convex because the matrix inverse is matrix convex. Moreover, f is twice continuously differentiable on its effective domain $\{\boldsymbol{\lambda} \in \mathbb{R}^n_+ : A(\boldsymbol{\lambda}) \in S^d_{++}\}$.

Scalarizations of positive semidefinite matrices and, in particular, of the covariance matrix $V(\boldsymbol{\lambda})$, are abundant in statistical inference literature in the context of experimental design [14–16]. We give two examples we use in our analysis below:

$$F_1(V) = \text{trace}(V), \qquad\qquad F_2(V) = \|V\|^2_F. \qquad (4)$$

Both scalarizations satisfy Assumption 2.

We denote by $\Gamma = \langle N, [0, 1/\sigma^2]^n, (J_i)_{i \in N} \rangle$ the game with set of players $N = \{1, \cdots, n\}$, where each each player $i \in N$ chooses her action λ_i in her action set $[0, 1/\sigma^2]$ to minimize her cost $J_i : [0, 1/\sigma^2]^n \to \mathbb{R}_+$, given by (3). We refer to a $\boldsymbol{\lambda} \in [0, 1/\sigma^2]^n$ as a *strategy profile* of the game Γ. We analyze the game as a *complete information game*, i.e., we assume that the set of players, the action sets and utilities are known by all players.

4   Nash Equilibria

We begin our analysis by characterizing the Nash equilibria of the game Γ. Observe first that Γ is a potential game [24]. Indeed, define the function $\Phi : [0, 1/\sigma^2]^n \to \bar{\mathbb{R}}$ such that

$$\Phi(\boldsymbol{\lambda}) = f(\boldsymbol{\lambda}) + \sum_{i \in N} c_i(\lambda_i), \quad (\boldsymbol{\lambda} \in [0, 1/\sigma^2]^n). \qquad (5)$$

Then for every $i \in N$ and for every $\lambda_{-i} \in [0, 1/\sigma^2]^{n-1}$, we have

$$J_i(\lambda_i, \lambda_{-i}) - J_i(\lambda_i', \lambda_{-i}) = \Phi(\lambda_i, \lambda_{-i}) - \Phi(\lambda_i', \lambda_{-i}), \quad \forall \lambda_i, \lambda_i' \in [0, 1/\sigma^2]. \quad (6)$$

Therefore, Γ is a potential game with potential function Φ.

In the game Γ, each player chooses her contribution λ_i to minimize her cost. A Nash equilibrium (in pure strategy) is a strategy profile $\boldsymbol{\lambda}^*$ satisfying

$$\lambda_i^* \in \arg\min_{\lambda_i} J_i(\lambda_i, \lambda_{-i}^*), \quad \text{for all } i \in N.$$

From (6), we see that (as for any potential game) the set of Nash equilibria coincides with the set of local minima of function Φ.

First note that there may exist Nash equilibria $\boldsymbol{\lambda}^*$ for which $f(\boldsymbol{\lambda}^*) = \infty$. For instance, if $d \geq 2$, $\boldsymbol{\lambda}^* = 0$ is a Nash equilibrium. Indeed, in that case, no individual has an incentive to deviate since a single $\lambda_i > 0$ still yields a non-invertible precision matrix $A(\boldsymbol{\lambda})$. In fact, any profile $\boldsymbol{\lambda}$ for which $A(\boldsymbol{\lambda})$ is non-invertible, and remains so under unilateral deviations, constitutes an equilibrium.

We call such Nash equilibria (at which the estimation cost is infinite) *trivial*. Existence of trivial equilibria can be avoided in practice using slight model adjustments. For instance, one can impose a finite upper bound on the variance σ_i of an individual i (or, equivalently, a positive lower bound on λ_i). Alternatively, the existence of d non-strategic individuals whose feature vectors span $\mathbb{R}^d$ is also sufficient to enforce a finite covariance at all $\boldsymbol{\lambda}$ across strategic individuals.

In the remainder, we focus on the more interesting *non-trivial* equilibria. Using the potential game structure of Γ, we derive the following result.

Theorem 1. *There exists a unique non-trivial equilibrium of the game Γ.*

Proof. Recall that the set of Nash equilibria coincides with the set of local minima of function Φ. To conclude the proof, we show that there exists a unique local minimum $\boldsymbol{\lambda}$ of Φ in the effective domain of f.

First note that, by Assumption 1, the privacy cost $c_i(\cdot)$ is finite on $[0, 1/\sigma^2]$ since it is continuous on a compact set. Therefore, $\Phi(\cdot)$ is finite *iif* $f(\cdot)$ is finite i.e., $\text{dom}\,\Phi \equiv \{\boldsymbol{\lambda} : \Phi(\boldsymbol{\lambda}) < \infty\} = \text{dom}\,f$, where dom is the effective domain. Recall that since X has rank d, $(0, 1/\sigma^2]^n \subset \text{dom}\,\Phi$, and $\text{dom}\,\Phi$ is non-empty.

By Assumptions 1 and 2, function Φ is strictly convex on its effective domain. Therefore it has at most one local minimum in $\text{dom}\,\Phi$. Since $\text{dom}\,\Phi$ is not compact, we still need to show that the minimum is achieved. By Assumption 1, the privacy cost derivatives are bounded and increasing. Let $M = \max_{i \in N} c'(1/\sigma^2) < \infty$ be the largest possible privacy cost derivative across all users and all λ_i's. On the other hand, the partial derivatives of the estimation cost can be written as

$$\frac{\partial f}{\partial \lambda_i}(\boldsymbol{\lambda}) = -\text{trace}\left(\frac{\partial F}{\partial V} \cdot (X^T \Lambda X)^{-1} \boldsymbol{x}_i \boldsymbol{x}_i^T (X^T \Lambda X)^{-1}\right)$$

$$= -\boldsymbol{x}_i^T (X^T \Lambda X)^{-1} \cdot \frac{\partial F}{\partial V} \cdot (X^T \Lambda X)^{-1} \boldsymbol{x}_i,$$

where

$$\frac{\partial F}{\partial V} = \begin{pmatrix} \frac{\partial F}{\partial V_{11}} & \cdots \\ \vdots & \ddots \end{pmatrix}.$$

Hence, since F is non-constant, there exists an $i \in N$ for which $\frac{\partial f}{\partial \lambda_i}(\boldsymbol{\lambda})$ is unbounded. Therefore, it is possible to define $\bar{\mathcal{E}} \equiv \{\boldsymbol{\lambda} \in [0, 1/\sigma^2]^n : f(\boldsymbol{\lambda}) > K\}$ with K large enough so that $\max_{i \in N} \left|\frac{\partial f}{\partial \lambda_i}(\boldsymbol{\lambda})\right| > M$ for all $\boldsymbol{\lambda} \in \bar{\mathcal{E}}$. Let $\mathcal{E}$ be the complement of $\bar{\mathcal{E}}$ in $[0, 1/\sigma^2]^n$. Let $\boldsymbol{\lambda} \in \bar{\mathcal{E}}$. Since $\max_{i \in N} \left|\frac{\partial f}{\partial \lambda_i}(\boldsymbol{\lambda})\right| > M$, there exists $\boldsymbol{\lambda}' \in \mathcal{E}$ such that $\Phi(\boldsymbol{\lambda}') < \Phi(\boldsymbol{\lambda})$. Therefore, there exists a point in $\mathcal{E}$ for which Φ is smaller than anywhere outside $\mathcal{E}$. Finally, by Assumption 2, $\mathcal{E}$ is compact. We deduce that Φ has a unique minimum on $\operatorname{dom} \Phi$ which concludes the proof. $\square$

The potential game structure of Γ has another interesting implication: if individuals start from an initial strategy profile $\boldsymbol{\lambda}$ such that $f(\boldsymbol{\lambda}) < \infty$, the so called *best-response dynamics* converge towards the unique non-trivial equilibrium (see, e.g., [25]). This implies that the non-trivial equilibrium is the only equilibrium reached when, e.g., all users start with non-infinite noise variance.

5   Price of Stability

Having established the uniqueness of a non-trivial equilibrium in our game, we turn our attention to issues of efficiency. We define the *social cost* function $C : \mathbb{R}^n \to \mathbb{R}_+$ as the sum of all individual costs, and say that a strategy profile $\boldsymbol{\lambda}^{\text{opt}}$ is *socially optimal* if it minimizes the social cost, i.e.,

$$C(\boldsymbol{\lambda}) = \sum_{i \in N} c_i(\lambda_i) + nf(\boldsymbol{\lambda}), \qquad \text{and} \qquad \boldsymbol{\lambda}^{\text{opt}} \in \underset{\boldsymbol{\lambda} \in [0, 1/\sigma^2]^n}{\arg\min} \ C(\boldsymbol{\lambda}).$$

Let $\text{opt} = C(\boldsymbol{\lambda}^{\text{opt}})$ be the minimal social cost. We define the *price of stability* (*price of anarchy*) as the ratio of the social cost of the best (worst) Nash equilibrium in Γ to opt, i.e.,

$$\text{PoS} = \min_{\boldsymbol{\lambda} \in \text{NE}} \frac{C(\boldsymbol{\lambda})}{\text{opt}}, \qquad \text{and} \qquad \text{PoA} = \max_{\boldsymbol{\lambda} \in \text{NE}} \frac{C(\boldsymbol{\lambda})}{\text{opt}},$$

where $\text{NE} \subset [0, 1/\sigma^2]^n$ is the set of Nash equilibria of Γ.

Clearly, in the presence of trivial equilibria, the price of anarchy is infinity. We thus turn our attention to determining the price of stability. Note however that since the non-trivial equilibrium is unique (Theorem 1), the price of stability and the price of anarchy coincide under slight model adjustments discussed in Section 4 that avoid existence of the trivial equilibria.

The fact that our game admits a potential function has the following immediate consequence (see, e.g., [25, 26]):

Theorem 2. *Under Assumptions 1 and 2, PoS $\leq n$.*

Proof. Under Assumptions 1 and 2, the unique non-trivial equilibrium $\boldsymbol{\lambda}^*$ minimizes the potential function $\Phi(\boldsymbol{\lambda}) = \sum_{i \in N} c_i(\lambda_i) + f(\boldsymbol{\lambda})$. Then, for $\boldsymbol{\lambda}^{\text{opt}}$ a minimizer of the social cost:

$$\Phi(\boldsymbol{\lambda}^*) \leq \Phi(\boldsymbol{\lambda}^{\text{opt}}) = \sum_{i \in N} c_i(\lambda_i^{\text{opt}}) + f(\boldsymbol{\lambda}^{\text{opt}}) \leq \sum_{i \in N} c_i(\lambda_i^{\text{opt}}) + nf(\boldsymbol{\lambda}^{\text{opt}}) = \text{opt}$$

by the positivity of f. On the other hand, $C(\boldsymbol{\lambda}^*) \leq n\Phi(\boldsymbol{\lambda}^*)$, by the positivity of c_i, and the theorem follows.   □

Improved bounds can be obtained for specific estimation and privacy cost functions. In what follows, we focus on the two inference cost functions given by (4). We make use of the following lemma, whose proof is in our technical report [1]:

Lemma 1. *If $A(\boldsymbol{\lambda}) \in S_{++}^d$, then for any $i \in N$,*

$$\frac{\partial \operatorname{trace}\left(A^{-1}(\boldsymbol{\lambda})\right)}{\partial \lambda_i} = -\boldsymbol{x}_i^T A^{-2}(\boldsymbol{\lambda}) \boldsymbol{x}_i, \quad and \quad \frac{\partial \|A^{-1}(\boldsymbol{\lambda})\|_F^2}{\partial \lambda_i} = -2\boldsymbol{x}_i^T A^{-3}(\boldsymbol{\lambda}) \boldsymbol{x}_i.$$

We begin by providing a bound on the price of stability when privacy costs are monomial functions, proved in our technical report [1]. The following theorem characterizes the PoS in these cases, improving on the linear bound of Theorem 2:

Theorem 3. *Assume that the cost functions are given by $c_i(\lambda) = c_i \lambda^k$, where $c_i > 0$ and $k \geq 1$. If the estimation cost is given by the extended-value extension of $F_1(V) = \operatorname{trace}(V)$, then $\mathsf{PoS} \leq n^{\frac{1}{k+1}}$. If the estimation cost is given by the extended-value extension of $F_2(V) = \|V\|_F^2$, then $\mathsf{PoS} \leq n^{\frac{2}{k+2}}$.*

The proof of Theorem 3 relies on characterizing explicitly the socially optimal profile under relaxed constraints, and showing it equals the Nash equilibrium $\boldsymbol{\lambda}^*$ multiplied by a scalar. Moreover, the theorem states that, among monomial privacy costs, the largest PoS is $n^{\frac{1}{2}}$ for $F = F_1$, and $n^{\frac{2}{3}}$ for $F = F_2$. Both are attained at linear privacy costs; in fact, the above "worst-case" bounds can be generalized to a class of functions beyond monomials.

Theorem 4. *Assume that for every $i \in N$ the privacy cost functions $c_i : \mathbb{R}_+ \to \mathbb{R}_+$ satisfy Assumption 1. If the estimation cost is the extended-value extension of $F_1(V) = \operatorname{trace}(V)$, and the derivatives c_i' satisfy*

$$nc_i'(\lambda) \leq c_i'(n^{\frac{1}{2}}\lambda) \tag{7}$$

then $\mathsf{PoS} \leq n^{\frac{1}{2}}$. Similarly, if the estimation cost is the extended-value extension of $F_2(V) = \|V\|_F^2$, and the derivatives c_i' satisfy

$$nc_i'(\lambda) \leq c_i'(n^{\frac{1}{3}}\lambda) \tag{8}$$

then $\mathsf{PoS} \leq n^{\frac{2}{3}}$.

Theorem 4, proved in our technical report [1], applies to privacy cost functions that have the "strong" convexity properties (7) and (8). Roughly speaking, such functions grow no slower than cubic and fourth-power monomials, respectively. In contrast to Theorem 3 , in the case of Theorem 4, we cannot characterize the social optimum precisely; as a result, the proof relies on Brouwer's fixed point theorem to relate λ^{opt} to the non-trivial Nash equilibrium λ^*.

We note that a similar worst-case efficiency of linear functions among convex cost families has also been observed in the context of other games, including routing [27] and resource allocation games [28]. As such, Theorems 3 and 4 indicate that this behavior emerges in our linear regression game as well.

6 An Aitken-Type Theorem for Nash Equilibria

Until this point, we have assumed that the analyst uses the generalized least-square estimator (2) to estimate model β. In the non-strategic case, where λ (and, equivalently, the added noise variance) is fixed, the generalized least-square estimator is known to satisfy a strong optimality property: the so-called Aitken/Gauss-Markov theorem, which we briefly review below, states that it is the best linear unbiased estimator, a property commonly refered to as BLUE. In this section, we give an extension of this theorem, in the strategic case where λ^* is not a priori fixed, but is the equilibrium reached by users, itself depending on the estimator used by the analyst.

For all technical results in this section, we restrict ourselves to the case where $F(V) = F_1(V) = \mathrm{trace}(V)$.

6.1 Linear Unbiased Estimators and the Aitken Theorem

A *linear* estimator $\hat{\beta}_L$ of the model β is a linear map of the perturbed variables $\tilde{y}$; i.e., it is an estimator that can be written as $\hat{\beta}_L = L\tilde{y}$ for some matrix $L \in \mathbb{R}^{d \times n}$. A linear estimator is called *unbiased* if $\mathbb{E}[L\tilde{y}] = \beta$ (the expectation taken over the inherent and added noise variables). Recall by (2) that the generalized least-square estimator $\hat{\beta}_{\mathrm{GLS}}$ is an unbiased linear estimator with $L = (X^T \Lambda X)^{-1} X^T \Lambda$ and covariance $Cov(\hat{\beta}_{\mathrm{GLS}}) = (X^T \Lambda X)^{-1}$.

Any linear estimator $\hat{\beta}_L = L\tilde{y}$ can be written without loss of generality as

$$L = (X^T \Lambda X)^{-1} X^T \Lambda + D^T \tag{9}$$

where $D = (L - (X^T \Lambda X)^{-1} X^T \Lambda)^T \in \mathbb{R}^{n \times d}$. It is easy to verify that $\hat{\beta}_L$ is unbiased if and only if $D^T X = 0$; in turn, using this result, the covariance of any linear unbiased estimator can be shown to be

$$Cov(\hat{\beta}_L) = (X^T \Lambda X)^{-1} + D^T \Lambda^{-1} D \succeq Cov(\hat{\beta}_{\mathrm{GLS}}).$$

In other words, the covariance of the generalized least-square estimator is minimal in the positive-semidefinite order among the covariances of *all linear unbiased estimators.* This optimality result is known as the Aitken theorem [29]. Applied specifically to homoschedastic noise (i.e., when all noise variances are identical), it is known as the Gauss-Markov theorem [23], which establishes the optimality of the ordinary least squares estimator. Both theorems provide a strong argument in favor of using least squares to estimate $\boldsymbol{\beta}$, in the presence of fixed noise variance.

6.2 Extension to a Non-cooperative Game

Suppose now that the data analyst uses a linear unbiased estimator $\hat{\boldsymbol{\beta}}_L$ of the form (9), with a given matrix $D \in \mathbb{R}^{n \times d}$, which may depend on X. As before, we can define a game Γ in which each individual i chooses her λ_i to minimize her cost; this time, however, the estimation cost depends on the variance of $\hat{\boldsymbol{\beta}}_L$. A natural question to ask is the following: it is possible that, despite the fact that the analyst is using an estimator that is "inferior" to $\hat{\boldsymbol{\beta}}_{\text{GLS}}$ in the BLUE sense, an equilibrium reached under $\hat{\boldsymbol{\beta}}_L$ is *better* than the equilibrium reached under $\hat{\boldsymbol{\beta}}_{\text{GLS}}$? If so, despite the Aitken theorem, the data analyst would clearly have an incentive to use $\hat{\boldsymbol{\beta}}_L$ instead.

In this section, we answer this question in the negative, in effect extending Aitken's theorem to the case of strategic individuals. Formally, we consider the game $\Gamma = \langle N, [0, 1/\sigma^2]^n, (J_i)_{i \in N} \rangle$ defined as in Section 3.3, except that the estimation cost is the extended-value extension of $F_1(V(\boldsymbol{\lambda}))$ with

$$V(\boldsymbol{\lambda}) \equiv (X^T \Lambda X)^{-1} + D^T \Lambda^{-1} D, \qquad (\Lambda = \text{diag}\,\boldsymbol{\lambda}). \qquad (10)$$

Γ is still a potential game with potential function given by (5). Moreover, Assumption 2 still holds since $V(\cdot)$ given by (10) is a matrix convex function, and the extended-value extension $f(\cdot)$ is still convex.

Since the proof of Theorem 1 relied on the convexity of the potential, a straightforward adaptation of the proof gives the following result.

Theorem 5. *For any matrix $D \in \mathbb{R}^{n \times d}$, there exists a unique non-trivial equilibrium of the game Γ under the corresponding linear unbiased estimator (9).*

As for the case of GLS, this result follows from the uniqueness of a minimizer of the potential function attained in the effective domain.

We are now ready to state our extension of Aitken Theorem, proved in our technical report [1].

Theorem 6. *The generalized least-square estimator gives an optimal covariance among linear unbiased estimators, in the strategic case, in the order given by the scalarization F_1 used in the estimation cost. That is, for any linear unbiased estimator $\hat{\boldsymbol{\beta}}_L$, we have*

$$f(\boldsymbol{\lambda}_L^*) \geq f(\boldsymbol{\lambda}_{\text{GLS}}^*),$$

where $\boldsymbol{\lambda}_L^*$ and $\boldsymbol{\lambda}_{\text{GLS}}^*$ are the non-trivial equilibria for the linear unbiased estimator and for the generalized least-square estimator respectively.

Theorem 6 therefore establishes the optimality of $\hat{\boldsymbol{\beta}}_{\text{GLS}}$ amongst linear unbiased estimators w.r.t. the scalarization F_1, in the presense of strategic individuals. The proof uses perturbative techniques similar to the ones used in [22].

7   Concluding Remarks

This paper studies linear regression in the presence of cost-minimizing individuals, modeling noise addition as a non-cooperative game. We establish existence of a unique non-trivial Nash equilibrium, and study its efficiency for several different classes of privacy and estimation cost functions. We also show an extension of the Aitken/Gauss-Markov theorem to this non-cooperative setup.

The efficiency result in Theorem 3 gives specific bounds on the price of stability for monomial privacy costs. These bounds are sub-linear. However, the efficiency result in Theorem 4 indicates that a sub-linear price of stability can be attained for a much wider class of privacy cost functions. Nevertheless, Theorem 3 includes functions not covered by Theorem 4, which leaves open the question of extending the bounds of Theorem 4, potentially to all privacy costs satisfying Assumption 1. Moreover, both of these theorems, as well as Theorem 6, are shown for specific scalarizations of the estimator variance. Going beyond these scalarizations is also an interesting open problem.

Our Aitken/Gauss-Markov-type theorem is weaker than these two classical results in two ways. First, the optimality of the generalized least squares estimator is shown w.r.t. the partial order imposed by the scalarization F_1, rather than the positive semidefinite order. It would be interesting to strenghten this result not only in this direction, but also in the case of the order imposed by other scalarizations used as estimation costs. Second, Theorem 6 applies to linear estimators whose difference from GLS does not depend on the actions λ. In the presence of arbitrary dependence on λ, the non-trivial equilibrium need not be unique (or even exist). Understanding when this occurs, and proving optimality results in this context, also remains open.

Finally, our model assumes that the variance added by each individual is known to the analyst. Amending this assumption brings issues of truthfulness into consideration: in particular, an important open question is whether there exists an estimator (viewed as a mechanism) that induces truthful noise reporting among individuals, at least in equilibrium. Again, an Aitken-type theorem seems instrumental in establishing such a result.

References

1. Ioannidis, S., Loiseau, P.: Linear regression as a non-cooperative game. Technical report, arXiv:1309.7824 (2013)
2. Vaidya, J., Clifton, C.W., Zhu, Y.M.: Privacy Preserving Data Mining. Springer (2006)
3. Domingo-Ferrer, J.: A survey of inference control methods for privacy-preserving data mining. In: Privacy-Preserving Data Mining, pp. 53–80. Springer (2008)
4. Traub, J.F., Yemini, Y., Woźniakowski, H.: The statistical security of a statistical database. ACM Transactions on Database Systems (TODS) 9(4), 672–679 (1984)
5. Duncan, G.T., Mukherjee, S.: Optimal disclosure limitation strategy in statistical databases: Deterring tracker attacks through additive noise. Journal of the American Statistical Association 95(451), 720–729 (2000)
6. Agrawal, R., Srikant, R.: Privacy-preserving data mining. In: ACM SIGMOD International Conference on Management of Data, pp. 439–450 (2000)

7. Oliveira, S.R., Zaiane, O.R.: Privacy preserving clustering by data transformation. In: SBBD, pp. 304–318 (2003)
8. Atallah, M., Bertino, E., Elmagarmid, A., Ibrahim, M., Verykios, V.: Disclosure limitation of sensitive rules. In: Workshop on Knowledge and Data Engineering Exchange (KDEX 1999), pp. 45–52 (1999)
9. Dwork, C.: Differential privacy. In: Bugliesi, M., Preneel, B., Sassone, V., Wegener, I. (eds.) ICALP 2006. LNCS, vol. 4052, pp. 1–12. Springer, Heidelberg (2006)
10. Kifer, D., Smith, A., Thakurta, A.: Private convex empirical risk minimization and high-dimensional regression. JMLR W&CP 23, 25.1–25.40 (2012); Proceedings of COLT 2012
11. Ghosh, A., Roth, A.: Selling privacy at auction. In: ACM EC, pp. 199–208 (2011)
12. Nissim, K., Smorodinsky, R., Tennenholtz, M.: Approximately optimal mechanism design via differential privacy. In: Innovations in Theoretical Computer Science (ITCS), pp. 203–213 (2012)
13. Ligett, K., Roth, A.: Take it or Leave it: Running a Survey when Privacy Comes at a Cost. In: Goldberg, P.W. (ed.) WINE 2012. LNCS, vol. 7695, pp. 378–391. Springer, Heidelberg (2012)
14. Pukelsheim, F.: Optimal design of experiments, vol. 50. Society for Industrial Mathematics (2006)
15. Atkinson, A., Donev, A., Tobias, R.: Optimum experimental designs, with SAS. Oxford University Press, New York (2007)
16. Boyd, S., Vandenberghe, L.: Convex Optimization. Cambridge University Press (2004)
17. Horel, T., Ioannidis, S., Muthukrishnan, S.: Budget feasible mechanisms for experimental design. arXiv preprint arXiv:1302.5724 (2013)
18. Dekel, O., Fischer, F., Procaccia, A.D.: Incentive compatible regression learning. Journal of Computer and System Sciences (76), 759–777 (2010)
19. Perote, J., Perote-Pena, J.: Strategy-proof estimators for simple regression. Mathematical Social Sciences (47), 153–176 (2004)
20. Morgan, J.: Financing public goods by means of lotteries. Review of Economic Studies 67(4), 761–784 (2000)
21. Loiseau, P., Schwartz, G., Musacchio, J., Amin, S., Sastry, S.S.: Congestion pricing using a raffle-based scheme. In: NetGCoop (October 2011)
22. Loiseau, P., Schwartz, G., Musacchio, J., Amin, S., Sastry, S.S.: Incentive mechanisms for internet congestion management: Fixed-budget rebate versus time-of-day pricing. IEEE/ACM Transactions on Networking (to appear, 2013)
23. Hastie, T., Tibshirani, R., Friedman, J.: The Elements of Statistical Learning: Data Mining, Inference and Prediction, 2nd edn. Springer (2009)
24. Monderer, D., Shapley, L.S.: Potential games. Games and Economic Behavior 14(1), 124–143 (1996)
25. Sandholm, W.H.: Population Games and Evolutionary Dynamics. MIT Press (2010)
26. Schäfer, G.: Online social networks and network economics. Lecture notes, Sapienza University of Rome (2011)
27. Roughgarden, T., Tardos, E.: How bad is selfish routing? Journal of the ACM 49(2), 236–259 (2002)
28. Johari, R., Tsitsiklis, J.N.: Efficiency loss in a network resource allocation game. Mathematics of Operations Research 29(3), 407–435 (2004)
29. Aitken, A.C.: On least squares and linear combinations of observations. Proceedings of the Royal Society of Edinburgh (1935)

Optimal Allocation for Chunked-Reward Advertising*

Weihao Kong[1], Jian Li[2], Tie-Yan Liu[3], and Tao Qin[3]

[1] Shanghai Jiao Tong University
kweihao@gmail.com
[2] Tsinghua University
lapordge@gmail.com
[3] Microsoft Research Asia
{taoqin,tyliu}@microsoft.com

Abstract. Chunked-reward advertising is commonly used in the industry, such as the guaranteed delivery in display advertising and the daily-deal services (e.g., Groupon) in online shopping. In chunked-reward advertising, the publisher promises to deliver at least a certain volume (a.k.a. tipping point or lower bound) of user traffic to an advertiser according to their mutual contract. At the same time, the advertiser may specify a maximum volume (upper bound) of traffic that he/she would like to pay for according to his/her budget constraint. The objective of the publisher is to design an appropriate mechanism to allocate the user traffic so as to maximize the overall revenue obtained from all such advertisers. In this paper, we perform a formal study on this problem, which we call Chunked-reward Allocation Problem (CAP). In particular, we formulate CAP as a knapsack-like problem with variable-sized items and majorization constraints. Our main results regarding CAP are as follows. (1) We first show that for a special case of CAP, in which the lower bound equals the upper bound for each contract, there is a simple dynamic programming-based algorithm that can find an optimal allocation in pseudo-polynomial time. (2) The general case of CAP is much more difficult than the special case. To solve the problem, we first discover several structural properties of the optimal allocation, and then design a two-layer dynamic programming-based algorithm that can find an optimal allocation in pseudo-polynomial time by leveraging these properties. (3) We convert the two-layer dynamic programming based algorithm to a fully polynomial time approximation scheme (FPTAS). Besides these results, we also investigate some natural generalizations of CAP, and propose effective algorithms to solve them.

1 Introduction

We study the traffic allocation problem for what we call "chunked-reward advertising". In chunked-reward advertising, an advertiser requests (and pays for)

---

* This work was conducted at Microsoft Research Asia.

Y. Chen and N. Immorlica (Eds.): WINE 2013, LNCS 8289, pp. 291–304, 2013.
© Springer-Verlag Berlin Heidelberg 2013

a chunk of advertising opportunities (e.g., user traffic, clicks, or transactions) from a publisher (or ad platform) instead of pursuing each individual advertising opportunity separately (which we call pay-per-opportunity advertising for ease of comparison). More precisely, when an advertiser i submits a request to the publisher, he/she specifies a tuple (l_i, u_i, p_i, b_i), where l_i is the lower bound of advertising opportunities he/she wants to obtain, u_i is the upper bound (which exists mainly due to the budget constraint of the advertiser), p_i is the per-opportunity price he/she is willing to pay, and b_i is a bias term that represents the base payment when the lower bound is achieved. If the number $\mathbf{x}_i$ of opportunities allocated to the advertiser is smaller than l_i, he/she does not need to pay anything because the lower bound is not met; if $\mathbf{x}_i > u_i$, the advertiser only needs to pay for the u_i opportunities and the over allocation is free to him/her. Mathematically, the revenue that the publisher extracts from advertiser i with $\mathbf{x}_i$ opportunities can be written as below:

$$r(\mathbf{x}_i; p_i, l_i, u_i, b_i) = \begin{cases} 0, & \text{if } \mathbf{x}_i < l_i, \\ p_i \mathbf{x}_i + b_i, & \text{if } l_i \leq \mathbf{x}_i \leq u_i, \\ p_i u_i + b_i, & \text{if } u_i < \mathbf{x}_i. \end{cases} \tag{1}$$

1.1   Examples of Chunked-Reward Advertising

Many problems in real applications can be formulated as chunked-reward advertising. Below we give two examples: daily-deal services in online shopping and guaranteed delivery in display advertising.

Daily-Deal Services. In daily-deal services, the publisher (or service provider, e.g., Groupon.com) shows (multiple) selected deals, with significant discount, to Web users every day. The following information of each deal is available for allocating user traffics to these deals.

- The discounted price w_i of the deal. This is the real price with which Web users purchase the deal.
- The tipping point L_i describes the minimum number of purchases that users are required to make in order for the discount of the deal to be invoked; otherwise, the deal fails and no one can get the deal and the discount.
- The purchase limit U_i denotes the maximum number of users that can purchase this deal. The purchase limit is constrained by the service capability of the advertiser/merchant. For example, a restaurant may be able to serve at most 200 customers during the lunch time.
- The revenue share ratio s_i represents the percentage of revenue that the publisher can get from each transaction of the item in the deal. That is, for each purchase of the deal, $w_i s_i$ goes to the publisher and $w_i(1 - s_i)$ goes to the merchant/advertiser. Note that the publisher can get the revenue from a deal i only if the deal is on (i.e., at least L_i purchases are achieved).
- Conversion probability λ_i denotes the likelihood that the i-th deal will be purchased by a web user given that he/she has noticed the deal.

It is straightforward to represent the daily-deal service in the language of chunked-reward advertising, by setting $l_i = L_i/\lambda_i$, $u_i = U_i/\lambda_i$, $p_i = w_i s_i \lambda_i$, and $b_i = 0$, where $\mathbf{x}_i$ is the number of effective impressions[1] allocated to ad i.

Guaranteed Delivery. Guaranteed delivery is a major form of display advertising: an advertiser makes a contract with the publisher (or ad platform) to describe his/her campaign goal that the publisher should guarantee. If the publisher can help achieve the campaign goal, it can extract a revenue higher than unguaranteed advertisements. However, if the publisher failed in doing so, a penalization will be imposed.

Specifically, in the contract, the advertiser will specify:

- The number of impressions (denoted by U_i) that he/she wants to achieve;
- The price P_i that he/she is willing to pay for each impression
- The penalty price Q_i that he/she would like to impose on the publisher for each undelivered impression.

If the publisher can successfully deliver U_i impressions for advertiser i, its revenue collected from the advertising is $P_i U_i$; on the other hand, if only $\mathbf{x}_i < U_i$ impressions are delivered, the publisher will be penalized for the undelivered impressions and can only extract a revenue of $\max\{0, P_i U_i - Q_i(U_i - \mathbf{x}_i)\}$.

Again, it is easy to express guaranteed delivery in the language of chunked-reward advertising, by setting $l_i = \frac{Q_i - P_i}{Q_i} U_i$, $u_i = U_i$, $p_i = Q_i$, and $b_i = (P_i - Q_i) u_i$, where $\mathbf{x}_i$ is the number of effective impressions allocated to ad i.

1.2 Chunked-Reward Allocation Problem

A central problem in chunked-reward advertising is how to efficiently allocate the user traffics to the advertisements (deals) so as to maximize the revenue of the publisher (ad platform). For ease of reference, we call such a problem the Chunked Allocation Problem (CAP), which is formulated in details in this subsection.

Suppose that a publisher has M candidate ads to show for a given period of time (e.g., the coming week) and N Web users that will visit its website during the time period. The publisher shows K ads at K slots to each Web user. Without loss of generality, we assume that the i-th slot is better (in terms of attracting the attention of a Web user) than the j-th slot if $i < j$, and use γ_k to denote the discount factor carried by each slot. Similar to the position bias of click probability in search advertising [4,1], we have that $1 \geq \gamma_1 > \gamma_2 > \ldots > \gamma_K \geq 0$.

If an ad is shown at slot k to x visitors, we say that the ad has $x\gamma_k$ effective impressions. We use N_k to denote the number of effective impressions of slot k: $N_k = N\gamma_k$. Therefore, $N_1 > N_2 > \cdots > N_K$. For simplicity and without much loss of accuracy, we assume that N_k is an integer. We can regard N_k as the expected number of visitors who have paid attention to the k-th slot. With

---

[1] Effective impressions means the real impressions adjusted by the slot discount factor, as shown in the following subsection.

the concept of effective impression, p_i denotes the (expected) revenue that the publisher can get from one effective impression of the i-th ad if the ad is tipped on.

When allocating user traffics to ads, one needs to consider quite a few constraints, which makes CAP a difficult task:

1. Multiple ads will be displayed, one at each slot, for one visitor; one ad cannot be shown at more than one slots for one visitor. This seemingly simple constraints combined with the fact that different slot positions have different discount factors will become challenging to deal with.
2. Each ad has both a lower bound and an upper bound. On one hand, to make money from an ad, the publisher must ensure the ad achieves the lower bound. On the other hand, to maximize revenue, the publisher needs to ensure the traffic allocated to an ad will not go beyond its upper bound. These two opposite forces make the allocation non-trivial.

In the next subsection, we will use formal language to characterize these constraints, and give a mathematical description of the CAP problem.

1.3   Problem Formulation

We use an integer vector $\mathbf{x}$ to denote an allocation, where the i-th element $\mathbf{x}_i$ denotes the number of effective impressions allocated to the i-th ad. For any vector $\mathbf{x} = \{\mathbf{x}_1, \mathbf{x}_2, \ldots, \mathbf{x}_n\}$, let $\mathbf{x}_{[1]} \geq \mathbf{x}_{[2]} \geq \ldots \mathbf{x}_{[n]}$ denotes the components of $\mathbf{x}$ in nonincreasing order (ties are broken in an arbitrary but fixed manner). Since there are multiple candidate ads and multiple slots, we need to ensure the feasibility of an allocation. As mentioned before, an allocation $\mathbf{x}$ is *feasible* if it satisfies that (i) no more than one ad is assigned to a slot for any visitor, and (ii) no ad is assigned to more than one slot for any visitor. Actually these constraints are essentially the same in preemptive scheduling of independent tasks on uniform machines. Consider M jobs with processing requirement $\mathbf{x}_i (i = 1, \ldots, M)$ to be processed on K parallel uniform machines with different speeds $N_j (j = 1, \ldots, K)$. Execution of job i on machine j requires $\mathbf{x}_i / N_j$ time units. $\mathbf{x}$ is a feasible allocation if and only if the minimum makespan of the preemptive scheduling problem is smaller or equal to 1. According to [2], the sufficient and necessary conditions for processing all jobs in the interval [0,1] are

$$\frac{\sum_{j=1}^{M} \mathbf{x}_{[j]}}{\sum_{j=1}^{K} N_j} \leq 1, \tag{2}$$

and

$$\frac{\sum_{j=1}^{i} \mathbf{x}_{[j]}}{\sum_{j=1}^{i} N_j} \leq 1 \text{ , for all } i \leq K. \tag{3}$$

Thus, a vector $\mathbf{x}$ is a feasible allocation for CAP if it satisfies the inequalities in Eqn. (2) and (3).

Based on the above notations, finding an optimal allocation means solving the following optimization problem.

$$\max_{\mathbf{x}} \sum_{i=1}^{M} r(\mathbf{x}_i; p_i, l_i, u_i, b_i)$$

$$s.t. \quad l_i \leq \mathbf{x}_i \leq u_i \quad \text{or} \quad \mathbf{x}_i = 0, \text{ for } i = 1, 2, \ldots M$$

$$\mathbf{x} \text{ is a feasible allocation.}$$

Note that $\mathbf{x}$ is a vector of integers, and we do not explicitly add it as a constraint when the context is clear. The first set of constraints says the number of effective impressions allocated to ad i should be between the lower bound l_i and upper bound u_i.

The feasibility conditions in Eqn. (2) and (3) can be exactly described by the majorization constraints.

Definition 1. *Majorization constraints*
The vector $\mathbf{x}$ is majorized[2] by vector $\mathbf{y}$ (denoted as $\mathbf{x} \preceq \mathbf{y}$) if the sum of the largest i entries in $\mathbf{x}$ is no larger than the sum of the largest i entries in $\mathbf{y}$ for all i, i.e.,

$$\sum_{j=1}^{i} \mathbf{x}_{[j]} \leq \sum_{j=1}^{i} \mathbf{y}_{[j]}. \tag{4}$$

In the above definition, $\mathbf{x}$ and $\mathbf{y}$ should contain the same number of elements. In Eqn. (2) and (3), N has less elements than $\mathbf{x}$; one can simply add $M - K$ zeros into N (i.e., $N_{[i]} = 0, \forall K < i \leq M$).

Now we are ready to abstract CAP as an combinatorial optimization problem as the following.

Definition 2. *Problem formulation for* **CAP**
There are M class of items, $\mathbb{C}_1, \ldots, \mathbb{C}_M$. Each class $\mathbb{C}_i$ is associated with a lower bound $l_i \in \mathbb{Z}^+$, an upper bound $u_i \in \mathbb{Z}^+$, and a bias term b_i. Each item of $\mathbb{C}_i$ has a profit p_i. We are also given a vector $\mathbf{N} = \{N_1, N_2, \ldots, N_K\}$, called the target vector, where $N_1 > N_2 > \cdots > N_K$. We use $|\mathbf{N}|$ to denote $\sum_{i=1}^{K} N_j$. Our goal is to choose $\mathbf{x}_i$ items from class $\mathbb{C}_i$ for each $i \in [M]$ such that the following three properties hold:

1. *Either $\mathbf{x}_i = 0$ (we do not choose any item of class $\mathbb{C}_i$ at all) or $l_i \leq \mathbf{x}_i \leq u_i$ (the number of items of class $\mathbb{C}_i$ must satisfy both the lower and upper bounds);*
2. *The vector $\mathbf{x} = \{\mathbf{x}_i\}_i$ is majorized by the target vector $\mathbf{N}$ (i.e., $\mathbf{x} \preceq \mathbf{N}$);*
3. *The total profit of chosen items (adjusted by the class bias term) is maximized.*

---

[2] In fact, the most rigorous term used here should be "sub-majorize" in mathematics and theoretical computer science literature (see e.g., [11,6]). Without causing any confusion, we omit the prefix for simplicity.

1.4   Relation to Scheduling and Knapsack Problems

CAP bears some similarity with the classic parallel machine scheduling problems [12]. The K slots can be viewed as K parallel machines with different speeds (commonly termed as the *uniformly related machines*, see, e.g., [3,7]). The M ads can be viewed as M jobs. One major difference between CAP and the scheduling problems lies in the objective functions. Most scheduling problems target to minimize some functions related to time given the constraint of finishing all the jobs, such as makespan minimization and total completion time minimization. In contrast, our objective is to maximize the revenue generated from the finished jobs (deals in our problem) given the constraint of limited time.

CAP is similar to the classic knapsack problem in which we want to maximize the total profit of the items that can be packed in a knapsack with a known capacity. Our FPTAS borrows the technique from [8] for the knapsack problem. Our work is also related to the *interval scheduling* problem [10,5] in which the goal is to schedule a subset of interval such that the total profit is maximized. CAP differs from these two problems in that the intervals/items (we can think each ad as an interval) have variable sizes.

1.5   Our Results

Our major results for CAP can be summarized as follows. Because of the space limitation, the last three items are included in the longer version of this work [9].

1. (Section 2) As a warmup, we start with a special case of the CAP problem: the lower bound of each class of items equals the upper bound. In this case, we can order the classes by decreasing lower bounds and the order enables us to design a nature dynamic programming-based algorithm which can find an optimal allocation in pseudo-polynomial running time.
2. (Section 3) We then consider the general case of the CAP problem where the lower bound can be smaller than the upper bound. The general case is considerably more difficult than the simple case in that there is no natural order to process the classes. Hence, it is not clear how to extend the previous dynamic program to the general case. To handle this difficulty, we discover several useful structural properties of the optimal allocation. In particular, we can show that the optimal allocation can be decomposed into multiple *blocks*, each of them has at most one *fractional* class (the number of allocated items for the class is less than the upper bound and larger than the lower bound). Moreover, in a block, we can determine for each class except the fractional class, whether the allocated number should be the upper bound or the lower bound. Hence, within each block, we can reduce the problem to the simpler case where the lower bound of every item equals the upper bound (with slight modifications). We still need a higher level dynamic program to assemble the blocks and need to show that no two different blocks use items from the same class. Our two level dynamic programming-based algorithm can find an optimal allocation in pseudo-polynomial time.

3. (Section 4 of [9]) Using the technique developed in [8], combined with some careful modifications, we can further convert the pseudo-polynomial time dynamic program to a fully polynomial time approximation scheme (FPTAS). We say there is an FPTAS for the problem, if for any fixed constant $\epsilon > 0$, we can find a solution with profit at least $(1-\epsilon)\mathcal{OPT}$ in $\text{poly}(M, K, \log |\mathbf{N}|, 1/\epsilon)$ time (See e.g., [13]).

4. (Section 5 of [9]) We consider the generalization from the strict decreasing target vector (i.e., $N_1 > N_2 > \cdots > N_K$) to the non-increasing target vector (i.e., $N_1 \geq N_2 \geq \ldots \geq N_K$), and briefly describe a pseudo-polynomial time dynamic programming-based algorithm for this setting based on the algorithm in Section 3.

5. (Section 6 of [9]) For theoretical completeness, we consider for a generalization of CAP where the target vector $\mathbf{N} = \{N_1, \ldots, N_K\}$ may be non-monotone. We provide a $\frac{1}{2} - \epsilon$ factor approximation algorithm for any constant $\epsilon > 0$. In this algorithm, we use somewhat different techniques to handle the majorization constraints, which may be useful in other variants of CAP.

2 Warmup: A Special Case

In this section, we investigate a special case of CAP, in which $l_i = u_i$ for every class. In other words, we either select a fixed number $(\mathbf{x}_i = l_i)$ of items from class $\mathbb{C}_i$, or nothing from the class. We present an algorithm that can find the optimal allocation in $\text{poly}(M, K, |\mathbf{N}|)$ time based on dynamic programming.

For simplicity, we assume that the M classes are indexed by the descending order of l_i in this section. That is, we have $l_1 \geq l_2 \geq l_3 \geq \cdots \geq l_M$.

Let $G(i, j, k)$ denote the maximal profit by selecting at most i items from extactly k of the first j classes, which can be expressed by the following integer optimization problem.

$$G(i, j, k) = \max_{\mathbf{x}} \sum_{t=1}^{j} r(\mathbf{x}_t; p_t, l_t, u_t, b_t)$$

$$\text{subject to} \quad \mathbf{x}_t = l_t \quad \text{or} \quad \mathbf{x}_t = 0, \quad \text{for } 1 \leq t \leq j \tag{5}$$

$$\sum_{t=1}^{r} \mathbf{x}_{[t]} \leq \sum_{t=1}^{r} N_t, \quad \text{for } r = 1, 2, ..., \min\{j, K\} \tag{6}$$

$$\sum_{t=1}^{j} \mathbf{x}_{[t]} \leq i \tag{7}$$

$$\mathbf{x}_{[k]} > 0, \quad \mathbf{x}_{[k+1]} = 0 \tag{8}$$

In the above formulation, $\mathbf{x} = \{\mathbf{x}_1, \mathbf{x}_2, ..., \mathbf{x}_j\}$ is a j dimensional allocation vector. $\mathbf{x}_{[t]}$ is the t-th largest element of vector $\mathbf{x}$. Eqn. (6) restates the majorization constraints. Eqn. (7) ensures that at most i items are selected and Eqn. (8) indicates that exactly k classes of items are selected. Further, we use $Z(i, j, k)$

to denote the optimal allocation vector of the above problem (a j-dimensional vector).

It is easy to see that the optimal profit of the special case is $\max_{1 \leq k \leq K} G(|\mathbf{N}|, M, k)$. In the following, we present an algorithm to compute the values of $G(i, j, k)$ for all i, j, k.

The Dynamic Program : Initially, we have the base cases that $G(i, j, k) = 0$ if i, j, k all equal zero. For each $1 \leq i \leq |\mathbf{N}|, 1 \leq j \leq M, 1 \leq k \leq j$, the recursion of the dynamic program for $G(i, j, k)$ is as follows.

$$G(i, j, k) =$$
$$\max \begin{cases} G(i, j - 1, k), & \text{if } j > 0 & \text{(A)} \\ G(i - 1, j, k), & \text{if } i > 0 & \text{(B)} \\ G(i - l_j, j - 1, k - 1) + l_j p_j + b_j, & \text{if } Z(i - l_j, j - 1, k - 1) \cup l_j \text{ is feasible} & \text{(C)} \end{cases}$$
$$\tag{9}$$

Note that for the case (C) of the above recursion, we need to check whether adding the j-th class in the optimal allocation vector $Z(i - l_j, j - 1, k - 1)$ is feasible, i.e., satisfying the majorization constraints in Eqn. (6). The allocation vector $Z(i, j, k)$ can be easily determined from the recursion as follows.

- If the maximum is achieved at case (A), we have $Z(i, j, k)_t = Z(i, j - 1, k)_t, \forall 1 \leq t \leq j - 1$, and $Z(i, j, k)_j = 0$.
- If the maximum is achieved at case (B), we have $Z(i, j, k)_t = Z(i - 1, j, k)_t, \forall 1 \leq t \leq j$.
- If the maximum is achieved at case (C), we have $Z(i, j, k)_t = Z(i - l_j, j - 1, k - 1)_t, \forall 1 \leq t \leq j - 1$, and $Z(i, j, k)_j = l_j$.

According to Eqn. (9), all $G(i, j, k)$ (and thus $Z(i, j, k)$) can be computed in the time[3] of $O(M^2 |\mathbf{N}|)$.

At the end of this section, we remark that the correctness of the dynamic program crucially relies on the fact that $u_i = l_i$ for all $\mathbb{C}_i$ and we can process the classes in descending order of their l_is. However, in the general case where $u_i \neq l_i$, we do not have such a natural order to process the classes and the current dynamic program does not work any more.

3   Algorithm for the General CAP

In this section, we consider the general case of CAP ($l_i \leq u_i$) and present an algorithm that can find the optimal allocation in poly($M, K, |\mathbf{N}|$) time based on dynamic programming. Even though the recursion of our dynamic program appears to be fairly simple, its correctness relies on several nontrivial structural properties of the optimal allocation of CAP. We first present these properties in

---

[3] One can further decrease the complexity of computing all the $G(i, j, k)$'s to $O(M|\mathbf{N}| \min(M, K))$ by using another recursion equation. We use the recursion equation as shown in Eqn. (9) considering its simplicity for presentation and understanding.

Section 3.1. Then we show the dynamic program in Section 3.2 and prove its correctness. Finally we discuss several extensions of CAP, for which the detailed algorithms are described in Appendix because of space limitations.

3.1   The Structure of the Optimal Solution

Before describing the structure of the optimal solution, we first define some notations.

For simplicity of description, we assume all p_is are distinct[4] and the M classes are indexed in the descending order of p_i. That is, we have that $p_1 > p_2 > \cdots > p_M$. Note that the order of classes in this section is different from that in Section 2.

For any allocation vector $\mathbf{x}$, $\mathbf{x}_i$ indicates the number of items selected from class i, and $\mathbf{x}_{[i]}$ indicates the i-th largest element in vector $\mathbf{x}$. For ease of notions, when we say "class $\mathbf{x}_{[i]}$", we actually refer to the class corresponding to $\mathbf{x}_{[i]}$. In a similar spirit, we slightly abuse the notation $p_{[i]}$ to denote the per-item profit of the class $\mathbf{x}_{[i]}$. For example, $p_{[1]}$ is the per-item profit of the class for which we allocate the most number of items in $\mathbf{x}$ (rather than the largest profit). Note that if $\mathbf{x}_{[i]} = \mathbf{x}_{[i+1]}$, then we put the class with the larger per-item profit before the one with the smaller per-item profit. In other words, if $\mathbf{x}_{[i]} = \mathbf{x}_{[i+1]}$, then we have $p_{[i]} > p_{[i+1]}$.

In an allocation $\mathbf{x}$, we call class $\mathbb{C}_i$ (or $\mathbf{x}_i$) *addable* (w.r.t. $\mathbf{x}$) if $\mathbf{x}_i < u_i$. Similarly, class $\mathbb{C}_i$ (or $\mathbf{x}_i$) is *deductible* (w.r.t. $\mathbf{x}$) if $\mathbf{x}_i > l_i$. A class $\mathbb{C}_i$ is *fractional* if it is both addable and deductible (i.e., $l_i < \mathbf{x}_i < u_i$).

Let $\mathbf{x}^\star$ be the optimal allocation vector. We start with a simple yet very useful lemma.

Lemma 1. *If a deductible class $\mathbb{C}_i$ and an addable class $\mathbb{C}_j$ satisfy $\mathbf{x}_i^\star > \mathbf{x}_j^\star$ in the optimal solution $\mathbf{x}^\star$, we must have $p_i > p_j$ (otherwise, we can get a better solution by setting $\mathbf{x}_i^\star = \mathbf{x}_i^\star - 1$ and $\mathbf{x}_j^\star = \mathbf{x}_j^\star + 1$).*

The proof of lemma is quite straightforward.

The following definition plays an essential role in this section.

Definition 3. *(Breaking Points and Tight Segments) Let the set of breaking points for the optimal allocation $\mathbf{x}^\star$ be*

$$P = \{t \mid \sum_{i=1}^{t} \mathbf{x}_{[i]}^\star = \sum_{i=1}^{t} N_i\} = \{t_1 < t_2 < \ldots < t_{|P|}\}.$$

To simplify the notations for the boundary cases, we let $t_0 = 0$ and $t_{|P|+1} = K$. We can partition $\mathbf{x}^\star$ into $|P| + 1$ tight segments, $S_1, \ldots, S_{|P|+1}$, where $S_i = \{\mathbf{x}_{[t_{i-1}+1]}^\star, \mathbf{x}_{[t_{i-1}+2]}^\star, \ldots, \mathbf{x}_{[t_i]}^\star\}$. We call $S_{|P|+1}$ the tail segment, and $S_1, \ldots, S_{|P|}$ non-tail tight segments. □

We have the following useful property about the number of items for each class in a non-tail tight segment.

---

[4] This is without loss of generality. If $p_i = p_j$ for some $i \neq j$, we can break tie by adding an infinitesimal value to p_i, which would not affect the optimality of our algorithm in any way.

Lemma 2. *Given a non-tail tight segment $S_k = \{\mathbf{x}^\star_{[t_{k-1}+1]}, \mathbf{x}^\star_{[t_{k-1}+2]}, \ldots, \mathbf{x}^\star_{[t_k]}\}$ which spans $N_{t_{k-1}+1}, \ldots, N_{t_k}$. For each class $\mathbb{C}_i$ that appears in S_k we must have $N_{t_{k-1}+1} \geq \mathbf{x}^\star_i \geq N_{t_k}$.*

Proof. From the definition $\sum_{i=1}^{t_k} \mathbf{x}^\star_{[i]} = \sum_{i=1}^{t_k} N_i$ and majorization constraint $\sum_{i=1}^{t_k-1} \mathbf{x}^\star_{[i]} \leq \sum_{i=1}^{t_k-1} N_i$ we know that $\mathbf{x}^\star_{[t_k]} \geq N_{t_k}$. As $\mathbf{x}^\star_{[t_k]}$ is the smallest in S_k, we proved $\mathbf{x}^\star_i \geq N_{t_k}$. Similarly form $\sum_{i=1}^{t_k-1} \mathbf{x}^\star_{[i]} = \sum_{i=1}^{t_k-1} N_i$ and $\sum_{i=1}^{t_{k-1}+1} \mathbf{x}^\star_{[i]} \leq \sum_{i=1}^{t_{k-1}+1} N_i$ we know that $N_{t_{k-1}+1} \geq \mathbf{x}^\star_{[t_{k-1}+1]}$. As $\mathbf{x}^\star_{[t_{k-1}+1]}$ is the biggest in S_k, we proved $N_{t_{k-1}+1} \geq \mathbf{x}^\star_i$. $\qquad\square$

Note that as we manually set $t_{|B|+1} = K$, the tail segment actually may not be tight. But we still have $N_{t_{k-1}+1} \geq \mathbf{x}^\star_i$.

Let us observe some simple facts about a tight segment S_k. First, there is at most one fractional class. Otherwise, we can get a better allocation by selecting one more item from the most profitable fractional class and removing one item from the least profitable fractional class. Second, in segment S_k, if $\mathbb{C}_i$ is deductible and $\mathbb{C}_j$ is addable, we must have $p_i > p_j$ (or equivalently $i < j$) . Suppose $\mathbb{C}_{\alpha(S_k)}$ is the per-item least profitable deductible class in S_k and $\mathbb{C}_{\beta(S_k)}$ is the per-item most profitable addable class in S_k. From the above discussion, we know $\alpha(S_k) \leq \beta(S_k)$. If $\alpha(S_k) = \beta(S_k)$, then $\alpha(S_k)$ is the only fractional class in S_k. If there is no deductible class in S_k, we let $\alpha(S_k) = 1$. Similarly, if there is no addable class in S_k, we let $\beta(S_k) = M$. Let us summarize the properties of tight segments in the lemma below.

Lemma 3. *Consider a particular tight segment S_k of the optimal allocation $\mathbf{x}^\star$. The following properties hold.*

1. *There is at most one fractional class.*
2. *For each class $\mathbb{C}_i$ that appears in S_k with $i < \beta(S_k)$, we must have $\mathbf{x}^\star_i = u_i$.*
3. *For each class $\mathbb{C}_i$ that appears in S_k with $i > \alpha(S_k)$, we must have $\mathbf{x}^\star_i = l_i$.*

Now, we perform the following greedy procedure to produce a coarser partition of $\mathbf{x}^\star$ into disjoint *blocks*, $B_1, B_2, \ldots, B_h$, where each block is the union of several consecutive tight segments. The purpose of this procedure here is to endow one more nice property to the blocks. We overload the definition of $\alpha(B_i)$ ($\beta(B_i)$ resp.) to denote the index of the per-item least (most resp.) profitable deductible (addable resp.) class in B_i. We start with $B_1 = \{S_1\}$. So, $\alpha(B_1) = \alpha(S_1)$ and $\beta(B_1) = \beta(S_1)$. Next we consider S_2. If $[\alpha(B_1), \beta(B_1)]$ intersects with $[\alpha(S_2), \beta(S_2)]$, we let $B_1 \leftarrow B_1 \cup S_2$. Otherwise, we are done with B_1 and start to create B_2 by letting $B_2 = S_2$. Generally, in the i-th step, suppose we are in the process of creating block B_j and proceed to S_i. If $[\alpha(B_j), \beta(B_j)]$ intersects with $[\alpha(S_i), \beta(S_i)]$, we let $B_j \leftarrow B_j \cup S_i$. Note that the new $[\alpha(B_j), \beta(B_j)]$ is the intersection of old $[\alpha(B_j), \beta(B_j)]$ and $[\alpha(S_i), \beta(S_i)]$. Otherwise, we finish creating B_j and let the initial value of B_{j+1} be S_i.

We list the useful properties in the following critical lemma. We can see that Property (2) is new (compared with Lemma 3).

Lemma 4. *Suppose $B_1, \ldots, B_h$ are the blocks created according to the above procedure from the optimal allocation $\mathbf{x}^*$, and $\alpha(B_i)$ and $\beta(B_i)$ are defined as above. The following properties hold.*

1. *Each block has at most one fractional class.*
2. $\alpha(B_1) \leq \beta(B_1) < \alpha(B_2) \leq \beta(B_2) < \ldots < \alpha(B_h) \leq \beta(B_h)$.
3. *For each class $\mathbb{C}_i$ that appears in any block B_k with $i < \beta(B_k)$, we must have*
   $x_i^* = u_i$.
4. *For each class $\mathbb{C}_i$ that appears in any block B_k with $i > \alpha(B_k)$, we must have*
   $x_i^* = l_i$.

Because of the space limitation, we omit the proof, which can be found in [9].

3.2   The Dynamic Program

Our algorithm for CAP has two levels, both based on dynamic programming. In the lower level, we attempt to find the optimal allocation for each block. Then in the higher level, we assemble multiple blocks together to form a global optimal solution. Lastly, we prove the optimal allocations for these individual blocks do not use one class of items multiple times, thus can be assembled together.

The Lower Level Dynamic Program: Let us first describe the lower level dynamic program. Denote $F(i, j, k), \forall 1 \leq i \leq j \leq K, 1 \leq k \leq M$ as the maximal profit generating from the block B which spans $N_i, N_{i+1}, \ldots, N_j$ and $\alpha(B) \leq k \leq \beta(B)$. Note here the block B is not one of the blocks created from the optimal allocation $\mathbf{x}^*$, but we still require that it satisfies the properties described in Lemma 4. More formally, $F(i, j, k)$ can be written as an integer program in the following form:

$$F(i, j, k) = \max \sum_{t=1}^{M} r(\mathbf{x}_t; p_t, l_t, u_t, b_t)$$

$$\text{subject to} \quad \mathbf{x}_t = u_t \quad \text{or} \quad \mathbf{x}_t = 0, \quad \text{for } t < k \tag{10}$$

$$\mathbf{x}_t = l_t \quad \text{or} \quad \mathbf{x}_t = 0, \quad \text{for } t > k \tag{11}$$

$$l_t \leq \mathbf{x}_t \leq u_t \quad \text{or} \quad \mathbf{x}_t = 0, \quad \text{for } t = k \tag{12}$$

$$\sum_{t=1}^{r} \mathbf{x}_{[t]} \leq \sum_{t=i}^{i+r-1} N_t, \quad \text{for } r = 1, 2, \ldots j - i \tag{13}$$

$$\sum_{t=1}^{j-i+1} \mathbf{x}_{[t]} = \sum_{t=i}^{j} N_t \tag{14}$$

$$\mathbf{x}_{[j-i+2]} = 0. \tag{15}$$

Constraints (10) and (11) correspond to Properties (3) and (4) in Lemma 4. The constraint (12) says $\mathbb{C}_k$ may be the only fractional constraint. The constraints (13) are the majorization constraints. Constraints (14) and (15) say B spans $N_i, \ldots, N_j$ with exactly $j - i + 1$ class of items. If $j = K$ (i.e., it is the last block), we do not have the last two constraints since we may not have to fill all slots, or with fixed number of classes.

To compute the value of $F(i, j, k)$, we can leverage the dynamic program developed in Section 2. The catch is that for any $x_k \in [l_k, u_k]$, according to Eqn. (10) and (11), $\mathbf{x}_i$ can only take 0 or a non-zero value (either u_i or l_i). This is the same as making $u_i = l_i$. Therefore, for a given $x_k \in [l_k, u_k]$, the optimal profit $F(i, j, k)$, denoted as $F_{x_k}(i, j, k)$, can be solved by the dynamic program in Section 2.[5] Finally, we have

$$F(i, j, k) = \max_{x_k = 0, l_k, l_k+1, l_k+2, \cdots, u_k} F_{x_k}(i, j, k).$$

The Higher Level Dynamic Program: We use $D(j, k)$ to denote the optimal allocation of the following subproblem: if $j < K$, we have to fill up exactly $N_1, N_2, \ldots, N_j$ (i.e., $\sum_i \mathbf{x}_i = \sum_{i=1}^{j} N_j$) and $\alpha(B) \leq k$ where B is the last block of the allocation; if $j = K$, we only require $\sum_i \mathbf{x}_i \leq \sum_{i=1}^{j} N_j$. Note that we still have the majorization constraints and want to maximize the profit. The recursion for computing $D(j, k)$ is as follows:

$$D(j, k) = \max \Big\{ \max_{i < j} \{ D(i, k-1) + F(i+1, j, k) \}, D(j, k-1) \Big\}. \tag{16}$$

We return $D(K, M)$ as the final optimal revenue of CAP.

As we can see from the recursion (16), the final value $D(K, M)$ is a sum of several F values, say $F(1, t_1, k_1), F(t_1 + 1, t_2, k_2), F(t_2 + 1, t_3, k_3), \ldots$, where $t_1 < t_2 < t_3 < \ldots$ and $k_1 < k_2 < k_3 < \ldots$. Each such F value corresponds to an optimal allocation of a block. Now, we answer the most critical question concerning the correctness of the dynamic program: whether the optimal allocations of the corresponding blocks together form a global feasible allocation? More specifically, the question is whether one class can appear in two different blocks? We answer this question negatively in the next lemma.

Lemma 5. *Consider the optimal allocations $\mathbf{x}^1$ and $\mathbf{x}^2$ corresponding to $F(i_1, j_1, k_1)$ and $F(i_2, j_2, k_2)$ respectively, where $i_1 \leq j_1 < i_2 \leq j_2$ and $k_1 < k_2$. For any class $\mathbb{C}_i$, it is impossible that both $\mathbf{x}_i^1 \neq 0$ and $\mathbf{x}_i^2 \neq 0$ are true.*

Proof. We distinguish a few cases. We will use Lemma 2 on blocks in the following proof.

1. $i \leq k_1$. Suppose by contradiction that $\mathbf{x}_i^1 \neq 0$ and $\mathbf{x}_i^2 \neq 0$. We always have $\mathbf{x}_i^1 \leq u_i$. Since $i \leq k_1 < k_2$, again by Lemma 4, we have also $\mathbf{x}_i^2 = u_i$. Moreover, from Lemma 2 we know that $\mathbf{x}_i^1 \geq N_{j_1} > N_{i_2} \geq \mathbf{x}_i^2$. This renders a contradiction.
2. $i \geq k_2$. Suppose by contradiction that $\mathbf{x}_i^1 \neq 0$ and $\mathbf{x}_i^2 \neq 0$. By Lemma 4, we know $\mathbf{x}_i^1 = l_i$ and $\mathbf{x}_i^2 \geq l_i$. We also have that $\mathbf{x}_i^1 > N_{i_2} \geq \mathbf{x}_i^2$ due to Lemma 2, which gives a contradiction again.
3. $k_1 < i < k_2$. Suppose by contradiction that $\mathbf{x}_i^1 \neq 0$ and $\mathbf{x}_i^2 \neq 0$. By Lemma 4, we know $\mathbf{x}_i^1 = l_i$ and $\mathbf{x}_i^2 = u_i$. We also have the contradiction by $\mathbf{x}_i^1 > \mathbf{x}_i^2$.

We have exhausted all cases and hence the proof is complete. $\square$

---

[5] The only extra constraint is (14), which is not hard to ensure at all since the dynamic program in Section 2 also keeps track of the number of slots used so far.

Theorem 1. *The dynamic program* (16) *computes the optimal revenue for* CAP *in time* $\text{poly}(M, K, |\mathbf{N}|)$.

Proof. By Lemma 4, the optimal allocation $\mathbf{x}^*$ can be decomposed into several blocks $B_1, B_2, \ldots, B_h$ for some h. Suppose B_k spans $N_{i_{k-1}+1}, \ldots, N_{i_k}$. Since the dynamic program computes the optimal value, we have $F(i_{k-1} + 1, i_k, \alpha(B_k)) \geq \sum_{i \in B_k} r(\mathbf{x}_i^*; p_i, l_i, u_i, b_i)$. Moreover, the higher level dynamic program guarantees that

$$D(K, M) \geq \sum_k F(i_{k-1} + 1, i_k, \alpha(B_k)) \geq \sum_k \sum_{i \in B_k} r(\mathbf{x}_i^*; p_i, l_i, u_i, b_i) = \mathcal{OPT}.$$

By Lemma 5, our dynamic program returns a feasible allocation. So, it holds that $D(K, M) \leq \mathcal{OPT}$. Hence, we have shown that $D(K, M) = \mathcal{OPT}$.   □

3.3   Extensions

We make some further investigations on three extensions for the CAP problem. Due to the space limitation, we only give some high level description in this subsection, and the details are given in the longer version [9] of this paper.

First, note the optimal algorithm developed in Section 3.2 runs in pseudo-polynomial time of $|\mathbf{N}|$. If $|\mathbf{N}|$ is very large, one may need some more efficient algorithm. In Section 4 of [9], we present a full polynomial time approximation scheme (FPTAS) for CAP, which can find a solution with profit at least $(1 - \epsilon)\mathcal{OPT}$ in time polynomial in the input size (i.e., $O(M + K \times \log |\mathbf{N}|)$) and $1/\epsilon$ for any fixed constant $\epsilon > 0$.

We further consider the general case where $N_1 \geq N_2 \geq \ldots \geq N_K$ and some inequalities hold with equality. Our previous algorithm does not work here since the proof of Lemma 5 relies on strict inequalities. In the general case, the lemma does not hold and we can not guarantee that no class is used in more than one blocks. We refine the algorithm proposed in Section 3.2 and present a new DP algorithm for this general setting in Section 5 of [9].

Third, in Section 6 of [9], we provide a $\frac{1}{2} - \epsilon$ factor approximation algorithm for a generalization of CAP where the target vector $\mathbf{N} = \{N_1, \ldots, N_K\}$ may not be monotone ($N_1 \geq N_2 \geq \ldots \geq N_K$ may not hold). We still require that $\sum_{t=1}^r \mathbf{x}_{[t]} \leq \sum_{t=1}^r N_t$ for all r. Although we are not aware of an application scenario that would require the full generality, the techniques developed here, which are quite different from those in Section 3.2, may be useful in handling other variants of CAP or problems with similar constraints. So we provide this approximation algorithm for theoretical completeness.

4   Conclusions

We have formulated and studied the traffic allocation problem for chunked-reward advertising, and designed a dynamic programming based algorithm, which can find an optimal allocation in pseudo-polynomial time. An FPTAS has also been derived based on the proposed algorithms, and two generalized settings have been further studied.

There are many research issues related to the CAP problem which need further investigations. (1) We have studied the offline allocation problem and assumed that the traffic N of a publisher is known in advance and all the ads are available before allocation. It is interesting to study the online allocation problem when the traffic is not known in advance and both website visitors and ads arrive online one by one. (2) We have assumed that the position discount γ_i of each slot. It is worthwhile to investigate how to maximize revenue with unknown position discount through online exploration. (3) We have not considered the strategic behaviors of advertisers. It is of great interest to study the allocation problem in the setting of auctions and analyze its equilibrium properties.

Acknowledgement. Jian Li was supported in part by the National Basic Research Program of China Grant 2011CBA00300, 2011CBA00301 and the National Natural Science Foundation of China Grant 61202009, 61033001, 61061130540, 61073174.

References

1. Aggarwal, G., Goel, A., Motwani, R.: Truthful auctions for pricing search keywords. In: Proceedings of the 7th ACM Conference on Electronic Commerce, pp. 1–7. ACM (2006)
2. Brucker, P.: Scheduling algorithms, 5th edn., pp. 124–125. Springer (2007)
3. Ebenlendr, T., Sgall, J.: Optimal and online preemptive scheduling on uniformly related machines. In: Diekert, V., Habib, M. (eds.) STACS 2004. LNCS, vol. 2996, pp. 199–210. Springer, Heidelberg (2004)
4. Edelman, B., Ostrovsky, M., Schwarz, M.: Internet advertising and the generalized second price auction: Selling billions of dollars worth of keywords. Technical report, National Bureau of Economic Research (2005)
5. Epstein, L., Jez, L., Sgall, J., van Stee, R.: Online interval scheduling on uniformly related machines (2012) (manuscript)
6. Goel, A., Meyerson, A.: Simultaneous optimization via approximate majorization for concave profits or convex costs. Algorithmica 44(4), 301–323 (2006)
7. Horvath, E.C., Lam, S., Sethi, R.: A level algorithm for preemptive scheduling. Journal of the ACM (JACM) 24(1), 32–43 (1977)
8. Ibarra, O.H., Kim, C.E.: Fast approximation algorithms for the knapsack and sum of subset problems. Journal of the ACM (JACM) 22(4), 463–468 (1975)
9. Kong, W., Li, J., Qin, T., Liu, T.-Y.: Optimal allocation for chunked-reward advertising. arXiv preprint arXiv:1305.5946 (2013)
10. Lipton, R.: Online interval scheduling. In: Proceedings of the Fifth Annual ACM-SIAM Symposium on Discrete Algorithms, pp. 302–311. Society for Industrial and Applied Mathematics (1994)
11. Nielsen, M.A.: An introduction to majorization and its applications to quantum mechanics (2002)
12. Pinedo, M.: Scheduling: theory, algorithms, and systems. Springer (2008)
13. Vazirani, V.V.: Approximation algorithms. Springer (2004)

Bicriteria Online Matching:
Maximizing Weight and Cardinality

Nitish Korula[1], Vahab S. Mirrokni[1], and Morteza Zadimoghaddam[2]

[1] Google Research, New York NY 10011, USA
{nitish,mirrokni}@google.com
[2] Massachusetts Institute of Technology, Cambridge MA 02139, USA
morteza@mit.edu

Abstract. Inspired by online ad allocation problems, many results have been developed for online matching problems. Most of the previous work deals with a single objective, but, in practice, there is a need to optimize multiple objectives. Here, as an illustrative example motivated by display ads allocation, we study a bi-objective online matching problem.

In particular, we consider a set of fixed nodes (ads) with capacity constraints, and a set of online items (pageviews) arriving one by one. Upon arrival of an online item i, a set of eligible fixed neighbors (ads) for the item is revealed, together with a weight w_{ia} for eligible neighbor a. The problem is to assign each item to an eligible neighbor online, while respecting the capacity constraints; the goal is to maximize both the total weight of the matching and the cardinality. In this paper, we present both approximation algorithms and hardness results for this problem.

An (α, β)-approximation for this problem is a matching with weight at least α fraction of the maximum weighted matching, and cardinality at least β fraction of maximum cardinality matching. We present a parametrized approximation algorithm that allows a smooth tradeoff curve between the two objectives: when the capacities of fixed nodes are large, we give a $p(1 - 1/e^{1/p}), (1 - p)(1 - 1/e^{1/1-p})$-approximation for any $0 \leq p \leq 1$, and prove a 'hardness curve' combining several inapproximability results. These upper and lower bounds are always close (with a maximum gap of 9%), and exactly coincide at the point $(0.43, 0.43)$. For small capacities, we present a smooth parametrized approximation curve for the problem between $(0, 1 - 1/e)$ and $(1/2, 0)$ passing through a $(1/3, 0.3698)$-approximation.

1   Introduction

In the past decade, there has been much progress in designing better algorithms for online matching problems. This line of research has been inspired by interesting combinatorial techniques that are applicable in this setting, and by online ad allocation problems. For example, the display advertising problem has been modeled as maximizing the weight of an online matching instance [11,10,8,2,19]. While weight is indeed important, this model ignores the fact that cardinality of the matching is also crucial in the display ad application. This example illustrates

Y. Chen and N. Immorlica (Eds.): WINE 2013, LNCS 8289, pp. 305–318, 2013.
© Springer-Verlag Berlin Heidelberg 2013

the fact that in many real applications of online allocation, one needs to optimize multiple objective functions, though most of the previous work in this area deals with only a single objective function. On the other hand, there is a large body of work exploring *offline* multi-objective optimization in the approximation algorithms literature. In this paper, we focus on simultaneously maximizing *online* two objectives which have been studied extensively in matching problems: cardinality and weight. Besides being a natural mathematical problem, this is motivated by online display advertising applications.

Applications in Display Advertising. In online display advertising, advertisers typically purchase bundles of millions of display ad impressions from web publishers. Display ad serving systems that assign ads to pages on behalf of publishers must satisfy the contracts with advertisers, respecting targeting criteria and delivery goals. Modulo this, publishers try to allocate ads intelligently to maximize overall quality (measured, for example, by clicks), and therefore a desirable property of an ad serving system is to maximize this quality while satisfying the contracts to deliver the purchased number $n(a)$ impressions to advertiser a. This has been modeled in the literature (e.g., [11,1,24,8,2,19]) as an online allocation problem, where quality is represented by edge weights, and contracts are enforced by overall delivery goals: While trying to maximize the *weight* of the allocation, the ad serving systems should deliver $n(a)$ impressions to advertiser a. However, online algorithms with adversarial input cannot guarantee the delivery of $n(a)$ impressions, and hence the goals $n(a)$ were previously modeled as upper bounds. But maximizing the *cardinality* subject to these upper bounds is identical to delivering as close to the targets as possible. This motivates our model of the display ad problem as simultaneously maximizing weight and cardinality.

Problem Formulation. More specifically, we study the following bicriteria online matching problem: consider a set of bins (also referred to as fixed nodes, or ads) A with capacity constraints $n(a) > 0$, and a set of online items (referred to as online nodes, or impressions or pageviews) I arriving one by one. Upon arrival of an online item i, a set S_i of eligible bins (fixed node neighbors) for the item is revealed, together with a weight w_{ia} for eligible bin $a \in S_i$. The problem is to assign each item i to an eligible bin in S_i or discard it online, while respecting the capacity constraints, so bin a gets at most $n(a)$ online items. The goal is to maximize both the cardinality of the allocation (i.e. the total number of assigned items) and the sum of the weights of the allocated online items.

It was shown in [11] that achieving any positive approximation guarantee for the total weight of the allocation requires the *free disposal* assumption, i.e. that there is no penalty for assigning more online nodes to a bin than its capacity, though these extra nodes do not count towards the objective. In the advertising application, this means that in the presence of a contract for $n(a)$ impressions, advertisers are only pleased by – or at least indifferent to – getting *more* than $n(a)$ impressions. More specifically, if a set I^a of online items are assigned to each bin a, and $I^a(k)$ denotes the set of k online nodes with maximum weight in I^a, the goal is to simultaneously maximize cardinality which is $\sum_{a \in A} \min(|I^a|, n(a))$, and total weight which is $\sum_{a \in A} \sum_{i \in I^a(n(a))} w_{ia}$.

Throughout this paper, we use W_{opt} to denote the maximum weight matching, and overload this notation to also refer to the weight of this matching. Similarly, we use C_{opt} to denote both the maximum cardinality matching and its cardinality. Note that C_{opt} and W_{opt} may be distinct matchings. We aim to find (α, β)-approximations for the bicriteria online matching problem: These are matchings with weight at least αW_{opt} and cardinality at least βC_{opt}. Our approach is to study parametrized approximation algorithms that allow a smooth tradeoff curve between the two objectives, and prove both approximation and hardness results in this framework. As an offline problem, the above bicriteria problem can be solved optimally in polynomial time, i.e., one can check if there exists an assignment of cardinality c and weight w respecting capacity constraints. (One can verify this by observing that the integer linear programming formulation for the offline problem is totally unimodular, and therefore the problem can be solved by solving the corresponding LP relaxation.) However in the online competitive setting, even maximizing one of these two objectives does not admit better than a $1 - 1/e$ approximation [18]. A naive greedy algorithm gives a $\frac{1}{2}$-approximation for maximizing a single objective, either for cardinality or for total weight under the free disposal assumption.

Results and Techniques. The seminal result of Karp, Vazirani and Vazirani [18] gives a simple randomized $(1 - 1/e)$-competitive algorithm for maximizing cardinality. For the weight objective, no algorithm better than the greedy $1/2$-approximation is known, but for the case of large capacities, a $1 - 1/e$-approximation has been developed [11] following the primal-dual analysis framework of Buchbinder et al. [5,21]. Using these results, one can easily get a $(\frac{p}{2}, (1 - p)(1 - \frac{1}{e}))$-approximation for the bicriteria online matching problem with small capacities, and a $(p(1 - \frac{1}{e}), (1 - p)(1 - \frac{1}{e}))$-approximation for large capacities. These factors are achieved by applying the online algorithm for weight, WeightAlg, and the online algorithm for cardinality, CardinalityAlg, as subroutines as follows: When an online item arrives, pass it to WeightAlg with probability p, and CardinalityAlg with probability $1-p$. As for a hardness result, it is easy to show that an approximation factor better than $(\alpha, 1 - \alpha)$ is not achievable for any $\alpha > 0$. There is a large gap between the above approximation factors and hardness results. For example, the naive algorithm gives a $(0.4, 0.23)$-approximation, but the hardness result does not preclude a $(0.4, 0.6)$-approximation. In this paper, we tighten the gap between these lower and upper bounds, and present new tradeoff curves for both algorithms and hardness results. Our lower and upper bound results are summarized in Figure 1. For the case of large capacities, these upper and lower bound curves are always close (with a maximum vertical gap of 9%), and exactly coincide at the point $(0.43, 0.43)$.

We first describe our hardness results. In fact, we prove three separate inapproximability results which can be combined to yield a 'hardness curve' for the problem. The first result gives better upper bounds for large values of β; this is based on structural properties of matchings, proving some invariants for any online algorithm on a family of instances, and writing a factor-revealing mathematical program (see Section 2.1). The second main result is an improved upper bound for large values of α, and is based on a new family of instances for which achieving a large value for α implies very small values of β (see Section 2.2). Finally, we show that for any achievable (α, β), we have $\alpha + \beta \leq 1 - \frac{1}{e^2}$ (see Theorem 3).

These hardness results show the limit of what can be achieved in this model. We next turn to algorithms, to see how close we can come to these limits. The key to our new algorithmic results lies in the fact that though each subroutine WeightAlg and CardinalityAlg only receives a fraction of the online items, it can use the entire set of bins. This may result in both subroutines filling up a bin, but if WeightAlg places t items in a bin, we can discard t of the items placed there by CardinalityAlg and still get at least the cardinality obtained by CardinalityAlg and the weight obtained by

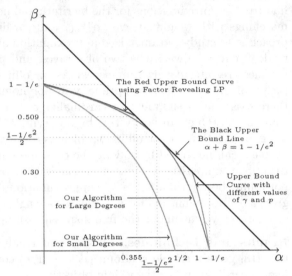

Fig. 1. New curves for upper and lower bounds

WeightAlg. Each subroutine therefore has access to the entire bin capacity, which is more than it 'needs' for those items passed to it. Thus, its competitive ratio can be made better than $1 - 1/e$. For large capacities, we prove the following theorem by extending the primal-dual analysis of Buchbinder *et al.* and Feldman *et al.* [11,21,5].

Theorem 1. *For all* $0 < p < 1$, *there is an algorithm for the bicriteria online matching problem with competitive ratios tending to* $\left(p(1 - \frac{1}{e^{1/p}}), (1 - p)(1 - \frac{1}{e^{1/(1-p)}})\right)$ *as* $\min_a\{n(a)\}$ *tends to infinity.*

For small capacities, our result is more technical and is based on studying structural properties of matchings, proving invariants for our online algorithm over any instance, and solving a factor-revealing LP that combines these new invariants and previously known combinatorial techniques by Karp, Vazirani, Vazirani, and Birnbaum and Mathieu [18,4]. Factor revealing LPs have been used in the context of online allocation problems [21,20]. In our setting, we need to prove new variants and introduce new inequalities to take into account and analyze the tradeoff between the two objective functions. This result can also be parametrized by p, the fraction of items sent to WeightAlg, but we do not have a closed form expression. Hence, we state the result for $p = 1/2$.

Theorem 2. *For all* $0 \le p \le 1$, *the approximation guarantee of our algorithm for the bicriteria online matching problem is lower bounded by the green curve of Figure 1. In particular, for* $p = 1/2$, *we have the point* $(1/3, 0.3698)$.

Related Work. Our work is related to online ad allocation problems, including the *Display Ads Allocation (DA)* problem [11,10,1,24], and the *AdWords (AW)* problem [21,7]. In both of these problems, the publisher must assign online impressions to an inventory of ads, optimizing efficiency or revenue of the allocation while respecting pre-specified contracts. The Display Ad (DA) problem is the online matching problem described above only considering the weight objective [11,2,19]. In the AdWords (AW) problem, the publisher allocates impressions resulting from search queries. Advertiser a has a budget $B(a)$ on the total spend instead of a bound $n(a)$ on the number of impressions. Assigning impression i to advertiser a consumes w_{ia} units of a's budget instead of 1 of the $n(a)$ slots, as in the DA problem. For both of these problems, $1 - \frac{1}{e}$-approximation algorithms have been designed under the assumption of large capacities [21,5,11]. None of the above papers for the adversarial model study multiple objectives at the same time.

Besides the adversarial model studied in this paper, online ad allocations have been studied extensively in various *stochastic models*. In particular, the problem has been studied in the *random order model*, where impressions arrive in a random order [7,10,1,24,17,20,23]; and the *iid* model in which impressions arrive iid according to a known (or unknown) distribution [12,22,16,8,9]. In such stochastic settings, primal and dual techniques have been applied to getting improved approximation algorithms. These techniques are based on computing offline optimal primal or dual solutions of an expected instance, and using this solution online [12,7]. It is not hard to generalize these techniques to the bicritera online matching problem. In this extended abstract, we focus on the adversarial model, and leave discussions of extensions of such techniques for the stochastic bicriteria problem to the full version of the paper. Note that in order to deal with traffic spikes, adversarial competitive analysis is important from a practical perspective, as discussed in [23].

Most previous work on online problems with multiple objectives has been in the domain of routing and scheduling, and with different models. Typically, goals are to maximize throughput and fairness; see the work of Goel *et al.* [15,14], Buchbinder and Naor [6], and Wang *et al.* [25]. In this literature, different objectives often come from applying different functions on the same set of inputs, such as processing times or bandwidth allocations. In a model more similar to ours, Bilò *et al.* [3] consider scheduling where each job has two different and unrelated requirements, processing time and memory; the goal is to minimize makespan while also minimizing maximum memory requirements on each machine. In another problem with distinct metrics, Flammini and Nicosia [13] consider the k-server problem with a distance metric and time metric defined on the set of service locations. However, unlike our algorithms, theirs do not compete simultaneously against the best solution for each objective; instead, they compete against offline solutions that must simultaneously do well on both objectives.

Further, the competitive ratio depends on the relative values of the two objectives. Such results are of limited use in advertising applications, for instance, where click-through rates per impression may vary by several orders of magnitude.

2   Hardness Instances

In this section for any $0 \le \alpha \le 1 - 1/e$, we prove upper bounds on β such that the bicriteria online matching problem admits an (α, β)-approximation. Note that it is not possible to achieve α-approximation guarantee for the total weight of the allocation for any $\alpha > 1 - 1/e$. We have two types of techniques to achieve upper bounds: a) Factor-Revealing Linear Programs, b) Super Exponential Weights Instances, which are discussed in Subsections 2.1, and 2.2 respectively. Factor revealing LP hardness instances give us the red upper bound curve in Figure 1. The orange upper bound curve in Figure 1 is proved by Super Exponential Weights Instances presented in Subsection 2.2, and the black upper bound line in Figure 1 is proved in Theorem 3.

2.1   Better Upper Bounds via Factor-Revealing Linear Programs

We construct an instance, and a linear program $LP_{\alpha,\beta}$ based on the instance where α and β are two parameters in the linear program. We prove that if there exists an (α, β)-approximation for the bicriteria online matching problem, we can find a feasible solution for $LP_{\alpha,\beta}$ based on the algorithm's allocation to the generated instance. Finally we find out for which pairs (α, β) the linear program $LP_{\alpha,\beta}$ is infeasible. These pairs (α, β) are upper bounds for the bicriteria online matching problem.

For any two integers C, l, and some large weight $W \gg 4l^2$, we construct the instance as follows. We have l phases, and each phase consists of l sets of C identical items, i.e. $l^2 C$ items in total. For any $1 \le t, i \le l$, we define $O_{t,i}$ to be the set i in phase t that has C identical items. In each phase, we observe the sets of items in increasing order of i. There are two types of bins: a) l weight bins $b_1, b_2, \cdots, b_l$ which are shared between different phases, b) l^2 cardinality bins $\{b'_{t,i}\}_{1 \le t, i \le l}$. For each phase $1 \le t \le l$, we have l separate bins $\{b'_{t,i}\}_{1 \le i \le l}$. The capacity of all bins is C. We pick two permutations $\pi_t, \sigma_t \in \mathbb{S}_n$ uniformly at random at the beginning of each phase t to construct edges. We note that these permutations are private knowledge, and they are not revealed to the algorithm. For any $1 \le i \le j \le l$, we put an edge between every item in set $O_{t,i}$ and bin $b'_{t,\sigma_t(j)}$ with weight 1 where $\sigma_t(j)$ is the jth number in permutation σ_t. We also put an edge between every item in set $O_{t,i}$ and bin $b_{\pi_t(j)}$ (for each $j \ge i$) with weight W^t.

Suppose there exists an (α, β)-approximation algorithm $A_{\alpha,\beta}$ for the bicriteria online matching problem. For any $1 \le t, i \le l$, let $x_{t,i}$ be the expected number of items in set $O_{t,i}$ that algorithm $A_{\alpha,\beta}$ assigns to weight bins $\{b_{\pi_t(j)}\}_{j=i}^{l}$. Similarly we define $y_{t,i}$ to be the expected number of items in set $O_{t,i}$ that algorithm $A_{\alpha,\beta}$

assigns to cardinality bins $\{b'_{t,\sigma_t(j)}\}_{j=i}^{l}$. We know that when set $O_{t,i}$ arrives, although the algorithm can distinguish between weight and cardinality bins, it sees no difference between the weight bins $\{b_{\pi_t(j)}\}_{j=i}^{l}$, and no difference between the cardinality bins $\{b'_{t,\sigma_t(j)}\}_{j=i}^{l}$. By uniform selection of π and σ, we ensure that in expectation the $x_{t,i}$ items are allocated equally to weight bins $\{b_{\pi_t(j)}\}_{j=i}^{l}$, and the $y_{t,i}$ items are allocated equally to cardinality bins $\{b'_{t,\sigma_t(j)}\}_{j=i}^{l}$. In other words, for $1 \le i \le j \le l$, in expectation $x_{t,i}/(l-i+1)$ and $y_{t,i}/(l-i+1)$ items of set $O_{t,i}$ is allocated to bins $b_{\pi_t(j)}$ and $b'_{t,\sigma_t(j)}$, respectively. It is worth noting that similar ideas have been used in previous papers on online matching [18,4].

Since weights of all edges to cardinality bins are 1, we can assume that the items assigned to cardinality bins are kept until the end of the algorithm, and they will not be thrown away. We can similarly say that the weights of all items for weight bins is the same in a single phase, so we can assume that an item that has been assigned some weight bin in a phase will not be thrown away at least until the end of the phase. However, the algorithm might use the free disposal assumption for weight bins in different phases. We have the following capacity constraints on bins $b_{\pi_t(j)}$ and $b'_{t,\sigma_t(j)}$:

$$\forall 1 \le t, j \le l: \sum_{i=1}^{j} x_{t,i}/(l-i+1) \le C \ \& \ \sum_{i=1}^{j} y_{t,i}/(l-i+1) \le C. \tag{1}$$

At any stage of phase t, the total weight assigned by the algorithm cannot be less than α times the optimal weight allocation up to that stage, or we would not have weight αW_{opt} if the input stopped at this point. After set $O_{t,i}$ arrives, the maximum weight allocation achieves at least total weight CiW^t which is achieved by assigning items in set $O_{t,i'}$ to weight bin $b_{\pi_t(i')}$ for each $1 \le i' \le i$. On the other hand, the expected weight in allocation of algorithm $A_{\alpha,\beta}$ is at most $C(tl + W^{t-1}l) + W^t \sum_{i'=1}^{i} x_{t,i'} \le W^t(C/\sqrt{W} + \sum_{i'=1}^{i} x_{t,i'})$. Therefore we have the following inequality for any $1 \le t, i \le l$:

$$\sum_{i'=1}^{i} x_{t,i'}/C \ge \alpha i - 1/\sqrt{W}. \tag{2}$$

We show in Lemma 1 that the linear program $LP_{\alpha,\beta}$ is feasible if there exists an algorithm $A_{\alpha,\beta}$ by defining $p_i = \sum_{t=1}^{l} x_{t,i}/lC$, and $q_i = \sum_{t=1}^{l} y_{t,i}/lC$. Now for any α, we can find the maximum β for which the $LP_{\alpha,\beta}$ has some feasible solution for large values of l and W. These factor-revealing linear programs yield the red upper bound curve in Figure 1.

$LP_{\alpha,\beta}$

C1:	$\sum_{i'=1}^{i} p_{i'}$	$\ge \alpha i - 1/\sqrt{W}$	$\forall 1 \le i \le l$
C2:	$\sum_{i=1}^{l} q_i$	$\ge l\beta - 1$	
C3:	$p_i + q_i$	≤ 1	$\forall 1 \le i \le l$
C4:	$\sum_{i=1}^{j} p_i/(l-i+1) \le 1$		$\forall 1 \le j \le l$
C5:	$\sum_{i=1}^{j} q_i/(l-i+1) \le 1$		$\forall 1 \le j \le l$

Lemma 1. *If there exists an (α, β)-approximation algorithm for the bicriteria online matching problem, there exists a feasible solution for $LP_{\alpha,\beta}$ as well.*

In addition to computational bounds for infeasibility of certain (α, β) pairs, we can theoretically prove in Theorem 3 that for any (α, β) with $\alpha + \beta > 1 - 1/e^2$, the $LP_{\alpha,\beta}$ is infeasible so there exists no (α, β) approximation for the problem. We note that Theorem 3 is a simple generalization of the $1 - 1/e$ hardness result for the classic online matching problem [18,4].

Theorem 3. *For any small $\epsilon > 0$, and $\alpha + \beta \geq 1 - 1/e^2 + \epsilon$, there exists no (α, β)-approximation algorithm for the bicriteria matching problem.*

Proof. We just need to show that $LP_{\alpha,\beta}$ is infeasible. Given a solution of $LP_{\alpha,\beta}$, we find a feasible solution for LP'_ϵ defined below by setting $r_i = p_i + q_i$ for any $1 \leq i \leq l$.

$$LP'_\epsilon \quad \sum_{i=1}^{l} r_i \qquad\qquad \geq (1 - 1/e^2 + \epsilon/2)l$$
$$r_i \qquad\qquad\qquad\quad \leq 1 \qquad\qquad \forall 1 \leq i \leq l$$
$$\sum_{i=1}^{j} r_i/(l - i + 1) \leq 2 \qquad\qquad \forall 1 \leq j \leq l$$

The first inequality in LP'_ϵ is implied by summing up the constraint C1 for $i = l$, and constraint C2 in $LP_{\alpha,\beta}$, and also using the fact that $\alpha + \beta \geq (1 - 1/e^2 + \epsilon/2) + \epsilon/2$. We note that the $\epsilon/2$ difference between the $\alpha + \beta$ and $1 - 1/e^2 + \epsilon/2$ takes care of $-1/\sqrt{W}$ and -1 in the right hand sides of constraints C1 and C2 for large enough values of l and W. Now we prove that LP'_ϵ is infeasible for any $\epsilon > 0$ and large enough l. Suppose there exists a feasible solution $r_1, r_2, \cdots, r_n$. For any pair $1 \leq i < j \leq n$, if we have $r_i < 1$ and $r_j > 0$, we update the values of r_i and r_j to $r_i^{new} = r_i + \min\{1 - r_i, r_j\}$, and $r_j^{new} = r_j - \min\{1 - r_i, r_j\}$. Since we are moving the same amount from r_j to r_i (for some $i < j$), all constraints still hold. If we do this operation iteratively until there is no pair r_i and r_j with the above properties, we reach a solution $\{r'_i\}_{i=1}^{l}$ of this form: $1, 1, \cdots, 1, x, 0, 0, \cdots, 0$ for some $0 \leq x \leq 1$. Let t be the maximum index for which r'_t is 1. Using the third inequality for $j = l$, we have that $\sum_{i=1}^{t} 1/(l - i + 1) \leq 2$ which means that $\ln(l/(l - t + 1)) \leq 2$. So t is not greater than $l(1 - 1/e^2)$, and consequently $\sum_{i=1}^{l} r'_i \leq t + 1 \leq (1 - 1/e^2)l + 1 < (1 - 1/e^2 + \epsilon/2)l$. This contradiction proves that LP'_ϵ is infeasible which completes the proof of theorem.

2.2 Hardness Results for Large Values of Weight Approximation Factor

The factor-revealing linear program $LP_{\alpha,\beta}$ gives almost tight bounds for small values of α. In particular, the gap between the the upper and lower bounds for the cardinality approximation ratio β is less than 0.025 for $\alpha \leq (1 - 1/e^2)/2$. But for large values of α ($\alpha > (1 - 1/e^2)/2$), this approach does not give anything better than the $\alpha + \beta \leq 1 - 1/e^2$ bound proved in Theorem 3 . This leaves a maximum gap of $1/e - 1/e^2 \approx 0.23$ between the upper and lower bounds at $\alpha = 1 - 1/e$. In order to close the gap at $\alpha = 1 - 1/e$, we present a different analysis based on a new set of instances, and reduce the maximum gap between lower and upper bounds from 0.23 to less than 0.09 for all values of $\alpha \geq (1 - 1/e^2)/2$.

The main idea is to construct a hardness instance I_γ for any $1/e \leq \gamma < 1$, and prove that for any $0 \leq p \leq 1 - \gamma$, the pair $(1 - 1/e - f(p), p/(1 - \gamma))$ is an upper bound on (α, β) where $f(p)$ is $\frac{p}{e(\gamma+p)}$. In other words, there exists no (α, β)-approximation algorithm for this problem with both $\alpha > 1 - 1/e - f(p)$ and $\beta > p/(1 - \gamma)$. By enumerating different pairs of γ and p, we find the orange upper bound curve in Figure 1.

For any $\gamma \geq 1/e$, we construct instance I_γ as follows: The instance is identical to the hardness instance in Subsection 2.1, but we change some of the edge weights. To keep the description short, we only describe the edges with modified weights here. Let r be $\lfloor 0.5 \log_{1/\gamma} l \rfloor$. In each phase $1 \leq t \leq l$, we partition the l sets of items $\{O_{t,i}\}_{i=1}^l$ into r groups. The first $l(1 - \gamma)$ sets are in the first group. From the remaining γl sets, we put the first $(1 - \gamma)$ fraction in the second group and so on. Formally, we put set $O_{t,i}$ in group $1 \leq z < r$ for any $i \in [l - l\gamma^{z-1} + 1, l - l\gamma^z]$. Group r of phase t contains the last $l\gamma^{r-1}$ sets of items in phase t. The weight of all edges from sets of items in group z in phase t is $W^{(t-1)r+z}$ for any $1 \leq z \leq r$ and $1 \leq t \leq l$.

Given an (α, β)-approximation algorithm $A_{\alpha,\beta}$, we similarly define $x_{t,i}$ and $y_{t,i}$ to be the expected number of items from set $O_{t,i}$ assigned to weight and cardinality bins by algorithm $A_{\alpha,\beta}$ respectively. We show in the following lemma that in order to have a high α, the algorithm should allocate a large fraction of sets of items in each group to the weight bins.

Lemma 2. *For any phase $1 \leq t \leq l$, and group $1 \leq z < r$, if the expected number of items assigned to cardinality bins in group z of phase t is at least $plC\gamma^{z-1}$ (which is p times the number of all items in groups $z, z + 1, \cdots, r$ of phase t), the weight approximation ratio cannot be greater than $1 - 1/e - f(p)$ where $f(p)$ is $\frac{p}{e(\gamma+p)}$.*

We conclude this part with the main result of this subsection:

Theorem 4. *For any small $\epsilon > 0$, $1/e \leq \gamma < 1$, and $0 \leq p \leq 1 - \gamma$, any algorithm for bicriteria online matching problem with weight approximation guarantee, α, at least $1 - 1/e - f(p)$ cannot have cardinality approximation guarantee, β, greater than $p/(1 - \gamma) + \epsilon$.*

Proof. Using Lemma 2, for any group $1 \leq z < r$ in any phase $1 \leq t \leq l$, we know that at most p fraction of items are assigned to cardinality bins, because $1 - 1/e - f(p)$ is a strictly increasing function in p. Since in each phase the number of items is decreasing with a factor of γ in consecutive groups, the total fraction of items assigned to cardinality bins is at most $p + p\gamma + p\gamma^2 + \cdots + p\gamma^{r-2}$ plus the fraction of items assigned to cardinality in the last group r of phase t. Even if the algorithm assigns all of group r to cardinality, it does not achieve more than fraction γ^{r-1} from these items in each phase. Since the optimal cardinality algorithm can match all items, the cardinality approximation guarantee is at most $p(1 + \gamma + \gamma^2 + \cdots + \gamma^{r-2}) + \gamma^{r-1}$. For large enough l (and consequently large enough r), this sum is not more than $p/(1 - \gamma) + \epsilon$.

One way to compute the best values for p and γ corresponding to the best upper bound curve is to solve complex equations explicitly. Instead, we compute these values numerically by trying different values of p and γ which, in turn, yield the orange upper bound curve in Figure 1.

3   Algorithm for Large Capacities

We now turn to algorithms, to see how close one can come to matching the upper bounds of the previous section. In this section, we assume that the capacity $n(a)$ of each bin $a \in A$ is "large", and give an algorithm with the guarantees in Theorem 1 as $\min_{a \in A} n(a) \to \infty$.

Recall that our algorithm Alg uses two subroutines WeightAlg and CardinalityAlg, each of which, if given an online item, suggests a bin to place it in. Each item i is independently passed to WeightAlg with probability p and CardinalityAlg with the remaining probability $1 - p$. First note that CardinalityAlg and WeightAlg are independent and unaware of each other; each of them thinks that the only items which exist are those passed to it. This allows us to analyze the two subroutines separately.

We now describe how Alg uses the subroutines. If WeightAlg suggests matching item i to a bin a, we match i to a. If a already has $n(a)$ items assigned to it in total, we remove any item assigned by CardinalityAlg arbitrarily; if all $n(a)$ were assigned by WeightAlg, we remove the item of lowest value for a. If CardinalityAlg suggests matching item i to a', we make this match unless a' has already had at least $n(a')$ total items assigned to it by both subroutines. In other words, the assignments of CardinalityAlg might be thrown away by some assignments of WeightAlg; however, the total number of items in a bin is always at least the the number assigned by CardinalityAlg. Items assigned by WeightAlg are never thrown away due to CardinalityAlg; they may only be replaced by later assignments of WeightAlg. Thus, we have proved the following proposition.

Proposition 1. *The weight and cardinality of the allocation of* Alg *are respectively at least as large as the weight of the allocation of* WeightAlg *and the cardinality of the allocation of* CardinalityAlg.

Note that the above proposition does not hold for any two arbitrary weight functions, and this is where we need one of the objectives to be cardinality. We now describe WeightAlg and CardinalityAlg, and prove Theorem 1. WeightAlg is essentially the exponentially-weighted primal-dual algorithm from [11], which was shown to achieve a $1 - \frac{1}{e}$ approximation for the weighted online matching problem with large degrees. For completeness, we present the primal and dual LP relaxations for weighted matching below, and then describe the algorithm. In the primal LP, for each item i and bin a, variable x_{ia} denotes whether impression i is one of the $n(a)$ most valuable items for bin a.

	Primal				**Dual**	

$$\max \quad \sum_{i,a} w_{ia} x_{ia}$$

$$\sum_a x_{ia} \leq 1 \quad (\forall\, i)$$
$$\sum_i x_{ia} \leq n(a) \quad (\forall\, a)$$
$$x_{ia} \geq 0 \quad (\forall\, i,a)$$

$$\min \sum_a n(a)\beta_a + \sum_i z_i$$
$$\beta_a + z_i \geq w_{ia} \quad (\forall i,a)$$
$$\beta_a, z_i \geq 0 \quad (\forall i,a)$$

Following the techniques of Buchbinder *et al.* [5], the algorithm of [11] simultaneously maintains a feasible solution to the primal LP, and provides a feasible solution to the dual LP after all online nodes arrive. Each dual variable β_a is initialized to 0. When item i arrives online:

- Assign i to the bin $a' = \arg\max_a \{w_{ia} - \beta_a\}$. (If this quantity is negative for all a, discard i.)
- Set $x_{ia'} = 1$. If a' previously had $n(a')$ items assigned to it, set $x_{i'a'} = 0$ for the least valuable item i' previously assigned to a'.
- In the dual solution, set $z_i = w_{ia'} - \beta_{a'}$ and *update* dual variable $\beta_{a'}$ as described below.

Definition 1 (Exponential Weighting). *Let $w_1, w_2, \ldots w_{n(a)}$ be the weights of the $n(a)$ items currently assigned to bin a, sorted in non-increasing order, and padded with 0s if necessary.*

Set $\beta_a = \frac{1}{p \cdot n(a) \cdot ((1 + 1/p \cdot n(a))^{n(a)} - 1)} \sum_{j=1}^{n(a)} w_j \left(1 + \frac{1}{p \cdot n(a)}\right)^{j-1}$.

Lemma 3. *If WeightAlg is the primal-dual algorithm, with dual variables β_a updated according to the exponential weighting rule defined above, the competitive ratio of WeightAlg regarding the weight objective is at least $p \cdot \left(1 - \frac{1}{k}\right)$ where $k = \left(1 + \frac{1}{p \cdot d}\right)^d$, and $d = \min_a\{n(a)\}$. Note that $\lim_{d\to\infty} k = e^{1/p}$.*

We provide some brief intuition here. If *all* items are passed to WeightAlg, it was proved in [11] that the algorithm has competitive ratio tending to $1 - 1/e$ as $d = \min_a\{n(a)\}$ tends to ∞; this is the statement of Lemma 3 when $p = 1$. Now, suppose each item is passed to WeightAlg with probability p. The expected value of the optimum matching induced by those items passed to WeightAlg is at least $p \cdot W_{\text{opt}}$, and this is *nearly* true (up to $o(1)$ terms) *even if we reduce the capacity of each bin a to $p \cdot n(a)$.* This follows since W_{opt} assigns at most $n(a)$ items to bin a, and as we are unlikely to sample more than $p \cdot n(a)$ of these items for the reduced instance, we do not lose much by reducing capacities. But note that WeightAlg can use the entire capacity $n(a)$, while there is a solution of value close to pW_{opt} even with capacities $p \cdot n(a)$. This extra capacity allows an improved competitive ratio of $1 - \frac{1}{e^{1/p}}$, proving the lemma.

Algorithm CardinalityAlg is identical to WeightAlg, except that it assumes all items have weight 1 for each bin. Since items are assigned to CardinalityAlg with probability $1 - p$, Lemma 3 implies the following corollary. This concludes the proof of Theorem 1.

Corollary 1. *The total cardinality of the allocation of* CardinalityAlg *is at least* $(1-p) \cdot (1 - \frac{1}{k})$, *where* $k = \left(1 + \frac{1}{(1-p)\cdot d}\right)^d$, *and* $d = \min_a\{n(a)\}$. *Note that* $\lim_{d\to\infty} k = e^{1/(1-p)}$.

4 Algorithm for Small Capacities

We now consider algorithms for the case when the capacities of bins are not large. Without loss of generality, we assume that the capacity of each bin is one, because we can think about a bin with capacity c as c identical bins with capacity one. So we have a set A of bins each with capacity one, and a set of items I arriving online. As before, we use two subroutines WeightAlg and CardinalityAlg, but the algorithms are slightly different from those in the previous section. Each item $i \in I$ is independently passed to WeightAlg with probability p and CardinalityAlg with the remaining probability $1 - p$.

In WeightAlg, we match item i (that has been passed to WeightAlg) to the bin that maximizes its marginal value. Formally we match i to bin $a = \arg\max_{a\in A}(w_{i,a} - w_{i',a})$ where i' is the last item assigned to a before item i.

In CardinalityAlg, we run the RANKING algorithm presented in [18]. So CardinalityAlg chooses a permutation π uniformly at random on the set of bins A, assigns an item i (that has been passed to it) to the bin a that is available, has the minimum rank in π, and there is also an edge between i and a.

4.1 $p/(p+1)$ Lower Bound on the Weight Approximation Ratio

Let $n = |I|$ be the number of items. We denote the ith arrived item by i. Let a_i be the bin that i is matched to in W_{opt} for any $1 \le i \le n$. One can assume that all unmatched items in the optimum weight allocation are matched with zero-weight edges to an imaginary bin. So W_{opt} is equal to $\sum_{i=1}^n w_{i,a_i}$. Let S be the set of items that have been passed to WeightAlg. If WeightAlg matches item i to bin a_j for some $j > i$, we call this a forwarding allocation (edge) because item j (the match of a_j in W_{opt}) has not arrived yet. We call it a selected forwarding edge if $j \in S$. We define the marginal value of assigning item i to bin a to be w_{ia} minus the value of any item previously assigned to a.

Lemma 4. *The weight of the allocation of* WeightAlg *is at least* $(p/(p+1))W_{opt}$.

Proof. Each forwarding edge will be a selected forwarding edge with probability p because $Pr[j \in S]$ is p for any $j \in I$. Let F be the total weight of forwarding edges of WeightAlg, where by weight of a forwarding edge, we mean its marginal value (not the actual weight of the edge). Similarly, we define F_s to be the sum of marginal values of selected forwarding edges. We have the simple equality that the expected value of F, $E(F)$, is $E(F_s)/p$. We define W' and W_s to be the total marginal values of allocation of WeightAlg, and the sum $\sum_{i\in S} w_{i,a_i}$. We know that $E(W_s)$ is pW_{opt} because $Pr[i \in S]$ is p. We prove that W' is at least $W_s - F_s$.

For every item i that has been selected to be matched by WeightAlg, we get at least marginal value w_{i,a_i} minus the sum of all marginal values of items that have been assigned to bin a by WeightAlg up to now. If we sum up all these lower bounds on our gains for all selected items, we get $W_s (= \sum_{i \in S} w_{i,a_i})$ minus the sum of all marginal values of items that has been assigned to a_i before item i arrives for all $i \in S$. The latter part is exactly the definition of F_s. Therefore W' is at least $W_s - F_s$. We also know that $W' \geq F$. Using $E[F] \geq E[F_s]/p$, we have that $E(W')$ is at least $E(W_s) - pE(W')$, and this yields the $p/(p+1)$ approximation factor.

Corollary 2. *The weight and cardinality approximation guarantees of* Alg *are at least $p/(p+1)$ and $(1-p)/(1-p+1)$ respectively.*

4.2 Factor Revealing Linear Program for CardinalityAlg

Due to limited space, we just mention that we get the lower bounds on the cardinality competitive ratio in the green curve of Figure 1 using the following LP. This LP gives a valid lower bound for any integer k for $p = 1/2$. A few simple adjustments generalize this factor revealing LP to the general case of arbitrary $0 < p < 1$.

$$
\begin{aligned}
\text{Minimize:} \quad & \beta \\
\forall 1 < i \leq k: & \; s_i \geq s_{i-1} \quad \& \; sf_i \geq sf_{i-1} \quad \& \; sb_i \geq sb_{i-1} \\
\forall 1 \leq i \leq k: & \; t_i \geq t_{i-1} \quad \& \; t_i \geq sf_i \quad \& \; s_i = sf_i + sb_i \\
& \beta \geq s_k + t_k \; \& \; \beta \geq 1/2 - sf_k \\
\forall 1 < i \leq k: & \; s_i - s_{i-1} \quad \geq 1/2k - (s_i + t_i)/k
\end{aligned}
$$

References

1. Agrawal, S., Wang, Z., Ye, Y.: A dynamic near-optimal algorithm for online linear programming. Working paper posted at http://www.stanford.edu/~yyye/
2. Bhalgat, A., Feldman, J., Mirrokni, V.S.: Online ad allocation with smooth delivery. In: ACM Conference on Knowledge Discovery, KDD (2012)
3. Bilò, V., Flammini, M., Moscardelli, L.: Pareto approximations for the bicriteria scheduling problem. Journal of Parallel and Distributed Computing 66(3), 393–402 (2006)
4. Birnbaum, B., Mathieu, C.: On-line bipartite matching made simple. SIGACT News 39(1), 80–87 (2008)
5. Buchbinder, N., Jain, K., Naor, J.(S.): Online Primal-Dual Algorithms for Maximizing Ad-Auctions Revenue. In: Arge, L., Hoffmann, M., Welzl, E. (eds.) ESA 2007. LNCS, vol. 4698, pp. 253–264. Springer, Heidelberg (2007)
6. Buchbinder, N., Naor, J.: Fair online load balancing. In: Proceedings of the Eighteenth Annual ACM Symposium on Parallelism in Algorithms and Architectures, pp. 291–298. ACM (2006)
7. Devanur, N., Hayes, T.: The adwords problem: Online keyword matching with budgeted bidders under random permutations. In: ACM EC (2009)

8. Devanur, N.R., Jain, K., Sivan, B., Wilkens, C.A.: Near optimal online algorithms and fast approximation algorithms for resource allocation problems. In: ACM Conference on Electronic Commerce, pp. 29–38 (2011)
9. Devanur, N.R., Sivan, B., Azar, Y.: Asymptotically optimal algorithm for stochastic adwords. In: ACM Conference on Electronic Commerce, pp. 388–404 (2012)
10. Feldman, J., Henzinger, M., Korula, N., Mirrokni, V.S., Stein, C.: Online stochastic packing applied to display ad allocation. In: de Berg, M., Meyer, U. (eds.) ESA 2010, Part I. LNCS, vol. 6346, pp. 182–194. Springer, Heidelberg (2010)
11. Feldman, J., Korula, N., Mirrokni, V., Muthukrishnan, S., Pál, M.: Online ad assignment with free disposal. In: Leonardi, S. (ed.) WINE 2009. LNCS, vol. 5929, pp. 374–385. Springer, Heidelberg (2009)
12. Feldman, J., Mehta, A., Mirrokni, V., Muthukrishnan, S.: Online stochastic matching: Beating 1 - 1/e. In: FOCS (2009)
13. Flammini, M., Nicosia, G.: On multicriteria online problems. In: Paterson, M. (ed.) ESA 2000. LNCS, vol. 1879, pp. 191–201. Springer, Heidelberg (2000)
14. Goel, A., Meyerson, A., Plotkin, S.: Approximate majorization and fair online load balancing. In: Proceedings of the Twelfth Annual ACM-SIAM Symposium on Discrete Algorithms, pp. 384–390. Society for Industrial and Applied Mathematics (2001)
15. Goel, A., Meyerson, A., Plotkin, S.: Combining fairness with throughput: Online routing with multiple objectives. Journal of Computer and System Sciences 63(1), 62–79 (2001)
16. Haeupler, B., Mirrokni, V.S., Zadimoghaddam, M.: Online stochastic weighted matching: Improved approximation algorithms. In: Chen, N., Elkind, E., Koutsoupias, E. (eds.) WINE 2011. LNCS, vol. 7090, pp. 170–181. Springer, Heidelberg (2011)
17. Karande, C., Mehta, A., Tripathi, P.: Online bipartite matching with unknown distributions. In: STOC (2011)
18. Karp, R.M., Vazirani, U.V., Vazirani, V.V.: An optimal algorithm for on-line bipartite matching. In: STOC, pp. 352–358 (1990)
19. Korula, N., Mirrokni, V.S., Yan, Q.: Whole-page ad allocation and. In: Ad Auctions Workshop (2012)
20. Mahdian, M., Yan, Q.: Online bipartite matching with random arrivals: A strongly factor revealing lp approach. In: STOC (2011)
21. Mehta, A., Saberi, A., Vazirani, U.V., Vazirani, V.V.: Adwords and generalized online matching. J. ACM 54(5) (2007)
22. Menshadi, H., OveisGharan, S., Saberi, A.: Offline optimization for online stochastic matching. In: SODA (2011)
23. Mirrokni, V., Gharan, S.O., ZadiMoghaddam, M.: Simultaneous approximations for adversarial and stochastic online budgeted allocation problems. In: SODA (2012)
24. Vee, E., Vassilvitskii, S., Shanmugasundaram, J.: Optimal online assignment with forecasts. In: ACM EC (2010)
25. Wang, C.-M., Huang, X.-W., Hsu, C.-C.: Bi-objective optimization: An online algorithm for job assignment. In: Abdennadher, N., Petcu, D. (eds.) GPC 2009. LNCS, vol. 5529, pp. 223–234. Springer, Heidelberg (2009)

Mitigating Covert Compromises

A Game-Theoretic Model of Targeted and Non-Targeted Covert Attacks

Aron Laszka[1,2], Benjamin Johnson[3], and Jens Grossklags[1]

[1] College of Information Sciences and Technology,
Pennsylvania State University, USA
[2] Department of Networked Systems and Services,
Budapest University of Technology and Economics, Hungary
[3] School of Information, University of California, Berkeley, USA

Abstract. Attackers of computing resources increasingly aim to keep security compromises hidden from defenders in order to extract more value over a longer period of time. These covert attacks come in multiple varieties, which can be categorized into two main types: targeted and non-targeted attacks. Targeted attacks include, for example, cyber-espionage, while non-targeted attacks include botnet recruitment.

We are concerned with the subclass of these attacks for which detection is too costly or technically infeasible given the capabilities of a typical organization. As a result, defenders have to mitigate potential damages under a regime of incomplete information. A primary mitigation strategy is to reset potentially compromised resources to a known safe state, for example, by reinstalling computer systems, and changing passwords or cryptographic private keys.

In a game-theoretic framework, we study the economically optimal mitigation strategies in the presence of targeted and non-targeted covert attacks. Our work has practical implications for the definition of security policies, in particular, for password and key renewal schedules.

Keywords: Game Theory, Computer Security, Covert Compromise, Targeted Attacks, Non-Targeted Attacks.

1 Introduction

Most organizations devote significant resources to prevent security compromises which may harm their financial bottomline or adversely affect their reputation. Security measures typically include technologies to detect known attack vectors. However, recent studies of anti-malware and anti-virus tools have demonstrated their ineffectiveness against novel attack approaches and even incrementally modified known malware.

At the same time, attackers prey upon opportunities to keep successful security compromises covert. The goal is to benefit from defenders' lack of awareness by exploiting resources, and extracting credentials and company secrets for as

Y. Chen and N. Immorlica (Eds.): WINE 2013, LNCS 8289, pp. 319–332, 2013.
© Springer-Verlag Berlin Heidelberg 2013

long as possible. In contrast to non-covert attacks and compromises that focus on short-term benefits, these long-lasting and (for typical organizations) undetectable attacks pose specific challenges to system adminstrators and creators of security policies. Discoveries of such attacks by sophisticated security companies provide evidence for damage caused over many months or years.

CDorked, a highly advanced and stealthy backdoor, was discovered in April 2013 [5]. The malware uses compromised webservers to infect visitors with common system configurations. To stay covert, the malware uses a number of different techniques, for example, not delivering malicious content if the visitor's IP address is in a customized blacklist. The operation has been active since at least December 2012, and has infected more than 400 webservers, including 50 from Alexa's top 100,000 most popular websites.

Another example is Gauss, a complex, nation-state sponsored cyber-espionage toolkit, which is closely related to the notorious Stuxnet [1,10]. Gauss was designed to steal sensitive financial data from targets primarily located in the Middle East, and was active for at least 10 months before it was discovered.

Such recently-revealed attack vectors as well as the suspected number of unknown attacks highlight the importance of developing mitigation strategies to minimize the resulting expected losses. Potentially effective mitigation approaches include resetting of passwords, changing cryptographic private keys, reinstalling servers, or reinstantiating virtual servers. These approaches are often effective at resetting the resource to a known safe state, but they reveal little about past compromises. For example, if a server is reinstalled, knowledge of if and when a compromise occured may be lost. Likewise, resetting a password does not reveal any information about the confidentiality of previous passwords.

Covert (and non-covert) attacks can be distinguished in another dimension by the extent to which the attack is targeted (or customized) for a particular organization [4,7]. Approaches related to cyber-espionage are important examples of targeted attacks, and require a high effort level customized to a specific target [14]. A typical example of a non-targeted covert attack is the recruitment of a computer into a botnet via drive-by-download. Such attacks are relatively low effort, and do not require a specific target. Further, they can often be scaled to affect many users for marginal additional cost [7]. See Table 1 for a comparison between targeted and non-targeted attacks.

Table 1. Comparison of Targeted and Non-Targeted Attacks

	Targeted	Non-Targeted
Number of attackers	low	high
Number of targets	low	high
Effort required for each attack	high	low
Success probability of each attack	high	low

The targeted nature of an attack also matters to the defender, because targeted and non-targeted attacks do different types of damage. For example, a targeting attacker might use a compromised computer system to access an organization's secret e-mails, which may potentially cause enormous economic damage; while a non-targeting attacker might use the same compromised machine to send out spam, causing different types of damage.

The presence of both targeted and non-targeted covert attacks presents an interesting dilemma for a medium-profile target to choose a mitigation strategy against covert attacks. Strategies which are optimal against non-targeted attacks may not be the best choice against targeted attacks. At the same time, mitigation strategies against targeted attacks may not be economically cost-effective against only non-targeting attackers.

To address this dichotomy, we present a game in which a defender must vie for a contested resource that is subject to both targeted attacks from a strategic attacker, and non-targeted covert attacks from a large set of non-strategic attackers. We identify Nash equilibria in the simultaneous game, and subgame perfect equilibria in the sequential game with defender leading. The optimal mitigation strategies for the defender against these combined attacks lend insights to policy makers regarding renewal requirements for passwords and cryptographic keys.

The rest of the paper is organized as follows. In Section 2, we review related work. We define our game-theoretic model in Section 3, and we give analytical results for this model in Section 4. In Section 5, we present numerical and graphical observations; we conclude in Section 6.

2 Related Work

2.1 Security Economics and Games of Timing

Research studies on the economics of security decision-making primarily investigate the optimal or bounded rational choice between different canonical options to secure a resource (i.e., protection, mitigation, risk-transfer), or the determination of the optimal level of investment in one of these security dimensions. In our own work, we have frequently contributed to the exploration of these research objectives (see, for example, [6,9,8]). Further, these studies have been thoroughly summarized in a recent review effort [11].

Another critical decision dimension for successfully securing resources is the consideration of *when* to act to successfully thwart attacks. Scholars have studied such time-related aspects of tactical security choices since the cold war era by primarily focusing on zero-sum games called *games of timing* [2]. The theoretical contributions on some subclasses of these games have been surveyed by [17].

2.2 FlipIt: Modeling Targeted Attacks

Closely related to our study is the FlipIt model which identifies optimal timing-related security choices under targeted attacks [19]. In FlipIt, two players compete for a resource that generates a payoff to the current owner. Players can make costly moves (i.e., "flips") to take ownership of the resource, however, they have to make moves under incomplete information about the current state of possession.

In the original FlipIt paper, equilibria and dominant strategies for simple cases of interaction are studied [19]. Other groups of researchers have worked on extensions [16,12]. For example, Laszka et al. extended the FlipIt game to the case with multiple resources. In addition, the usefulness of the FlipIt game has been investigated for various application scenarios [3,19]. We detail the difference of our model to FlipIt in Section 3.3. The current study generalizes our previous work which was restricted to exponential distributions for the attack time [13].

FlipIt has been studied in experiments in which human participants were matched with computerized opponents [15]. This work has also been extended to consider different interface feedback modalities [18]. The results complement the theoretical work by providing evidence for the difficulty to identify optimal choices when timing is the critical decision dimension.

3 Model Definition

We model the covert compromise scenario as a randomized, one-shot, non-zero-sum game. For a list of symbols used in our model, see Table 2. The player who is the rightful owner of the resource is called the defender, while the other players are called attackers. The game starts at time $t = 0$ with the resource being uncompromised, and it is played indefinitely as $t \to \infty$. We assume that time is continuous.

Table 2. List of Symbols

C_D	move cost for the defender
C_A	move cost for the targeting attacker
B_A	benefit received per unit of time for the targeting attacker
B_N	benefit received per unit of time for the non-targeting attackers
F_A	cumulative distribution function of the attack time for the targeting attacker
λ_N	rate of the non-targeted attacks' arrival

We let D, A, and N denote the defender, the targeting attacker, and the non-targeting attackers, respectively. At any time instance, player i may make a move, which costs her C_i. (Note that, for attackers, we will use the words attack and move synonymously). When the defender makes a move, the resource becomes uncompromised immediately for every attacker. When the targeting attacker makes a move, she starts her attack, which takes some random amount of time. If the defender makes a move while an attack is in progress, the attack fails. We assume that the time required by a targeted attack follows the same distribution every time. Its cumulative distribution function is denoted by F_A, and is subject to $F_{A_i}(0) = 0$. In practice, this distribution can be based on industry-wide beliefs, statistics of previous attacks, etc.

The attackers' moves are stealthy; i.e., the defender does not know when the resource became compromised or if it is compromised at all. On the other hand, the defender's moves are non-stealthy. In other words, the attackers learn immediately when the defender has made a move.

The cost rate for player i up to time t, denoted by $c_i(t)$, is the number of moves per unit of time made by player i up to time t, multiplied by the cost per move C_i.

For attacker $i \in \{A, N\}$, the benefit rate $b_i(t)$ up to time t is the fraction of time up to t that the resource has been compromised by i, multiplied by B_i. Note that if multiple attackers have compromised the resource, they all receive benefits until the defender's next move. For the defender D, the benefit rate $b_D(t)$ up to time t is $-\sum_{i \in \{A,N\}} b_i(t)$. The relation between defender and attacker benefits implies that the game would be zero-sum if we only considered the players' benefits. Because our players' payoffs also consider move costs, our game is *not* zero-sum. Player i's payoff is defined as

$$\liminf_{t \to \infty} \; b_i(t) - c_i(t) \; . \tag{1}$$

It is important to note that the asymptotic benefit rate $\liminf_{t \to \infty} b_i(t)$ of attacker i is equal to the probability that i has the resource compromised at a random time instance, multiplied by B_i. For a discussion on computing the payoffs for the key strategy profiles in this paper, see the extended version of this paper available on the authors' websites.

3.1   Types of Strategies for the Defender and the Targeting Attacker

Not Moving. A player can choose to *never move*. While this might seem counter-intuitive, it is actually a best-response if the expected benefit from making a move is always less than the cost of moving.

Adaptive Strategies for the Targeting Attacker. Let $\mathcal{T}(n) = \{T_0, T_1, \ldots, T_n\}$ denote the move times of the defender up to her nth move (or in the case of $T_0 = 0$, the start of the game). The attacker uses an *adaptive strategy* if she waits for $W(\mathcal{T}(n))$ time until making a move after the defender's nth move (or after the start of the game), where W is a non-deterministic function. If the defender makes her $n+1$st move before the chosen wait time is up, the attacker chooses a new wait time $W(\mathcal{T}(n+1))$, which also considers the new information that is the defender's $n+1$st move time. This class is a simple representation of all the rational strategies available to an attacker, since W can depend on all the information that the attacker has, and we do not have any constraints on W.

Renewal Strategies. Player i uses a *renewal strategy* if the time intervals between consecutive moves are identically distributed independent random variables, whose distribution is given by the cumulative function F_{R_i}. Renewal strategies are well-motivated by the fact that the defender is playing blindly; thus, she has the same information available after each move. So it makes sense to use a strategy which always chooses the time until her next flip according to the same distribution.

Periodic Strategies. Player i uses a *periodic strategy* if the time intervals between consecutive moves are identical. This period is denoted by δ_i. Periodic strategies are a special case of renewal strategies.

3.2 Non-Targeted Attacks

Suppose that there are N non-targeting attackers. In practice, N is very large, but the expected number of attacks in any time interval is finite. Hence, as N goes to infinity, the probability that a given non-targeting attacker targets the defender approaches zero. Since non-targeting attackers operate independently of each other, the number of successful attacks in any time interval depends solely on the length of the interval. Thus, the arrival of *successful non-targeted attacks* follows a *Poisson process*.

Furthermore, since the economic decisions of the non-targeting attackers depend on a very large pool of possible targets, the effect of the defender's behaviour on the non-targeting attackers' strategies is negligible. Thus, the non-targeting attackers' strategies (that is, the arrival rate of the Poisson process) can be considered exogenously given. We let λ_N denote the expected number of arrivals that occur per unit of time; and we model all the non-targeting attackers together as a single attacker whose benefit per unit of time is B_N.

3.3 Comparison to FlipIt

Even though our game-theoretic model resembles FlipIt in many ways, it differs in three key assumptions. First, we assume that the defender's moves are *non-stealthy*. The motivation for this is that, when the attacker receives benefits from continuously exploiting the compromised resource, she should know whether she has the resource compromised or not. For example, if the attacker uses the compromised password of an account to regularly spy on its e-mails, she will learn of a password reset immediately when she tries to access the account. Second, we assume that the targeting attacker's moves are *not instantaneous*, but take some time. The motivation for this is that an attack requires some effort to be carried out in practice. Furthermore, the time required for a successful attack may vary, which we model using a random variable for the attack time. Third, we assume that the defender faces *multiple attackers*, not only a single one.

Moreover, to the authors' best knowledge, papers published on FlipIt so far give analytical results only on a very restricted set of strategies. In contrast, we completely describe our game's equilibria and give optimal defender strategies based on very mild assumptions, which effectively do not limit the power of players (see the introduction of Section 4).

4 Analytical Results

In this section, we provide analytical results based on our model. We start with a discussion of the players' strategies.

Recall that the defender has to play blindly, which means that she has the same information available after each one of her moves. Consequently, it makes sense for her to choose the time until her next flip according to the same distribution each time. In other words, a rational defender can restrict herself to using only renewal strategies.

Now, if the defender uses a renewal strategy, the time of her next move depends only on the time elapsed since her last move T_n, and the times of her previous moves (including T_n) are irrelevant to the future of the game. Therefore, it is reasonable to assume that the attacker's response strategy to a renewal strategy also does not depend on $T_0, T_1, \ldots, T_n$. For the remainder of the paper, when the defender plays a renewal strategy, the attacker uses a fixed probability distribution – given by the density function f_W – over her wait times for when to begin her attack. Note that it is clear that there always exists a best-response strategy of this form for the attacker against a renewal strategy.

Since the attacker always waits an amount of time that is chosen according to a fixed probability distribution after the defender's each move, the amount of time until the resource would be successfully compromised after the defender's move also follows a fixed probability distribution. Let S be the random variable measuring the time after the defender has moved until the attacker's attack would finish. The probability density function f_S of S can be computed as

$$f_S(s) = \int_{w=0}^{s} f_W(w) \int_{a=0}^{(s-w)} f_A(a) \, da \, dw \ . \tag{2}$$

Finally, we let F_S denote the cumulative distribution function of S.

4.1   Best Responses

Defender's Best Response. We begin our analysis with finding the defender's best-response strategy.

Lemma 1. *Suppose that the non-targeted attacks arrive according to a Poisson process with rate λ_N, and the targeting attacker uses an adaptive strategy with a fixed wait time distribution F_W. Then,*
  – not moving is the only best response if $C_D = \mathcal{D}(l)$ has no solution for $l > 0$, where

$$\mathcal{D}(l) = B_A \left(l F_S(l) - \int_{s=0}^{l} F_S(s) \, ds \right) + B_N \left(-l e^{-\lambda_N l} + \frac{1 - e^{-\lambda_N l}}{\lambda_N} \right) \ ; \tag{3}$$

  – the periodic strategy whose period is the unique solution to $C_D = \mathcal{D}(l)$ is the only best response otherwise.

The proof is available in the paper's extended version on the authors' websites.

Even though we cannot express the solution of $C_D = \mathcal{D}(l)$ in closed form, it can be easily found using numerical methods, as the right hand side is continuous and increasing.[1] Note that all the equations presented in the subsequent lemmas and theorems of this paper can also be solved by applying numerical methods.

Lemma 2. *Suppose that the non-targeted attacks arrive according to a Poisson process with rate λ_N, and the targeting attacker never attacks. Then,*

---

[1] We show in the proof of the lemma that the right hand side is increasing in l.

– *not moving is the only best response if $C_D = \mathcal{D}^N(l)$ has no solution for $l > 0$, where*

$$\mathcal{D}^N(l) = B_N \left(-le^{-\lambda_N l} + \frac{1 - e^{-\lambda_N l}}{\lambda_N} \right) ; \qquad (4)$$

– *the periodic strategy whose period is the unique solution to $C_D = \mathcal{D}^N(l)$ is the only best response otherwise.*

Proof. Follows readily from the proof of Lemma 1 with the terms belonging to the targeting attacker omitted everywhere. □

Observe that $\mathcal{D}(0) = \mathcal{D}^N(0) < 0$ and $\mathcal{D}(l) \geq \mathcal{D}^N(l)$. Consequently, $C_D = \mathcal{D}(l)$ has a solution whenever $C_D = \mathcal{D}^N(l)$ has one. Furthermore, if both have solutions, the solution of $C_D = \mathcal{D}(l)$ is less than or equal to the solution of $C_D = \mathcal{D}^N(l)$. In other words, the *defender is more likely to keep moving if there is a threat of targeted attacks*, and she *will move at least as frequently as she would if there was no targeting attacker.*

Attacker's Best Response. We continue our analysis with finding the attacker's best-response strategy.

Lemma 3. *Against a defender who uses a periodic strategy with period δ_D,*
– *never attacking is the only best response if $C_A > \mathcal{A}(\delta_D)$, where*

$$\mathcal{A}(\delta) = B_A \int_{a=0}^{\delta} F_A(a) da ; \qquad (5)$$

– *attacking immediately after the defender has moved is the only best response if $C_A < \mathcal{A}(\delta_D)$;*
– *both not attacking and attacking immediately are best responses otherwise.*

The proof is available in the paper's extended version on the authors' websites.

The lemma shows that the targeting attacker should either attack immediately or not attack at all, but she should never wait to attack. For the never attack strategy, we already have the defender's best response from Lemma 2. For the attacking immediately strategy, the defender can determine the optimal period of her strategy *solely based on the distribution of A*, which is an exogenous parameter of the game. More formally, the defender's best response is not to move if $C_D = \mathcal{D}^A(l)$ has no solution, and it is a periodic strategy whose period is the unique solution to $C_D = \mathcal{D}^A(l)$ otherwise, where

$$\mathcal{D}^A(l) = B_A \left(lF_A(l) - \int_{a=0}^{l} F_A(a)\, da \right) + B_N \left(-le^{-\lambda_N l} + \frac{1 - e^{-\lambda_N l}}{\lambda_N} \right) . \qquad (6)$$

This follows readily from Lemma 1 by substituting F_S for F_A.[2]

---

[2] Recall that S was defined as the sum of the waiting time W, which is always zero in this case, and the attack time A.

4.2   Nash Equilibria

Based on the previous lemmas, we can describe all the equilibria of the game (if there are any) as follows.

Theorem 1. *Suppose that the defender uses a renewal strategy, the targeting attacker uses an adaptive strategy, and the non-targeted attacks arrive according to a Poisson process. Then, the game's equilibria can be described as follows.*

1. *If $C_D = \mathcal{D}^A(l)$ does not have a solution for l, then there is a unique equilibrium in which the defender does not move and in which the targeting attacker moves once at the beginning of the game.*
2. *If $C_D = \mathcal{D}^A(l)$ does have a solution δ_D for l:*
    (a) *If $C_A \leq \mathcal{A}(\delta_D)$, then there is a unique equilibrium in which the defender plays a periodic strategy with period δ_D, and the targeting attacker moves immediately after each of the defender's moves.*
    (b) *If $C_A > \mathcal{A}(\delta_D)$,*
        i. *if $C_D = \mathcal{D}^N(l)$ has a solution δ'_D for l, and $C_A \geq \mathcal{A}(\delta'_D)$, then there is a unique equilibrium in which the defender plays a periodic strategy with period δ_D, and the targeting attacker never moves;*
        ii. *otherwise, there is no equilibrium.*

The proof is available in the paper's extended version on the authors' websites. For an illustration of the hierarchy of the theorem's criteria, see Figure 1. Finally, recall that a discussion on the payoffs can also be found in the extended version.

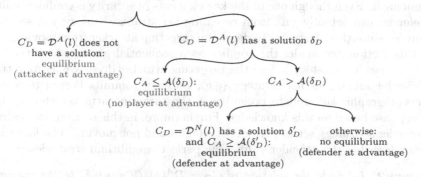

$C_D = \mathcal{D}^A(l)$ does not have a solution: equilibrium (attacker at advantage)

$C_D = \mathcal{D}^A(l)$ has a solution δ_D

$C_A \leq \mathcal{A}(\delta_D)$: equilibrium (no player at advantage)

$C_A > \mathcal{A}(\delta_D)$

$C_D = \mathcal{D}^N(l)$ has a solution δ'_D and $C_A \geq \mathcal{A}(\delta'_D)$: equilibrium (defender at advantage)

otherwise: no equilibrium (defender at advantage)

Fig. 1. Illustration for the hierarchy of criteria in Theorem 1

In the first case, the attacker is at an overwhelming advantage, as the relative cost of defending the resource is prohibitively high. Consequently, the defender simply "gives up," as any effort to protect the resource is not profitable, and the attacker will eventually have the resource compromised indefinitely (see Figure 2 for an illustration). In the second case, no player is at an overwhelming advantage. Both players are actively moving, and the resource gets compromised and uncompromised from time to time. In the third and fourth cases, the defender is at an overwhelming advantage. However, this does not necessarily lead to an equilibrium. If the defender moves with a sufficiently high rate, she makes moving unprofitable for the targeting attacker. But if the targeting attacker decides

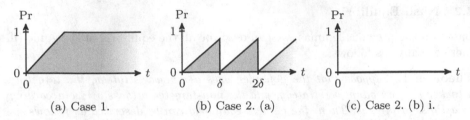

(a) Case 1.   (b) Case 2. (a)   (c) Case 2. (b) i.

Fig. 2. The probability that the targeting attacker has compromised the resource (vertical axis) as a function of time (horizontal axis) in various equilibria (see Theorem 1 for each case). Note that these are just examples, the actual shapes of the functions depend on F_A.

not to move, then the defender switches to a lower move rate, which is optimal against only non-targeted attacks. However, once the defender switches to the lower move rate, it might again be profitable for the targeting attacker to move, which would in turn trigger the defender to switch back to the higher move rate.

4.3   Sequential Game: Deterrence by Committing to a Strategy

So far, we have modeled the mitigation of covert compromises as a simultaneous game. This is realistic for scenarios where neither the defender nor the targeting attacker can learn the opponent's strategy choice in advance. However, in practice, the defender can easily let the targeting attacker know her strategy by publicly announcing it. Even though one of the key elements of security is confidentiality, the defender can actually gain from revealing her strategy – as we will show in Section 5 – since this allows her to deter the targeting attacker from moving.

In this section, we model the conflict as a sequential game, where the defender chooses her strategy before the targeting attacker does. We assume that the defender announces her strategy (e.g., publicly commits herself to a certain cryptographic-key update policy) and the targeting attacker chooses her best response based on this knowledge. Furthermore, in this section, we restrict the defender's strategy set to periodic strategies and not moving. The following theorem describes the defender's subgame perfect equilibrium strategies.

Theorem 2. *Let δ_1 be the solution of $C_D = \mathcal{D}^A(\delta)$ (if any), δ_2 be the maximal period δ for which $C_A = \mathcal{A}(\delta)$, and δ_3 be the solution of $C_D = \mathcal{D}^N(\delta)$ (if any). In a subgame perfect equilibrium, the defender's strategy is one of the following:*
– not moving,
– periodic strategies with periods $\{\delta_1, \delta_2, \delta_3\}$.

The proof is available in the paper's extended version on the authors' websites.

Based on the above theorem, one can easily find all subgame perfect equilibria by iterating over the above strategies and, for each strategy, computing the targeting attacker's best response using Lemma 3, and finally comparing the defender's payoffs to find her equilibrium strategy (or strategies). Note that, for each case of Theorem 1, the set of possible equilibrium strategies in Theorem 2 could be restricted further. For example, in Case 2. (b) i., the only subgame perfect equilibrium is the defender moving periodically with δ'_D and the targeting attacker never moving. We defer the remaining cases to future work.

5   Numerical Illustrations

In this section, we present numerical results on our game. For the illustrations, we instantiate our model with the *exponential distribution* as the distribution of the attack time. For rate parameter λ_A, the cumulative distribution function of the exponential distribution is $F_A(a) = 1 - e^{-\lambda_A a}$. For the remainder of this section, unless indicated otherwise, the parameters of the game are $C_D = C_A = B_A = \lambda_A = \lambda_N = 1$ and $B_N = 0.1$. Finally, we refer to the simultaneous-game Nash equilibria simply as equilibria, and we refer to the defender's subgame perfect equilibrium strategies as optimal strategies (because they maximize the defender's payoff given that the targeting attacker will play her best response).

First, in Figure 3, we study the effects of varying the value of the resource, that is, the unit benefit B_A received by the targeting attacker. Figure 3a shows the equilibrium payoffs as functions of B_A (the defender's period for the same

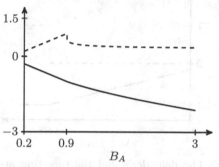

(a) The defender's and the targeting attacker's payoffs (solid and dashed lines, respectively) in equilibria as functions of B_A

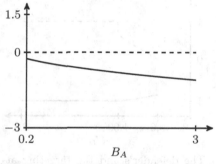

(b) The defender's and the targeting attacker's payoffs for the defender's optimal strategy as functions of B_A

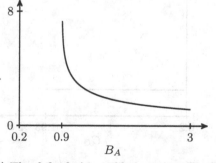

(c) The defender's equilibrium period as a function of B_A

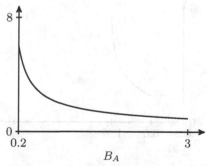

(d) The defender's optimal period as a function of B_A

Fig. 3. The effects of varying the unit benefit B_A of the targeting attacker

setup is shown by Figure 3c). The defender's payoff is strictly decreasing, which is not surprising: the more valuable the resource is, the higher the cost of security is. The targeting attacker's payoff, on the other hand, starts growing linearly, but then suffers a sharp drop, and finally converges to a finite positive value.

For lower values ($B_A < 0.9$), the defender does not protect the resource, as it is not valuable enough. Accordingly, Figure 3c shows no period for this range, and the targeting attacker's payoff is the value of the resource B_A. However, once the value reaches about 0.9, the defender starts protecting the resource. Hence, the targeting attacker's payoff drops as she no longer always has the resource. For higher values, the defender balances between losses and costs, which means that the time the resource is compromised decreases as its value increases.

Figure 3b shows the payoffs for the defender's optimal strategy as functions of B_A (the optimal period is shown by Figure 3d). The figure shows that the defender's strategy for this range of B_A is always to deter the targeting attacker (hence, the targeting attacker's payoff is zero). To achieve this, the defender is using a strictly shorter period than her equilibrium period. Interestingly, the defender's payoff is much higher compared to her equilibrium payoff.

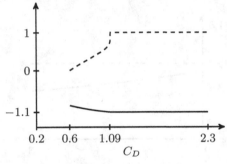

(a) The defender's and the targeting attacker's payoffs (solid and dashed lines, respectively) in equilibria as functions of C_D

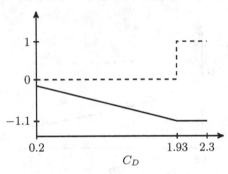

(b) The defender's and the targeting attacker's payoffs for the defender's optimal strategy as functions of C_D

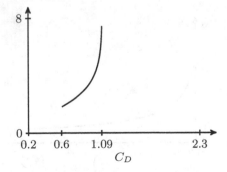

(c) The defender's equilibrium period as a function of C_D

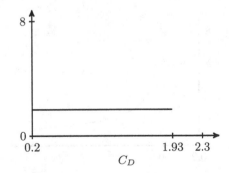

(d) The defender's optimal period as a function of C_D

Fig. 4. The effects of varying the defender's move cost C_D

In Figure 4, we study the effects of varying the defender's move cost C_D. Figure 4a shows the equilibrium payoffs as functions of C_D (the defender's period for the same setup is shown by Figure 4c). The figure shows that the defender's payoff is decreasing, while the targeting attacker's payoff is increasing, which is unsurprising: the more costly it is to defend, the greater the attacker's advantage.

For lower costs ($C_D < 0.6$), the defender is at an overwhelming advantage, but there is no equilibrium (Case 2. (b) ii. of Theorem 1). For costs between 0.6 and 1.09, no player is at an overwhelming advantage; hence, both players move from time to time. For higher costs, the targeting attacker is at an overwhelming advantage. In this case, the defender never moves, while the attacker moves once. Hence, their payoffs are $B_A + B_N = -1.1$ and $B_A = 1$, respectively.

Figure 4b shows the payoffs for the defender's optimal strategy as a function of C_D (the optimal period is shown by Figure 4d). The defender's optimal strategy for move costs lower than 1.93 is to deter the targeting attacker. Hence, the targeting attacker's payoff is zero. The defender's payoff decreases linearly as the cost of deterrence increases. Again, we see that the defender's payoff is much higher than her equilibrium payoff. However, for higher move costs, she must give up defending the resource, as in her equilibrium strategy for this range.

6 Conclusions

In this paper, we studied the mitigation of both targeted and non-targeted covert attacks. As our main result, we found that periodic mitigation is the most effective strategy against both types of attacks and their combinations. Considering the simplicity of this strategy, our result can be surprising, but it also serves as a theoretical justification for the prevalent periodic password and cryptographic-key renewal practices. Moreover, this result contradicts the lesson learned from the FlipIt model [19], which suggests that a defender facing an adaptive attacker should use an unpredictable, randomized strategy.

Further, a defender is more willing to commit resources to defensive moves when being threatened by non-targeted and targeted attacks at the same time. This stands in contrast to the result that a high level of either threat type can force the defender to abandon defensive activities altogether.

Finally, we observed that there is an important difference between the defender's simultaneous and sequential (i.e., optimal) equilibrium strategies, both in the lengths of the periods and the resulting payoffs. Thus, a defender should not try to keep her strategy secret, but rather publicly commit to it.

Acknowledgements. We gratefully acknowledge the support of the Penn State Institute for CyberScience and the National Science Foundation under ITR award CCF-0424422 (TRUST). We also thank the reviewers for their comments on an earlier draft of the paper.

References

1. Bencsath, B., Pek, G., Buttyán, L., Felegyhazi, M.: The cousins of Stuxnet: Duqu, Flame, and Gauss. Future Internet 4(4), 971–1003 (2012)

2. Blackwell, D.: The noisy duel, one bullet each, arbitrary accuracy. Technical report, The RAND Corporation, D-442 (1949)
3. Bowers, K., van Dijk, M., Griffin, R., Juels, A., Oprea, A., Rivest, R., Triandopoulos, N.: Defending against the unknown enemy: Applying FLIPIT to system security. In: Grossklags, J., Walrand, J. (eds.) GameSec 2012. LNCS, vol. 7638, pp. 248–263. Springer, Heidelberg (2012)
4. Casey, E.: Determining intent - Opportunistic vs. targeted attacks. Computer Fraud & Security 2003(4), 8–11 (2003)
5. ESET Press Center. ESET and Sucuri uncover Linux/Cdorked.A: The most sophisticated Apache backdoor (2013), http://www.eset.com/int/about/press/articles/article/eset-and-sucuri-uncover-linuxcdorkeda-apache-webserver-backdoor-the-most-sophisticated-ever-affecting-thousands-of-web-sites/
6. Grossklags, J., Christin, N., Chuang, J.: Secure or insure? A game-theoretic analysis of information security games. In: Proc. of the 17th International World Wide Web Conference (WWW), pp. 209–218 (2008)
7. Herley, C.: The plight of the targeted attacker in a world of scale. In: 9th Workshop on the Economics of Information Security, WEIS (2010)
8. Johnson, B., Böhme, R., Grossklags, J.: Security games with market insurance. In: Baras, J.S., Katz, J., Altman, E. (eds.) GameSec 2011. LNCS, vol. 7037, pp. 117–130. Springer, Heidelberg (2011)
9. Johnson, B., Grossklags, J., Christin, N., Chuang, J.: Are security experts useful? Bayesian nash equilibria for network security games with limited information. In: Gritzalis, D., Preneel, B., Theoharidou, M. (eds.) ESORICS 2010. LNCS, vol. 6345, pp. 588–606. Springer, Heidelberg (2010)
10. Kaspersky Lab. Gauss (2012), http://www.kaspersky.com/gauss
11. Laszka, A., Felegyhazi, M., Buttyán, L.: A survey of interdependent security games. Technical Report CRYSYS-TR-2012-11-15, CrySyS Lab, Budapest University of Technology and Economics (November 2012)
12. Laszka, A., Horvath, G., Felegyhazi, M., Buttyan, L.: FlipThem: Modeling targeted attacks with FlipIt for multiple resources. Technical report, Budapest University of Technology and Economics (2013)
13. Laszka, A., Johnson, B., Grossklags, J.: Mitigation of targeted and non-targeted covert attacks as a timing game. In: Proc. of GameSec 2013 (2013)
14. Laszka, A., Johnson, B., Schöttle, P., Grossklags, J., Böhme, R.: Managing the weakest link: A game-theoretic approach for the mitigation of insider threats. In: Crampton, J., Jajodia, S., Mayes, K. (eds.) ESORICS 2013. LNCS, vol. 8134, pp. 273–290. Springer, Heidelberg (2013)
15. Nochenson, A., Grossklags, J.: A behavioral investigation of the FlipIt game. In: 12th Workshop on the Economics of Information Security, WEIS (2013)
16. Pham, V., Cid, C.: Are we compromised? Modelling security assessment games. In: Grossklags, J., Walrand, J. (eds.) GameSec 2012. LNCS, vol. 7638, pp. 234–247. Springer, Heidelberg (2012)
17. Radzik, T.: Results and problems in games of timing. In: Statistics, Probability and Game Theory: Papers in Honor of David Blackwell. Lecture Notes-Monograph Series, Statistics, vol. 30, pp. 269–292 (1996)
18. Reitter, D., Grossklags, J., Nochenson, A.: Risk-seeking in a continuous game of timing. In: Proc. of the 13th International Conference on Cognitive Modeling (ICCM), pp. 397–403 (2013)
19. van Dijk, M., Juels, A., Oprea, A., Rivest, R.: FlipIt: The game of "stealthy takeover". Journal of Cryptology 26, 655–713 (2013)

Characterization of Truthful Mechanisms for One-Dimensional Single Facility Location Game with Payments

Pinyan Lu[1] and Lan Yu[2],[*]

[1] Microsoft Research, China
pinyanl@microsoft.com
[2] Division of Mathematical Sciences, School of Physical and Mathematical Sciences,
Nanyang Technological University, Singapore
YULA0001@ntu.edu.sg

Abstract. In a one-dimensional single facility location game, each player resides at a point on a straight line (his location); the task is to decide the location of a single public facility on the line. Each player derives a nonnegative cost, which is a monotonically increasing function of the distance between the location of the facility and himself, so he may misreport his location to minimize his cost. It is desirable to design an incentive compatible allocation mechanism, in which no player has an incentive to misreport.

Offering/Charging payments to players is a usual tool for a mechanism to adjust incentives. Our game setting without payment is equivalent to the voting setting where voters have single-peaked preferences. A complete parametric characterization of incentive compatible allocation mechanisms in this setting was given by [17], while the problem for games with payments is left open. We give a characterization for the case of linear and strictly convex cost functions by showing the sufficiency of weak-monotonicity, which, more importantly, implies an interesting monotone triangular structure on every single-player subfunction.

1 Introduction

People live in communities, where public facilities need to be built to serve the residents. The location of a public facility is one of the most important decisions to make since the convenience of accessing a facility is mainly affected by the distance to the facility. Although the social goal is to provide convenient service to the whole community, tradeoffs have to be made as people reside at different locations. Hence the well-known public facility location problem has been a long-lasting attraction to researchers.

Conventionally, the convenience of accessing a facility is quantified by the negation of a cost, indicating the effort needed to reach the facility. The cost can be calculated through a cost function C on the distance d to the facility, which grows with the distance. For different people or circumstances, the growth

* Supported by National Research Foundation (Singapore) under grant RF2009-08.

Y. Chen and N. Immorlica (Eds.): WINE 2013, LNCS 8289, pp. 333–346, 2013.

rate may vary; the cost function may be linear, convex, concave, or a combination of the three. A linear cost function $C(d) = \alpha d$ is the simplest case, which has a stable growth rate α. A convex cost function has higher growth rate for longer distances (such as the taxi fare), which captures the nature that people become exhausted after long distances. In contrast, a concave cost function has lower growth rate for longer distances (such as the subway fare). This represents situations where people become adapted after long distances.

In algorithm design, people solve optimization problems, such as to minimize the total/maximum cost of the people served in the community. There are also interesting variations in the number of facilities, the cost function, and the location space (discrete or continuous). In our paper, we consider one of the simplest variations: one-dimensional single facility location, where the location space is the one-dimensional real line $\mathbb{R}$ and only a single facility needs to be built. All three types of cost functions are investigated.

However, optimization is not the focus of our paper: We investigate the game-theoretic setting where each resident is modeled as a player in the facility location game intending to maximize his *utility*, i.e., his overall benefit. Each player i's true location r_i is his private information, and the location of the facility is chosen based on the reported locations from all players $\mathbf{x} = (x_1, \ldots, x_n)$. Hence a player may have an incentive to misreport his location to make the facility closer to him. Naturally, to achieve good public service, we would like a solution where no player has an incentive to misreport. This property is called *incentive compatibility*, or simply *truthfulness*, which is the main solution concept of the field of *mechanism design* [21,20].

Offering/Charging payments to players is a usual tool in mechanism design to adjust incentives. Our work allows solutions with payments. Thus, an *allocation mechanism*, i.e., a solution to our facility location game, is composed of an *allocation function* and a *payment function* vector. The allocation function f takes the reported locations of all players $\mathbf{x}$ as input and outputs a location y of the public facility; The payment function vector $\mathbf{p}$ contains a payment function p_i as its ith component for each player i. Function p_i takes the same input as f, and assigns a (positive or negative) payment to player i. The setting without payments restricts $\mathbf{p} \equiv \mathbf{0}$.

Under a mechanism $(f, \mathbf{p})$, the utility u_i of player i is his payment under reported locations $p_i(\mathbf{x})$ minus his cost under true location $C(|f(\mathbf{x}) - r_i|)$. The mechanism is public knowledge, and a truthful mechanism ensures truth-telling in the following sense: No matter what other players may report, for each player, given the mechanism and other players' reports, reporting his true location always maximizes his utility.

The goal of our work is to characterize the set of truthful allocation functions, i.e., allocation functions f for which there exists a payment function vector $\mathbf{p}$ such that $(f, \mathbf{p})$ is truthful. Characterization of truthful functions is meaningful in mechanism design, since it allows mechanism designers to focus on the function and not to worry about payments, whose existence is already guaranteed by the characterization. Furthermore, in most applications, there are other desirable properties the allocation function should also satisfy, such as fairness or efficiency. A good characterization provides a useful description of the set of truthful functions for a designer to start with to work further on the other properties, or to prove the impossibility to satisfy other properties simultaneously. A great number of results in mechanism design follow this path [3,10,22].

For the game setting without payment, a complete parametric characterization of truthful allocation mechanisms was given by Moulin [17]. (One unnecessary assumption in the proof is dropped by Barberà and Jackson [5], and Sprumont [26].) Observe that, without payment, a player's utility is simply the negation of his cost, and the definition of truthfulness is only concerned with the comparison of the cost of two locations. Moreover, the cost is *single-peaked*: it reaches its minimum 0 at the player's true location and increases monotonically on both sides; the formula of the cost function becomes irrelevant. In fact, this setting is essentially equivalent to the voting setting where voters have single-peaked preferences. Moulin considered the voting setting, and hence his characterization, a parametric representation of truthful allocations is called a *generalized median voter scheme*. It is an extension of the function that selects the median voter's preference peak, which, interpreted into our setting, is the median location out of the locations of all players.

The characterization for games with payments is left open, which is what we studied in this paper. This question is interesting in its own right: In real life, some facility builders are willing to provide payments. For example, when a company chooses the location of its office, employees living far away from the office are subsidized. On the other hand, the generalized median voter scheme is very restricted, and does not satisfy certain other desirable properties, such as fairness or cost minimization. For example, the average function of all agents' locations minimizes the sum of squares of agents' cost, which is a widely used objective function in operational research to balance the social welfare and fairness. This nice function is not truthfully implementable without payment. However, by our characterization, it can be made truthful with payment, so designers may want to consider investing some money to realize this allocation function.

1.1   Our Contributions

It turns out that the set of truthful allocation functions with payments is a much wider class than the generalized median voter scheme. We show that weak monotonicity, an easily proven necessary condition for truthfulness [20], is also sufficient in this setting for linear and strictly convex cost functions. There has been a series of works on characterization of truthfulness for various settings [23,19,3,15,24,27,18,16,11,16,14,9,7,1,4,2], and most of them involve weak monotonicity, or some other kind of monotonicity properties. It turns out that the characterization results are closely related to the *domain* of the problem setting, i.e., the set of all possible *valuation functions* on the set of outcomes. In our setting, the domain of player i is $\{-C(|y - r_i|) : r_i \in \mathbb{R}\}$, where each element $-C(|y - r_i|)$ is a single-peaked function mapping each location of facility $y \in \mathbb{R}$ to the valuation (convenience) player i derives when his true location is r_i.

Evidently, the domain of our setting is restricted, so Roberts' Theorem that every truthful function in an unrestricted domain is an affine maximizer [23] does not apply. Furthermore, our domain is not a special case of the convex domain or single-parameter domain for which the sufficiency of weak monotonicity is proved [19,3,24]. Clearly, our domain is not convex, and an easy way to see this is that the average of two valuation functions no longer has peak of value 0. On the other hand, though the domain of each player i is associated with a single parameter r_i, the single-peaked function does not conform to the function in the definition of single-parameter domain. It is interesting that none of the previous characterization results covers our setting although it is very simple and realistic. In particular, most of the previous result involves some kind of convexity: either the valuation is convex or the type space is convex. The fact that there are infinite (uncountable) many alternatives also makes the result interesting since most of previous results assume a finite set of alternatives.

On the other hand, from the mechanism design point of view, weak monotonicity, as a condition on any two locations, is not directly applicable; it is more desirable to derive its equivalent properties that describe global features of the allocation function (usually on every single player subfunction), from which truthful payments can also be described. The characterization of the single-parameter domain in [19,3] is successful in this aspect: Various mechanisms for specific settings with different objectives are derived based on this characterization [19,3,12,10,25,6]. For our problem, we also succeed in providing a characterization of this kind. In fact, the sufficiency of weak monotonicity is shown indirectly through the correctness of this characterization.

More specifically, our characterization results are presented in three steps: In Section 3, we derive some properties from weak-monotonicity on every single player subfunction:[1] For strictly convex cost functions, the allocation function is simply monotonically non-decreasing in the usual sense; Linear cost functions imply a weaker condition, which we call *partially monotonically non-decreasing*. As shown in Section 4, this condition implies a *monotone triangular partition*, which graphically divides the allocation function into pieces each within a triangle, and the set of triangles obeys some "monotone" property. For strictly convex cost functions, this part is evident as the allocation function is monotone. Finally we provide a payment function with respect to a monotone triangular partition and prove truthfulness in Section 5 for linear and strictly convex cost functions respectively.

In summary, here are our main characterization results for one-dimensional single facility location game with payments (which also apply to the setting where the location space is a closed interval):

Theorem 1. *For linear and strictly convex cost functions, an allocation function is truthful if and only if it satisfies weak-monotonicity.*

---

[1] A single player subfunction on player i is the allocation function restricted to some fixed reported locations of players other than i. See Section 2 for a formal definition.

Theorem 2. *For linear cost functions, an allocation function is truthful if and only if each of its single player subfunctions is partially monotonically non-decreasing.*

Theorem 3. *For strictly convex cost functions, an allocation function is truthful if and only if each of its single player subfunctions is monotonically non-decreasing.*

Although Theorem 1 has its own theoretical significance (there are domains for which weak monotonicity is not sufficient [15]), Theorems 2 and 3 are more informative: they provide a global monotone structure on every single player subfunction, which is more intuitive and easier to verify for practical mechanism design. Strictly convex cost functions enforce a simple monotone structure; the linear cost function case is more intriguing: here monotonicity is required in a hidden (partial) way, captured in our notion of monotone triangular partition.

Consider a single player subfunction f. Since the distance to the facility $|f(x) - x|$ switches sign at $f(x) = x$, the sign of $f(x) - x$ (thus the line $y = x$) is important. Our monotone triangular partition is a partition of the real line into intervals such that, for each interval I, all $f(x)$ are within the closure $\tilde{I}$ of the interval and on the same side of line $y = x$. Hence f is monotonically non-decreasing between different intervals (i.e., f on a right interval is never below f on a left interval), but need not be monotone within an interval I. Graphically, each interval I corresponds to one of the two triangles generated by dividing $I \times \tilde{I}$ with line $y = x$. The sign of the interval, i.e., the uniform sign of $f(x) - x$, corresponds to which side of $y = x$ the triangle resides. Therefore, f is contained in these monotone triangles, and we call this nice interesting structure a monotone triangular partition. For intervals where $f(x) \equiv x$, we allow the corresponding triangle to degenerate into the line segment on $y = x$ intersecting $I \times \tilde{I}$.

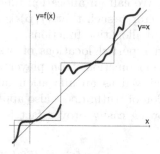

Fig. 1. A monotone triangular partition when C is linear

Unsurprisingly, the payment function in Section 5 is closely related to the triangular partition. Since truthfulness is unaffected by shifting the payment function by any arbitrary constant, we pick an arbitrary reference point and set its payment 0. Then to find out the payment for any point x, imagine taking a walk from the reference point towards the allocated facility location $f(x)$ and counting throughout the way. For cost function $C(d) = d$, we simply count the distance we have walked, but with a sign according to the sign of the interval we are walking at (We take negation of it if $f(x)$ is to the left of the reference point). Hence in the formula, the payment is a directed summation of lengths of intervals corresponding to a monotone triangular partition. For linear cost functions with slope $\alpha \neq 1$ or strictly convex cost functions, we need to adjust the quantity (not as easy as distance here) counted into the payment, but the idea is the same.

2   Preliminaries and Notation

Now we formally define a one-dimensional single facility location game. Suppose there are n players and player i's location is represented by a real number $x_i \in \mathbb{R}$. Given a location vector $\mathbf{x} = (x_1, \ldots, x_n)$ of n players, an allocation mechanism chooses a location $y = f(\mathbf{x}) \in \mathbb{R}$ for the single facility and assigns payments $\mathbf{p}(\mathbf{x}) = (p_1(\mathbf{x}), \ldots, p_n(\mathbf{x}))$ to players where player i gets payment $p_i(\mathbf{x})$. The setting without payments restricts $\mathbf{p}(\mathbf{x}) \equiv \mathbf{0}$.

Let $C(d)$ denote the cost function of all players, which is a smooth monotonically-increasing function on nonnegative distances, and can always be normalized to satisfy $C(0) = 0$. Let $\mathbf{r} = (r_1, \ldots, r_n)$ be the true location vector of the n players, in which r_i is player i's private information. Then the utility of player i is $u_i(\mathbf{x}) = -C(|f(\mathbf{x}) - r_i|) + p_i(\mathbf{x})$.

In the game-theoretic model, each player intends to maximize his utility. An allocation mechanism $(f, \mathbf{p})$ is incentive compatible, or truthful, if for each player i, reporting his true location r_i always maximizes his utility. Formally, it requires that, for each player i, for each fixed reported locations of players other than i, written as $\mathbf{x}_{-i} = (x_1, \ldots, x_{i-1}, x_{i+1}, \ldots, x_n)$, and for any r_i and x_i, we have $u_i((r_i, \mathbf{x}_{-i})) \geq u_i((x_i, \mathbf{x}_{-i}))$. We call an allocation function f truthful if there exists a payment function vector $\mathbf{p}$ such that $(f, \mathbf{p})$ is truthful. Our goal is to characterize the set of truthful allocation functions.

For a player i, each fixed reported locations of other players $\mathbf{x}_{-i}$ induces a subfunction of the allocation function f on player i's location: $f^i_{\mathbf{x}_{-i}}(x_i) = f((x_i, \mathbf{x}_{-i}))$, which can be viewed as an allocation function for a game of a single player i. Thus the notion of truthfulness also applies to such single player subfunctions of f. The following easily proved fact is used extensively in the literature:

Proposition 4. *The allocation function f is truthful if and only if every single player subfunction of f is truthful.*

By Proposition 4, it is meaningful to characterize the set of truthful allocation functions of one player: now an allocation function $f : \mathbb{R} \to \mathbb{R}$ maps a location x to location y of the facility. We want to know for which f there exists payment function $p : \mathbb{R} \to \mathbb{R}$ such that (f, p) is truthful.

Weak monotonicity is a well-known necessary condition for a truthful function. In our setting, it translates to the following: for any $x, x' \in \mathbb{R}$, $C(|f(x) - x|) - C(|f(x) - x'|) \le C(|f(x') - x|) - C(|f(x') - x'|)$. We obtain a nice characterization by showing weak monotonicity is also sufficient and providing more illustrative conditions equivalent to weak monotonicity on the allocation function f.

It turns out that the shape of the cost function C plays an important role. In our work, we investigate linear, strictly convex and strictly concave cost functions. $C(d) = \alpha d$ where $\alpha > 0$ is a linear cost function; C is strictly convex if for any two points $d_1 \ne d_2$ and $t \in (0, 1)$, it holds that $C(td_1 + (1 - t)d_2) < tC(d_1) + (1 - t)C(d_2)$. Symmetrically C is strictly concave if for any two points $d_1 \ne d_2$ and $t \in (0, 1)$, it holds that $C(td_1 + (1 - t)d_2) > tC(d_1) + (1 - t)C(d_2)$.

3 Implication of Weak-Monotonicity

In this section, for linear and strictly convex cost functions respectively, derive from weak monotonicity an equivalent condition on every single player subfunction.

3.1 Convex Cost Functions

Lemma 5. *If the cost function is strictly convex, a single player allocation function f satisfies weak monotonicity if and only if it is monotonically non-decreasing, i.e., $f(x_1) \le f(x_2)$ for any $x_1 < x_2$.*

Proof. We use the following property of strictly convex functions:

Proposition 6. *If function C is strictly convex, $C(d_1) + C(d_4) > C(d_2) + C(d_3)$ holds for any $d_1 < d_2 \le d_3 < d_4$ satisfying $d_1 + d_4 = d_2 + d_3$.*

Now given a strictly convex cost function C, for any $x_1 < x_2$, we claim that function $\Delta(z) = C(|z - x_1|) - C(|z - x_2|)$ is a monotonically increasing function: For any $z_1 < z_2 \le x_1$, set $d_1 = x_1 - z_2$, $d_4 = x_2 - z_1$, $d_2 = \min(x_1 - z_1, x_2 - z_2)$ and $d_3 = \max(x_1 - z_1, x_2 - z_2)$. We can easily check $d_1 < d_2 \le d_3 < d_4$ and $d_1 + d_4 = d_2 + d_3$. By Proposition 6, $C(x_1 - z_2) + C(x_2 - z_1) > C(x_1 - z_1) + C(x_2 - z_2)$. We rearrange the terms and change the distances to the form of absolute values to get $C(|z_1 - x_1|) - C(|z_1 - x_2|) < C(|z_2 - x_1|) - C(|z_2 - x_2|)$, i.e., $\Delta(z_1) < \Delta(z_2)$.

The case $x_2 \le z_1 < z_2$ is symmetric. For $x_1 \le z_1 < z_2 \le x_2$, we have $C(|z_1 - x_1|) = C(z_1 - x_1) < C(z_2 - x_1) = C(|z_2 - x_1|)$ and $C(|z_1 - x_2|) = C(x_2 - z_1) > C(x_2 - z_2) = C(|z_2 - x_1|)$. Taking the difference of the two inequalities gives $C(|z_1 - x_1|) - C(|z_1 - x_2|) < C(|z_2 - x_1|) - C(|z_2 - x_2|)$, i.e., $\Delta(z_1) < \Delta(z_2)$.

The monotonicity of the entire function Δ can be easily derived by its monotonicity on the three closed intervals $z \le x_1$, $x_1 \le z \le x_2$, $z \ge x_2$ derived above. The condition of weak monotonicity can be rewritten as $\Delta(f(x_1)) \le \Delta(f(x_2))$, which holds if and only if $f(x_1) \le f(x_2)$ since function Δ is strictly monotonically increasing.

3.2   Linear Cost Functions

Lemma 7. *If the cost function is linear, a single player allocation function f satisfies weak monotonicity if and only if for any $x_1 < x_2$, $f(x_1) > x_1$ implies $f(x_2) \geq \min(x_2, f(x_1))$.*

This property is weaker than being monotonically non-decreasing, which we call *partially monotonically non-decreasing*.

Proof. If the cost function $C(d) = \alpha d$ ($\alpha > 0$), for any $x_1 < x_2$, function $\Delta(z) = C(|z - x_1|) - C(|z - x_2|)$ is a continuous non-decreasing piecewise linear function: For $z \leq x_1$, $\Delta(z)$ is a negative constant $\alpha(x_1 - x_2)$, whereas it constantly equals its negation $\alpha(x_2 - x_1)$ for $z \geq x_2$. Between $z = x_1$ and $z = x_2$ is a linear piece of slope $2\alpha > 0$.

The condition of weak monotonicity can be rewritten as $\Delta(f(x_1)) \leq \Delta(f(x_2))$. This always holds for $f(x_1) \leq x_1$ since $\Delta(f(x_1))$ reaches the minimum. If $f(x_1) > x_1$, there are two cases: for $f(x_1) < x_2$, $f(x_1)$ belongs to the linearly increasing piece, so $\Delta(f(x_1)) \leq \Delta(f(x_2))$ if and only if $f(x_1) \leq f(x_2)$; otherwise, $f(x_1) \geq x_2$, $\Delta(f(x_1))$ reaches the maximum, thus $\Delta(f(x_2))$ is also the maximum, i.e., $f(x_2) \geq x_2$. The summary of the two cases is exactly $f(x_2) \geq \min(x_2, f(x_1))$.

4   Monotone Triangular Partition

In this section, we show that weak monotonicity implies a monotone triangular partition. We start with the following key separation theorem:

Theorem 8. *If f is partially monotonically non-decreasing, then for any $x_1 < x_2$ satisfying $(f(x_1) - x_1)(f(x_2) - x_2) < 0$, there exists $x^* \in [x_1, x_2]$ such that $f(x) \leq x^*$ for $x < x^*$ and $f(x) \geq x^*$ for $x > x^*$. In particular, $f(x^*) = x^*$ for the case $f(x_1) > x_1$ and $f(x_2) < x_2$.*

Proof. $(f(x_1) - x_1)(f(x_2) - x_2) < 0$ implies that $f(x_1) - x_1$ and $f(x_2) - x_2$ have different signs. There are two cases:

If $f(x_1) < x_1$ and $f(x_2) > x_2$, we take $x^* = \inf\{x : f(x) > x, x \geq x_1\}$, where the infimum exists since the set is non-empty (contains x_2) and bounded below by x_1. Clearly $x^* \in [x_1, x_2]$.

First, we show $f(x) \leq x^*$ for $x < x^*$ in this case. This is immediate for $x \geq x_1$ by the definition of x^*. For $x < x_1$, suppose $f(x) > x^*$ for contradiction. Then since f is partially monotonically non-decreasing, $x < x_1$ and $f(x) > x$ implies $f(x_1) \geq \min(x_1, f(x)) \geq \min(x_1, x^*) = x_1$, contradicting that $f(x_1) < x_1$.

Next, for $x > x^*$, we want to show $f(x) \geq x^*$. By the definition of x^*, there exists $x' \in [x^*, x)$ satisfying $f(x') > x'$. Now we apply the partial monotonicity condition again with $x' < x$: $f(x') > x'$ implies $f(x) \geq \min(x, f(x')) > x^*$.

Interestingly, the second case $f(x_1) > x_1$ and $f(x_2) < x_2$ is not symmetric. Here we take $x^* = \inf\{f(x) : f(x) < x, x \geq x_1\}$. Again the set is non-empty since it contains $f(x_2)$. It is bounded below by x_1 since we can apply the partial monotonicity condition with $x_1 < x$, $f(x_1) > x_1$ and get $f(x) \geq \min(x, f(x_1)) > x_1$.

Hence x^* is also well-defined in this case. Moreover, the above argument plus $f(x_2) < x_2$ guarantees $x^* \in [x_1, x_2]$. For this case, we need to show a slightly stronger statement: $f(x) \leq x^*$ for $x \leq x^*$ and $f(x) \geq x^*$ for $x \geq x^*$, which at $x = x^*$ implies $f(x^*) = x^*$.

First we prove $f(x) \leq x^*$ for $x \leq x^*$. For contradiction, suppose $f(x) > x^*$. By the definition of x^*, there exists $x'(\geq x_1)$ such that $x^* \leq f(x') < f(x)$ and $f(x') < x'$: This immediately implies $f(x') < \min(x', f(x))$; On the other hand, $x \leq x^* \leq f(x') < x'$ and $f(x) > x^* \geq x$ allows us to apply the partial monotonicity condition and get $f(x') \geq \min(x', f(x))$, which is a contradiction.

Now $f(x) \geq x^*$ for $x \geq x^*$. For those x satisfying $f(x) \geq x$, $x \geq x^*$ immediately gives $f(x) \geq x^*$; otherwise, $f(x) < x$, then the definition of x^* implies that $f(x)$ is at least the infimum x^*.

Theorem 8 enables us to repeatedly partition the real line into intervals: as long as there exist two points $x_1 < x_2$ within the same interval I whose signs of $f(x) - x$ are different, we disect the interval at x^*. Point x^* belongs to the left subinterval I_1 if $f(x^*) < x^*$, to the right subinterval I_2 if $f(x^*) > x^*$, and to either one of the two if $f(x^*) = x^*$ (Note that I_1 and I_2 are both nonempty but may only contain a single point). This disection at the same time disects the allocation function by line $y = x^*$: for x in I_1 and all intervals left to I_1, $f(x) \leq x^*$, the allocation function does not exceed this line; symmetrically for x in I_2 and all intervals right to I_2, $f(x) \geq x^*$, the allocation function never goes below the line. Graphically, f appears within the region $x \leq x^*, y \leq x^*$ and $x \geq x^*, y \geq x^*$.

Eventually, we get a partition $\mathcal{P} = \{I\}$ of $\mathbb{R}$ satisfying the following:

- Within each interval I, the sign of $f(x) - x$ is uniformly $\delta_I \in \{-1, 1\}$, i.e., $\delta_I(f(x) - x) \geq 0$ for all $x \in I$.
- For each interval I, $f(x) \in \tilde{I}$ for all $x \in I$, where $\tilde{I}$ is the closure of I.
- Between different intervals I_1 and I_2, if I_1 is to the left of I_2, $f(x_1) \leq f(x_2)$ for any $x_1 \in I_1, x_2 \in I_2$.

The second property is immediate from the disecting argument in the description of our partition process; and the last property immediately follows from the second.

Graphically, the partition $\mathcal{P}$ defines a triangular structure: each interval $I \in \mathcal{P}$ corresponds to a triangle $T_I : T_I = \{(x, y) : x \in I, y \in \tilde{I}, y \leq x\}$ for $\delta_I = -1$ and $T_I = \{(x, y) : x \in I, y \in \tilde{I}, y \geq x\}$ for $\delta_I = 1$. And the allocation function f only appears within the set of triangles. Moreover, the triangular structure is "monotonic" in the sense that "a triangle to the right is always above". Therefore, we call such a partition $\mathcal{P}$ a *monotone triangular partition*.

Combining Theorem 8 with Lemma 7 in Subsection 3.2, we derive that weak monotonicity guarantees the existence of such a partition for linear cost functions. For convex cost functions, Lemma 5 says that weak monotonicity implies that the allocation function f is monotonically non-decreasing, which is stronger than the condition of partially monotonically non-decreasing required in Theorem 8. Thus a monotone triangular partition exists for convex cost functions as well. This can also be derived directly from the monotonicity of the allocation function.

5   Incentive Compatible Payments

In this section, for any allocation function f that admits a monotone triangular partition, we would like to provide a payment function p such that (f, p) is truthful. We have an explicit formula of p for partitions where any finite range $[a, b]$ only intersects finitely many intervals, i.e., $\mathcal{P}$ can be written as $\{I_i : b_\ell \leq i \leq b_r\}$, where the I_i's are ordered from left to right (possibly $b_\ell = -\infty$, or $b_r = +\infty$, or both). For general f that does not admit a partition of this form, the same idea works; yet it involves infinite summations and makes our argument notationally much more complicated. Handling such technical details is not the focus of our paper here.

Now given a monotone triangular partition $\mathcal{P} = \{I_i : b_\ell \leq i \leq b_r\}$, let $\{a_i : b_\ell \leq i \leq b_r + 1\}$ be the set of boundary points, where $a_i \leq a_{i+1}$ and the left/right endpoint of I_i is a_i/a_{i+1}. $\mathcal{P}$ may contain only finitely many intervals, including the very special case $|\mathcal{P}| = 1$ where $b_\ell = b_r$; otherwise, there is an infinite sequence of intervals to the left end of the real line ($b_\ell = -\infty$), or to the right end ($b_r = +\infty$), or both. If b_ℓ is finite, $a_{b_\ell} = -\infty$; If b_r is finite, $a_{b_r+1} = +\infty$. Other than these two, all a_i's are finite.

A monotone triangular partition $\mathcal{P} = \{I_i : b_\ell \leq i \leq b_r\}$ of $\mathbb{R}$ satisfies the following three properties:

- Each interval I_i is associated with $\delta_i \in \{-1, 0, 1\}$, which denotes the uniform sign of $f(x) - x$. We have $\delta_i(f(x) - x) \geq 0$ for all $x \in I_i$, and in particular, $\delta_i = 0$ requires $f(x) \equiv x$ for all $x \in I_i$.
- For each i, $f(x) \in \tilde{I}_i$ for all $x \in I_i$, where $\tilde{I}_i$ is the closure of I_i.
- For any $i < j$ and $x \in I_i, x' \in I_j$, we have $f(x) \leq f(x')$.

Here we allow $\delta_i = 0$ for an interval I_i where $f(x) \equiv x$, while for such an interval, the other two choices -1 and 1 are also allowed. Graphically $\delta_i = 0$ indicates that the corresponding triangle of I_i shrinks to the line segment $\{(x, y) : x \in I_i, y = x\}$. This extra freedom does not add any difficulty to our proofs, but as shown in Subsection 5.2, now our payment function includes the no-payment case, i.e., for an allocation function that is truthful without payments, there is a monotone triangular partition with associated δ's under which our payment function is exactly $p(x) \equiv 0$.

For linear and strictly convex cost functions respectively, we present a formula of the payment function p and show its incentive compatibility based on the above properties. This, combined with Section 4, and the necessity of weak monotonicity, completes the proof of Theorem 1-3. Due to the space limit, we defer the convex cost function part to the full paper.

5.1   Linear Cost Functions

Given a monotone triangular partition $\mathcal{P} = \{I_i : b_\ell \leq i \leq b_r\}$, we define a function $q : \mathbb{R} \to \mathbb{R}$ on the location $y \in \mathbb{R}$ of the public facility as follows:

$q(y) = \delta_0 y$ if $|\mathcal{P}| = 1$; otherwise, choose a reference boundary point a_{b_0}, where $b_\ell < b_0 \le b_r$.

$$q(y) = \begin{cases} \delta_k(y - a_k) + \sum_{i=b_0}^{k-1} \delta_i(a_{i+1} - a_i), & y \in I_k, b_0 \le k \le b_r \\ -\delta_k(a_{k+1} - y) - \sum_{i=k+1}^{b_0-1} \delta_i(a_{i+1} - a_i), & y \in I_k, b_\ell \le k < b_0 \end{cases}$$

Our definition of q only involves a_i with $b_l + 1 \le i \le b_r$, which are all finite, thus function q is well-defined. Moreover, observe that for any finite boundary point a_k, the value of $q(a_k)$ is the same no matter whether a_k belongs to interval I_{k-1} or I_k. Hence the above formula holds for any $y \in \tilde{I}_k$ as well.

In particular, the value of q at the reference boundary point a_{b_0} is set to 0. Each interval I_i, or part of an interval, contributes to the payment if and only if it is between a_{b_0} and y. The contribution equals its sign δ_i times the length of the interval if it is to the right of a_{b_0}, and its negation if it is to the left of a_{b_0}. Under this summarization, the difference of the function value of any two points y and y' is irrelevant to the choice of the reference point a_{b_0}. The following lemma can be easily proven:

Lemma 9. *Suppose $y \in \tilde{I}_k$ and $y' \in \tilde{I}_{k'}$. For $k < k'$,*

$$q(y') - q(y) = \delta_{k'}(y' - a_{k'}) + \sum_{i=k+1}^{k'-1} \delta_i(a_{i+1} - a_i) + \delta_k(a_{k+1} - y);$$

For $k = k'$, $q(y') - q(y) = \delta_k(y' - y)$.

Theorem 10. *Let f be an allocation function that admits a monotone triangular partition $\mathcal{P} = \{I_i : b_\ell \le i \le b_r\}$, and $C(d) = \alpha d$ ($\alpha > 0$) is the cost function. Then mechanism (f, p) is truthful where the payment function is defined as $p(x) = \alpha q(f(x))$.*

Proof. To prove (f, p) is truthful, we need to show that for any true location x and $x' \ne x$,

$$-C(|f(x) - x|) + p(x) = u(x) \ge u(x') = -C(|f(x') - x|) + p(x'),$$

i.e., reporting true location x always maximizes the player's utility. Substituting $C(d) = \alpha d$, $p(x) = \alpha q(f(x))$ and $f(x) = y$, $f(x') = y'$, the inequality simplifies to

$$q(y) - |y - x| \ge q(y') - |y' - x|.$$

Now we verify this inequality in three cases as follows. Throughout our proof, we repeatedly use the simple fact that, for $x \in I_k$ and $y = f(x)$, $|y - x| = \delta_k(y - x)$. This is immediate from the first property of the partition.

Case 1: x and x' are in the same interval I_k.

In this case, $y, y' \in \tilde{I}_k$ from the second property of the partition. By Lemma 9, we have $q(y') - q(y) = \delta_k(y' - y)$. We substitute this and $|y - x| = \delta_k(y - x)$ and the inequality simplifies to $|y' - x| \geq \delta_k(y' - y) + \delta_k(y - x) = \delta_k(y' - x)$, which always holds given $\delta_k \in \{-1, 0, 1\}$.

Case 2 : $x \in I_k$, $x' \in I_{k'}$ and $k < k'$.

In this case, $y \in \tilde{I}_k$ and $y' \in \tilde{I}_{k'}$. Applying Lemma 9 gives

$$q(y') - q(y) = \delta_{k'}(y' - a_{k'}) + \sum_{i=k+1}^{k'-1} \delta_i(a_{i+1} - a_i) + \delta_k(a_{k+1} - y).$$

On the other hand, $|y' - x| = y' - x = (y' - a_{k'}) + \sum_{i=k+1}^{k'-1}(a_{i+1} - a_i) + (a_{k+1} - x)$.

Putting all equalities together, we get

$$q(y') - q(y) - |y' - x| + |y - x| = (\delta_{k'} - 1)(y' - a_{k'}) + \sum_{i=k+1}^{k'-1} (\delta_i - 1)(a_{i+1} - a_i)$$
$$+ \delta_k(a_{k+1} - y) - (a_{k+1} - x) + \delta_k(y - x)$$
$$\leq \delta_k(a_{k+1} - y) - (a_{k+1} - x) + \delta_k(y - x)$$
$$= (\delta_k - 1)(a_{k+1} - x) \leq 0,$$

given $\delta_k, \delta_{k'} \in \{-1, 0, 1\}$. Rearranging the terms gives exactly the inequality we want to prove.

Case 3 : $x \in I_k$, $x' \in I_{k'}$ and $k > k'$. This case is symmetric to Case 2.

5.2   Generality and Non-uniqueness of Our Payment

As mentioned before, by allowing degenerated triangles (allowing $\delta_I = 0$ for interval I where $f(x) \equiv x$) in our monotone triangular partition, we make our payment formula include the all-zero payment function for truthful allocation functions in the no payment setting.

For games without payment, every single player subfunction behaves as follows: as player's location x grows, the facility location $y = f(x)$ either remains the same, or jumps to a symmetric (higher point) with respect to x, or continues to equal x. Formally, for any single player subfunction of a truthful allocation function, there exists a monotone triangular partition satisfying the following properties:

- For any I with $\delta_I = 0$, $f(x) \equiv x$. This is always required by a monotone triangular partition. We state it here for completeness.
- For any I with $\delta_I = 1$, $f(x)$ always equals to the right endpoint of I.
- For any I with $\delta_I = -1$, $f(x)$ always equals to the left endpoint of I.
- For any I_1 adjacent to I_2 and to the left of I_2, $\delta_{I_1} = -1$ implies $\delta_{I_2} = 1$ and the lengths of the two intervals are equal.

It can be verified that our payment function p in Subsection 5.1 based on the above monotone triangular partition is constant, thus can be made all-zero by a constant shift.

On the other hand, for an interval I where $f(x) \equiv x$, we can still set $\delta_I = -1$ or 1, or even divide it into more intervals and set different δ's. This freedom results in different monotone triangular partitions, which, plugged into our payment formula, results in payment functions that differ more than a constant shift. Therefore, the payment function for a truthful allocation function may not be unique. In contrast, the classic unique-payment theorem [20] states that the payment function is unique for a truthful mechanism when the domain is connected; and the domain of our setting is connected. The inconsistency comes from the fact that our outcome set (the set of possible facility locations) is uncountable, while the theorem assumes the outcome set to be finite. This is called the revenue equivalence in economics literature [13,8].

6   Conclusion and Open Questions

In this paper, we characterize the set of truthful allocation functions for one-dimensional single facility location game with payments: we show the sufficiency of weak monotonicity, and its equivalent global monotone structure on every single player subfunction for linear and strictly convex cost functions respectively.

When investigating concave cost functions, we observe certain anti-monotone feature implied by weak monotonicity, which makes this case greatly different from the cases we have solved. We would love to see characterization results of this case: it is not known yet whether weak monotonicity is sufficient or not. We note here that, when the cost function is concave, the global utility function is still not convex (it is convex in both sides of its true location but not if we view it globally).

Another direction is to consider the game for more facilities, say, two facilities. In this case, the valuation domain for each agent is more complicated. Even the characterization for truthful mechanisms without payment is still open.

References

1. André, B., Rudolf, M., Hossein, N.S.: Path-monotonicity and incentive compatibility. Technical report (2010)
2. Archer, A., Kleinberg, R.: Truthful germs are contagious: A local to global characterization of truthfulness. In: Proceedings of the 9th ACM Conference on Electronic Commerce, pp. 21–30 (2008)
3. Archer, A., Tardos, É.: Truthful mechanisms for one-parameter agents. In: FOCS, pp. 482–491. IEEE Computer Society (2001)
4. Ashlagi, I., Braverman, M., Hassidim, A., Monderer, D.: Monotonicity and implementability. Econometrica 78(5), 1749–1772 (2010)
5. Barberà, S., Jackson, M.: A characterization of strategy-proof social choice functions for economies with pure public goods. Social Choice and Welfare 11(3), 241–252 (1994)

6. Bei, X., Chen, N., Gravin, N., Lu, P.: Budget feasible mechanism design: from prior-free to bayesian. In: STOC, pp. 449–458 (2012)
7. Bikhchandani, S., Chatterji, S., Lavi, R., Mu'alem, A., Nisan, N., Sen, A.: Weak monotonicity characterizes deterministic dominant-strategy implementation. Econometrica 74(4), 1109–1132 (2006)
8. Carbajal, J.C., Ely, J.C.: Mechanism design without revenue equivalence. Journal of Economic Theory (2012)
9. Cuff, K., Hong, S., Schwartz, J.A., Wen, Q., Weymark, J.A.: Dominant strategy implementation with a convex product space of valuations. Social Choice and Welfare 39(2-3), 567–597 (2012)
10. Dhangwatnotai, P., Dobzinski, S., Dughmi, S., Roughgarden, T.: Truthful approximation schemes for single-parameter agents. In: FOCS, pp. 15–24 (2008)
11. Frongillo, R.M., Kash, I.A.: General truthfulness characterizations via convex analysis. arXiv preprint arXiv:1211.3043 (2012)
12. Goldberg, A.V., Hartline, J.D., Karlin, A.R., Saks, M., Wright, A.: Competitive auctions. Games and Economic Behavior 55(2), 242–269 (2006)
13. Heydenreich, B., Müller, R., Uetz, M., Vohra, R.V.: Characterization of revenue equivalence. Econometrica 77(1), 307–316 (2009)
14. Jehiel, P., Moldovanu, B., Stacchetti, E.: Multidimensional mechanism design for auctions with externalities. Journal of Economic Theory 85(2), 258–293 (1999)
15. Lavi, R., Mu'alem, A., Nisan, N.: Towards a characterization of truthful combinatorial auctions. In: FOCS, p. 574 (2003)
16. Mishra, D., Roy, S.: Implementation in multidimensional dichotomous domains. Theoretical Economics 8(2), 431–466 (2013)
17. Moulin, H.: On strategy-proofness and single peakedness. Public Choice 35(4), 437–455 (1980)
18. Müller, R., Perea, A., Wolf, S.: Weak monotonicity and bayes–nash incentive compatibility. Games and Economic Behavior 61(2), 344–358 (2007)
19. Myerson, R.B.: Optimal auction design. Mathematics of Operations Research 6(1), 58–73 (1981)
20. Nisan, N.: Introduction to mechanism design. In: Nisan, N., Roughgarden, T., Tardos, E., Vazirani, V. (eds.) Algorithmic Game Theory, ch. 9, pp. 209–242 (2007)
21. Nisan, N., Ronen, A.: Algorithmic mechanism design. Games and Economic Behavior 35(1-2), 166–196 (2001)
22. Papadimitriou, C.H., Schapira, M., Singer, Y.: On the hardness of being truthful. In: FOCS, pp. 250–259 (2008)
23. Roberts, K.: The characterization of implementable choice rules. In: Aggregation and Revelation of Preferences, pp. 321–348 (1979)
24. Saks, M., Yu, L.: Weak monotonicity suffices for truthfulness on convex domains. In: Proceedings of ACM EC, pp. 286–293. ACM (2005)
25. Singer, Y.: Budget feasible mechanisms. In: FOCS, pp. 765–774 (2010)
26. Sprumont, Y.: The division problem with single-peaked preferences: A characterization of the uniform allocation rule. Econometrica, 509–519 (1991)
27. Vohra, R.V.: Mechanism Design: A linear programming approach, vol. 47. Cambridge Univ. Pr. (2011)

Equilibrium in Combinatorial Public Projects

Brendan Lucier[1], Yaron Singer[2,*], Vasilis Syrgkanis[3,**], and Éva Tardos[3,***]

[1] Microsoft Research
brlucier@microsoft.com
[2] School of Engineering and Applied Sciences, Harvard University
yaron@seas.harvard.edu
[3] Dept. of Computer Science, Cornell University, Ithaca, NY, USA
{vasilis,eva}@cs.cornell.edu

Abstract. We study simple item bidding mechanisms for the combinatorial public project problem and explore their efficiency guarantees in various well-known solution concepts. We first study sequential mechanisms where each agent, in sequence, reports her bid for every item in a predefined order on the agents determined by the mechanism. We show that if agents' valuations are unit-demand any subgame perfect equilibrium of a sequential mechanism achieves the optimal social welfare. For the simultaneous bidding equivalent of the above auction we show that for any class of bidder valuations, all Strong Nash Equilibria achieve at least a $O(\log n)$ factor of the optimal social welfare. For Pure Nash Equilibria we show that the worst-case loss in efficiency is proportional to the number of agents. For public projects in which only one item is selected we show constructively that there always exists a Pure Nash Equilibrium that guarantees at least $\frac{1}{2}(1 - \frac{1}{n})$ of the optimum. We also show efficiency bounds for Correlated Equilibria and Bayes-Nash Equilibria, via the recent smooth mechanism framework [26].

1 Introduction

In recent years considerable attention has been devoted to the design and analysis of "simple" mechanisms: algorithms for strategic environments that yield provable guarantees yet are simple enough to run in practice. This trend is motivated by the realization that mechanisms which are implemented in practice and encourage participation cannot be arbitrarily complex. Since simplicity often comes at the price of lowered economic efficiency, the goal in analyzing simple mechanisms is to quantify this loss in comparison to some theoretical optimum.

Simple mechanisms have been explored thus far primarily in auction domains. Some examples include posted price mechanisms that approximate revenue-optimal auctions [5], and simultaneous single item auctions that have good social efficiency relative to the fully efficient combinatorial auctions [6,1,13,10].

* Part of this work was done while the author was visiting Cornell University.
** Supported in part by ONR grant N00014-98-1-0589, NSF grant CCF-0729006 and a Simons Graduate Fellowship.
*** Supported in part by NSF grants CCF-0910940 and CCF-1215994, ONR grant N00014-08-1-0031, a Yahoo! Research Alliance Grant, and a Google Research Grant.

In contrast to auctions where agents compete for allocated resources, *public projects* require agents to coordinate on resources that are collectively allocated. This is captured in the *combinatorial public projects* model introduced in [21], where there are multiple resources (items), each agent has a combinatorial valuation function on subsets of the items, and the goal is to select some fixed number of items that maximizes the sum of the agents' valuations. This problem provides the first evidence of the computational hardness of truthful implementation: for agents with nondecreasing submodular valuations, there are constant-factor approximation algorithms when agents' valuations are known, but there is no computationally efficient truthful mechanism that can obtain a reasonable approximation under standard complexity-theoretic assumptions [21].

As a canonically hard mechanism design problem, there has been an ongoing investigation of mechanisms for combinatorial public projects under various valuation classes and solution concepts [3,12,8,14,7]. When considering simple mechanisms, the *item-bidding with first-prices* mechanism, which is the analogue of those used in combinatorial auctions [6,1,13], is arguably the simplest non-trivial mechanism for combinatorial public projects: the mechanism asks each agent, simultaneously, to report her valuation separately for each item, then chooses the k items whose sum of reported valuations is maximal and charges agents first prices, i.e. each agent pays her reported valuation for every item selected by the mechanism. Despite its appealing simplicity, it turns out that achieving desirable efficiency guarantees at equilibrium is not trivial in this mechanism. In evidence, consider an instance with n agents and 3 items A, B, C, in which a single item is to be selected and agents valuations are as those summarized in the following $n \times 3$ matrix:

$$[v_{ij}] = \begin{bmatrix} n-1 & 1 & 0 \\ \vdots & \vdots & \vdots \\ n-1 & 1 & 0 \\ 0 & 0 & n-1 \end{bmatrix}$$

In this instance all the agents except agent n have a valuation $v_{iA} = n-1$ for A, $v_{iB} = 1$ for item B, and $v_{iC} = 0$ for item C, and agent n has a valuation of 0 for items A and B and a valuation of $n-1$ for project C. Obviously the optimal outcome is for A to be chosen leading to a social welfare of $(n-1)^2$. However, there exists a Nash Equilibrium where project B is selected: if all agents except agent n bid 1 for B and 0 for A and agent n bids truthfully, then in this profile the bids on B total to $(n-1)$ (assuming tie-breaking is chosen in favor of B) and B is selected. To see that this an equilibrium, note that for any agent except n the only two ways to alter the allocation is either by only reducing the current bid for B to 0 and letting C be selected, or reducing the bid for B to 0 *and* increasing his bid on A to $n-1$. For both deviations the utility (valuation minus payment) would be 0, which is exactly what the agent is currently getting.

The above example shows that even when agents have very simple valuation functions (unit-demand), the *price of anarchy* – the ratio between the optimal solution and that achievable in equilibrium – can be as bad as linear in the number of agents in the system. It may therefore seem like simple mechanisms for combinatorial public projects are of little interest to anyone interested in reasonable efficiency guarantees. However, a more careful observation at the above example leaves some hope. The major difficulty in designing simple and efficient mechanisms for combinatorial public projects is the inability of the bidders to coordinate on the right equilibrium. Thus, to achieve worst case efficiency guarantees, either the mechanism should allow for agents to signal among each other, or the solution concept should allow for such signaling. Alternatively, an optimistic designer could be interested in best-case guarantees by studying the existence of good equilibria (Price of Stability), rather than that any equilibrium will be good, since the designer himself could somehow signal which equilibrium should be chosen. In this work we address all these three different routes. We study both *sequential* and *simultaneous* bidding mechanisms. In sequential bidding, the mechanism determines some order in which agents place their bids, while in the simultaneous case all agents bid simultaneously.

1.1   Main Results

To understand outcome quality in our item bidding mechanisms, we consider a number of standard solution concepts.

Sequential Item-Bidding Mechanism. We start by analyzing a sequential version of the item-bidding first-price mechanism, where the agents are asked sequentially (in an arbitrary order) to report their willingness-to-pay separately for each item. We focus on the subgame-perfect equilibria of this extensive form game, which is the most well-established concept for sequential games. We show that when bidders have unit-demand valuations (i.e. their valuation when a set S of items is chosen is their maximum valued item in the set) then every subgame-perfect equilibrium achieves optimal social welfare. We reiterate that in the simultaneous mechanism the inefficiency can grow linearly with the number of players even for unit-demand valuations, as illustrated in the example given above, rendering the design of good mechanisms for such valuations a non-trivial task. The intuition behind our result is that the sequentiality of the moves allows the agents to signal their preferences and to coordinate on a specific equilibrium by pre-committing on their declared valuations.

Strong-Nash Equilibria of the Simultaneous Item-Bidding Mechanism. Subsequently we analyze the quality of Strong Nash Equilibria (equilibria that are stable under coalitional deviations) of the simultaneous item-bidding mechanism, where all agents are asked to simultaneously submit their willingness-to-pay separately for each item. Our simple mechanism for the public project problem can be thought of as a coordination game, where agents need to coordinate on their best set of items. In this context the Strong Nash Equilibrium is a very natural solution concept. We show that the loss in efficiency (the strong price of anarchy) is no more than $O(\log n)$. Essentially, Strong Nash Equilibria alleviate the coordination problem inherent in public project auctions by allowing the agents to coallitionally deviate if they found themselves stuck at a bad equilibrium where no agent unilaterally could affect the set of chosen items.

Nash Equilibria of the Simultaneous Item-Bidding Mechanism. Next we consider the quality of Pure Nash Equilibria of the simultaneous auction, and show that the worst-case loss in efficiency is proportional to n, the number of agents[1]. Our upper bound requires no assumption on bidders' valuations. Note that while the n bound on the price of anarchy seems weak, it is better than any deterministic truthful mechanisms: [21] shows that computationally efficient deterministic dominant strategy mechanisms cannot do better than $\sqrt{m}$ (where m is the number of resources). For the special case of unit-demand agents (whose valuation for a set of items is the value of the best item selected) we give an improved bound of n/k, where k is the number of items that need to be chosen.

The high inefficiency of the worst pure nash equilibrium is due to the fact that certain bad equilibria survive due to the lack of good unilateral deviations. However, such equilibria tend to be unreasonable and unnatural. Thus it is interesting to study the existence of good equilibria of the auction. For the case when one item is to be chosen we show constructively that the *best* Pure Nash Equilibrium is guaranteed to obtain at least a $\frac{1}{2}(1 - \frac{1}{n})$ fraction of the optimal welfare.

Learning Behavior and Incomplete Information. The equilibrium analysis so far assumed that agents will reach a stable solution of the bidding game, i.e., an equilibrium. We also explore the quality of solution achieved in a repeated version of the simultaneous game under the weaker assumption that all agents employ no-regret learning strategies. If all agents use no-regret learning strategies [2], than the resulting outcome distribution is a coarse correlated equilibrium of the game. We show that the loss in efficiency of any coarse correlated equilibrium is no more than $2 \cdot n \cdot k$ for arbitrary valuations, $\frac{2e}{e-1}n$ for fractionally subadditive valuations and $\frac{e}{e-1}n/k$ for unit-demand valuations. The latter bounds are given via the smooth-mechanism framework [26] and thereby also carry over to the set of Bayes-Nash equilibria of the incomplete information setting, where valuations are private and drawn from commonly known distributions.

1.2   Related Work

There is a long recent literature on combinatorial public projects that mainly tries to find truthful mechanisms with good efficiency guarantees [21,24,3,8]. Specifically, as mentioned above, in [21] it is shown that under standard assumptions, no tractable truthful mechanism can achieve an approximation factor better than $\sqrt{m}$ for agents with nondecreasing submodular valuations.

There has been a long line of research on quantifying inefficiency of equilibria starting from [15] who introduced the notion of the price of anarchy. Several

---

[1] We also show that this bound is essentially tight, by giving a lower bound of $n - 1$ on the price of anarchy.

recent papers have studied the efficiency of simple mechanisms. A series of papers, Christodoulou, Kovacs and Schapira [6], Bhawalkar and Roughgarden [1] Hassidim, Kaplan, Mansour, Nisan [13], and Feldman et al. [10], studied the inefficiency of Bayes-Nash equilibria of non-truthful combinatorial auctions that are based on running simultaneous separate single-item auctions. Lucier and Borodin studied Bayes-Nash Equilibria of non-truthful auctions based on greedy allocation algorithms [16]. Paes Leme and Tardos [20], Lucier and Paes Leme [17] and Caragiannis et al. [4] studied the ineffficiency of Bayes-Nash equilibria of the generalized second price auction. Roughgarden [22] showed that many price of anarchy bounds carry over to imply bounds also for learning outcomes. Roughgarden [23] and Syrgkanis [25] showed that such bounds also extend to bound the inefficiency of games of incomplete information. Recently, in [26] we give a more specialized framework for the case of non-truthful mechanisms in settings with quasi-linear preferences, showing how to capture several of the previous results. In this work we show that our upper bounds for coarse correlated and Bayes-Nash equilibria of the simultaneous auction fall into the framework of [26].

The quality of subgame-perfect equilibria of sequential versions of simultaneous games, was introduced in [19] and has been applied to cost-sharing games, cut and consensus games, load balancing games. Our result on the sequential item-bidding mechanism is of similar flavor to this line of work and gives another interesting application of the latter approach.

2   Model and Notation

In combinatorial public projects there is a set of m items and n agents. Each agent $i \in [n]$ has a valuation function $v_i : 2^{[m]} \to \mathbb{R}_{\geq 0}$ for each set of chosen items. Given some fixed parameter k, the goal of the designer is to select a set S of size k that maximizes the total valuation of the agents $V(S) = \sum_i v_i(S)$. Since agents' valuations are considered private information, the mechanism enforces payments to help achieve good equilibria. For a profile of agents' bids b, when the selected subset by the mechanism is S and the payments profile is $p = (p_1, p_2, \ldots, p_n)$, the *utility* of an agent i denoted $u_i(b)$ is $v_i(S) - p_i$.

Valuation Classes. In some cases we state results over arbitrary valuation classes, and in others our results are stated for valuation classes that have particular combinatorial structure. A valuation v is *additive* if $v(S) = \sum_{j \in S} v(\{j\})$ for all $S \subseteq [m]$. A valuation v is (nondecreasing) *submodular* if it has a decreasing marginal utilities property: $v(S) - v(S \cup \{j\}) \geq v(T) - v(T \cup \{j\})$ for $S \subseteq T$ and all $j \in [m]$. A valuation v is *unit-demand* if $v(S) = \max_{j \in S} v(\{j\})$. It is easy to see that every unit-demand valuation is submodular.[2]

---

[2] In our case, when agents have submodular valuations, the algorithmic problem becomes that of maximizing a submodular function under a cardinality constraint for which there is a computationally efficient greedy algorithm that is well-known to be within a factor of $1 - 1/e$ of the optimum (assuming that agents comply with the protocol and reveal their true valuations) [18].

A generalization of submodular valuations which we will use in this work is that of *fractionally subadditive* (or *XOS*) valuations (see [9]): a valuation v is fractionally subadditive if and only if there exist a set of additive valuations $(v^\ell)_{\ell \in \mathcal{L}}$ such that $v(S) = \max_{\ell \in \mathcal{L}} v^\ell(S)$.

The First-Price Item-Bidding Mechanism. We consider the following simple item-bidding mechanism. Each agent $i \in [n]$ submits a bid b_{ij} for each item $j \in [m]$. For an item $j \in [m]$ let $B_j = \sum_i b_{ij}$ be the *total bid* placed on j. The mechanism chooses the k items with the highest total bids. For profile b let $S(b)$ be the chosen set. Each agent is charged her bids for the chosen items: $p_i = \sum_{j \in S(b)} b_{ij}$. We consider two variants: in the *simultaneous* mechanism, all agents submit their bids simultaneously. In the *sequential* mechanism, the agents submit their bids sequentially in some order, with each agent seeing the bids of those who came before. We define solution concepts for both mechanisms.

Solution Concepts for Simultaneous Games. We now define the main solution concepts that we will use in the context of the simultaneous move mechanism. A *Pure Nash Equilibrium* (PNE) is a set of bids $(b_{ij})_{i \in [n], j \in [m]}$ such that, for each agent i, there is no bid vector b'_i such that

$$u_i(b'_i, b_{-i}) > u_i(b). \tag{1}$$

If we allow the agents to make coalitional deviations then the appropriate equilibrium concept is *Strong Nash Equilibrium* (SNE). A set of bids $(b_{ij})_{i \in [n], j \in [m]}$ constitutes a SNE if, for each set of agents $S \subseteq [n]$, there is no bid vector $b'_S = (b'_i)_{i \in S}$ such that

$$\forall i \in S : u_i(b'_S, b_{-S}) > u_i(b). \tag{2}$$

A relaxed notion of equilibrium corresponds to no-regret learning outcomes (due to space limitations see [2] for a survey). It is known that such learning outcomes correspond to Coarse Correlated Equilibria of a game. A (possibly correlated) distribution on bids $b \sim D$ is a *Coarse Correlated Equilibrium* if, for every agent i and for every bid b'_i,

$$\mathbb{E}_{b \sim \mathcal{D}}[u_i(b)] \geq \mathbb{E}_{b \sim \mathcal{D}}[u_i(b'_i, b_{-i})] \tag{3}$$

that is, no agent i can improve his expected utility by unilaterally deviating.

All of the above equilibrium notions implicitly assume a full-information model, where agent valuations are commonly known. In the alternative model of incomplete information, the valuation profile v is drawn from distribution F, where this distribution is common knowledge. In the Item-Bidding Mechanism under incomplete information, each agent's strategy is a function $b_i(v_i)$ that outputs an agent's bids given her realized valuation. A *Bayes-Nash Equilibrium* of this game is a profile of strategies $b(v) = (b_i(v_i))_{i \in N}$ such that

$$\forall i \in N, \forall v_i : E_{\mathbf{v}_{-i}|v_i}[u_i(b_i(v_i), b_{-i}(\mathbf{v_{-i}}))] \geq E_{\mathbf{v}_{-i}|v_i}[u_i(b'_i(v_i), b_{-i}(\mathbf{v_{-i}}))]$$

The Sequential Item-Bidding Mechanism. In the sequential item-bidding mechanism, the agents are ordered under some arbitrary but commonly-known predefined sequence. Each agent is asked sequentially, in this order, to report a bid b_{ij} for each item $j \in [m]$. After all the agents have reported their bids, the mechanism chooses the set of k items with the highest total bid $B_j = \sum_i b_{ij}$ and charges each agent her bid on each selected item. In this sequential game the strategy of each agent is not simply a set of bids b_{ij} for each item $j \in [m]$ but rather it is a contingency of plans describing how the agent will bid conditioned on the history of play up to her turn. If we denote with h_i the history of play (i.e. reported bids) up to agent i then the strategy of an agent is a set of functions $b_{ij}(h_i)$ that maps each history to a bid vector.

Subgame-Perfect Equilibrium. A natural solution concept for sequential games is Subgame-Perfect Equilibrium, a refinement of the Nash Equilibrium. A profile of strategies is a subgame-perfect equilibrium if it constitutes an equilibrium on any subgame induced for any possible history of play. Note that this definition restricts the behavior of agents outside the equilibrium path, ruling out non-viable threats (for detailed discussion of subgame-perfection see [11]).

Measure of Efficiency. For each equilibrium notion above, we can measure worst-case efficiency by way of the *price of anarchy*. For a given equilibrium concept, the corresponding price of anarchy is the ratio between the minimum expected welfare of *any* equilibrium (with expectation over randomness in the strategies and/or realizations of bidders' valuations) and the expected optimal social welfare (over randomness in the bidders' valuations).

3   Sequential Item-Bidding Mechanism

We begin by considering outcomes of the sequential item-bidding mechanism at subgame-perfect equilibrium. We will focus on the case that agents have unit-demand valuations, where we find that the price of anarchy is 1. That is, the agents always select an optimal outcome. Note that this is in contrast to the example discussed in the introduction which shows that the price of anarchy (of Nash equilibrium) for the simultaneous item-bidding mechanism can be as large as $n - 1$.

Theorem 1. *For unit-demand valuations and any $k \geq 1$, the unique subgame perfect equilibrium of the sequential item-bidding mechanism selects a welfare-optimal outcome. Moreover, at this equilibrium each agent bids on a single item.*

Proof. For any value profile V, let $OPT(V) = \text{argmax}_{S:\, |S|=k}\{\sum_i v_i(S)\}$, breaking ties arbitrarily. Throughout the proof we will think of a bid vector $\mathbf{b}_i$ as an additive valuation function b_i, so that (in particular) $OPT(\mathbf{b})$ is the outcome selected by the mechanism when agents submit bids $\mathbf{b}$. We will also write $V(S) = \sum_i v_i(S)$ for the social welfare of an outcome S under valuations V.

Let $\mathbf{b}$ be an arbitrary bid profile, and for each i consider the valuation profile $\mathbf{b}^{(i)} = (b_1, b_2, \ldots, b_i, v_{i+1}, \ldots, v_n)$. Note $\mathbf{b}^{(0)} = V$ and $\mathbf{b}^{(n)} = \mathbf{b}$. For each $i \geq 1$, consider the subgame that occurs just before agent i is about to bid, after agents 1 through $i-1$ bid according to $\mathbf{b}$. We will show by backward induction that the unique equilibrium of this subgame selects outcome $OPT(\mathbf{b}^{(i-1)})$, and each agent bids on at most one item in this equilibrium. Taking $i = 1$ will then prove our theorem, since then $OPT(\mathbf{b}^{(0)}) = OPT(V)$ is the outcome of the mechanism.

The base case $i = n + 1$ is trivial, since by definition the mechanism selects outcome $OPT(\mathbf{b}) = OPT(\mathbf{b}^{(n)})$. For $i \leq n$, we know by induction that, for any bid b_i made by agent i, the mechanism returns $OPT(b_i, \mathbf{b}_{-i}^{(i)})$. We must show that, for the utility-maximizing bid b_i for agent i, $OPT(b_i, \mathbf{b}_{-i}^{(i)}) = OPT(\mathbf{b}^{(i-1)})$.

One potential strategy for agent i is to bid nothing (i.e., the zero bid $\mathbf{0}$), obtaining utility $v_i(OPT(\mathbf{0}, \mathbf{b}_{-i}^{(i-1)}))$. Since v_i is unit-demand, the only way to obtain higher utility is to choose some j with $v_i(j) > v_i(OPT(\mathbf{0}, \mathbf{b}_{-i}^{(i-1)}))$ and bid some (minimal) b_i so that $j \in OPT(b_i, \mathbf{b}_{-i}^{(i)})$. Note that this b_i will place a positive bid only on item j; let x_j be the minimal value such that $j \in OPT(b_i, \mathbf{b}_{-i}^{(i)})$ when $b_i(j) = x_j$. If agent i makes this minimal bid for j, he obtains utility $v_i(j) - x_j$.

We have argued that agent i maximizes utility by bidding on at most one item, so (by induction) each agent's bid at equilibrium is unit-demand.

Recalling the definition of x_j, let $S_j \ni j$ be the set selected by the mechanism when i bids x_j on j. Since all valuations in $\mathbf{b}_{-i}^{(i-1)}$ are unit demand, we have

$$x_j = \mathbf{b}_{-i}^{(i-1)}\left(OPT\left(\mathbf{0}, \mathbf{b}_{-i}^{(i-1)}\right)\right) - \mathbf{b}_{-i}^{(i-1)}(S_j). \tag{4}$$

We now consider two cases. First, suppose i maximizes utility by bidding nothing, so $b_i = \mathbf{0}$. Then $OPT(\mathbf{b}^{(i)}) = OPT(\mathbf{0}, \mathbf{b}_{-i}^{(i-1)})$. We will show $OPT(\mathbf{b}^{(i-1)}) = OPT(\mathbf{0}, \mathbf{b}_{-i}^{(i-1)})$, by showing that each set S_j achieves lower social welfare than $OPT(\mathbf{0}, \mathbf{b}_{-i}^{(i-1)})$ under profile $\mathbf{b}^{(i-1)}$. (This suffices because, as argued above, $OPT(\mathbf{b}^{(i-1)})$ must be either $OPT(\mathbf{0}, \mathbf{b}_{-i}^{(i-1)})$ or S_j for some item j, since v_i is unit-demand). Pick any $j \notin OPT(\mathbf{0}, \mathbf{b}_{-i}^{(i-1)})$. Since b_i is utility-maximal, $v_i(OPT(\mathbf{0}, \mathbf{b}_{-i}^{(i-1)})) \geq v_i(j) - x_j$. Substituting (4) and rearranging, we get that $\mathbf{b}^{(i-1)}(OPT(\mathbf{0}, \mathbf{b}_{-i}^{(i-1)})) \geq v_i(j) + \mathbf{b}_{-i}^{(i-1)}(S_j) = \mathbf{b}^{(i-1)}(S_j)$ and hence we conclude $\mathbf{b}^{(i-1)}(OPT(\mathbf{0}, \mathbf{b}_{-i}^{(i-1)})) \geq \mathbf{b}^{(i-1)}(S_j)$ as required.

Next suppose that i maximizes his utility by choosing b_i to be a bid of x_j on item j. We will show that $OPT(\mathbf{b}^{(i-1)}) = S_j$, by showing that neither $OPT(\mathbf{0}, \mathbf{b}_{-i}^{(i-1)})$ nor $S_{j'}$ for $j' \neq j$ can achieve higher welfare under valuation profile $\mathbf{b}^{(i-1)}$. Since i maximized utility by bidding on j, we have $v_i(OPT(\mathbf{0}, \mathbf{b}_{-i}^{(i-1)})) \leq v_i(j) - x_j$ and $v_i(j) - x_j \geq v_i(j') - x_{j'}$ for all $j' \neq j$. Rearranging these inequalities implies $\mathbf{b}^{(i-1)}(OPT(\mathbf{0}, \mathbf{b}_{-i}^{(i-1)})) \leq \mathbf{b}^{(i-1)}(S_j)$ and $\mathbf{b}^{(i-1)}(S_{j'}) \leq \mathbf{b}^{(i-1)}(S_j)$ for all $j' \neq j$, and hence $OPT(\mathbf{b}^{(i-1)}) = S_j$ as required.

In either case, we have that the equilibrium at this subgame selects outcome $OPT(\mathbf{b}^{(i-1)})$. The theorem now follows by induction.

4   Strong Nash Equilibrium of the Simultaneous Item-Bidding Mechanism

In this section we give efficiency bounds for strong Nash equilibria of the simultaneous item-bidding mechanism. A state of a game is a strong Nash equilibrium (SNE) if there is no coalition of agents that can each individually benefit by deviating as a group. Despite being a strong requirement, SNE is a natural solution concept in public projects as allocations of resources are collectively shared by agents.

Theorem 2. *Any Strong Nash Equilibrium of the first-price item bidding mechanism has efficiency at least* $\log(n)$ *of the optimal.*

Proof. Let $B_i(A)$ be the sum of bids of agent i for set A. Let S be the set that is selected at a strong nash equilibrium and OPT be the optimal set.

First we show that at any Strong Nash Equilibrium all the chosen projects receive the same B_j, i.e. $\forall j \in S : B_j = p$. Suppose that some chosen project has $B_j > p$. Then a agent i who is bidding positively on this project could just decrease his bid by some ϵ. The selected set would remain unchanged and agent i would be paying ϵ less than before. Hence, his utility would increase.

Now, suppose that all the agents deviate to bidding some small ϵ only on the optimal set OPT. The definition of a strong Nash equilibrium states that there exists an agent that doesn't prefer the utility at the deviation. W.l.o.g. rearrange the agents such that it is agent 1; then $v_1(S) - B_1(S) \geq v_1(OPT)$. Now suppose that the agents $\{2, \ldots, n\}$ deviate to bidding each $\frac{p}{n-1}$ on each item in OPT. By definition of SNE there exists an agent (w.l.o.g., agent 2) that doesn't prefer this deviation; that is, $v_2(S) - B_2(S) \geq v_2(OPT) - \frac{kp}{n-1}$. By similar reasoning we can reorder the agents such that, for each i,

$$v_i(S) - B_i(S) \geq v_i(OPT) - \frac{kp}{n-i+1} \tag{5}$$

Summing all the above inequalities we get:

$$V(S) - \sum_i B_i(S) \geq V(OPT) - kp \sum_{i=1}^{n} \frac{1}{n-i+1} \implies$$
$$V(S) - kp \geq V(OPT) - kp\log(n)$$

Since $kp < V(S)$ we get that $V(S) \geq \frac{1}{\log(n)}V(OPT)$.   ∎

The above result gives a reasonable bound on the efficiency loss in such equilibria. Regarding existence of equilibrium, the non-existence for unit-demand agents when choosing two items and of PNE in Section 5 applies here as well, since SNE is a stronger solution concept.

5 Pure Nash Equilibria of the Simultaneous Item-Bidding Mechanism

We now examine the efficiency and existence of Pure Nash Equilibria for the simultaneous item-bidding mechanism. For brevity we defer some proofs to the appendix. In [21] the $\sqrt{m}$ lower bound on truthful mechanisms applies to two agents with submodular valuations, and thus grows with the number of resources in the problem. In contrast, we show here that for the item bidding mechanism the loss in efficiency at any Pure Nash Equilibrium (whenever it exists) and for any type of bidder valuations is at most proportional to the number of agents.

Theorem 3. *Any PNE of the item bidding mechanism has PoA $\leq n$.*

Proof. As in the proof of Theorem 2 it is easy to see that at any Pure Nash Equilibrium all the chosen items receive the same B_j, i.e. $\forall j \in S : B_j = p$.

Let b be a Nash Equilibrium and S the chosen set. Let OPT be the optimal set of items for the true valuations of the agents. Each agent i could change the chosen set to OPT by bidding $p + \epsilon$ on every item $j \in OPT$. Since we are at a Nash Equilibrium this deviation wouldn't be profitable:

$$v_i(S) - \sum_{j \in S} b_{ij} \geq v_i(OPT) - kp$$

Summing over all agents and using the fact that $\sum_{i \in [n]} \sum_{j \in S} b_{ij} = k \cdot p$ we get:

$$V(S) - k \cdot p \geq V(OPT) - n \cdot k \cdot p$$

Due to individual rationality no agent is paying above his total value. Hence, $k \cdot p \leq V(S)$. Thus: $nV(S) \geq V(OPT)$. ∎

As shown in the Introduction, even when $k = 1$ the PoA of unit-demand agents can be as bad as $n - 1$, implying that our PoA upper bound is nearly tight. Note that when $k = 1$ unit-demand and additive valuations coincide. Hence, our example proves that the PoA bound is tight even for additive agents.

Theorem 4. *For unit-demand agents the PoA of the item bidding mechanism can be at least $n - 1$, even when choosing a single item ($k = 1$).*

Price of Stability. We now investigate existence of good pure Nash equilibria.

Theorem 5. *There always exists a pure Nash equilibrium of the item bidding mechanism when $k = 1$ and arbitrary number of agents. Moreover, it achieves at least $\frac{1}{2}(1 - \frac{1}{n})$ of the optimal social welfare.*

Proof. For a set of agents S and an item j let: $V_j(S) = \sum_{i \in S} v_{ij}$. Moreover, let $a_{j,j'}^S = V_j(S) - V_{j'}(S)$. Let $a_{A,B}^{S^*} = \max_{S \subset N} \max_{j \in M} \max_{j' \in M - j} a_{j,j'}^S$, that is among all possible quantities $a_{j,j'}^S$, $a_{A,B}^{S^*}$ is the maximum one. Observe that in the above maximum we take maximum only among sets that are strict subsets of N. In other words $S^* \subset N$ and $N - S^* \neq \emptyset$. The reason is that we need at least one agent to price set the "winners". We claim that the following outcome is an equilibrium:

- $\forall i \in S^* : b_i(A) = v_{iA} - v_{iB}$ and $\forall j \neq A : b_i(j) = 0$
- $\forall i \notin S^* : b_i(B) = v_{iB} - v_{iA} + \frac{a_{A,B}^{S^*} - a_{B,A}^{N-S^*}}{n}$ and $\forall j \neq B : b_i(j) = 0$

We denote with $p = \sum_{i \in S^*} b_i(A)$. Notice that by the definition of the equilibrium $p = \sum_{i \in S^*} b_i(A) = V_A(S^*) - V_B(S^*) = a_{A,B}^{S^*}$. Moreover, $p = \sum_{i \notin S^*} b_i(B) = V_B(N - S^*) - v_A(N - S^*) + a_{A,B}^{S^*} - a_{B,A}^{N-S^*} = a_{A,B}^{S^*}$.

We first focus on an agent $i \in S^*$. We take cases on his possible deviations and show that none is profitable:

- Drop bid on A and let B win: To show this is not profitable we need to show that $v_{iA} - b_i(A) \geq v_{iB} \Leftrightarrow v_{iA} - v_{iB} \geq b_i(A)$. From the equilibrium definition this is satisfied with equality.
- Drop bid on A and bid p on an item $j \neq A, B$ to make it win: We need to show that $v_{iA} - b_i(A) \geq v_{ij} - p \Leftrightarrow v_{iA} - v_{iA} + v_{iB} \geq v_{ij} - a_{A,B}^{S^*} \Leftrightarrow a_{A,B}^{S^*} \geq v_{ij} - v_{iB} = a_{j,B}^{\{i\}}$. Which holds by the maximality of $a_{A,B}^{S^*}$.

Now we focus on a agent $i \notin S^*$.

- Slightly increase his bid on B to make B win: We need to show that $v_{iA} \geq v_{iB} - b_i(B) \Leftrightarrow b_i(B) \geq v_{iB} - v_{iA}$. By the maximality of $a_{A,B}^{S^*}$, we have that $a_{A,B}^{S^*} \geq a_{B,A}^{N-S^*}$. Hence, the inequality holds by the definition of equilibrium.
- Drop bid on B and bid p on item $j \neq A, B$ to make it win: We need to show that $v_{iA} \geq v_{ij} - p \Leftrightarrow v_{iA} \geq v_{ij} - a_{A,B}^{S^*} \Leftrightarrow a_{A,B}^{S^*} \geq v_{ij} - v_{iA} = a_{j,A}^{\{i\}}$, which holds by the maximality of $a_{A,B}^{S^*}$.

In fact, above we gave a specific equilibrium where the price setting agents split equally the excess $a_{A,B}^{S^*} - a_{B,A}^{N-S^*}$, one can easily see that any splitting of that excess among the price setting agents is an equilibrium.

Efficiency. For the equilibrium constructed above we know that:

$$\forall j \neq j', S' \subset N : V_A(S^*) - V_B(S^*) \geq V_j(S') - V_{j'}(S')$$

Let j^* be the optimal item. Consider the above property for $j = j^*$, $j' = A$, and $S' = N - \{\arg\min_{i \in N} v_{ij^*}\}$. The condition gives us:

$$V_A(S^*) - V_B(S^*) \geq V_{j^*}(S') - V_A(S')$$

By the definition of S' we know that $V_{j^*}(S') \geq (1 - \frac{1}{n})V_{j^*}(N)$. In addition $V_A(S^*) \leq V_A(N)$ and $V_A(S') \leq V_A(N)$. Combining all the above together we get:

$$2V_A(N) \geq V_A(S^*) + V_A(S') - V_B(S^*) \geq V_{j^*}(S') \geq \left(1 - \frac{1}{n}\right)V_{j^*}(N)$$

∎

From this point onwards, due to lack of space we defer all proofs to the full version of the paper. For the case of two agents we can show existence of an optimal equilibrium.

Theorem 6. For $k = 1$, $n = 2$ there exists an optimal equilibrium.

Existence and Complexity. In contrast to case where one project is chosen, we show that even when $k = 2$ there may not exist a PNE when agents valuations are additive.

Theorem 7. *For additive agents and $k = 2$ there may not be a PNE.*

Regarding complexity, the computational hardness can be shown by reducing the problem of finding a Pure Nash Equilibrium to that of the well-studied problem of finding maximal coverage of a universe of elements.

Theorem 8. *It is NP-hard to compute a Pure Nash Equilibrium of the item bidding mechanism even for two agents with coverage valuations.*

6   Smoothness of the Item-Bidding Mechanism

In this section we study the efficiency achieved at learning outcomes and also when players have incomplete information about the valuations of the rest of the players. We defer proofs to the appendix. We give efficiency bounds by utilizing the recently proposed *Smooth Mechanism* framework [26]. For completeness, we present the basic definition of smoothness and the theorem that we will utilize.

Definition 1 (Syrgkanis, Tardos [26]). *A mechanism is (λ, μ)-smooth if for any valuation profile v, there exist strategies $b'_i(b_i, v)$, such that for any strategies $\boldsymbol{b}_{-i}$ of the rest of the players:*

$$\sum_i u_i(b'_i(b_i, v), \boldsymbol{b}_{-i}) \geq \lambda OPT(v) - \mu \sum_i P_i(\boldsymbol{b}) \tag{6}$$

where $OPT(v)$ is the optimal social welfare for valuation profile v, and $P_i(\boldsymbol{b})$ is the payment of player i under bid profile $\boldsymbol{b}$.

Theorem 9 (Syrgkanis, Tardos [26]). *If a mechanism (λ, μ)-smooth then the efficiency at any correlated equilibrium of the complete information game and at any Bayes-Nash equilibrium of the incomplete information game where v_i's are drawn from commonly known independent distributions F_i, is at least $\frac{\lambda}{\max\{\mu, 1\}}$. If the deviations $b'_i(b_i, v)$ in Definition 1 are independent of b_i then the latter holds also at coarse-correlated equilibria.*

We show that for any class of bidder valuations and for any k the Item-Bidding Mechanism is a $(\frac{1}{2}, n \cdot k)$-smooth mechanism, thereby implying a Price of Anarchy of at most $2nk$ for the aforementioned solution concepts.

Theorem 10. *For agents with arbitrary monotone valuations the Item-Bidding Mechanism is $\left(\frac{1}{2}, n \cdot k\right)$-smooth.*

When the bidders are fractionally subadditive, we show that the Item-Bidding Mechanism is $\left(\frac{1}{2}\left(1 - \frac{1}{e}\right), n\right)$-smooth, implying a Price of Anarchy of at most $\frac{2e}{e-1}n$, independent of the number k of projects to be chosen.

Theorem 11. *For agents with fractionally subadditive monotone valuations the Item-Bidding Mechanism is $\left(\frac{1}{2}\left(1 - \frac{1}{e}\right), n\right)$-smooth.*

When players are unit-demand we are able to show that the Item-Bidding Mechanism actually satisfies the *semi-smoothness* property of Lucier et al. [17] which is essentially the special case of Definition 1 where the deviating bids depend only on a players own valuation. Such a stronger property allows for the efficiency guarantee of theorem 9 to carry over to incomplete information settings where the bidder distributions are correlated. We show that the mechanism is $\left(1 - e^{-1}, \frac{n}{k}\right)$ semi-smooth for the case of unit-demand bidders implying a Price of Anarchy bound of at most $\frac{e}{e-1}\frac{n}{k}$, which decreases as the number of chosen project increases (i.e. more players are satisfied by building more projects).

Theorem 12. *When bidder valuations are unit-demand then the Item-Bidding Mechanism is $\left(1 - e^{-1}, \frac{n}{k}\right)$ semi-smooth.*

7 Discussion

The work presented in this paper is a first step towards the broader understanding of equilibria induced in combinatorial public projects. More generally, we explore the tools for mechanism design under solution concepts other than dominant strategy truthfulness.

While our bounds for pure Nash equilibria are nearly tight, better efficiency bounds may perhaps be achieved for the other solution concepts we explored in this work. It may be possible that sublogarithmic bounds can be shown for Strong Nash Equilibria, or that constant factor bounds may be achieveable by subgame perfect equilibria beyond the case of unit-demand bidders.

The simple mechanism for public projects we analyzed here is the item bidding mechanism with first prices. This is arguably the simplest non-trivial mechanism in this setting, and can be extended in multiple ways; similar allocation rules with second prices, or including constraints on the allocation rule, may lead to substantially different results than the ones presented here. In particular, we believe there is a simple mechanism where agents can reach an efficient equilibrium through natural dynamics for public projects, even in settings where no reasonable Maximal-In-Range mechanisms exist.

References

1. Bhawalkar, K., Roughgarden, T.: Welfare Guarantees for Combinatorial Auctions with Item Bidding. In: SODA 2011 (2011)
2. Blum, A., Mansour, Y.: Learning, regret minimization, and equilibria. In: Nisan, N., Roughgarden, T., Tardos, E., Vazirani, V. (eds.) Algorithmic Game Theory, ch. 4, pp. 4–30. Cambridge University Press (2007)
3. Buchfuhrer, D., Schapira, M., Singer, Y.: Computation and incentives in combinatorial public projects. In: ACM Conference on Electronic Commerce, pp. 33–42 (2010)

4. Caragiannis, I., Kaklamanis, C., Kanellopoulos, P., Kyropoulou, M.: On the Efficiency of Equilibria in Generalized Second Price Auctions. In: EC (2011)
5. Chawla, S., Hartline, J.D., Malec, D.L., Sivan, B.: Multi-parameter mechanism design and sequential posted pricing. In: Proceedings of the 42nd ACM Symposium on Theory of Computing, STOC 2010, pp. 311–320. ACM, New York (2010)
6. Christodoulou, G., Kovács, A., Schapira, M.: Bayesian combinatorial auctions. In: Aceto, L., Damgård, I., Goldberg, L.A., Halldórsson, M.M., Ingólfsdóttir, A., Walukiewicz, I. (eds.) ICALP 2008, Part I. LNCS, vol. 5125, pp. 820–832. Springer, Heidelberg (2008)
7. Dobzinski, S.: An impossibility result for truthful combinatorial auctions with submodular valuations. In: STOC, pp. 139–148 (2011)
8. Dughmi, S.: A truthful randomized mechanism for combinatorial public projects via convex optimization. In: ACM Conference on Electronic Commerce, pp. 263–272 (2011)
9. Feige, U.: On maximizing welfare when utility functions are subadditive. In: STOC (2006)
10. Feldman, M., Fu, H., Gravin, N., Lucier, B.: Simultaneous auctions are (almost) efficient. In: STOC (2013)
11. Fudenberg, D., Tirole, J.: Game Theory. MIT Press (1991)
12. Gupta, A., Ligett, K., McSherry, F., Roth, A., Talwar, K.: Differentially private combinatorial optimization. In: SODA, pp. 1106–1125 (2010)
13. Hassidim, A., Kaplan, H., Mansour, Y., Nisan, N.: Non-price equilibria in markets of discrete goods. In: EC 2011, p. 295. ACM Press, New York (2011)
14. Huang, Z., Kannan, S.: The exponential mechanism for social welfare: Private, truthful, and nearly optimal. In: FOCS, pp. 140–149 (2012)
15. Koutsoupias, E., Papadimitriou, C.: Worst-case equilibria. In: Meinel, C., Tison, S. (eds.) STACS 1999. LNCS, vol. 1563, pp. 404–413. Springer, Heidelberg (1999)
16. Lucier, B., Borodin, A.: Price of anarchy for greedy auctions. In: SODA 2010, p. 20 (2010)
17. Lucier, B., Leme, R.: GSP auctions with correlated types. In: EC 2011 (2011)
18. Nemhauser, G.L., Wolsey, L.A., Fisher, M.L.: An analysis of approximations for maximizing submodular set functions ii. Math. Programming Study 8, 73–87 (1978)
19. Paes Leme, R., Syrgkanis, V., Tardos, E.: The curse of simultaneity. In: ITCS (2012)
20. Paes Leme, R., Tardos, E.: Pure and Bayes-Nash price of anarchy for generalized second price auction. In: FOCS (2010)
21. Papadimitriou, C.H., Schapira, M., Singer, Y.: On the hardness of being truthful. In: FOCS, pp. 250–259 (2008)
22. Roughgarden, T.: Intrinsic robustness of the price of anarchy. In: STOC 2009 (2009)
23. Roughgarden, T.: The price of anarchy in games of incomplete information. In: Proceedings of the 13th ACM Conference on Electronic Commerce, EC 2012, pp. 862–879. ACM, New York (2012)
24. Schapira, M., Singer, Y.: Inapproximability of combinatorial public projects. In: Papadimitriou, C., Zhang, S. (eds.) WINE 2008. LNCS, vol. 5385, pp. 351–361. Springer, Heidelberg (2008)
25. Syrgkanis, V.: Bayesian games and the smoothness framework. CoRR, abs/1203.5155 (2012)
26. Syrgkanis, V., Tardos, E.: Composable and efficient mechanisms. In: STOC (2013)

Exchange Markets: Strategy Meets Supply-Awareness *

(Abstract)

Ruta Mehta[1],[**] and Milind Sohoni[2]

[1] College of Computing, Georgia Tech
rmehta@cc.gatech.edu
[2] Dept. of CSE, IIT, Bombay
sohoni@cse.iitb.ac.in

The exchange market model, proposed by Leon Walras (1874), has been studied extensively since more than a century due to its immense practical relevance [8,14]. The two implicit assumptions in this model are that agents behave truthfully, and are unaware of the total supply of goods in the market. In this paper we study exchange markets, with each of these assumptions dropped separately, and establish a surprising connection between their solutions.

The strategic behavior of agents is well known; many different types of market games have been formulated and analyzed for its Nash equilibria [1,2,5,6,7,13]. Generalizing the Fisher market[1] game of [1], we define the *exchange market game*, as where agents are the players and strategies are the utility functions that they may pose. We derive a complete characterization of the symmetric Nash equilibria (SNE) of this game, for the case when utility functions are *linear*.

Using the characterization of SNE we obtain: (i) the payoffs at SNE are always Pareto-optimal, and (ii) every competitive equilibrium allocation can be achieved at a SNE. Apart from these, we also obtain structural properties for the SNE set, like (iii) connectedness, and (iv) the necessary and sufficient conditions for uniqueness. These properties are important in equilibrium theory, both competitive and Nash, and a lot of work has been done to characterize such instances [2,7,9,10,11,12,13].

The other assumption that agents are unaware of the total supply of goods in the market, may not hold in many rural and informal markets where supplies are visible. Given that agents know the supply of all the goods, it is rational for them to take the supplies into consideration while calculating their demand bundles. This will change the demand dynamics, and as a consequence the set of competitive equilibria. Such a setting has been analyzed for auction markets [3,4], however to the best of our knowledge no such work for exchange markets is known.

We make significant progress towards understanding the effects of supply-aware agents in exchange markets. We show that the set of competitive equilibria

---

* A full version of this paper is available at http://www.cse.iitb.ac.in/~sohoni/supplyaware.pdf

** Research supported by NSF Grant CCF-1216019.
[1] Fisher market is a special case of exchange market model.

Y. Chen and N. Immorlica (Eds.): WINE 2013, LNCS 8289, pp. 361–362, 2013.

(CE) of such a market is equivalent to the set of SNE of the corresponding exchange market game. Through this equivalence, we obtain both the welfare theorems, and connectedness and uniqueness conditions of CE for the supply-aware markets with *linear* utilities.

Finally, for markets with arbitrary concave utilities, we derive sufficiency conditions for a strategy to be a symmetric Nash equilibrium, while restricting strategies of the agents to linear functions in the game. Using these conditions we obtain the first two properties, namely, Pareto-optimality and achieving CE allocations at SNEs, for this general setting. Further, we extend the connection between CE and SNE to markets with concave utility functions, and as a consequence obtain both the welfare theorems for the *supply-aware* markets in general.

We note that even though supply-awareness may be thought of as exchange markets with concave utility functions, where no more utility is obtained after the available supply of goods, the welfare theorems do not follow directly as they require non-satiated utility functions [15].

References

1. Adsul, B., Babu, C.S., Garg, J., Mehta, R., Sohoni, M.: Nash equilibria in fisher market. In: Kontogiannis, S., Koutsoupias, E., Spirakis, P.G. (eds.) SAGT 2010. LNCS, vol. 6386, pp. 30–41. Springer, Heidelberg (2010)
2. Amir, R., Sahi, S., Shubik, M., Yao, S.: A strategic market game with complete markets. Journal of Economic Theory 51, 126–143 (1990)
3. Bhattacharya, S.: Auctions, equilibria, and budgets. PhD Thesis (2012)
4. Borgs, C., Chayes, J., Etesami, O., Immorlica, N., Jain, K., Mahdian, M.: Dynamics of bid optimization in online advertisement auctions. In: WWW (2007)
5. Chen, N., Deng, X., Zhang, H., Zhang, J.: Incentive ratios of fisher markets. In: Czumaj, A., Mehlhorn, K., Pitts, A., Wattenhofer, R. (eds.) ICALP 2012, Part II. LNCS, vol. 7392, pp. 464–475. Springer, Heidelberg (2012)
6. Chen, N., Deng, X., Zhang, J.: How profitable are strategic behaviors in a market? In: European Symposium on Algorithms (2011)
7. Dubey, P., Geanakoplos, J.: From Nash to Walras via Shapley-Shubik. Journal of Mathematical Economics 39, 391–400 (2003)
8. Ellickson, B.: Competitive equilibrium: Theory and applications. Cambridge University Press (1994)
9. Florig, M.: Equilibrium correspondence of linear exchange economies. Journal of Optimization Theory and Applications 120(1), 97–109 (2004)
10. Mas-Colell, A.: On the uniqueness of equilibrium once again. In: Equilibrium Theory and Applications, pp. 275–296 (1991)
11. Pearce, I.F., Wise, J.: On the uniqueness of competitive equilibrium: Part ii, bounded demand. Econometrica 42(5), 921–932 (1974)
12. Rosen, J.B.: Existence and uniqueness of equilibrium points for concave n-person games. Econometrica 33(3), 520–534 (1965)
13. Shapley, L., Shubik, M.: Trade using one commodity as a means of payment. Journal of Political Economy 85(5), 937–968 (1977)
14. Shoven, J.B., Whalley, J.: Applying general equilibrium. Cambridge University Press (1992)
15. Varian, H.: Microeconomic Analysis. W. W. Norton & Company (1992)

A Lemke-Like Algorithm for the Multiclass Network Equilibrium Problem

Frédéric Meunier and Thomas Pradeau

Université Paris Est, CERMICS (ENPC)
F-77455 Marne-la-Vallée
{frederic.meunier,thomas.pradeau}@enpc.fr

Abstract. We consider a nonatomic congestion game on a connected graph, with several classes of players. Each player wants to go from its origin vertex to its destination vertex at the minimum cost and all players of a given class share the same characteristics: cost functions on each arc, and origin-destination pair. Under some mild conditions, it is known that a Nash equilibrium exists, but the computation of an equilibrium in the multiclass case is an open problem for general functions. We consider the specific case where the cost functions are affine and propose an extension of Lemke's algorithm able to solve this problem. At the same time, it provides a constructive proof of the existence of an equilibrium in this case.

Keywords: affine cost functions, congestion externalities, constructive proof, Lemke algorithm, nonatomic games, transportation network.

1   Introduction

Context. Being able to predict the impact of a new infrastructure on the traffic in a transportation network is an old but still important objective for transport planners. In 1952, Wardrop [28] noted that after some while the traffic arranges itself to form an equilibrium and formalized principles characterizing this equilibrium. With the terminology of game theory, the equilibrium is a Nash equilibrium for a congestion game with nonatomic players. In 1956, Beckmann [4] translated these principles as a mathematical program which turned out to be convex, opening the door to the tools from convex optimization. The currently most commonly used algorithm for such convex programs is probably the Frank-Wolfe algorithm [15], because of its simplicity and its efficiency, but many other algorithms with excellent behaviors have been proposed, designed, and experimented.

One of the main assumptions used by Beckmann to derive his program is the fact that all users are equally impacted by congestion. With the transportation terminology, it means that there is only one *class*. In order to improve the prediction of traffic patterns, researchers started in the 70s to study the *multiclass* situation where each class has its own way of being impacted by the congestion. Each class models a distinct mode of transportation, such as cars, trucks,

Y. Chen and N. Immorlica (Eds.): WINE 2013, LNCS 8289, pp. 363–376, 2013.
© Springer-Verlag Berlin Heidelberg 2013

or motorbikes. Dafermos [8,9] and Smith [27] are probably the first who proposed a mathematical formulation of the equilibrium problem in the multiclass case. However, even if this problem has been the topic of many research works, an efficient algorithm for solving it remains to be designed, except in some special cases [13,16,20,22]. In particular, there is no general algorithm in the literature for solving the problem when the cost of each arc is in an affine dependence with the flow on it.

Our main purpose is to propose such an algorithm.

Model. We are given a directed graph $D = (V, A)$ modeling the transportation network. A *route* is an s-t path of D and is called an s-t route. The set of all routes (resp. s-t routes) is denoted by $\mathcal{R}$ (resp. $\mathcal{R}_{(s,t)}$). The population of *players* is modeled as a bounded real interval I endowed with the Lebesgue measure λ, the *population measure*. The set I is partitioned into a finite number of measurable subsets $(I^k)_{k \in K}$ – the *classes* – modeling the players with same characteristics: they share a same collection of cost functions $(c_a^k : \mathbb{R}_+ \to \mathbb{R}_+)_{a \in A}$, a same origin s^k, and a same destination t^k. A player in I^k is said to be of *class k*. The set of vertices (resp. arcs) reachable from s^k is denoted V^k (resp. A^k).

A *strategy profile* is a measurable mapping $\sigma : I \to \mathcal{R}$ such that $\sigma(i) \in \mathcal{R}_{(s^k,t^k)}$ for all $k \in K$ and all $i \in I^k$. For each arc $a \in A$, the measure x_a^k of the set of all class k players i such that a is in $\sigma(i)$ is the *class k flow* on a in σ:

$$x_a^k = \lambda \{i \in I^k : a \in \sigma(i)\} .$$

The *total flow* on a is $x_a = \sum_{k \in K} x_a^k$. The cost of arc a for a class k player is then $c_a^k(x_a)$. For a class k player, the cost of a route r is defined as the sum of the costs of the arcs contained in r. Each player wants to select a minimum-cost route.

A strategy profile is a (pure) Nash equilibrium if each route is only chosen by players for whom it is a minimum-cost route. In other words, a strategy profile σ is a Nash equilibrium if for each class $k \in K$ and each player $i \in I^k$ we have

$$\sum_{a \in \sigma(i)} c_a^k(x_a) = \min_{r \in \mathcal{R}_{(s^k,t^k)}} \sum_{a \in r} c_a^k(x_a) . \tag{1}$$

This game enters in the category of *nonatomic congestion games with player-specific cost functions*, see Milchtaich [21]. The problem of finding a Nash equilibrium for such a game is called the *Multiclass Network Equilibrium Problem*.

Contribution. Our results concern the case when the cost functions are affine and strictly increasing: for all $k \in K$ and $a \in A^k$, there exist $\alpha_a^k > 0$ and $\beta_a^k \geq 0$ such that $c_a^k(x) = \alpha_a^k x + \beta_a^k$ for all $x \in \mathbb{R}_+$. In this case, the Multiclass Network Equilibrium Problem can be written as a linear complementarity problem. In 1965, Lemke [19] designed a pivoting algorithm for solving a linear complementarity problem under a quite general form. This algorithm has been adapted and extended several times – see for instance [1,3,5,7,12,25] – to be able to deal with linear complementarity problems that do not directly fit in the required framework of the original Lemke algorithm.

We show that there exists a pivoting Lemke-like algorithm solving the Multiclass Network Equilibrium Problem when the costs are affine. To our knowledge, it is the first algorithm solving this problem. We prove its efficiency through computational experiments. Moreover, the algorithm provides the first constructive proof of the existence of an equilibrium for this problem. The initial proof of the existence from Schmeidler [26] uses a non-constructive approach with the help of a general fixed point theorem.

On our track, we extend slightly the notion of basis used in linear programming and linear complementarity programming to deal directly with unsigned variables. Even if it is natural, we are not aware of previous use of such an approach. An unsigned variable can be replaced by two variables – one for the nonnegative part and one for the nonpositive part. Such an operation considerably increases the size of the matrices, while, in our approach, we are able to deal directly with the unsigned variables.

Related Works. We already gave some references of works related to ours with respect to the linear complementarity. The work by Schiro *et al.* [25] is one of them and deals actually with a problem more general than ours. They propose a pivotal algorithm to solve it. However, our problem is not covered by their termination results (the condition of their Proposition 5 is not satisfied by our problem). Another close work is the one by Eaves [12], which allows additional affine constraints on the variables, but the constraints we need – flow constraints – do not enter in this framework. Note also the work by De Schutter and De Moor [11], devoted to the "Extended Linear Complementarity Problem" which contains our problem. They propose a method that exhaustively enumerates all solutions and all extreme rays, without giving *a priori* guarantee for the existence of a solution.

Papers dealing with algorithms for solving the Multiclass Network Equilibrium Problems propose in general a Gauss-Seidel type diagonalization method, which consists in sequentially fixing the flows for all classes but one and solving the resulting single-class problem by methods of convex programming, see [13,14,16,20] for instance. For this method, a condition ensuring the convergence to an equilibrium is not always stated, and, when there is one, it requires that "the interaction between the various users classes be relatively weak compared to the main effects (the latter translates a requirement that a complicated matrix norm be less than unity)" [20]. Such a condition does clearly not cover the case with affine cost functions. Another approach is proposed by Marcotte and Wynter [22]. For cost functions satisfying the "nested monotonicity" condition – a notion developed by Cohen and Chaplais [6] – they design a descent method for which they are able to prove the convergence to a solution of the problem. However, we were not able to find any paper with an algorithm solving the problem when the costs are polynomial functions, or even affine functions.

Structure of the Paper. In Section 2, we explain how to write the Multiclass Network Equilibrium Problem as a linear complementarity problem. We get the formulation $(AMNEP(e))$ on which the remaining of the paper focuses. Section 3 presents the notions that underly the Lemke-like algorithm. All these notions, likes basis, secondary ray, pivot, and so on, are classical in the context of the Lemke algorithm. They require however to be redefined in order to be able to deal with the features of $(AMNEP(e))$. The algorithm is then described in Section 4. We also explain why it provides a constructive proof of the existence of an equilibrium. Section 5 is devoted to the experiments and shows the efficiency of the proposed approach.

Due to page limitations, all proofs are omitted. The interested reader may find them in the full version of the paper available on

http://cermics.enpc.fr/~pradeath/Research.html.

2   Formulation as a Linear Complementarity Problem

In this section, we formulate the Multiclass Network Equilibrium Problem as a complementarity problem which turns out to be linear when the cost functions are affine.

From now on, we assume that the cost functions are increasing. In the single-class case, i.e. $|K| = 1$, the equilibrium flows are optimal solutions of a convex optimization problem, see Beckmann [4]. If the flows $\boldsymbol{x}^{k'}$ for $k' \neq k$ are fixed, finding the equilibrium flows for the class k is again a single-class problem which can be formulated as a convex optimization problem. With the help of the Karush-Kuhn-Tucker conditions, we get that the equilibrium flows (x_a^k) coincide with the solutions of a system of the following form, where $\boldsymbol{b} = (b_v^k)$ is a given vector with $\sum_{v \in V^k} b_v^k = 0$ for all k.

$$\sum_{a \in \delta^+(v)} x_a^k = \sum_{a \in \delta^-(v)} x_a^k + b_v^k \qquad k \in K, v \in V^k$$

$$c_{uv}^k(x_{uv}) + \pi_u^k - \pi_v^k - \mu_{uv}^k = 0 \qquad k \in K, (u,v) \in A^k \qquad (MNEP_{gen})$$

$$x_a^k \mu_a^k = 0 \qquad k \in K, a \in A^k$$

$$x_a^k \geq 0, \mu_a^k \geq 0, \pi_v^k \in \mathbb{R} \qquad k \in K, a \in A^k, v \in V^k \,.$$

Actually in our model, we should have moreover $b_v^k = 0$ for $v \notin \{s^k, t^k\}$, and the inequalities $b_{s^k}^k > 0$ and $b_{t^k}^k < 0$, but we relax this condition to deal with a slightly more general problem. Moreover, in this more general form, we can easily require the problem to be non-degenerate, see Section 3.2.

Finding solutions for such systems is a *complementarity problem*, the word "complementarity" coming from the condition $x_a^k \mu_a^k = 0$ for all (a, k) such that $a \in A^k$.

We have thus the following proposition.

Proposition 1. $(\boldsymbol{x}^k)_{k \in K}$ *is an equilibrium flow if and only if there exist* $\boldsymbol{\mu}^k \in \mathbb{R}_+^{A^k}$ *and* $\boldsymbol{\pi}^k \in \mathbb{R}^{V^k}$ *for all k such that* $(\boldsymbol{x}^k, \boldsymbol{\mu}^k, \boldsymbol{\pi}^k)_{k \in K}$ *is a solution of the complementarity problem* $(MNEP_{gen})$.

When the cost functions are affine $c_a^k(x) = \alpha_a^k x + \beta_a^k$, solving the Multiclass Network Equilibrium Problem amounts thus to solve the following linear complementarity problem

$$\sum_{a \in \delta^+(v)} x_a^k = \sum_{a \in \delta^-(v)} x_a^k + b_v^k \qquad\qquad k \in K, v \in V^k$$

$$\alpha_{uv}^k \sum_{k' \in K} x_{uv}^{k'} + \pi_u^k - \pi_v^k - \mu_{uv}^k = -\beta_{uv}^k \qquad k \in K, (u,v) \in A^k \qquad (MNEP)$$

$$x_a^k \mu_a^k = 0 \qquad\qquad k \in K, a \in A^k$$

$$x_a^k \geq 0, \mu_a^k \geq 0, \pi_v^k \in \mathbb{R} \qquad\qquad k \in K, a \in A^k, v \in V^k \ .$$

Similarly as for the Lemke algorithm, we rewrite the problem as an optimization problem. It will be convenient for the exposure of the algorithm, see Section 3. This problem is called the *Augmented Multiclass Network Equilibrium Problem*. It uses a vector $\boldsymbol{e} = (e_a^k)$ defined for all $k \in K$ and $a \in A^k$. Problem $(AMNEP(\boldsymbol{e}))$ is

$$\min \omega$$
$$\text{s.t.} \sum_{a \in \delta^+(v)} x_a^k = \sum_{a \in \delta^-(v)} x_a^k + b_v^k \qquad\qquad k \in K, v \in V^k$$

$$\alpha_{uv}^k \sum_{k' \in K} x_{uv}^{k'} + \pi_u^k - \pi_v^k - \mu_{uv}^k + e_{uv}^k \omega = -\beta_{uv}^k \qquad k \in K, (u,v) \in A^k$$

$$x_a^k \mu_a^k = 0 \qquad\qquad k \in K, a \in A^k$$

$$x_a^k \geq 0, \mu_a^k \geq 0, \omega \geq 0, \pi_v^k \in \mathbb{R} \qquad\qquad k \in K, a \in A^k, v \in V^k \ .$$
$$(AMNEP(\boldsymbol{e}))$$

Some choices of $\boldsymbol{e}$ allow to find easily feasible solutions to problem $(AMNEP(\boldsymbol{e}))$. In Section 3, $\boldsymbol{e}$ will be chosen in such a way. A key remark is that solving $(MNEP)$ amounts to find an optimal solution for $(AMNEP(\boldsymbol{e}))$ with $\omega = 0$.

Without loss of generality, we impose that $\pi_{s^k}^k = 0$ for all $k \in K$ and it holds throughout the paper. It allows to rewrite problem $(AMNEP(\boldsymbol{e}))$ under the form

$$\min \omega$$

$$\text{s.t. } \overline{M}^e \begin{pmatrix} x \\ \mu \\ \omega \end{pmatrix} + \begin{pmatrix} 0 \\ M^T \end{pmatrix} \pi = \begin{pmatrix} b \\ -\beta \end{pmatrix}$$

$$x \cdot \mu = 0$$

$$x \geq 0, \ \mu \geq 0, \ \omega \geq 0, \ \pi \in \mathbb{R}^{\Sigma_k V^k \setminus \{s^k\}},$$

where $\overline{M}^e$ and C are defined as follows. (The matrix $\overline{M}^e$ is denoted with a superscript e in order to emphasize its dependency on e).

We define $M = \text{diag}((M^k)_{k \in K})$ where M^k is the incidence matrix of the directed graph (V^k, A^k) from which the s^k-row has been removed:

$$M^k_{v,a} = \begin{cases} 1 & \text{if } a \in \delta^+(v), \\ -1 & \text{if } a \in \delta^-(v), \\ 0 & \text{otherwise}. \end{cases}$$

We also define $C^k = \text{diag}((\alpha^k_a)_{a \in A^k})$ for $k \in K$, and then C the real matrix $C = (\underbrace{(C^k, \cdots, C^k)}_{|K| \text{ times}})_{k \in K}$. Then let

$$\overline{M}^e = \begin{pmatrix} M & 0 & 0 \\ C & -I & e \end{pmatrix}.$$

For $k \in K$, the matrix M^k has $|V^k| - 1$ rows and $|A^k|$ columns, while C^k is a square matrix with $|A^k|$ rows and columns. Then the whole matrix $\overline{M}^e$ has $\sum_{k \in K}(|A^k| + |V^k| - 1)$ rows and $2\left(\sum_{k \in K} |A^k|\right) + 1$ columns.

3   Bases, Pivots, and Rays

3.1   Bases

We define $\mathcal{X}$ and $\mathcal{M}$ to be two disjoint copies of $\{(a, k) : k \in K, a \in A^k\}$. We denote by $\phi^x(a, k)$ (resp. $\phi^\mu(a, k)$) the element of $\mathcal{X}$ (resp. $\mathcal{M}$) corresponding to (a, k). The set $\mathcal{X}$ models the set of all possible indices for the 'x' variables and $\mathcal{M}$ the set of all possible indices for the 'μ' variables for problem $(AMNEP(e))$. We consider moreover a dummy element o as the index for the 'ω' variable.

We define a *basis* for problem $(AMNEP(e))$ to be a subset B of the set $\mathcal{X} \cup \mathcal{M} \cup \{o\}$ such that the square matrix of size $\sum_{k \in K} (|A^k| + |V^k| - 1)$ defined by

$$\left(\overline{M}^e_B \middle| \begin{matrix} 0 \\ M^T \end{matrix} \right)$$

is nonsingular. Note that this definition is not standard. In general, a basis is defined in this way but without the submatrix $\begin{pmatrix} 0 \\ M^T \end{pmatrix}$ corresponding to the 'π' columns. We use this definition in order to be able to deal directly with the unsigned variables 'π'. We will see that this approach is natural (and could be used for linear programming as well). However, we are not aware of a previous use of such an approach.

As a consequence of this definition, since M^T has $\sum_{k \in K}(|V^k| - 1)$ columns, a basis is always of cardinality $\sum_{k \in K} |A^k|$.

The following additional notation is useful: given a subset $Z \subseteq \mathcal{X} \cup \mathcal{M} \cup \{o\}$, we denote by Z^x the set $(\phi^x)^{-1}(Z \cap \mathcal{X})$ and by Z^μ the set $(\phi^\mu)^{-1}(Z \cap \mathcal{M})$. In other words, (a, k) is in Z^x if and only if $\phi^x(a, k)$ is in Z, and similarly for Z^μ.

3.2 Basic Solutions and Non-degeneracy

Let B a basis. If it contains o, the unique solution $(\bar{x}, \bar{\mu}, \bar{\omega}, \bar{\pi})$ of

$$
\begin{cases}
\left(\overline{M}^e_B \,\middle|\, \begin{matrix} \mathbf{0} \\ M^T \end{matrix} \right)
\begin{pmatrix} x_{B^x} \\ \mu_{B^\mu} \\ \omega \\ \pi \end{pmatrix}
= \begin{pmatrix} b \\ -\beta \end{pmatrix} \\[6pt]
x^k_a = 0 \quad \text{for all } (a, k) \notin B^x \\
\mu^k_a = 0 \quad \text{for all } (a, k) \notin B^\mu .
\end{cases}
\tag{2}
$$

is called the *basic solution* associated to B.

If B does not contain o, we define similarly its associated *basic solution*. It is the unique solution $(\bar{x}, \bar{\mu}, \bar{\omega}, \bar{\pi})$ of

$$
\begin{cases}
\left(\overline{M}^e_B \,\middle|\, \begin{matrix} \mathbf{0} \\ M^T \end{matrix} \right)
\begin{pmatrix} x_{B^x} \\ \mu_{B^\mu} \\ \pi \end{pmatrix}
= \begin{pmatrix} b \\ -\beta \end{pmatrix} \\[6pt]
x^k_a = 0 \quad \text{for all } (a, k) \notin B^x \\
\mu^k_a = 0 \quad \text{for all } (a, k) \notin B^\mu \\
\omega = 0 .
\end{cases}
\tag{3}
$$

A basis is said to be *feasible* if the associated basic solution is such that $\bar{x}, \bar{\mu}, \bar{\omega} \geq 0$.

The problem $(AMNEP(e))$ is said to *satisfy the non-degeneracy assumption* if, for any feasible basis B, the associated basic solution $(\bar{x}, \bar{\mu}, \bar{\omega}, \bar{\pi})$ is such that

$$
\left((a, k) \in B^x \Rightarrow \bar{x}^k_a > 0 \right) \text{ and } \left((a, k) \in B^\mu \Rightarrow \bar{\mu}^k_a > 0 \right) .
$$

Note that if we had defined the vector b to be 0 on all vertices $v \notin \{s^k, t^k\}$, the problem would not in general satisfy the non-degeneracy assumption. An example of a basis for which the condition fails to be satisfied is the basis B^{ini} defined in Section 3.5. Remark 1 in that section details the example.

3.3 Pivots and Polytope

The following lemmas are key results that will eventually lead to the Lemke-like algorithm. They are classical for the usual definition of bases. Since we have extended the definition, we have to prove that they still hold.

Lemma 1. *Let B be a feasible basis for problem $(AMNEP(e))$ and assume non-degeneracy. Let i be an index in $\mathcal{X} \cup \mathcal{M} \cup \{o\} \setminus B$. Then there is at most one feasible basis $B' \neq B$ in the set $B \cup \{i\}$.*

The operation consisting in computing B' given B and the *entering index i* is called the *pivot operation*.

If we are able to determine an index in $\mathcal{X} \cup \mathcal{M} \cup \{o\} \setminus B$ for any basis B, Lemma 1 leads to a "pivoting" algorithm. At each step, we have a current basis B^{curr}, we determine the entering index i, and we compute the new basis in $B^{curr} \cup \{i\}$, if it exists, which becomes the new current basis B^{curr}; and so on. Next lemma allows to characterize situations where there is no new basis, i.e. situations for which the algorithm gets stuck.

The feasible solutions of $(AMNEP(e))$ belong to the polytope

$$P(e) = \left\{ (x, \mu, \omega, \pi) : \overline{M}^e \begin{pmatrix} x \\ \mu \\ \omega \end{pmatrix} + \begin{pmatrix} 0 \\ M^T \end{pmatrix} \pi = \begin{pmatrix} b \\ -\beta \end{pmatrix}, \right.$$

$$\left. x \geq 0, \mu \geq 0, \pi \geq 0, \omega \in \mathbb{R}_+ \right\}.$$

Lemma 2. *Let B be a feasible basis for problem $(AMNEP(e))$ and assume non-degeneracy. Let i be an index in $\mathcal{X} \cup \mathcal{M} \cup \{o\} \setminus B$. If there is no feasible basis $B' \neq B$ in the set $B \cup \{i\}$, then the polytope $P(e)$ contains an infinite ray originating at the basic solution associated to B.*

3.4   Complementarity and Twin Indices

A basis B is said to be *complementary* if for every (a, k) with $a \in A^k$, we have $(a, k) \notin B^x$ or $(a, k) \notin B^\mu$: for each (a, k), one of the components x_a^k or μ_a^k is not activated in the basic solution. In case of non-degeneracy, it coincides with the condition $x \cdot \mu = 0$. An important point to be noted for a complementary basis B is that if $o \in B$, then there is (a_0, k_0) with $a_0 \in A^{k_0}$ such that

- $(a_0, k_0) \notin B^x$ and $(a_0, k_0) \notin B^\mu$, and
- for all $(a, k) \neq (a_0, k_0)$ with $a \in A^k$, exactly one of the relations $(a, k) \in B^x$ and $(a, k) \in B^\mu$ is satisfied.

This is a direct consequence of the fact that there are exactly $\sum_{k \in K} |A^k|$ elements in a basis and that each (a, k) is not present in at least one of B^x and B^μ. In case of non-degeneracy, this point amounts to say that $x_a^k = 0$ or $\mu_a^k = 0$ for all (a, k) with $a \in A^k$ and that there is exactly one such pair, denoted (a_0, k_0), such that both are equal to 0.

We say that $\phi^x(a_0, k_0)$ and $\phi^\mu(a_0, k_0)$ for such (a_0, k_0) are the *twin indices*.

3.5   Initial Feasible Basis

A good choice of e gives an easily computable initial feasible complementary basis to problem $(AMNEP(e))$.

An *s-arborescence* in a directed graph is a spanning tree rooted at s that has a directed path from s to any vertex of the graph. We arbitrarily define a collection $\mathcal{T} = (T^k)_{k \in K}$ where $T^k \subseteq A^k$ is an s^k-arborescence of (V^k, A^k). Then the vector $e = (e_a^k)_{k \in K}$ is chosen with the help of $\mathcal{T}$ by

$$e_a^k = \begin{cases} 1 \text{ if } a \notin T^k \\ 0 \text{ otherwise} . \end{cases} \tag{4}$$

Lemma 3. *Let the set of indices $Y \subseteq \mathcal{X} \cup \mathcal{M} \cup \{o\}$ be defined by*

$$Y = \{\phi^x(a, k) : a \in T^k, k \in K\} \cup \{\phi^\mu(a, k) : a \in A^k \setminus T^k, k \in K\} \cup \{o\} .$$

Then, one of the following situations occurs:

- *$Y \setminus \{o\}$ is a complementary feasible basis providing an optimal solution of problem $(AMNEP(e))$ with $\omega = 0$.*
- *There exists (a_0, k_0) such that $B^{ini} = Y \setminus \{\phi^\mu(a_0, k_0)\}$ is a feasible complementary basis for problem $(AMNEP(e))$.*

We emphasize that B^{ini} depends on the chosen collection $\mathcal{T}$ of arborescences. Note that the basis B^{ini} is polynomially computable.

Remark 1. As already announced in Section 3.2, if we had defined the vector b to be 0 on all vertices $v \notin \{s^k, t^k\}$, the problem would not satisfy the non-degeneracy assumption as soon as there is $k \in K$ such that T^k has a vertex of degree 3 (which happens when (V^k, A^k) has no Hamiltonian path). In this case, the basis B^{ini} shows that the problem is degenerate. Since the unique solution $x_{T^k}^k$ of $M_{T^k}^k x_{T^k}^k = b^k$ consists in sending the whole demand on the unique route in T^k from s^k to t^k, we have for all arcs $a \in T^k$ not belonging to this route $x_a^k = 0$ while $(a, k) \in B^{ini,x}$.

3.6   No Secondary Ray

Let $(\bar{x}^{ini}, \bar{\mu}^{ini}, \bar{\omega}^{ini}, \bar{\pi}^{ini})$ be the feasible basic solution associated to the initial basis B^{ini}, computed according to Lemma 3 and with e given by Equation (4). The following inifinite ray

$$\rho^{ini} = \left\{(\bar{x}^{ini}, \bar{\mu}^{ini}, \bar{\omega}^{ini}, \bar{\pi}^{ini}) + t(0, e, 1, 0) : t \geq 0\right\},$$

has all its points in $\mathcal{P}(e)$. This ray with direction $(0, e, 1, 0)$ is called the *primary ray*. In the terminology of the Lemke algorithm, another infinite ray originating at a solution associated to a feasible complementary basis is called a *secondary ray*. Recall that we defined $\pi_{s^k}^k = 0$ for all $k \in K$ in Section 2 (otherwise we would have a trivial secondary ray). System $(AMNEP(e))$ has no secondary ray for the chosen e.

Lemma 4. *Let e be defined by Equation (4). Under the non-degeneracy assumption, there is no secondary ray in $\mathcal{P}(e)$.*

3.7   A Lemke-Like Algorithm

Assuming non-degeneracy, the combination of Lemma 1 and the point explicited
in Section 3.4 give raise to a Lemke-like algorithm. Two feasible complementary
bases B and B' are said to be *neighbors* if B' can be obtained from B by a pivot
operation using one of the twin indices as an entering index, see Section 3.4.
Note that this is a symmetrical notion: B can then also be obtained from B'
by a similar pivot operation. The abstract graph whose vertices are the feasible
complementary bases and whose edges connect neighbor bases is thus a collection
of paths and cycles. According to Lemma 3, we can find in polynomial time an
initial feasible complementary basis for $(AMNEP(e))$ with the chosen vector e.
This initial basis has exactly one neighbor according to Lemma 2 since there is
a primary ray and no secondary ray (Lemma 4).

Algorithm 1 explains how to follow the path starting at this initial feasible
complementary basis. Function $\texttt{EnteringIndex}(B, i')$ is defined for a feasible
complementary basis B and an index $i' \notin B$ being a twin index of B and
computes the other twin index $i \neq i'$. Function $\texttt{LeavingIndex}(B, i)$ is defined
for a feasible complementary basis B and an index $i \notin B$ and computes the
unique index $j \neq i$ such that $B \cup \{i\} \setminus \{j\}$ is a feasible complementary basis (see
Lemma 1).

Since there is no secondary ray (Lemma 4), a pivot operation is possible because of Lemma 2 as long as there are twin indices. By finiteness, a component in
the abstract graph having an endpoint necessarily has another endpoint. It implies that the algorithm reaches at some moment a basis B without twin indices.
Such a basis is such that $o \notin B$ (Section 3.4), which implies that we have a solution of problem $(AMNEP(e))$ with $\omega = 0$, i.e. a solution of problem $(MNEP)$,
and thus a solution of our initial problem.

---

input : The matrix $\overline{M}^e$, the matrix M, the vectors $\boldsymbol{b}$ and β, an initial feasible
        complementary basis B^{ini}

output: A feasible basis B^{end} with $o \notin B^{end}$.

$\phi^\mu(a_0, k_0) \leftarrow$ twin index in $\mathcal{M}$;

$i \leftarrow \texttt{EnteringIndex}(B^{ini}, \phi^\mu(a_0, k_0))$;

$j \leftarrow \texttt{LeavingIndex}(B^{ini}, i)$;

$B^{curr} \leftarrow B^{ini} \cup \{i\} \setminus \{j\}$;

while *There are twin indices* **do**

    $i \leftarrow \texttt{EnteringIndex}(B^{curr}, j)$;

    $j \leftarrow \texttt{LeavingIndex}(B^{curr}, i)$;

    $B^{curr} \leftarrow B^{curr} \cup \{i\} \setminus \{j\}$;

end

$B^{end} \leftarrow B^{curr}$;

return B^{end};

Algorithm 1. Lemke-like algorithm

4 Algorithm and Main Result

We are now in a position to describe the full algorithm under the non-degeneracy assumption.

1. For each $k \in K$, compute a collection $\mathcal{T} = (T^k)$ where $T^k \subseteq A^k$ is an s^k-arborescence of (V^k, A^k).
2. Define e as in Equation (4) (which depends on $\mathcal{T}$).
3. Define $Y = \{\phi^x(a,k) : a \in T^k, k \in K\} \cup \{\phi^\mu(a,k) : a \in A^k \setminus T^k, k \in K\} \cup \{o\}$.
4. If $Y \setminus \{o\}$ is a complementary feasible basis providing an optimal solution of problem $(AMNEP(e))$ with $\omega = 0$, then we have a solution of problem $(MNEP)$, see Lemma 3.
5. Otherwise, let B^{ini} be defined as in Lemma 3 and apply Algorithm 1, which returns a basis B^{end}.
6. Compute the basic solution associated to B^{end}.

All the elements proved in Section 3 lead finally to the following result.

Theorem 1. *Under the non-degeneracy assumption, this algorithm solves problem $(MNEP)$, i.e. the Multiclass Network Equilibrium Problem with affine costs.*

This result provides actually a constructive proof of the existence of an equilibrium for the Multiclass Network Equilibrium Problem when the cost are affine and strictly increasing, even if the non-degeneracy assumption is not satisfied. If we compute $b = (b_v^k)$ strictly according to the model, we have

$$b_v^k = \begin{cases} \lambda(I^k) & \text{if } v = s^k \\ -\lambda(I^k) & \text{if } v = t^k \\ 0 & \text{otherwise}. \end{cases} \tag{5}$$

In this case, the non-degeneracy assumption is not satisfied as it has been noted at the end of Section 3.5 (Remark 1). Anyway, we can slightly perturb b and $-\beta$ in such a way that any feasible complementary basis of the perturbated problem is still a feasible complementary basis for the original problem. Such a perturbation exists by standard arguments, see [7]. Theorem 1 ensures then the termination of the algorithm on a feasible complementary basis B whose basic solution is such that $\omega = 0$. It provides thus a solution for the original problem.

It shows also that the problem of finding such an equilibrium belongs to the PPAD complexity class. The PPAD class – defined by Papadimitriou [24] in 1994 – is the complexity class of functional problems for which we know the existence of the object to be found because of a (oriented) path-following argument. There are PPAD-complete problems, i.e. PPAD problems as hard as any problem in the PPAD class, see [18] for examples of such problems. A natural question would be whether the Multiclass Network Equilibrium Problem with affine costs is PPAD-complete. We do not know the answer. Another natural question is whether the problem belongs to other complexity classes often met in the context of congestion games, such as the PLS class [17] or the CLS class [10]. However, these latter classes require the existence of some potential functions which is not likely to be the case for our problem.

Another consequence of Theorem 1 is that if the demands $\lambda(I^k)$ and the cost parameters α_a^k, β_a^k are rational numbers, then there exists an equilibrium inducing rational flows on each arc and for each class k. It is reminiscent of a similar result for two players matrix games: if the matrices involve only rational entries, there is an equilibrium involving only rational numbers [23].

5   Computational Experiments

5.1   Instances

The experiments are made on $n \times n$ grid graphs (Manhattan instances). For each pair of adjacent vertices u and v, both arcs (u, v) and (v, u) are present. We built several instances on these graphs with various sizes n, various numbers of classes, and various cost parameters α_a^k, β_a^k. The cost parameters were chosen uniformly at random such that for all a and all k

$$\alpha_a^k \in [1, 10] \quad \text{and} \quad \beta_a^k \in [0, 100] \, .$$

5.2   Results

The algorithm has been coded in C++ and tested on a PC Intel® Core™ i5-2520M clocked at 2.5 GHz, with 4 GB RAM. The experiments are currently in progress. However, some preliminary computational results are given in Table 1. Each row of the table contains average figures obtained on five instances on the same graph and with the same number classes, but with various origins, destinations, and costs parameters.

The columns "Classes", "Vertices", and "Arcs" contain respectively the number of classes, the number of vertices, and the number of arcs. The column "Pivots" contains the number of pivots performed by the algorithm. They are done during Step 5 in the description of the algorithm in Section 4 (application of Algorithm 1). The column "Algorithm 1" provides the time needed for the whole execution of this pivoting step. The preparation of this pivoting step requires a first matrix inversion, and the final computation of the solution requires such an inversion as well. The times needed to perform these inversions are given in the column "Inversion". The total time needed by the complete algorithm to solve the problem is the sum of the "Algorithm 1" time and twice the "Inversion" time, the other steps of the algorithm taking a negligible time.

It seems that the number of pivots remains always reasonable. Even if the time needed to solve large instances is sometimes important with respect to the size of the graph, the essential computation time is spent on the two matrix inversions. The program has not been optimized. Since there are several efficient techniques known for inverting matrices, the results can be considered as very positive.

Table 1. Performances of the complete algorithm for various instance sizes

Classes	Grid	Vertices	Arcs	Pivots	Algorithm 1 (seconds)	Inversion (seconds)
2	2 × 2	4	8	2	<0.01	<0.01
	4 × 4	16	48	21	0.01	0.03
	6 × 6	36	120	54	0.08	0.5
	8 × 8	64	224	129	0.9	4.0
3	2 × 2	4	8	4	<0.01	<0.01
	4 × 4	16	48	33	0.03	0.1
	6 × 6	36	120	97	0.4	1.9
	8 × 8	64	224	183	2.6	12
4	2 × 2	4	8	3	<0.01	<0.01
	4 × 4	16	48	41	0.06	0.3
	6 × 6	36	120	126	0.9	4.7
	8 × 8	64	224	249	5.4	25
10	2 × 2	4	8	11	<0.01	0.02
	4 × 4	16	48	107	0.7	4.1
	6 × 6	36	120	322	15	70
	8 × 8	64	224	638	87	385
50	2 × 2	4	8	56	0.3	2.6
	4 × 4	16	48	636	105	511

References

1. Adler, I., Verma, S.: The Linear Complementarity Problem, Lemke Algorithm, Perturbation, and the Complexity Class PPAD, Technical Report (2011)
2. Ahuja, R.K., Magnanti, T.L., Orlin, J.B.: Network Flows: Theory, Algorithms, and Applications. Prentice-Hall, Englewood Cliffs (1993)
3. Asmuth, R., Eaves, B.C., Peterson, E.L.: Computing Economic Equilibria on Affine Networks with Lemke's Algorithm. Math. Oper. Res. 4, 209–214 (1979)
4. Beckmann, M., McGuire, C.B., Winsten, C.B.: Studies in Economics of Transportation. Yale University Press, New Haven (1956)
5. Cao, M., Ferris, M.C.: A pivotal method for affine variational inequalities. Math. Oper. Res. 21, 44–64 (1996)
6. Cohen, G., Chaplais, F.: Nested monotonicity for variational inequalities over product of spaces and convergence of iterative algorithms. J. Optim. Theory Appl. 59, 369–390 (1988)
7. Cottle, R.W., Pang, J.S., Stone, R.E.: The linear complementarity problem. Academic Press, New York (1992)
8. Dafermos, S.: The Traffic Assignment Problem for Multiclass-User Transportation Networks. Transportation Sci. 6, 73–87 (1972)
9. Dafermos, S.: Traffic equilibrium and variational inequalities. Transportation Sci. 14, 42–54 (1980)
10. Daskalakis, C., Papadimitriou, C.: Continuous Local Search. In: 22nd Annual ACM-SIAM Symposium on Discrete Algorithms (SODA), San Francisco (2011)
11. De Schutter, B., De Moor, B.: The Extended Linear Complementarity Problem. Tech. Report (1995)

12. Eaves, B.C.: Polymatrix games with joint constraints. SIAM J. Appl. Math. 24, 418–423 (1973)
13. Florian, M.: A traffic equilibrium model of travel by car and public transit modes. Transportation Sci. 11, 166–179 (1977)
14. Florian, M., Spiess, H.: The convergence of diagonalisation algorithms for asymmetric network equilibrium problems. Transportation Res. Part B 16, 477–483 (1982)
15. Frank, M., Wolfe, P.: An algorithm for quadratic programming. Naval Research Logistics Quarterly 3, 95–110 (1956)
16. Harker, P.T.: Accelerating the convergence of the diagonalization and projection algorithms for finite-dimensional variational inequalities. Math. Programming 41, 29–59 (1988)
17. Johnson, D.S., Papadimitriou, C., Talwar, K.: How easy is Local Search? Journal of Computer and System Sciences 37, 79–100 (1988)
18. Kintali, S., Poplawski, L.J., Rajaraman, R., Sundaram, R., Teng, S.-H.: Reducibility among fractional stability problems. In: 50th IEEE Symposium on Foundations of Computer Science (FOCS), Atlanta (2009)
19. Lemke, C.E.: Bimatrix equilibrium points and equilibrium programming. Management Science 11, 681–689 (1965)
20. Mahmassani, H.S., Mouskos, K.C.: Some numerical results on the diagonalization algorithm for network assignment with asymmetric interactions between cars and trucks. Transportation Res. Part B 22, 275–290 (1988)
21. Milchtaich, I.: Congestion games with player-specific payoff functions. Games Econom. Behavior 13, 111–124 (1996)
22. Marcotte, P., Wynter, L.: A new look at the multiclass network equilibrium problem. Transportation Sci. 38, 282–292 (2004)
23. Nash, J.F.: Non-Cooperative games. Annals of Mathematics 54, 286–295 (1951)
24. Papadimitriou, C.: On the complexity of the parity argument and other inefficient proofs of existence. Journal of Computer and System Sciences 48, 498–532 (1994)
25. Schiro, D.A., Pang, J.-S., Shanbhag, U.V.: On the solution of affine generalized Nash equilibrium problems with shared constraints by Lemke's method. Math. Program. (2012)
26. Schmeidler, D.: Equilibrium points on nonatomic games. J. Statist. Phys. 7, 295–300 (1970)
27. Smith, M.J.: The existence, uniqueness, and stability of traffic equilibria. Transportation Res. Part B. 15, 443–451 (1979)
28. Wardrop, J.G.: Some theoretical aspects of road traffic research. Proc. Inst. Civil Engineers 2, 325–378 (1952)

Near-Optimal and Robust Mechanism Design for Covering Problems with Correlated Players*

Hadi Minooei** and Chaitanya Swamy**

Combinatorics and Optimization, Univ. Waterloo, Waterloo, ON N2L 3G1
{hminooei,cswamy}@math.uwaterloo.ca

Abstract. We consider the problem of designing incentive-compatible, ex-post individually rational (IR) mechanisms for covering problems in the Bayesian setting, where players' types are drawn from an underlying distribution and may be correlated, and the goal is to minimize the expected total payment made by the mechanism. We formulate a notion of incentive compatibility (IC) that we call *robust Bayesian IC* (robust BIC) that is substantially more robust than BIC, and develop black-box reductions from robust-BIC mechanism design to algorithm design. For single-dimensional settings, this black-box reduction applies even when we only have an LP-relative *approximation algorithm* for the algorithmic problem. Thus, we obtain near-optimal mechanisms for various covering settings including single-dimensional covering problems, multi-item procurement auctions, and multidimensional facility location.

1 Introduction

We consider the problem of designing incentive-compatible, ex-post individually rational (IR) mechanisms for *covering problems* (also called *procurement auctions*) in the *Bayesian setting*, where players' types are drawn from an underlying distribution and may be *correlated*, and the goal is to *minimize the expected total payment made by the mechanism*. Consider the simplest such setting of a *single-item procurement auction*, where a buyer wants to buy an item from any one of n sellers. Each seller incurs a private cost, which we refer to as his *type*, for supplying the item and the sellers must therefore be incentivized via a suitable payment scheme. Myerson's seminal result [16] solves this problem (and other single-dimensional problems) when players' private types are *independent*. However, no such result (or characterization) is known when players' types are correlated. This is the question that motivates our work.

Whereas the analogous revenue-maximization problem for packing domains, such as combinatorial auctions (CAs), has been extensively studied in the algorithmic mechanism design (AMD) literature, both in the case of independent and correlated (even interdependent) player-types (see, e.g., [3,4,2,1,5,9,8,17,2,19] and the references therein), surprisingly, there are *almost no results* on the

---

* A full version is available on the CS arXiv.
** Research supported partially by NSERC grant 327620-09 and the second author's Discovery Accelerator Supplement Award, and Ontario Early Researcher Award.

Y. Chen and N. Immorlica (Eds.): WINE 2013, LNCS 8289, pp. 377–390, 2013.

payment-minimization problem in the AMD literature (see however [4]). The economics literature does contain various general results that apply to both covering and packing problems. However much of this work focuses on characterizing special cases; see, e.g., [12]. An exception is the work of Crémer and McLean [6,7], which shows that under certain conditions, one can devise a Bayesian-incentive-compatible (BIC) mechanism whose expected total payment exactly equal to the expected cost incurred by the players, albeit one where players may incur negative utility under certain type-profile realizations.

Our contributions. We initiate a study of payment-minimization (PayM) problems from the AMD perspective of designing computationally efficient, near-optimal mechanisms. We develop black-box reductions from mechanism design to algorithm design whose application yields a variety of optimal and near-optimal mechanisms. As we elaborate below, covering problems turn out to behave quite differently in certain respects from packing problems, which necessitates new approaches (and solution concepts).

Formally, we consider the setting of *correlated players* in the explicit model, that is, where we have an explicitly-specified *arbitrary discrete* joint distribution of players' types. The most common solution concept in Bayesian settings is Bayesian incentive compatibility (BIC) and interim individual rationality (interim IR), wherein at the *interim* stage when a player knows his type but is oblivious of the random choice of other players' types, truthful participation in the mechanism by all players forms a Bayes-Nash equilibrium. Two serious drawbacks of this solution concept (which are exploited strikingly and elegantly in [6,7]) are that: (i) a player may regret his decision of participating and/or truthtelling *ex post*, that is, after observing the realization of other players' types; and (ii) it is overly-reliant on having precise knowledge of the true underlying distribution making this a rather *non-robust* concept: if the true distribution differs, possibly even slightly, from the mechanism designer and/or players' beliefs or information about it, then the mechanism could lose its IC and IR properties.

We formulate a notion of incentive compatibility (IC) that we call *robust Bayesian IC* (robust BIC) that on the one hand is substantially more robust than BIC, and on the other is flexible enough that it allows one to obtain various polytime near-optimal mechanisms satisfying this notion. A *robust-(BIC, IR)* mechanism (see Section 2) ensures that truthful participation in the mechanism is in the best interest of every player (i.e. a "no-regret" choice) *even at the ex-post stage* when the other players' (randomly-chosen) types are revealed to him. Thus, a robust-(BIC, IR) mechanism is significantly more robust than a (BIC, interim-IR) mechanism since it retains its IC and IR properties for a wide variety of distributions, including those having the same support as the actual distribution. In other words, in keeping with Wilson's doctrine of detail-free mechanisms, the mechanism functions robustly even under fairly limited information about the type-distribution.

We show that for a variety of settings, one can reduce the robust-(BIC, IR) payment-minimization (PayM) mechanism-design problem to the algorithmic cost-minimization (CM) problem of finding an outcome that minimizes the total cost incurred. Moreover, this black-box reduction applies to: (a) single-dimensional settings even when we only have an LP-relative *approximation algorithm* for the PCM problem (that is required to work only with nonnegative costs) (Theorem 9); and (b) *multidimensional problems* with additive types (Corollary 6). We emphasize that our definition of additive types (see Section 2) should not be confused with, and is more general than, additive valuations in combinatorial auctions (CAs). (For example, in a CA with m items, valuation functions of the form $v(S) = \sum_{T \subseteq S} a_T$ for all $S \subseteq [m]$ form an additive type: if $v, v' \in V$, then $v + v'$ defined by $(v + v')(S) = v(S) + v'(S)$ also has the above form. But v is not an additive valuation as there need not exist item values v_j for $j \in [m]$ such that one can express $v(S) = \sum_{j \in S} v_j$.)

Our reduction yields near-optimal robust-(BIC-in-expectation, IR) mechanisms for a variety of covering settings such as (a) various single-dimensional covering problems including single-item procurement auctions (Table 1); (b) multi-item procurement auctions (Theorem 10); and (c) multidimensional facility location (Theorem 12). (Robust BIC-in-expectation means that the robust-BIC guarantee holds for the *expected utility* of a player, where the expectation is over the random coin tosses of the mechanism.) Our techniques can be adapted to yield *truthful-in-expectation* mechanisms with the same guarantees for single-dimensional problems with a constant number of players. These are the *first* results for the PayM mechanism-design problem with correlated players under a notion stronger than (BIC, interim IR). To our knowledge, our results are new even for the simplest covering setting of single-item procurement auctions.

On a side note, we note that we can leverage our ideas to also expand upon the results in [8] for revenue-maximization with correlated players and make significant progress on a research direction proposed in [8]. In the full version, we show that any "integrality-gap verifying" ρ-approximation algorithm for the SWM problem (as defined in [11]) can be used to obtain a truthful-in-expectation mechanism whose revenue is at least a ρ-fraction of the optimum revenue.

Our techniques. The starting point for our construction is the observation that the problem of designing an optimal robust-(BIC, IR)-in-expectation mechanism can be encoded via an LP (P). This was also observed by [8] in the context of the revenue-maximization problem for CAs, but the covering nature of the problem renders various techniques utilized successfully in the context of packing problems inapplicable, and therefore from here on our techniques diverge.

We show that an optimal solution to (P) can be computed given an optimal algorithm $\mathcal{A}$ for the CM problem since $\mathcal{A}$ can be used to obtain a separation oracle for the dual LP. Next, we prove that a feasible solution to (P) yields a robust-(BIC-in-expectation, IR) mechanism with no larger objective value.

For single-dimensional problems, we show that even LP-relative ρ-approximation algorithms for the CM problem can be utilized, as follows. We move to a relaxation of (P), where we replace the set of allocations with the feasible region of the CM-LP. This can be solved efficiently, since the separation oracle for the dual can be obtained by optimizing over the feasible region of CM-LP, which can be done efficiently! But now we need to work harder to "round" an optimal solution (x, p) to the relaxation of (P) and obtain a robust-(BIC-in-expectation, IR) mechanism. Here, we exploit the Lavi-Swamy [11] convex-decomposition procedure, using which we can show (roughly speaking) that we can decompose ρx into a convex combination of allocations. This allows us to obtain a robust-(BIC-in-expectation, IR) mechanism while blowing up the payment by a ρ-factor.

In comparison with [8], which is the work most closely-related to ours, our reduction from robust-BIC mechanism design to the algorithmic CM problem is stronger than the reduction in [8] in two ways. First, for single-dimensional settings, it applies even with LP-relative *approximation algorithms*, and the approximation algorithm is required to work only for "proper inputs" with nonnegative costs. (Note that whereas for packing problems, allowing negative-value inputs can be benign, this can change the character of a covering problem considerably; in particular, the standard notion of approximation becomes meaningless since the optimum could be negative.) In contrast, Dobzinski et al. [8] require an exact algorithm for the analogous social-welfare-maximization (SWM) problem. Second, our reduction also applies to multidimensional settings with additive types (see Section 2), albeit we now require an exact algorithm for the CM problem.

Differences with respect to packing problems. Note that [8] obtain (DSIC-in-expectation, IR)-mechanisms, which is a subtly stronger notion than the robust-(BIC-in-expectation, IR) solution concept that our mechanisms satisfy. This difference arises due to the different nature of covering and packing problems. [8] also first obtain a robust-(BIC, IR)-in-expectation mechanism. The key difference is that for combinatorial auctions, one can show that *any* robust-(BIC, IR)-in-expectation mechanism—in particular, the one obtained from the optimal LP solution—can be converted into a (DSIC-in-expectation, IR) mechanism without any loss in expected revenue. Intuitively, this works because one can focus on a single player by allocating no items to the other players. Clearly, one cannot mimic this for covering problems: dropping players may render the problem infeasible, and it is not clear how to extend an LP-solution to a (DSIC-in-expectation, IR) mechanism for covering problems. We suspect that there is a gap between the optimal expected total payments of robust-(BIC-in-expectation, IR) and (DSIC, IR) mechanisms; we leave this as an open problem. Due to this complication, we sacrifice a modicum of the IC, IR properties in favor of obtaining polytime near-optimal mechanisms and settle for the weaker, but still quite robust notion of robust (BIC-in-expectation, IR). We consider this to be a reasonable starting point for exploring mechanism-design solutions for covering problems, which leads to various interesting research directions.

A more-stunning aspect where covering and packing problems diverge can be seen when one considers the idea of a k-lookahead auction [18,8]. This was used by [8] to convert their results in the explicit model to the oracle model introduced by [18]. This however fails spectacularly in the covering setting. One can show that even for single-item procurement auctions, dropping even a single player can lead to an arbitrarily large payment compared to the optimum.

Other related work. In the economics literature, the classical results of Crémer and McLean [6,7] and McAfee and Reny [14], also apply to covering problems, and show that one can devise a (BIC, interim IR) mechanism with correlated players whose expected total payment is at most the expected total cost incurred provided the underlying type-distribution satisfies a certain full-rank assumption. These mechanisms may however cause a player to have negative utility under certain realizations of the random type profile.

The AMD literature has concentrated mostly on the independent-players setting [3,4,2,1,5,9]. There has been some, mostly recent, work that also considers correlated players [18,8,17,2,19]; as noted earlier, all of this work pertains to the revenue-maximization setting. Ronen [18] considers the single-item auction setting in the *oracle model*, where one sample from the distribution conditioned on some players' values. He proposes the (1-) lookahead auction and shows that it achieves a $\frac{1}{2}$-approximation. [17] shows that the optimal (DSIC, IR) mechanism for the single-item auction can be computed efficiently with at most 2 players, and is *NP*-hard otherwise. Cai et al. [2] give a characterization of the optimal auction under certain settings. [19] considers interdependent types, which generalizes the correlated type-distribution setting, and develop an analog of Myerson's theory for certain such settings.

Various reductions from revenue-maximization to SWM are given in [3,4,2]. These reductions also apply to covering problems and the PayM objective, but they are incomparable to our results. These works focus on the (BIC, interim-IR) solution concept, which is a rather weak/liberal notion for correlated distributions. Most (but not all) of these consider independent players and additive valuations, and often require that the SWM-algorithm also work with negative values, which is a benign requirement for downwards-closed environments such as CAs but is quite problematic for covering problems when only has an approximation algorithm. [2] considers correlated players and obtains mechanisms having running time polynomial in the maximum support-size of the marginal distribution of a player, which could be substantially smaller than the support-size of the entire distribution. This savings can be traced to the use of the (BIC, interim-IR) notion which allows [2] to work with a compact description of the mechanism. It is unclear if these ideas are applicable when one considers robust-(BIC, IR) mechanisms. A very interesting open question is whether one can design robust-(BIC-in-expectation, IR) mechanisms having running time polynomial in the support-sizes of the marginal player distributions (as in [2,8]).

2   Preliminaries

Covering mechanism-design problems. We adopt the formulation in [15] to describe general covering mechanism-design problems. There are m items that need to be covered, and n players who provide covering objects. Let $[k]$ denote the set $\{1, \ldots, k\}$. Each player i provides a set $\mathcal{T}_i$ of covering objects. All this information is public knowledge. Player i has a *private cost* or *type* vector $c_i = \{c_{i,v}\}_{v \in \mathcal{T}_i}$, where $c_{i,v} \geq 0$ is the cost he incurs for providing object $v \in \mathcal{T}_i$; for $T \subseteq \mathcal{T}_i$, we use $c_i(T)$ to denote $\sum_{v \in T} c_{i,v}$. A feasible solution or allocation selects a subset

$T_i \subseteq \mathcal{T}_i$ for each agent i, denoting that i provides the objects in T_i, such that $\bigcup_i T_i$ covers all the items. Given this solution, each agent i incurs the private cost $c_i(T_i)$, and the mechanism designer incurs a publicly known cost $pub(\bigcup_i T_i) \geq 0$, which may be used to encode any feasibility constraints in the covering problem.

Let C_i denote the set of all possible types of agent i, and $C = \prod_{i=1}^n C_i$. Let $\Omega := \{(T_1, \ldots, T_n) : pub(\bigcup_i T_i) < \infty\}$ be the (finite) set of all possible feasible allocations. For a tuple $x = (x_1, \ldots, x_n)$, we use x_{-i} to denote $(x_1, \ldots, x_{i-1}, x_{i+1}, \ldots, x_n)$. Similarly, let $C_{-i} = \prod_{j \neq i} C_j$. For an allocation $\omega = (T_1, \ldots, T_n)$, we sometimes use ω_i to denote T_i, $c_i(\omega)$ to denote $c_i(\omega_i) = c_i(T_i)$, and $pub(\omega)$ to denote $pub(\bigcup_i T_i)$. We make the mild assumption that $pub(\omega') \leq pub(\omega)$ if $\omega_i \subseteq \omega'_i$ for all i; so in particular, if ω is feasible, then adding covering objects to the ω_is preserves feasibility.

A (direct revelation) *mechanism* $M = (\mathcal{A}, p_1, \ldots, p_n)$ for a covering problem consists of an allocation algorithm $\mathcal{A} : C \mapsto \Omega$ and a payment function $p_i : C \mapsto \mathbb{R}$ for each agent i. Each agent i reports a cost function c_i (that might be different from his true cost function). The mechanism computes the allocation $\mathcal{A}(c) = (T_1, \ldots, T_n) = \omega \in \Omega$, and pays $p_i(c)$ to each agent i. The *utility* $u_i(c_i, c_{-i}; \bar{c}_i)$ that player i derives when he reports c_i and the others report c_{-i} is $p_i(c) - \bar{c}_i(\omega_i)$ where $\bar{c}_i$ is his true cost function, and each agent i aims to maximize his own utility. We refer to $\max_i |\mathcal{T}_i|$ as the dimension of a covering problem. Thus, for a *single-dimensional* problem, each player i's cost can be specified as $c_i(\omega) = c_i \alpha_{i,\omega}$, where $c_i \in \mathbb{R}_+$ is his private type and $\alpha_{i,\omega} = 1$ if $\omega_i \neq \emptyset$ and 0 otherwise.

The above setup yields a *multidimensional covering mechanism-design problem* with *additive types*, where additivity is the property that if $c_i, c'_i \in C_i$, then the type $c_i + c'_i$ defined by $(c_i + c'_i)(\omega) = c_i(\omega) + c'_i(\omega)$ for all $\omega \in \Omega$, is also in C_i. It is possible to define more general multidimensional settings, but additive type spaces is a reasonable starting point to explore the multidimensional covering mechanism-design setting. (As noted earlier, there has been almost no work on designing polytime, near-optimal mechanisms for covering problems.)

The Bayesian setting. We consider *Bayesian* settings where there is an underlying publicly-known *discrete* and possibly *correlated* joint type-distribution on C from which the players' types are drawn. We consider the so-called explicit model, where the players' type distribution is explicitly specified. We use $\mathcal{D} \subseteq C$ to denote the support of the type distribution, and $\Pr_{\mathcal{D}}(c)$ to denote the probability of realization of $c \in C$. Also, we define $\mathcal{D}_i := \{c_i \in C_i : \exists c_{-i} \text{ s.t. } (c_i, c_{-i}) \in \mathcal{D}\}$, and $\mathcal{D}_{-i}$ to be $\{c_{-i} : \exists c_i \text{ s.t. } (c_i, c_{-i}) \in \mathcal{D}\}$.

Solution concepts. A mechanism sets up a game between players, and the solution concept dictates certain desirable properties that this game should satisfy, so that one can reason about the outcome when rational players' are presented with a mechanism satisfying the solution concept. The two chief properties that one seeks to capture relate to incentive compatibility (IC), which (roughly speaking) means that every agent's best interest is to reveal his type truthfully, and individual rationality (IR), which is the notion that no agent is harmed by participating in the mechanism. Differences and subtleties arise in Bayesian settings depending on the stage at which we impose these properties and how robust we would like these properties to be with respect to the underlying type distribution.

Definition 1. *A mechanism* $M = (\mathcal{A}, \{p_i\})$ *is* Bayesian incentive compatible *(BIC) and* interim IR *if for every player i and every $\overline{c}_i, c_i \in C_i$, we have* $\mathrm{E}_{c_{-i}}[u_i(\overline{c}_i, c_{-i}; \overline{c}_i)|\overline{c}_i] \geq \mathrm{E}_{c_{-i}}[u_i(c_i, c_{-i}; \overline{c}_i)|\overline{c}_i]$ *(BIC) and* $\mathrm{E}_{c_{-i}}[u_i(\overline{c}_i, c_{-i}; \overline{c}_i)|\overline{c}_i] \geq 0$ *(interim IR), where* $\mathrm{E}_{c_{-i}}[.|\overline{c}_i]$ *denotes the expectation over the other players' types conditioned on i's type being $\overline{c}_i$.*

As mentioned earlier, the (BIC, interim-IR) solution concept may yet lead to ex-post "regret", and is quite non-robust in the sense that the mechanism's IC and IR properties rely on having detailed knowledge of the distribution; thus, in order to be confident that a BIC mechanism achieves its intended functionality, one must be confident about the "correctness" of the underlying distribution, and learning this information might entail significant cost. To remedy these weaknesses, we propose and investigate the following stronger IC and IR notions.

Definition 2. *A mechanism* $M = (\mathcal{A}, \{p_i\})$ *is* robust BIC *and* robust IR, *if for every player i, every $\overline{c}_i, c_i \in C_i$, and every $c_{-i} \in \mathcal{D}_{-i}$, we have* $u_i(\overline{c}_i, c_{-i}; \overline{c}_i) \geq u_i(c_i, c_{-i}; \overline{c}_i)$ *(robust BIC) and* $u_i(\overline{c}_i, c_{-i}; \overline{c}_i) \geq 0$ *(robust IR).*

Robust (BIC, IR) ensures that participating truthfully in the mechanism is in the best interest of every player even at the ex-post stage when he knows the realized types of all players. To ensure that robust BIC and robust IR are compatible, we focus on *monopoly-free* settings: for every player i, there is some $\omega \in \Omega$ with $\omega_i = \emptyset$. Notice that robust (BIC, IR) is subtly weaker than the notion of (dominant-strategy IC (DSIC), IR), wherein the IC and IR conditions of Definition 2 must hold for all $c_{-i} \in C_{-i}$, ensuring that truthtelling and participation are no-regret choices for a player even if the other players' reports are outside the support of the underlying type-distribution. We focus on robust BIC because it forms a suitable middle-ground between BIC and DSIC: it inherits the desirable robustness properties of DSIC, making it much more robust than BIC (and closer to a worst-case notion), and yet is flexible enough that one can devise polytime mechanisms satisfying this solution concept.

The above definitions are stated for a deterministic mechanism, but they have analogous extensions to a randomized mechanism M; the only change is that each $u_i(.)$ and $p_i(.)$ term is now replaced by the expected utility $\mathrm{E}_M[u_i(.)]$ and expected price $\mathrm{E}_M[p_i(.)]$ over the random coin tosses of M. We denote the analogous solution concept for a randomized mechanism by appending "in expectation" to the solution concept, e.g., a (BIC, interim IR)-in-expectation mechanism denotes a randomized mechanism whose expected utility satisfies the BIC and interim-IR requirements stated in Definition 1.

A robust-(BIC, IR)-in-expectation mechanism $M = (\mathcal{A}, \{p_i\})$ can be easily modified so that the IR condition holds *with probability 1* (with respect to M's coin tosses) while the expected payment to a player (again over M's coin tosses) is unchanged: on input c, if $\mathcal{A}(c) = \omega \in \Omega$ with probability q, the new mechanism returns, with probability q, the allocation ω, and payment $c_i(\omega) \cdot \frac{\mathrm{E}_M[p_i(c)]}{\mathrm{E}_M[c_i(\omega)]}$ to each player i (where we take $0/0$ to be 0, so if $c_i(\omega) = 0$, the payment to i is 0). Thus, we obtain a mechanism whose expected utility satisfies the robust-BIC condition, and IR holds with probability 1 for all $\bar{c}_i \in C_i, c_{-i} \in \mathcal{D}_{-i}$, A similar transformation can be applied to a (DSIC, IR)-in-expectation mechanism.

Optimization problems. Our main consideration is to minimize the expected total payment of the mechanism. It is natural to also incorporate the mechanism-designer's cost into the objective. Define the *disutility* of a mechanism $M = (f, \{p_i\})$ under input v to be $\sum_i p_i(v) + \kappa \cdot pub(f(v))$, where $\kappa \geq 0$ is a scaling factor. Our objective is to devise a polynomial-time robust (BIC (in-expectation), IR)-mechanism with minimum expected disutility. Since most problems we consider have $pub(\omega) = 0$ for all feasible allocations, in which case disutility equals the total payment, abusing terminology slightly, we refer to the above mechanism-design problem as the *payment-minimization* (PayM) problem. (An exception is metric *uncapacitated facility location* (UFL), where players provide facilities and the underlying metric is public knowledge; here, $pub(\omega)$ is the total client-assignment cost of the solution ω.) We always use O^* to denote the expected disutility of an optimal mechanism for the PayM problem under consideration.

We define the *cost minimization* (CM) problem to be the *algorithmic* problem of finding $\omega \in \Omega$ that minimizes the total cost $\sum_i c_i(\omega) + pub(\omega)$ incurred.

The following technical lemma will prove quite useful since it allows us to restrict the domain to a bounded set, which is essential to achieve IR with finite prices. (For example, in the single-dimensional setting, the payment is equal to the integral to ∞ of a certain quantity; a bounded domain ensures that this is well defined.) Note that such complications do not arise for packing problems. Let $\mathbf{1}_{\mathcal{T}_i}$ be the $|\mathcal{T}_i|$-dimensional all 1s vector. Let $\mathcal{I}$ denote the input size.

Lemma 3. *We can efficiently compute an estimate $m_i > \max_{c_i \in \mathcal{D}_i, v \in \mathcal{T}_i} c_{i,v}$ with $\log m_i = \text{poly}(\mathcal{I})$ for all i such that there is an optimal robust-(BIC-in-expectation, IR) mechanism $M^* = (\mathcal{A}^*, \{p_i^*\})$ where $\mathcal{A}^*(m_i \mathbf{1}_{\mathcal{T}_i}, c_{-i}) = \emptyset$ with probability 1 (over the random choices of M^*) for all i and all $c_{-i} \in \mathcal{D}_{-i}$.*

It is easy to obtain the stated estimates if we consider only *deterministic* mechanisms, but it turns out to be tricky to obtain this when one allows randomized mechanisms due to the artifact that a randomized mechanism may choose arbitrarily high-cost solutions as long as they are chosen with small enough probability. In the sequel, we set $\overline{\mathcal{D}}_i := \mathcal{D}_i \cup \{m_i \mathbf{1}_{\mathcal{T}_i}\}$ for all $i \in [n]$, and $\overline{\mathcal{D}} := \bigcup_i (\overline{\mathcal{D}}_i \times \mathcal{D}_{-i})$. Note that $|\overline{\mathcal{D}}| = O(n|\mathcal{D}|^2)$.

3   LP-Relaxations for the Payment-Minimization Problem

The starting point for our results is the LP (P) that essentially encodes the payment-minimization problem. Throughout, we use i to index players, c to index type-profiles in $\overline{\mathcal{D}}$, and ω to index Ω. We use variables $x_{c,\omega}$ to denote the probability of choosing ω, and $p_{i,c}$ to denote the expected payment to player i, for input c. For $c \in \overline{\mathcal{D}}$, let $\Omega(c) = \Omega$ if $c \in \bigcup_i (\mathcal{D}_i \times \mathcal{D}_{-i})$, and otherwise if

$c = (m_i \mathbf{1}_{\mathcal{T}_i}, c_{-i})$, let $\Omega(c) = \{\omega \in \Omega : \omega_i = \emptyset\}$ (which is non-empty since we are in a monopoly-free setting).

$$\min \quad \sum_{c \in \mathcal{D}} \mathrm{Pr}_{\mathcal{D}}(c) \Big(\sum_i p_{i,c} + \kappa \sum_\omega x_{c,\omega} pub(\omega) \Big) \tag{P}$$

$$\text{s.t.} \quad \sum_\omega x_{c,\omega} = 1 \qquad \forall c \in \overline{\mathcal{D}} \tag{1}$$

$$p_{i,(c_i,c_{-i})} - \sum_\omega c_i(\omega) x_{(c_i,c_{-i}),\omega} \geq$$
$$p_{i,(c_i',c_{-i})} - \sum_\omega c_i(\omega) x_{(c_i',c_{-i}),\omega} \qquad \forall i, c_i, c_i' \in \overline{\mathcal{D}}_i, c_{-i} \in \mathcal{D}_{-i} \tag{2}$$

$$p_{i,(c_i,c_{-i})} - \sum_\omega c_i(\omega) x_{(c_i,c_{-i}),\omega} \geq 0 \qquad \forall i, c_i \in \overline{\mathcal{D}}_i, c_{-i} \in \mathcal{D}_{-i} \tag{3}$$

$$p, x \geq 0, \quad x_{c,\omega} = 0 \qquad \forall c, \omega \notin \Omega(c). \tag{4}$$

(1) encodes that an allocation is chosen for every $c \in \overline{\mathcal{D}}$, and (2) and (3) encode the robust BIC and robust IR conditions respectively. Lemma 3 ensures that (P) correctly encodes PayM, so that $OPT := OPT_{\mathrm{P}}$ is a lower bound on the expected disutility of an optimal mechanism.

Our results are obtained by computing an optimal solution to (P), or a further relaxation of it, and translating this to a near-optimal robust (BIC-in-expectation, IR) mechanism. Both steps come with their own challenges. Except in very simple settings (such as single-item procurement auctions), $|\Omega|$ is typically exponential in the input size, and therefore it is not clear how to solve (P) efficiently. We therefore consider the dual LP (D), which has variables γ_c, $y_{i,(c_i,c_{-i}),c_i'}$ and $\beta_{i,(c_i,c_{-i})}$ corresponding to (1), (2) and (3) respectively.

$$\max \quad \sum_c \gamma_c \tag{D}$$

$$\text{s.t.} \sum_{i : c \in \overline{\mathcal{D}}_i \times \mathcal{D}_{-i}} \Big(\sum_{c_i' \in \overline{\mathcal{D}}_i} \big(c_i(\omega) y_{i,(c_i,c_{-i}),c_i'} - c_i'(\omega) y_{i,(c_i',c_{-i}),c_i} \big) + c_i(\omega) \beta_{i,c} \Big)$$
$$+ \kappa \cdot \mathrm{Pr}_{\mathcal{D}}(c) pub(\omega) \geq \gamma_c \qquad \forall c \in \overline{\mathcal{D}}, \omega \in \Omega(c) \tag{5}$$

$$\sum_{c_i' \in \overline{\mathcal{D}}_i} \big(y_{i,(c_i,c_{-i}),c_i'} - y_{i,(c_i',c_{-i}),c_i} \big) + \beta_{i,c_i,c_{-i}} \leq \mathrm{Pr}_{\mathcal{D}}(c) \quad \forall i, c_i \in \overline{\mathcal{D}}_i, c_{-i} \in \mathcal{D}_{-i}$$

$$y, \beta \geq 0.$$

With additive types, the separation problem for constraints (5) amounts to determining if the optimal value of the CM problem defined by a certain input with possibly negative costs, is at least γ_c. Hence, an optimal algorithm for the CM problem can be used to solve (D), and hence, (P) efficiently.

Theorem 4. *With additive types, one can efficiently solve* (P) *given an optimal algorithm for the CM problem.*

Complementing Theorem 4, we argue that a feasible solution (x, p) to (P) can be "rounded" to a robust-(BIC-in-expectation, IR) mechanism having expected disutility at most the value of (x, p) (Theorem 5). Combining this with Theorem 4 yields the corollary that an optimal algorithm for the CM problem can be used to obtain an optimal mechanism for the PayM problem (Corollary 6).

Theorem 5. *We can extend a feasible solution (x, p) to (P) to a robust-(BIC-in-expectation, IR) mechanism with expected disutility $\sum_c \Pr_{\mathcal{D}}(c)(\sum_i p_{i,c} + \kappa \sum_\omega x_{c,\omega} pub(\omega))$.*

Proof. Let $\Omega' = \{\omega : x_{c,\omega} > 0 \text{ for some } c \in \overline{\mathcal{D}}\}$. We use x_c to denote the vector $\{x_{c,\omega}\}_{\omega \in \Omega'}$. Consider a player i, $c_{-i} \in \mathcal{D}_{-i}$, and $c_i, c_i' \in \overline{\mathcal{D}}_i$. Note that (2) implies that if $x_{(c_i, c_{-i})} = x_{(c_i', c_{-i})}$, then $p_{i,(c_i, c_{-i})} = p_{i,(c_i', c_{-i})}$. For $c_{-i} \in \mathcal{D}_{-i}$, define $R(i, c_{-i}) = \{x_{(c_i, c_{-i})} : (c_i, c_{-i}) \in \overline{\mathcal{D}}\}$, and for $y = x_{(c_i, c_{-i})} \in R(i, c_{-i})$ define $p_{i,y}$ to be $p_{i,(c_i, c_{-i})}$ (which is well defined by the above argument).

We now define the randomized mechanism $M = (\mathcal{A}, \{q_i\})$, where $\mathcal{A}(c)$ and $q_i(c)$ denote respectively the probability distribution over allocations and the expected payment to player i, on input c. We sometimes view $\mathcal{A}(c)$ equivalently as the random variable specifying the allocation chosen for input c. Fix an allocation $\omega_0 \in \Omega$. Consider an input c. If $c \in \overline{\mathcal{D}}$, we set $\mathcal{A}(c) = x(c)$, and $q_i(c) = p_{i,c}$ for all i. So consider $c \notin \overline{\mathcal{D}}$. If there is no i such that $c_{-i} \in \mathcal{D}_{-i}$, we simply set $\mathcal{A}(c) = \omega_0$, $q_i(c) = c_i(\omega_0)$ for all i; such a c does not figure in the robust (BIC, IR) conditions. Otherwise there is a unique i such that $c_{-i} \in \mathcal{D}_{-i}$, $c_i \in C_i \setminus \overline{\mathcal{D}}_i$. Set $\mathcal{A}(c) = \arg\max_{y \in R(i, c_{-i})} (p_{i,y} - \sum_{\omega \in \Omega'} c_i(\omega) y_\omega)$ and $q_j(c) = p_{j,\mathcal{A}(c)}$ for all players j. Note that (c_i, c_{-i}) figures in (2) *only* for player i. Crucially, note that since $y = x_{(m_i, c_{-i})} \in R(i, c_{-i})$ and $\sum_{\omega \in \Omega} c_i(\omega) y_\omega = 0$ by definition, we always have $q_i(c) - \mathbb{E}_{\mathcal{A}}[c_i(\mathcal{A}(c))] \geq 0$. Thus, by definition, and by (2), we have ensured that M is robust (BIC, IR)-in-expectation and its expected disutility is exactly the value of (x, p). This can be modified so that IR holds with probability 1. $\square$

Corollary 6. *Given an optimal algorithm for the CM problem, we can obtain an optimal robust-(BIC-in-expectation, IR) mechanism for the PayM problem in multidimensional settings with additive types.*

The CM problem is however often *NP*-hard (e.g., for vertex cover), and we would like to be able to exploit approximation algorithms for the CM problem to obtain near-optimal mechanisms. The usual approach is use an approximation algorithm to "approximately" separate over constraints (5). However, this does not work since the CM problem that one needs to solve in the separation problem involves negative costs, which renders the usual notion of approximation meaningless. Instead, if the CM problem admits a certain type of LP-relaxation (C-P), then we argue that one can solve a relaxation of (P) where the allocation-set is the set of extreme points of (C-P) (Theorem 7). For single-dimensional problems (Section 4), we leverage this to obtain strong and far-reaching results. We show that a ρ-approximation algorithm relative to (C-P) can be used to "round" the optimal solution to this relaxation to a robust-(BIC-in-expectation, IR)-mechanism losing a ρ-factor in the disutility (Theorem 9). Thus, we obtain near-optimal mechanisms for a variety of single-dimensional problems.

Suppose that the CM problem admits an LP-relaxation of the following form, where $c = \{c_{i,v}\}_{i \in [n], v \in \mathcal{T}_i}$ is the input type-profile.

$$\min \quad c^T x + d^T z \quad \text{s.t.} \quad Ax + Bz \geq b, \quad x, z \geq 0. \tag{C-P}$$

Intuitively x encodes the allocation chosen, and $d^T z$ encodes $pub(.)$. For $x \geq 0$, define $z(x) := \arg\min\{d^T z : (x, z) \text{ is feasible to (C-P)}\}$; if there is no z such that (x, z) is feasible to (C-P), set $z(x) := \bot$. Define $\Omega_{\mathrm{LP}} := \{x : z(x) \neq \bot, \ x_{i,v} \leq 1 \ \forall i, v \in \mathcal{T}_i\}$. We require that: (a) a $\{0, 1\}$-vector x is in Ω_{LP} iff it is the characteristic vector of an allocation $\omega \in \Omega$, and in this case, we have $d^T z(x) = pub(\omega)$; (b) $A \geq 0$; (c) for any input $c \geq 0$ to the covering problem, (C-P) is not unbounded, and if it has an optimal solution, it has one where $x \in \Omega_{\mathrm{LP}}$; (d) for any c, we can efficiently find an optimal solution to (C-P) or detect that it is unbounded or infeasible.

We extend the type c_i of each player i and pub to assign values also to points in Ω_{LP}: define $c_i(x) = \sum_{v \in \mathcal{T}_i} c_{i,v} x_{i,v}$ and $pub(x) = d^T z(x)$ for $x \in \Omega_{\mathrm{LP}}$. Let Ω_{ext} denote the finite set of extreme points of Ω_{LP}. Condition (a) ensures that Ω_{ext} contains the characteristic vectors of all feasible allocations. Let (P') denote the relaxation of (P), where we replace the set of feasible allocations Ω with Ω_{ext} (so ω indexes Ω_{ext} now), and for $c \in \overline{\mathcal{D}}$ with $c_i = m_i \mathbf{1}(\mathcal{T}_i)$, we now define $\Omega(c) := \{\omega \in \Omega_{\mathrm{ext}} : \omega_{i,v} = 0 \ \forall v \in \mathcal{T}_i\}$. Since one can optimize efficiently over Ω_{LP}, and hence Ω_{ext}, even for negative type-profiles, we have the following.

Theorem 7. *We can efficiently compute an optimal solution to* (P').

4   Single-Dimensional Problems

Corollary 6 immediately yields results for certain single-dimensional problems (see Table 1), most notably, *single-item procurement auctions*. We now substantially expand the scope of PayM problems for which one can obtain near-optimal mechanisms by showing how to leverage "LP-relative" approximation algorithms for the CM problem. Suppose that the CM problem can be encoded as (C-P). An *LP-relative ρ-approximation algorithm* for the CM problem is a polytime algorithm that for any input $c \geq 0$ to the covering problem, returns a $\{0, 1\}$-vector $x \in \Omega_{\mathrm{LP}}$ such that $c^T x + d^T z(x) \leq \rho OPT_{\mathrm{C-P}}$. Using the convex-decomposition procedure in [11] (see Section 5.1 of [11]), one can show the following.

Lemma 8. *Let $x \in \Omega_{\mathrm{LP}}$. Given an LP-relative ρ-approximation algorithm for the CM problem, one can efficiently obtain $(\lambda^{(1)}, x^{(1)}), \ldots, (\lambda^{(k)}, x^{(k)})$, where $\sum_\ell \lambda^{(\ell)} = 1, \lambda \geq 0$, and $x^{(\ell)}$ is a $\{0, 1\}$-vector in Ω_{LP} for all ℓ, such that $\sum_\ell \lambda^{(\ell)} x_{i,v}^{(\ell)} = \min(\rho x_{i,v}, 1)$ for all $i, v \in \mathcal{T}_i$, and $\sum_\ell \lambda^{(\ell)} d^T z(x^{(\ell)}) \leq \rho d^T z(x)$.*

Theorem 9. *Given an LP-relative ρ-approximation algorithm for the CM problem, one can obtain a polytime ρ-approximation robust-(BIC-in-expectation, IR) mechanism for the PayM problem.*

Proof. We solve (P') to obtain an optimal solution (x, p). Since $|\mathcal{T}_i| = 1$ for all i, it will be convenient to view $\omega \in \Omega_{\mathrm{LP}}$ as a vector $\{\omega_i\}_{i \in [n]}$, where $\omega_i \equiv \omega_{i,v}$ for the single covering object $v \in \mathcal{T}_i$. Fix $c \in \overline{\mathcal{D}}$. Define $y_c = \sum_{\omega \in \Omega_{\mathrm{ext}}} x_{c,\omega} \omega$ (which can be efficiently computed). Then, $\sum_{\omega \in \Omega_{\mathrm{ext}}} c_i(\omega) x_{c,\omega} = c_i y_{c,i}$ and $d^T z(y) \leq$

$\sum_{\omega \in \Omega_{\text{ext}}} pub(\omega)x_{c,\omega}$. By Lemma 8, we can efficiently find a point $\tilde{y}_c = \sum_{\omega \in \Omega} \tilde{x}_{c,\omega}\omega$, where $\tilde{x}_c \geq 0$, $\sum_{w \in \Omega} \tilde{x}_{c,\omega} = 1$, in the convex hull of the $\{0,1\}$-vectors in Ω_{LP} such that $\tilde{y}_{c,i} = \min(\rho y_{c,i}, 1)$ for all i, and $\sum_{w \in \Omega} \tilde{x}_{c,\omega} pub(\omega) \leq \rho d^T z(y)$.

We now argue that one can obtain payments $\{q_{i,c}\}$ such that $(\tilde{x}, q)$ is feasible to (P) and $q_{i,c} \leq \rho p_{i,c}$ for all $i, c \in \overline{\mathcal{D}}$. Thus, the value of $(\tilde{x}, q)$ is at most ρ times the value of (x, p). Applying Theorem 5 to $(\tilde{x}, q)$ yields the desired result.

Fix i and $c_{-i} \in \mathcal{D}_{-i}$. Constraints (4) and (2) and ensure that $y_{(m_i, c_{-i}),i} = 0$, and $y_{(c_i, c_{-i}),i} \geq y_{(c'_i, c_{-i}),i}$ for all $c_i, c'_i \in \overline{\mathcal{D}}_i$ s.t. $c_i < c'_i$. Hence, $\tilde{y}_{(m_i, c_{-i})mi} = 0$, $\tilde{y}_{(c_i, c_{-i}),i} \geq \tilde{y}_{(c'_i, c_{-i}),i}$ for $c_i, c'_i \in \overline{\mathcal{D}}_i$, $c_i > c'_i$. Define $q_{i,(m_i, c_{-i})} = 0$. Let $0 \leq c_i^1 < c_i^2 < \ldots < c_i^{k_i}$ be the values in $\mathcal{D}_i$. For $c_i = c_i^\ell$, define $q_{i,(c_i, c_{-i})} = c_i \tilde{y}_{(c_i, c_{-i}),i} + \sum_{t=\ell+1}^{k_i} (c_{i^t} - c_{i^{t-1}}) \tilde{y}_{(c_{i^t}, c_{-i}),i}$. Since $\sum_{\omega \in \Omega} c_i(\omega) \tilde{x}_{(c_i, c_{-i}),\omega} = c_i \tilde{y}_{(c_i, c_{-i}),i}$, (3) holds. By construction, for consecutive values $c_i = c_i^\ell$, $c'_i = c_i^{\ell+1}$, we have $q_{i,(c_i, c_{-i})} - q_{i,(c'_i, c_{-i})} = c_i(\tilde{y}_{(c_i, c_{-i}),i} - \tilde{y}_{(c'_i, c_{-i}),i}) \leq \rho \cdot c_i(y_{(c_i, c_{-i}),i} - y_{(c'_i, c_{-i}),i}) \leq \rho(p_{i,(c_i, c_{-i})} - p_{i,(c'_i, c_{-i})})$. Since $q_{i,(m_i, c_{-i})} = 0 \leq \rho p_{i,(m_i, c_{-i})}$, this implies that $q_{i,(c_i, c_{-i})} \leq \rho p_{i,(c_i, c_{-i})}$. Finally, it is easy to verify that for any $c_i, c'_i \in \mathcal{D}_i$, we have $q_{i,(c_i, c_{-i})} - q_{i,(c'_i, c_{-i})} \geq c_i(\tilde{y}_{(c_i, c_{-i}),i} - \tilde{y}_{(c'_i, c_{-i}),i})$, so $(\tilde{x}, q)$ satisfies (2). □

Corollary 6 and Theorem 9 yield polytime near-optimal mechanisms for a host of single-dimensional PayM problems, as summarized by Table 1. Even for single-item procurement auctions, these are the *first* results for PayM problems with correlated players satisfying a notion stronger than (BIC, interim IR).

Table 1. Results for some representative single-dimensional PayM problems

Problem	Approximation	Due to
Single-item procurement auction: buy one item provided by n players	1	Corollary 6
Metric UFL: players are facilities, output should be a UFL solution	1.488 using [13]	Theorem 9
Vertex cover: players are nodes, output should be a vertex cover	2	Theorem 9
Set cover: players are sets, output should be a set cover	$O(\log n)$	Theorem 9
Steiner forest: players are edges, output should be a Steiner forest	2	Theorem 9
Multiway cut (a), *Multicut* (b): players are edges, output should be a multiway cut in (a), or a multicut in (b)	2 for (a) $O(\log n)$ for (b)	Theorem 9

5   Multidimensional Problems

We obtain results for multidimensional PayM problems via two distinct approaches. One is by directly applying Corollary 6 (e.g., Theorem 10). The other approach is based on again moving to an LP-relaxation of the CM problem and utilizing Theorem 7 in conjunction with a *stronger* LP-rounding approach. This yields results for multidimensional (metric) UFL and its variants (Theorem 12).

Multi-item procurement auctions. Here, we have n sellers and k (heterogeneous) items. Each seller i has a *supply vector* $s_i \in \mathbb{Z}_+^k$ denoting his supply for the various items, and the buyer has a *demand vector* $d \in \mathbb{Z}_+^k$ specifying his demand for the various items. This is public knowledge. Each seller i has a *private* cost-vector $c_i \in \mathbb{R}_+^k$, where $c_{i,\ell}$ is the cost he incurs for supplying *one unit* of item ℓ. A feasible solution is an allocation specifying how many units of each item each seller supplies to the buyer such that for each item ℓ, each seller i provides at most $s_{i,\ell}$ units of ℓ and the buyer obtains d_ℓ total units of ℓ. The corresponding CM problem is a min-cost flow problem (in a bipartite graph), which can be efficiently solved optimally, thus we obtain a polytime optimal mechanism.

Theorem 10. *There is a polytime optimal robust-(BIC-in-expectation, IR) mechanism for multi-unit procurement auctions.*

Multidimensional budgeted (metric) uncapacitated facility location (UFL). Here, we have a set $\mathcal{E}$ of clients that need to be serviced by facilities, and a set $\mathcal{F}$ of locations where facilities may be opened. Each player i may provide facilities at the locations in $\mathcal{T}_i \subseteq \mathcal{F}$. We may assume that the $\mathcal{T}_i$s are disjoint. For each facility $\ell \in \mathcal{T}_i$ that is opened, i incurs a private opening cost $f_\ell \equiv f_{i,\ell}$, and assigning client j to an open facility ℓ incurs a publicly-known assignment cost $d_{\ell j}$, where the $d_{\ell j}$s form a metric. We are also given a public assignment-cost budget B. The goal in **Budget-UFL** is to open a subset $F \subseteq \mathcal{F}$ of facilities and assign each client j to an open facility $\sigma(j) \in F$ so as to minimize $\sum_{\ell \in F} f_\ell + \sum_{j \in \mathcal{E}} d_{\sigma(j)j}$ subject to $\sum_{j \in \mathcal{E}} d_{\sigma(j)j} \leq B$; UFL is the special case where $B = \infty$. We can define $pub(T)$ to be the total assignment cost if this is at most B, and ∞ otherwise.

Let O^* denote the expected disutility of an optimal mechanism for **Budget-UFL**. We obtain a mechanism with expected disutility at most $2O^*$ that always returns a solution with expected assignment cost at most $2B$. Consider the following LP-relaxation for **Budget-UFL**.

$$\min \sum_{\ell \in \mathcal{F}} f_\ell x_\ell + \sum_{j \in \mathcal{E}, \ell \in \mathcal{F}} d_{\ell j} z_{\ell j} \quad \text{s.t.} \qquad \text{(BFL-P)}$$

$$\sum_{j \in \mathcal{E}, \ell \in \mathcal{F}} d_{\ell j} z_{\ell j} \leq B, \quad \sum_{\ell \in \mathcal{F}} z_{\ell j} \geq 1 \ \forall j \in \mathcal{E}, \quad 0 \leq z_{\ell j} \leq x_\ell \ \forall \ell \in \mathcal{F}, j \in \mathcal{E}. \quad (6)$$

Let (FL-P) denote (BFL-P) with $B = \infty$, and $OPT_{\text{FL-P}}$ denote its optimal value. We say that an algorithm $\mathcal{A}$ is a *Lagrangian multiplier preserving* (LMP) ρ-approximation algorithm for UFL if for every instance, it returns a solution (F, σ) such that $\rho \sum_{\ell \in F} f_\ell + \sum_{j \in \mathcal{E}} d_{\sigma(j)j} \leq \rho \cdot OPT_{\text{FL-P}}$. In [15], it is shown that given such an algorithm $\mathcal{A}$, one can take any solution (x, z) to (FL-P) and obtain a convex combination of *UFL solutions* $(\lambda^{(1)}; F^{(1)}, \sigma^{(1)}), \ldots, (\lambda^{(k)}; F^{(k)}, \sigma^{(k)})$, so $\lambda \geq 0$, $\sum_r \lambda^{(r)} = 1$, such that $\sum_{r: \ell \in F^{(r)}} \lambda^{(r)} = x_\ell$ for all ℓ and $\sum_r \lambda^{(r)} \left(\sum_j d_{\sigma^{(r)}(j)j} \right) \leq \rho \sum_{j,\ell} d_{\ell j} z_{\ell j}$. An LMP 2-approximation algorithm for UFL is known [10].

Lemma 11. *Given an LMP ρ-approximation algorithm for UFL, one can design a polytime robust-(BIC-in-expectation, IR) mechanism for **Budget-UFL** whose expected disutility is at most ρO^* while violating the budget by at most a ρ-factor.*

Theorem 12. *There is a polytime robust-(BIC-in-expectation, IR) mechanism for **Budget-UFL** with expected disutility at most $2O^*$, which always returns a solution with expected assignment cost at most $2B$.*

6   Extension: DSIC Mechanisms

We can strengthen our results from Section 4 to obtain (near-) optimal DSIC mechanisms for single-dimensional problems in time exponential in n. The key change is in the LP (P) (or (P')), where we now enforce (1)—(4) for all type profiles in $\prod_i \overline{\mathcal{D}}_i$. The rounding procedure and arguments in Theorem 9 proceed essentially identically to yield a near-optimal solution to this LP. But we can now argue that in single-dimensional settings, a feasible solution to the LP can be rounded to a (DSIC-in-expectation, IR) mechanism without increasing the expected disutility. Thus, we obtain the same guarantees as in Table 1, but under the stronger solution concept of DSIC-in-expectation and IR.

References

1. Alaei, S., Fu, H., Haghpanah, N., Hartline, J., Malekian, A.: Bayesian optimal auctions via multi- to single-agent reduction. In: Proc. EC, p. 17 (2012)
2. Cai, Y., Daskalakis, C., Weinberg, S.: Optimal multi-dimensional mechanism design: reducing revenue to welfare maximization. In: Proc. FOCS, pp. 130–139 (2012)
3. Cai, Y., Daskalakis, C., Weinberg, S.: Reducing revenue to welfare maximization: approximation algorithms and other generalizations. In: Proc. SODA (2013)
4. Cai, Y., Daskalakis, C., Weinberg, S.M.: Understanding incentives: mechanism design becomes algorithm design. In: Proc. FOCS (2013)
5. Chawla, S., Hartline, J., Malec, D., Sivan, B.: Multi-parameter mechanism design and sequential posted pricing. In: Proc. STOC, pp. 311–320 (2010)
6. Crémer, J., McLean, R.: Optimal selling strategies under uncertainty for a discriminating monopolist when demands are interdependent. Econometrica 53(2), 345–361 (1985)
7. Crémer, J., McLean, R.: Full extraction of the surplus in Bayesian and dominant strategy auctions. Econometrica 56(6), 1247–1257 (1988)
8. Dobzinski, S., Fu, H., Kleinberg, R.: Optimal auctions with correlated bidders are easy. In: Proc. STOC, pp. 129–138 (2011)
9. Hart, S., Nisan, N.: Approximate revenue maximization with multiple items. In: Proc. EC, p. 656 (2012)
10. Jain, K., Mahdian, M., Markakis, E., Saberi, A., Vazirani, V.: Greedy facility location algorithms analyzed using dual fitting with factor-revealing LP. J. ACM 50, 795–824 (2003)
11. Lavi, R., Swamy, C.: Truthful and near-optimal mechanism design via linear programming. Journal of the ACM 58(6), 25 (2011)
12. Krishna, V.: Auction Theory. Academic Press (2010)
13. Li, S.: A 1.488 approximation algorithm for the uncapacitated facility location problem. Inf. Comput. 222, 45–58 (2013)
14. McAfee, R., Reny, P.: Correlated information and mechanism design. Econometrica 60(2), 395–421 (1992)
15. Minooei, H., Swamy, C.: Truthful mechanism design for multidimensional covering problems. In: Goldberg, P.W. (ed.) WINE 2012. LNCS, vol. 7695, pp. 448–461. Springer, Heidelberg (2012)
16. Myerson, R.: Optimal auction design. Math. of Operations Research 6, 58–73 (1981)
17. Papadimitriou, C., Pierrakos, G.: On optimal single-item auctions. In: Proc. STOC, pp. 119–128 (2011)
18. Ronen, A.: On approximating optimal auctions. In: Proc. EC, pp. 11–17 (2001)
19. Roughgarden, T., Talgam-Cohen, I.: Optimal and near-optimal mechanism design with interdependent values. In: Proc. EC, pp. 767–784 (2013)

Bounding the Inefficiency of Altruism through Social Contribution Games

Mona Rahn[1] and Guido Schäfer[1,2]

[1] CWI Amsterdam, The Netherlands
[2] VU University Amsterdam, The Netherlands
{rahn,g.schaefer}@cwi.nl

Abstract. We introduce a new class of games, called *social contribution games (SCGs)*, where each player's individual cost is equal to the cost he induces on society because of his presence. Our results reveal that SCGs constitute useful abstractions of altruistic games when it comes to the analysis of the robust price of anarchy. We first show that SCGs are *altruism-independently smooth*, i.e., the robust price of anarchy of these games remains the same under arbitrary altruistic extensions. We then devise a general reduction technique that enables us to reduce the problem of establishing smoothness for an altruistic extension of a base game to a corresponding SCG. Our reduction applies whenever the base game relates to a canonical SCG by satisfying a simple *social contribution boundedness* property. As it turns out, several well-known games satisfy this property and are thus amenable to our reduction technique. Examples include min-sum scheduling games, congestion games, second-price auctions and valid utility games. Using our technique, we derive mostly tight bounds on the robust price of anarchy of their altruistic extensions. For the majority of the mentioned game classes, the results extend to the more differentiated friendship setting. As we show, our reduction technique covers this model if the base game satisfies three additional natural properties.

1 Introduction

The study of the inefficiency of equilibria in strategic games has been one of main research streams in algorithmic game theory in the last decade and contributed to the explanation of several phenomena observed in real life. More recently, researchers have also started to incorporate more complex social relationships among the players in such studies, accounting for the fact that players cannot always be regarded as isolated entities that merely act on their own behalf (see also [12]). In particular, the extent by which other-regarding preferences such as *altruism* and *spite* impact the inefficiency of equilibria has been studied intensively; see, e.g., [1, 4–7, 11, 15, 14, 16].

In this context, some counterintuitive results have been shown that are still not well-understood. For example, in a series of papers [4, 5, 7] it was observed that for congestion games the inefficiency of equilibria gets worse as players

Y. Chen and N. Immorlica (Eds.): WINE 2013, LNCS 8289, pp. 391–404, 2013.
© Springer-Verlag Berlin Heidelberg 2013

become more altruistic, therefore suggesting that altruistic behavior can actually be harmful for society. On the other hand, valid utility games turn out to be unaffected by altruism as their inefficiency remains unaltered under altruistic behavior [7]. These discrepancies triggered our interest in the research conducted in this paper. The basic question that we are asking here is: What is it that impacts the inefficiency of equilibria of games with altruistic players?

To this aim, we consider two different models that have previously been studied in the literature: the *altruism model* [7] and the *friendship model* [1]. In both models, one starts from a strategic game (called the *base game*) specifying the *direct cost* of each player and then extends this game by defining the *perceived cost* of each player as a function of his neighbors' direct costs. In the altruism model, player i's perceived cost is a convex combination of his direct cost and the overall social cost. In the more general friendship model, player i's perceived cost is a linear combination of his direct cost and his friends' costs.

In order to quantify the inefficiency of equilibria in our games we resort to the concept of the *price of anarchy (PoA)* [18], which is defined as the worst-case relative gap between the cost of a Nash equilibrium and a social optimum (over all instances of the game). By now, a standard approach to prove upper bounds on the PoA is through the use of the *smoothness framework* introduced by Roughgarden [19]. Basically, this framework allows us to derive bounds on the *robust price of anarchy* by showing that the underlying game satisfies a certain (λ, μ)-*smoothness property* for some parameters λ and μ. The robust PoA holds for various solution concepts, ranging from pure Nash equilibria to coarse correlated equilibria (see, e.g., Young [24]).

The original smoothness framework [19] has been extended to both the altruism and the friendship model in [7] and [1], respectively. Applying these adapted smoothness frameworks to bound the robust PoA is often technically involved because of the altruistic terms that need to be taken into account additionally (see also the analyses in [1, 7]).

Instead, we take a different approach here. As we will show, there is a natural class of games, which we term *social contribution games (SCGs)*, that is intimately connected with our altruism and friendship games. We establish a general reduction technique that enables us to reduce the problem of establishing smoothness for our altruism or friendship game to the problem of proving smoothness for a corresponding SCG. The latter is usually much simpler than proving smoothness for the altruism or friendship game directly. This also opens up the possibility to derive better bounds on the robust PoA of these games through the usage of our new reduction technique.

Our Contributions. Our main contributions are as follows:

- We introduce a new class of games, which we term *social contribution games (SCGs)*, where each player's individual cost is defined as the cost he incurs on society because of his presence. Said differently, player i's cost is equal to the difference in social cost if player i is present/absent in the game. We show that SCGs are *altruism-independently smooth*, i.e., if the SCG is (λ, μ)-smooth then every altruistic extension is (λ, μ)-smooth as well.

Table 1. Robust PoA bounds derived in this paper for the friendship model

Games	Robust PoA our results	Robust PoA previous best	Remarks
$R\|\|\sum_j w_j C_j$	$= 4^\star$	$\leq 23.31^\S$ [1]	RPoA $= 4$ (selfish players) [10]
$P\|\|\sum_j C_j$	≤ 2		RPoA $= \frac{3}{2} - \frac{1}{2m}$ (selfish players)
linear congestion games	$= \frac{17}{3}$	≤ 7 [1]	$5 \leq$ PoA $\leq \frac{17}{3}$ (special case) [2]
p-poly. congestion games	$\leq (1+p)\gamma(p)^\dagger$		PoA $= \gamma(p)^\dagger$ (selfish players) [8]
second-price auctions	$= 2$		RPoA $= 2$ (selfish players) [22]
valid utility games	$= 2^\ddagger$	$= 2^\ddagger$ [7]	RPoA $= 2$ (selfish players) [19]

* holds only if a certain weight condition is satisfied
§ for the special case $R\|\|\sum_j C_j$ only
† $\gamma(p) = p^{p(1-o(1))}$
‡ for the altruism model only

- We derive a general reduction technique to bound the robust PoA of both altruism and friendship games. Basically, the reduction can be applied whenever the underlying base game is *social contribution bounded*, meaning that the direct cost of each player is bounded by his respective cost in the corresponding SCG (for the friendship model a slightly stronger condition needs to hold). It is worth mentioning that this reduction preserves the (λ, μ)-smoothness parameters, i.e., the altruism or friendship game inherits the (λ, μ)-smoothness parameters of the SCG.
- We generalize smoothness for friendship extensions to *weight-bounded* social cost functions. In previous papers, the used techniques usually required sum-boundedness, which is a stronger condition [1]. Applying this definition to scheduling games with weighted sum as social cost, we derive a nice characterization of those scheduling games whose robust PoA does not grow for friendship extensions.
- We show that social contribution boundedness is satisfied by several well-known games, like min-sum scheduling games, congestion games, second-price auctions and valid utility games. Using our reduction technique, we then derive upper bounds on the robust PoA of their friendship/altruism extensions. In most cases we prove matching lower bounds. The results are summarized in Table 1.

Even though we focus on the complete information setting in this paper, our results extend to the incomplete information setting in which players are uncertain about the friendship levels of the other players. More details will be given in the full version of the paper.

Related Work. Several articles propose models of altruism and spite [2, 4–7, 11, 14–16]. Among these articles, the inefficiency of equilibria in the presence of altruism and spite was studied for various games in [2, 4–7, 11]. After its introduction in [19], the smoothness framework has been extended to incomplete information settings [20, 22] and altruism/spite settings [1, 7].

The robust PoA for minsum scheduling (not taking altruism or friendship into account) was studied in various papers. In [17] the authors show that it does not exceed 2 for $Q||\sum_j C_j$ (here we improve this bound to $\frac{3}{2} - \frac{1}{2m}$ for the special case $P||\sum_j C_j$). A robust PoA of 4 for $R||\sum_j w_j C_j$ has been proven in [9]. Our work on linear congestion games generalizes a result in [2]. They show that the pure price of anarchy does not exceed 17/3 in a restricted friendship setting ($\alpha_{ij} \in \{0,1\}$).

As indicated above, most related to our work are the articles [1, 7]. We significantly improve the bounds on the robust price of anarchy for congestion games and unrelated machine scheduling games in [1] and at the same time simplify the analysis by using our reduction technique.

2   Preliminaries

Let $G = (N, \{\Sigma_i\}_{i \in N}, \{C_i\}_{i \in N})$ be a *cost-minimization game*, where $N = [n]$ is the set of players, Σ_i is player i's strategy space, $\Sigma = \prod_{i \in N} \Sigma_i$ is the set of strategy profiles, and $C_i : \Sigma \to \mathbb{R}$ denotes the cost player i must pay for a given strategy profile. We assume that each player seeks to minimize his cost. A *social cost function* $C : \Sigma \to \mathbb{R}$ assigns a social cost to each strategy profile. We usually require C to be *sum-bounded*, i.e., $C(s) \leq \sum_{i \in N} C_i(s)$ for all $s \in \Sigma$.

We denote *payoff-maximization games* as $G = (N, \{\Sigma_i\}_{i \in N}, \{\Pi_i\}_{i \in N})$ with *social welfare* $\Pi : \Sigma \to \mathbb{R}$. In this case, each player i tries to maximize his *utility (or payoff)* Π_i. Again, we usually assume that Π is *sum-bounded*, i.e. $\Pi(s) \geq \sum_{i \in N} \Pi_i(s)$ for all $s \in \Sigma$.

Subsequently, we state most of the definitions and theorems only for cost-minimization games. The payoff-maximization case works similarly by reversing all inequalities. So, unless stated otherwise, G denotes a cost-minimization game with social cost function C.

Definition 1. *A* coarse equilibrium *is a probability distribution σ over Σ such that the following holds: If s is a random variable with distribution σ, then for all players i and all strategies $s_i^* \in \Sigma_i$, $\mathbf{E}_{s \sim \sigma}[C_i(s)] \leq \mathbf{E}_{s_{-i} \sim \sigma_{-i}}[C_i(s_i^*, s_{-i})]$, where σ_{-i} is the projection of σ on $\Sigma_{-i} = \prod_{j \neq i} \Sigma_j$. A* mixed Nash equilibrium *is a coarse equilibrium σ that is the product of independent probability distributions σ_i on Σ_i. A (pure)* Nash equilibrium *(NE) is a strategy profile $s \in \Sigma$ such that for all $s^* \in \Sigma$, $C_i(s) \leq C_i(s_i^*, s_{-i})$, where $s_{-i} = s|_{\Sigma_{-i}}$.*

The coarse (resp. correlated, mixed, pure) price of anarchy *(PoA) is defined as $\sup_s C(s)/C(s^*)$, where s^* minimizes C and s runs over the coarse (resp. correlated, mixed, pure) Nash equilibria of G.[1] The* coarse (resp. correlated, mixed, pure) PoA *of a class $\mathcal{G}$ of games is defined as the supremum of the respective PoA values of games in $\mathcal{G}$.*

Note that pure Nash equilibria constitute a subset of mixed Nash equilibria which constitute a subset of coarse equilibria. This implies that the respective prices of anarchy are non-decreasing (in this order).

---

[1] Similarly, we define the respective types of PoA for a payoff-maximization game as $\sup \Pi(s^*)/\Pi(s)$, where s and s^* are as above.

Due to lack of space, several proofs were omitted from this extended abstract and will be given in the full version of the paper.

2.1   The Altruism Model

Definition 2 ([7]). *Let $\alpha \in [0,1]^N$. The α-altruistic extension of G is defined as the cost-minimization game $G^\alpha = (N, \{\Sigma_i\}_{i \in N}, \{C_i^\alpha\}_{i \in N})$, where for any $i \in N$ the perceived cost is the convex combination $C_i^\alpha = (1 - \alpha_i)C_i + \alpha_i C$. We call G the* base game. *The social cost function of G^α is again C, i.e., the cost of the base game.*

The higher the 'altruism level' α_i, the more i cares about the society in general.

Definition 3. *Let G have sum-bounded social cost and let $\alpha \in [0,1]^N$. Define $C_{-i} := C - C_i$. G^α is (λ, μ)-smooth if there exists an optimal strategy s^* such that for any strategy $s \in \Sigma$,*

$$\sum_{i \in N} \left(C_i(s_i^*, s_{-i}) + \alpha_i(C_{-i}(s_i^*, s_{-i}) - C_{-i}(s)) \right) \leq \lambda C(s^*) + \mu C(s),$$

The robust PoA *of G^α is defined as $\inf\{\frac{\lambda}{1-\mu} \mid G^\alpha$ is (λ, μ)-smooth, $\mu < 1\}$.*

Theorem 1 ([7]). *Let G^α be an α-altruistic extension of G. Then the coarse (and thus the correlated, mixed and pure) PoA of G^α is bounded from above by the robust PoA of G^α.*

2.2   The Friendship Model

Definition 4 ([1]). *Let $\alpha \in [0,1]^{N \times N}$ such that $\alpha_{ii} = 1$ for all $i \in N$. The α-friendship extension of G is defined as $G^\alpha = (N, \{\Sigma_i\}_{i \in N}, \{C_i^\alpha\}_{i \in N})$, where for any $i \in N$ the perceived cost is defined as $C_i^\alpha = \sum_j \alpha_{ij} C_j$. Like in the altruism model, we consider C, the social cost function of the base game, as the social cost for G^α.*

For players i and j, α_{ij} can be interpreted as the level of affection i feels towards j. Note that if $C = \sum_j C_j$, then the altruism model is a special case of the friendship model because in this case, $C_i^\alpha = C_i + \sum_{j \neq i} \alpha_i C_j$ (for $\alpha \in [0,1]^N$).

Next we adapt the smoothness definition in [1] for the friendship model to the weighted player case.

Definition 5. *Let G^α be friendship extension of a cost-minimization game with a weight-bounded social cost function, i.e., $C \leq \sum_i w_i C_i$ for some $w \in \mathbb{R}_+^N$. G^α is (λ, μ)-smooth if there exists a (possibly randomized) strategy profile $\bar{s}$ such that for all strategy profiles s and all optima s^*,*

$$\sum_{i \in N} w_i \big(C_i(\bar{s}_i, s_{-i}) + \sum_{j \neq i} \alpha_{ij}(C_j(\bar{s}_i, s_{-i}) - C_j(s)) \big) \leq \lambda C(s^*) + \mu C(s).$$

We define the robust PoA of G^α as $\inf\{\frac{\lambda}{1-\mu} \mid G^\alpha$ is (λ, μ)-smooth, $\mu < 1\}$.

Theorem 2. *Let G^α be a friendship extension of a cost-minimization game with weight-bounded social cost function C. If G^α is (λ, μ)-smooth with $\mu < 1$, then the coarse PoA of G^α is at most $\frac{\lambda}{1-\mu}$.*

In both models, we can replace the deterministic factor α by a stochastic variable that is distributed with respect to some probability distribution over $[0,1]^N$ (in the altruism model) or $[0,1]^{N \times N}$ (in the friendship model). Thus, we can incorporate *incomplete information* into our model, reflecting the fact that often players are uncertain about other players' feelings. The bounds on the PoA continue to hold in this case. We defer the details to the full version of the paper.

3 Social Contribution Games

Definition 6. *We call G a (cost-minimization) social contribution game (SCG) if for all players i there exists a default strategy $\emptyset_i$ such that for all $s \in \Sigma$, $C_i(s) = C(s) - C(\emptyset_i, s_{-i})$.*

The strategy $\emptyset_i$ is often interpreted as 'refusing to participate in the game'. In that sense, i pays exactly the social cost he causes by choosing to play; in the payoff-maximization case, he gets exactly what he contributes to the social welfare. So social contribution games are 'fair' in some sense.

Basic utility games [23] satisfy the definition of an SCG (see also Section 7). In particular, the competitive facility location game (which is a basic utility game by [23]) is an SCG.

We now show that social contribution games satisfy the following invariance property with respect to their α-altruistic extensions.

Lemma 1. *Any social contribution game is altruism-independently smooth, i.e., for all $\alpha = (\alpha_i)_{i \in N}$ and corresponding altruistic extensions G^α of G, the robust price of anarchy in G and G^α is the same.*

Proof. For all players i, $C_{-i}(s) = C(s) - C_i(s)$ is independent of s_i since $C(s) - C_i(s) = C(\emptyset_i, s_{-i})$. Thus for all strategy profiles s, s^*, and all $\alpha \in \mathbb{R}^N$,

$$\sum_i \big(C_i(s_i^*, s_{-i}) + \alpha_i(C_{-i}(s_i^*, s_{-i}) - C_{-i}(s)) \big) = \sum_i C_i(s_i^*, s_{-i}).$$

It follows that for all $(\lambda, \mu) \in \mathbb{R}^2$, G^α is (λ, μ)-smooth iff G is. □

The notions of α-altruistic extensions and α-independent smoothness can be easily extended to $\alpha \in \mathbb{R}^N$. The above lemma continues to hold in this case. So even if a player wants to *hurt* society, the robust PoA stays the same for SCGs.

3.1   Social Contribution Bounded Games

Definition 7. *Assume C is sum-bounded. We call G social contribution bounded (SC-bounded) if for all players i there exists a default strategy $\emptyset_i$ such that for all $s \in \Sigma$, $C_i(s) \leq C(s) - C(\emptyset_i, s_{-i})$. In this case, we define the corresponding social contribution game $\bar{G} = (N, \{\Sigma_i\}_{i \in N}, \{\bar{C}_i\}_{i \in N})$ by setting $\bar{C}_i(s) = C(s) - C(\emptyset_i, s_{-i})$.*

As before, we think of $\emptyset_i$ as the option that i does not participate.[2]

The following theorem shows that if we want to get a bound on the PoA of α-altruistic extensions of an SC-bounded game, we might as well consider the corresponding SCG *regardless of α*.

Theorem 3. *Let G be SC-bounded and suppose that the robust PoA of the corresponding SCG $\bar{G}$ is ξ. Then for all altruistic extensions G^α of G, the robust PoA is at most ξ.*

In order to be able to derive our results for the *friendship* extensions, we need a slightly stronger definition.

Definition 8. *A cost minimization game G with weight-bounded social cost is strongly SC-bounded if for all $s \in \Sigma$ and every player i:*

1. $C_i(\emptyset_i, s_{-i}) = 0$ *(if i does not participate, he pays nothing)*
2. $\forall j \neq i : C_j(\emptyset_i, s_{-i}) \leq C_j(s)$ *(other players' costs can only increase if i participates)*
3. $w_i \sum_j (C_j(s) - C_j(\emptyset_i, s_{-i})) \leq C(s) - C(\emptyset_i, s_{-i})$ *(the weighted impact of i's participation on the players' costs is bounded by his impact on the social cost)*

If all weights are 1, then assumption (3) easily follows from
3b. $C(s) = \sum_j C_j(s)$ *(social cost is sum of individual costs).*

Theorem 4. *Let G be strongly SC-bounded. Suppose the robust PoA of $\bar{G}$ is ξ. Then for all friendship extensions G^α, the robust PoA is at most ξ.*

Proof. We have for every player i,

$$w_i\left(C_i(\bar{s}_i, s_{-i}) + \sum_{j \neq i} \alpha_{ij}(C_j(\bar{s}_i, s_{-i}) - C_j(s))\right)$$

$$\overset{(2)}{\leq} w_i\left(C_i(\bar{s}_i, s_{-i}) + \sum_{j \neq i} \alpha_{ij}(C_j(\bar{s}_i, s_{-i}) - C_j(\emptyset_i, s_{-i}))\right)$$

---

[2] Note that $\emptyset_i$ need not actually be an element of Σ_i. In many games (such as scheduling or congestion games) it is not an option to not participate. So, formally we should require that there exists a function $\mathfrak{C} : \prod_i(\Sigma_i \cup \{\emptyset_i\}) \to \mathbb{R}$ such that $\mathfrak{C}|_\Sigma = C$ and $C_i(s) \leq \mathfrak{C}(s) - \mathfrak{C}(\emptyset_i, s_{-i})$ for all i and s. However, there is a natural way to extend C (and C_i) on $\prod_i(\Sigma_i \cup \{\emptyset_i\})$, as we will see later. For notational convenience, we write C instead of $\mathfrak{C}$.

$$\overset{(2)}{\leq} w_i\big(C_i(\bar{s}_i, s_{-i}) + \sum_{j\neq i}(C_j(\bar{s}_i, s_{-i}) - C_j(\emptyset_i, s_{-i}))\big)$$

$$\overset{(1)}{=} w_i \sum_j (C_j(\bar{s}_i, s_{-i}) - C_j(\emptyset_i, s_{-i})) \overset{(3)}{\leq} C(\bar{s}_i, s_{-i}) - C(\emptyset_i, s_{-i}) = \bar{C}_i(\bar{s}_i, s_{-i}).$$

Summing over all i, it follows that if $\bar{G}$ is (λ, μ)-smooth[3], then so is G^α. $\quad\square$

If all weights are 1, then SC-boundedness follows from strong SC-boundedness. To see this, consider the case where $\alpha = 0$ and carry out the proof of Theorem 4 for s instead of $(\bar{s}_i, s_{-i})$.

4   Minsum Machine Scheduling

A *scheduling game* $G = (m, n, (p_{ij})_{i\in M, j\in N}, (w_j)_{j\in N})$ consists of a set of jobs (players) $[n] = \{1, \ldots, n\}$ and a set of machines $[m] = \{1, \ldots, m\}$. For each machine i and job j, $p_{ij} \in \mathbb{R}_+$ denotes the *processing time* of j on i. Furthermore, w_j is the *weight* of job j. The strategy space Σ_i of a job j is simply the set of machines. By $\emptyset_i = \emptyset$ we mean the strategy where i uses no machine.

Let x be a strategy profile. For a machine i, we denote by X_i the set of jobs that are scheduled on i. Furthermore, x_j denotes the machine j is assigned to. Following the notation by Cole et al. [9], we define $\rho_{ij} = p_{ij}/w_j$. We assume that the jobs on a machine are scheduled in increasing order of ρ_{ij}, which is known as *Smith's rule* [21]; if two jobs on a machine have the same time-to-weight ratio, we use a tie-breaking rule. The cost C_j of job j which it seeks to minimize is simply its completion time. In the following, we assume for simplicity that the ρ_{ij} are pairwise distinct (but the results continue to hold without this assumption). Then we can write $C_j(x) = \sum_{k\in X_i : \rho_{ik} \leq \rho_{ij}} p_{ik}$. The social cost C we consider is the weighted sum of the players' completion times, i.e., $C = \sum_j w_j C_j$.

In the following, we use the three-field notation by Graham et al [13]. In this notation, the problem we described is denoted by $R || \sum_j w_j C_j$. If all weights are 1, we write $\sum_j C_j$ instead of $\sum_j w_j C_j$. Furthermore, if there are *speeds* s_i for each machine i and *fixed* processing times p_j for each job such that $p_{ij} = p_j/s_i$, we write Q instead of R. Finally, if we have in addition identical speeds $s_i = 1$ for all machines i, the problem is denoted by P.

4.1 $R || \sum_j w_j C_j$

Lemma 2 ([9]). *For all strategy profiles x and x^*,*

$$\sum_{i\in[m]} \sum_{j\in X_i^*} w_j p_{ij} + \sum_{i\in[m]} \sum_{j\in X_i^*} \sum_{k\in X_i} w_j w_k \min\{\rho_{ij}, \rho_{ik}\} \leq 2C(x^*) + \frac{1}{2}C(x),$$

where X_i^* is defined similarly to X_i as $X_i^* = \{j \in J \mid x_j^* = i\}$.

---

[3] in the sense that there exist $\bar{s} \in \Sigma$ and an optimal $s^* \in \Sigma$ such that for all $s \in \Sigma$ it holds that $\sum_i C_i(\bar{s}_i, s_{-i}) \leq \lambda C(s) + \mu C(s^*)$, generalizing Roughgarden's definition of smoothness [19].

Proof. The claim is shown in the proof of [9, Theorem 3.2].   □

Theorem 5. *Let G be an instance of $R||\sum_j w_j C_j$ that satisfies the following condition for all jobs j, k and all machines i: $\rho_{ij} \leq \rho_{ik}$ implies $w_j \leq w_k$ (i.e., if k gets scheduled after j on i, then it is because of its processing time, not its weight). Then the robust PoA of all friendship extensions G^α of G is at most 4.*

For jobs j and k, α_{jk} has an influence on j's strategy in an equilibrium only if there is a machine i such that k gets scheduled after j on i because j cannot influence k's costs otherwise. Hence the weight condition tells us that the only jobs that could potentially have an influence on j are in fact the jobs that are at least equally important as j. Hence j cannot 'misplace his affections' and care too much about unimportant jobs.

Proof. First we show that G is strongly SC-bounded. Clearly, (1) and (2) are satisfied. For (3), note that for all jobs j, strategy profiles x, and $i = x_j$,

$$w_j \sum_k (C_k(x) - C_k(\emptyset, x_{-i})) = w_j \left(C_j(x) + \sum_{k \in X_i: \, \rho_{ik} > \rho_{ij}} p_{ij} \right)$$

$$\leq w_j C_j(x) + \sum_{k \in X_i: \, \rho_{ik} > \rho_{ij}} w_k p_{ij} = \bar{C}_j(x),$$

where the inequality follows from the condition on the weights. We calculate

$$\bar{C}_j(x_j^*, x_{-j}) = w_j C_j(x_j^*, x_{-j}) + \sum_{k \in X_i: \, \rho_{ik} > \rho_{ij}} w_k p_{ij}$$

$$= w_j p_{ij} + \sum_{k \in X_i: \, \rho_{ik} < \rho_{ij}} w_k w_j \rho_{ik} + \sum_{k \in X_i: \, \rho_{ik} > \rho_{ij}} w_k w_j \rho_{ij}$$

$$\leq w_j p_{ij} + \sum_{k \in X_i} w_j w_k \min\{\rho_{ij}, \rho_{ik}\}.$$

Summing over all machines i and $j \in X_i^*$, this is the same expression as in Lemma 2. Hence $\sum_j \bar{C}_j(x_j^*, x_{-j}) \leq 2C(x^*) + \frac{1}{2}C(x)$ and $\bar{G}$ is $(2, \frac{1}{2})$-smooth. It follows by Theorem 4 that the robust PoA in G^α is at most 4.   □

This bound is tight and the weight condition is necessary. In fact, if we drop it, the pure PoA is unbounded even for $P||\sum_j w_j C_j$ instances with unit-size jobs. We defer these results to the full version.

4.2 $P||\sum_j C_j$

Fix an ordering of the jobs such that $p_j > p_{j'}$ implies $j > j'$. We use the same notation as in [17]: For a schedule x, a job j and a machine i, let $h_i^x(j) = |\{j' > j | x_{j'} = x_j\}|$. This is the number of jobs that are scheduled after j on i. Using this notation, we can write $\bar{C}_j(x) = C_j(x) + h_{x_j}^x(j) \cdot p_j$ for instances with unit speeds. Throughout this section, let $\bar{x}$ denote the randomized schedule that assigns each job to each machine with probability $\frac{1}{m}$.

The following theorem will be helpful to establish an upper bound on the robust PoA for the friendship model and might be of independent interest.

Theorem 6. *For any schedule x and any optimal x^*, $\sum_j C_j(\bar{x}_j, x_{-j}) \leq C(x^*) + (\frac{1}{2} - \frac{1}{2m})\sum_j p_j$. In particular, the robust price of anarchy of $P||\sum_j C_j$ is at most $\frac{3}{2} - \frac{1}{2m}$. This bound is tight.*

Theorem 7. *Let G be an instance of $P||\sum_j C_j$. Then the robust PoA for any friendship extension G^α is at most 2.*

Proof. Let x be arbitrary. Then by linearity of expectation,

$$\mathbf{E}\Big[\sum_j \bar{C}_j(\bar{x}_j, x_{-j})\Big] = \sum_j \mathbf{E}[C_j(\bar{x}_j, x_{-j})] + \sum_j \mathbf{E}[h^x_{\bar{x}_j}(j)] \cdot p_j.$$

We know that

$$\mathbf{E}[h^x_{\bar{x}_j}(j)] = \frac{1}{m}\sum_i h^x_i(j) = \frac{1}{m}|\{j' \in J \mid j' > j\}| = \mathbf{E}[h^{\bar{x}}_{x_j}(j)].$$

Hence the second term evaluates as

$$\sum_j \mathbf{E}[h^x_{\bar{x}_j}(j)] \cdot p_j = \sum_j \mathbf{E}[h^{\bar{x}}_{x_j}(j)] \cdot p_j = \sum_j \mathbf{E}[C_j(\bar{x}_j, x_j)] - \sum_j p_j.$$

We know by Theorem 6 that $\sum_j \mathbf{E}[C_j(\bar{x}_j, x_{-j})] \leq C(x^*) + (\frac{1}{2} - \frac{1}{2m})\sum_j p_j$. Hence

$$\sum_j \mathbf{E}[\bar{C}_j(\bar{x}_j, x_{-j})] = 2\sum_j \mathbf{E}[C_j(\bar{x}_j, x_{-j})] - \sum_j p_j \leq 2C(x^*) - \frac{1}{m}\sum_j p_j \leq 2C(x^*),$$

for any schedule x^*. Hence the robust PoA for the friendship extension is at most 2. $\qquad\square$

5   Congestion Games

An *atomic congestion game* $G = (N, E, \{\Sigma_i\}_{i \in N}, (d_e)_{e \in E})$ is given by a set E of *resources* together with *delay functions* $d_e : \mathbb{N} \to \mathbb{R}_+$ indicating the delay on e for a given number of players using e. Each player's strategy set consists of subsets of E; $\Sigma_i \subseteq \mathcal{P}(E)$ for all i. For $s \in \Sigma$, let $x_e(s) = |\{i \in N \mid e \in s_i\}|$. The cost of each player i under s is given by $C_i(s) = \sum_{e \in s_i} d_e(x_e(s))$. If all delay functions are linear, we say that G is *linear*. Further, if all delay functions are polynomials of maximum degree p with non-negative coefficients, we say that G is *p-polynomial*. The social cost C is simply the sum over all individual cost. By $\emptyset_i = \emptyset$ we mean the strategy where player i uses no machine.

It is known that we can without loss of generality assume that all latency functions are of the form $l_e(x) = x$. This was first mentioned in [8]; for a proof see [7]. The following lemma is shown in the proof of [8, Theorem 1].

Lemma 3 ([8]). *Let G be a linear congestion game and $s, s^* \in \Sigma$. Then $\sum_i C_i(s^*_i, s_{-i}) \leq \sum_e x_e(s^*)(x_e(s) + 1)$.*

Lemma 4 ([2]). *For any pair $\alpha, \beta \in \mathbb{N}$, it holds that $\frac{2}{5}\alpha^2 + \frac{17}{5}\beta^2 \geq \beta(2\alpha + 1)$.*

Bilò et al. show in their paper [2] that the *pure* PoA lies between 5 and 17/3 for a restricted friendship setting, where $\alpha_{ij} \in \{0, 1\}$ for all i, j. We generalize their result to the *robust* PoA for arbitrary $\alpha_{ij} \in [0, 1]$ and show tightness.

Theorem 8. *Let G be a linear congestion game. Then the robust PoA of all friendship extensions G^α is bounded by $\frac{17}{3} \approx 5.67$. This bound is tight.*

Proof. We have

$$\bar{C}_i(s) = C_i(s) + \sum_{e \in s_i} |\{j \neq i \mid e \in s_j\}| = C_i(s) + \sum_{e \in s_i} x_e(\emptyset, s_{-i}) \geq C_i(s),$$

so G is SC-bounded. Also G is strongly SC-bounded: If i does not use any resource, he experiences no cost; the other's costs can only increase if another player enters; and finally, $C = \sum_j C_j$.

Let $s, s^* \in \Sigma$. We abbreviate $x_e(s)$ and $x_e(s^*)$ by x_e and x_e^*, respectively. The calculation of the robust PoA for $\bar{G}$ yields

$$\sum_i \bar{C}_i(s_i^*, s_{-i}) = \sum_i C_i(s_i^*, s_{-i}) + \sum_i \sum_{e \in s_i^*} x_e(\emptyset, s_{-i}).$$

The first term is at most $\sum_e x_e^*(x_e + 1)$ by Lemma 3. The second term is bounded by $\sum_i \sum_{e \in s_i^*} x_e(s) = \sum_{e \in E} x_e x_e^*$. Hence we get in total by Lemma 4

$$\sum_e x_e^*(2x_e + 1) \leq \sum_e \left(\frac{17}{5}(x_e^*)^2 + \frac{2}{5}x_e^2 \right) = \frac{17}{5}C(s^*) + \frac{2}{5}C(s).$$

It follows that the robust PoA of $\bar{G}$ is at most $\frac{17}{5} / (1 - \frac{2}{5}) = \frac{17}{3}$.

We show now that the bound of $\frac{17}{3}$ is asymptotically tight. Let $n \geq 0$. Consider an instance with $n + 3$ blocks of players $B_0, \ldots, B_{n+2}$ consisting of three players each: $B_k = \{a_k, b_k, c_k\}$. We construct a NE s and an optimal strategy profile s^* as follows. For all resources e, we set $l_e(x) = x$. For $0 \leq k \leq n$, the pattern of strategies repeats (see Figure 1). Here player $i = a_k$ has two strategies $s_i = \{3k, 3k+1, 3k+2\}$ and $s_i^* = \{3k+6\}$. Player $i = b_k$ has two strategies $s_i = \{3k+2, 3k+3\}$ and $s_i^* = \{3k+7\}$. Player $i = c_k$ has two strategies $s_i = \{3k+3, 3k+4\}$ and $s_i^* = \{3k+8\}$.

The strategies s_i of players in the final blocks B_{n+1} and B_{n+2} are defined as above. However, we need to change the definition of s_i^* because otherwise, s is not a Nash equilibrium. So for each $i \in B_{n+1} \cup B_{n+2}$, we insert sets of new, previously unused resources s_i^* such that $C_i(s_i) = |s_i^*|$.

We define $\alpha_{ij} = 1$ for the following pairs of players: (a_k, b_{k+1}), (a_k, c_{k+1}), (a_k, a_{k+2}) as well as (b_k, c_{k+1}), (b_k, a_{k+2}) and (c_k, a_{k+2}), (c_k, b_{k+2}), where $0 \leq k \leq n$. All other α_{ij} are zero. Hence $\alpha_{ij} = 1$ iff s_i^* intersects s_j. Note that if $s_i \cap s_j \neq \emptyset$, then $\alpha_{ij} = 0$.

Now, we claim that s is a NE. In fact, for all $0 \leq k \leq n$ and $i = a_k$, $C^\alpha(s) = C(s) + \sum_{j \neq i} \alpha_{ij} C_j(s) = 7 + 5 + 5 + 7 = 24$, which equals $C_i^\alpha(s^*, s_{-i}) = 4 + 6 + 6 + 8$.

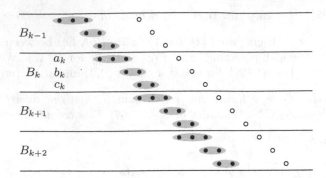

Fig. 1. The strategy profiles s (grey) and s^* (white). Columns correspond to resources

A similar calculation shows $C_i(s) = C_i(s_i, s_{-i})$ for $i = b_k, c_k$. Observe that for $k = n + 1, n + 2$, and $i \in B_k$, $C_i^\alpha(s) = C(s) = |s_i^*| = C(s_i^*, s_{-i})$ by our construction of s_i^*. Hence s is indeed a NE.

For $k = 1, \ldots, n$, block B_k has the same cost: $C(B_k) := \sum_{i \in B_k} C_i(s) = 17$ and $C^*(B_k) := \sum_{i \in B_k} C_i(s^*) = 3$. Let $X = C(B_0) + C(B_{n+1}) + C(B_{n+2})$ and $X^* = C^*(B_0) + C^*(B_{n+1}) + C^*(B_{n+2})$ and observe that these are constants independent of n. It follows that

$$\frac{C(s)}{C(s^*)} = \frac{17n + X}{3n + X^*} = \frac{17 + o(n)}{3 + o(n)}. \qquad \square$$

We obtain the following result for friendship extensions of p-polynomial congestion games. Note that the pure PoA of the base game is $\gamma(p) := p^{p(1-o(1))}$ [8]. That is, altruism increases the PoA by at most a factor of $(1 + p)$ in this case.

Theorem 9. *Let G be a p-polynomial congestion game. Then the robust PoA of all friendship extensions G^α is bounded by $(1 + p) \cdot p^{p(1-o(1))}$.*

6   Second-Price Auctions

A single-item auction G consists of an *allocation rule* $a : \Sigma \to N$ which determines which bidder gets the item and a *pricing rule* $p : \Sigma \to \mathbb{R}^N$ indicating how much each player should pay. Each bidder i is assumed to have a certain *valuation* $v_i \in \mathbb{R}_+$ for the item. For a given bidding profile $b \in \mathbb{R}_+^N$, the social welfare is $\Pi(b) = v_{a(b)}$. Player i's utility is given by $\Pi_i(b) = v_i - p_i(b)$ if he gets the object and $-p_i(b)$ otherwise. In a *second-price auction*, the highest bidder gets the item and pays the second highest bid, while everybody else pays nothing.

We do not allow *overbidding*, i.e., for all bidders i, $b_i \leq v_i$. This is a standard assumption because overbidding is a dominated strategy. We denote by $\beta(b, i)$ the name of the player who places the i-th highest bid in b. We write $\beta(i)$ instead of $\beta(b, i)$ if the bidding profile is clear from the context. $\emptyset_i = 0$ denotes the strategy where bidder i bids nothing.

Note that here the friendship model is *not* a generalization of the altruism model because $\Pi \neq \sum_i \Pi_i$. We summarize our results in the following theorem.

Theorem 10. *Let G be a second-price auction. Then the robust PoA of all altruism extensions G^α is at most 2. Further, the coarse PoA of the class of friendship extensions of G is exactly 2.*

7   Valid Utility Games

A *valid utility game* [23] is defined as a payoff-maximization game $G = (N, E, \{\Sigma_i\}_{i \in N}, \{\Pi\}_{i \in N}, V)$, where E is a ground set of resources, $\Sigma_i \subseteq \mathcal{P}(E)$ and V is a submodular and non-negative function on E. The social welfare Π is given by $\Pi(s) = V(\bigcup_{i \in N} s_i)$ and is assumed to be sum-bounded. Furthermore, we require G to satisfy $\Pi_i(s) \geq \Pi(s) - \Pi(\emptyset, s_{-i})$ for all $s \in \Sigma$. If G additionally satisfies the last inequation with equality, it is called *basic utility game* [23]. For all players i, set $\emptyset_i = \emptyset$.

Theorem 11 ([19]). *The robust PoA of valid utility games with non-decreasing[4] set function V is bounded by 2.*

An example for valid utility games with non-decreasing set functions are competitive facility location games without fixed costs [23].

The following theorem has already been proven in [7] and tightness of this bound has been shown in [3] for the base game. We now use our framework to provide a shorter proof that illustrates why the robust PoA does not increase for altruistic extensions: The corresponding SCG falls into the same category of games.

Theorem 12. *Let G be a valid utility game with non-decreasing V. Then the robust price of anarchy of every altruistic extension G^α of G is bounded by 2.*

Proof. It follows directly from the definition that G is SC-bounded. It is easy to verify that the corresponding SCG $\bar{G} = (N, E, \{\Sigma_i\}_{i \in N}, \{\bar{\Pi}\}_{i \in N}, V)$ is again a valid utility game: $\sum_i \bar{\Pi}_i(s) \leq \sum_i \Pi_i(s) \leq \Pi(s)$ and $\bar{\Pi}_i(s) = \Pi(s) - \Pi(\emptyset, s_{-i})$. So the robust PoA of $\bar{G}$ is at most 2. Our claim follows by Theorem 3. □

References

1. Anagnostopoulos, A., Becchetti, L., de Keijzer, B., Schäfer, G.: Inefficiency of games with social context. In: Vöcking, B. (ed.) SAGT 2013. LNCS, vol. 8146, pp. 219–230. Springer, Heidelberg (2013)
2. Bilò, V., Celi, A., Flammini, M., Gallotti, V.: Social context congestion games. In: Kosowski, A., Yamashita, M. (eds.) SIROCCO 2011. LNCS, vol. 6796, pp. 282–293. Springer, Heidelberg (2011)
3. Blum, A., Hajiaghayi, M.T., Ligett, K., Roth, A.: Regret minimization and the price of total anarchy. In: Proc. 40th ACM Symp. on Theory of Computing (2008)

---

[4] where non-decreasing means that for all $A \subseteq B \subseteq E$ it holds that $V(A) \leq V(B)$.

4. Buehler, R., et al.: The price of civil society. In: Chen, N., Elkind, E., Koutsoupias, E. (eds.) WINE 2011. LNCS, vol. 7090, pp. 375–382. Springer, Heidelberg (2011)
5. Caragiannis, I., Kaklamanis, C., Kanellopoulos, P., Kyropoulou, M., Papaioannou, E.: The impact of altruism on the efficiency of atomic congestion games. In: Wirsing, M., Hofmann, M., Rauschmayer, A. (eds.) TGC 2010, LNCS, vol. 6084, pp. 172–188. Springer, Heidelberg (2010)
6. Chen, P.A., Kempe, D.: Altruism, selfishness, and spite in traffic routing. In: Proc. 9th ACM Conf. on Electronic Commerce, pp. 140–149 (2008)
7. Chen, P.-A., de Keijzer, B., Kempe, D., Schäfer, G.: The robust price of anarchy of altruistic games. In: Chen, N., Elkind, E., Koutsoupias, E. (eds.) WINE 2011. LNCS, vol. 7090, pp. 383–390. Springer, Heidelberg (2011)
8. Christodoulou, G., Koutsoupias, E.: The price of anarchy of finite congestion games. In: Proc. 37th ACM Symp. on Theory of Computing (2005)
9. Cole, R., Correa, J.R., Gkatzelis, V., Mirrokni, V., Olver, N.: Inner product spaces for minsum coordination mechanisms. In: Proc. 43rd ACM Symp. on Theory of Computing, pp. 539–548 (2011)
10. Correa, J.R., Queyranne, M.: Efficiency of equilibria in restricted uniform machine scheduling with total weighted completion time as social cost. Naval Research Logistics (NRL) 59(5), 384–395 (2012)
11. Elias, J., Martignon, F., Avrachenkov, K., Neglia, G.: Socially-aware network design games. In: Proc. 29th Conf. on Information Communications, pp. 41–45 (2010)
12. Fehr, E., Schmidt, K.M.: The Economics of Fairness, Reciprocity and Altruism: Experimental Evidence and New Theories. In: Handbook on the Economics of Giving, Reciprocity and Altruism, vol. 1, ch. 8, pp. 615–691. Elsevier (2006)
13. Graham, R., Lawler, E., Lenstra, J., Kan, A.R.: Optimization and approximation in deterministic sequencing and scheduling: A survey. Ann. of Discrete Math. (1979)
14. Hoefer, M., Skopalik, A.: Stability and convergence in selfish scheduling with altruistic agents. In: Leonardi, S. (ed.) WINE 2009. LNCS, vol. 5929, pp. 616–622. Springer, Heidelberg (2009)
15. Hoefer, M., Skopalik, A.: Altruism in atomic congestion games. In: Fiat, A., Sanders, P. (eds.) ESA 2009. LNCS, vol. 5757, pp. 179–189. Springer, Heidelberg (2009)
16. Hoefer, M., Skopalik, A.: Social context in potential games. In: Goldberg, P.W. (ed.) WINE 2012. LNCS, vol. 7695, pp. 364–377. Springer, Heidelberg (2012)
17. Hoeksma, R., Uetz, M.: The price of anarchy for minsum related machine scheduling. In: Solis-Oba, R., Persiano, G. (eds.) WAOA 2011. LNCS, vol. 7164, pp. 261–273. Springer, Heidelberg (2012)
18. Koutsoupias, E., Papadimitriou, C.: Worst-case equilibria. Computer Science Review 3(2), 65–69 (2009)
19. Roughgarden, T.: Intrinsic robustness of the price of anarchy. In: Proc. 41st ACM Symp. on Theory of Computing, pp. 513–522 (2009)
20. Roughgarden, T.: The price of anarchy in games of incomplete information. In: Proc. 13th ACM Conf. on Electronic Commerce, pp. 862–879 (2012)
21. Smith, W.: Various optimizers for single stage production. Naval Res. Logist. Quart. (1956)
22. Syrgkanis, V., Tardos, É.: Composable and efficient mechanisms. In: Proc. Symp. on the Theory of Computing (2013)
23. Vetta, A.: Nash equilibria in competitive societies. In: Proc. 43rd Symp. on Found. of Comp. Science (2002)
24. Young, H.P.: Strategic Learning and its Limits. Oxford University Press (1995)

Welfare-Improving Cascades
and the Effect of Noisy Reviews

Nick Arnosti* and Daniel Russo**

Stanford University

Abstract. We study a setting in which firms produce items whose quality is ex-ante unobservable, but learned by customers over time. Firms take customer learning into account when making production decisions. We focus on the effect that the review process has on product quality. Specifically, we compare equilibrium quality levels in the setting described above to the quality that would be produced if customers could observe item quality directly. We find that in many cases, customers are better off when relying on reviews, i.e. better off in the world where they have less information. The idea behind our result is that the risk of losing future profits due to bad initial reviews may drive firms to produce an exceptional product. This intuitive insight contrasts sharply with much of the previous academic literature on the subject.

1 Introduction

It is often impossible to directly determine the quality of an item before buying it. When this is the case, potential customers try to learn from the experiences of others. Increasingly, they rely on online reviews to help with their decisions. These reviews can significantly influence a firm's profitability: a Harvard Business School study [11] recently concluded that each additional star on Yelp generates (on average) a 5-9% increase in revenue for small businesses. Not only does firm success depend on reviews, but evidence is mounting that business owners are *aware* of this, and take it into account when making decisions. Some businesses respond directly to customers who leave negative reviews, and try to make amends. Others change their business practices: one Chicago bookstore "totally revamped our customer service approach" due to reviews left on Yelp [17].

This paper studies a setting in which customers learn from reviews left by others, and firms make choices with this fact in mind. There are many interesting questions that one could ask about such a market. We focus on the quality of items produced. Intuition suggests that reviews are an imperfect substitute for directly observing item quality. Formalizing this idea, the literature on this topic almost universally reaches the conclusion that product quality is highest when

* Supported in part by DARPA under the GRAPHS program.
** Supported by a Burt and Deedee McMurty Stanford Graduate Fellowship.

Y. Chen and N. Immorlica (Eds.): WINE 2013, LNCS 8289, pp. 405–420, 2013.
© Springer-Verlag Berlin Heidelberg 2013

customers can observe it. We illustrate that when item quality is endogenously chosen by firms, and firms take customer learning into account, the *opposite* might be true.

In our model, firms begin by making an irreversible decision about the quality of items that they will produce. Customers cannot observe this decision, but whenever a customer patronizes the firm they leave behind a review. Each review provides a noisy signal of the firms' item quality, which future customers use to draw inferences. Even high quality firms are at risk of losing business due to a bad review. To guard against this possibility, firms may produce a product that is better than customers would demand if they could directly observe its quality.

We clarify this intuition through two stylized models. In each model, firms are distinguished by their cost structures: some firms can provide high quality items more cheaply than others. We first consider a one-shot model, in which customer decisions are based on a single review. We then extend this to an infinite horizon model in which firms are visited by a sequence of customers. Customers only purchase from firms with a sufficiently favorable review history, and the only way that a firm can signal its quality is through customer reviews. This can lead to a cascade-like phenomenon: if early customers leave bad reviews, later ones may choose not to purchase. When this occurs, customers stop learning about the firm, and so the firm may go out of business even if they are producing high quality goods. The threat of this cascade gives firms an incentive to set higher initial quality levels.

The remainder of the paper is organized as follows. Section 2 places our work in the context of related academic literature. We introduce and discuss our general framework in Section 3. From there, we describe and analyze a single-period model in Section 4 before considering the infinite-horizon case in Section 5. We close with a summary of our results and a discussion of possibilities for future work.

2   Literature Review

The topic of signaling and reputation in markets has a rich history in economics. In a groundbreaking paper, Spence [16] introduces a simple model of job market signaling. In his model, intrinsically high quality workers put in effort in school in order to signal their ability to potential employers and, by doing so, differentiate themselves from low quality workers for whom sending positive signals is more costly. The firms in our model could be naturally re-interpreted as workers attempting to maintain a good reputation, and the customers as potential employers deciding whether to hire the worker. Seen in this light, our work differs from Spence's in at least two ways. The first is that we model the observation of individual performance as a noisy process. The second is that the effort employees exert in order to signal their quality directly benefits their employers.

In our model a potentially long-lived firm interacts with a series of short-lived customers. Thus, though we speak of a firm's "reputation," our setting differs substantially from the large literature that studies the role of reputation in long run strategic relationships (see for example Rubinstein [13], Fudenberg and Maskin [7], Fudenberg et al. [8]).

One distinctive component of our model is that firms receive a review only if a customer purchases their product. Because of this, a firm could be unable to repair its reputation after receiving several negative reviews, since it must attract customers in order to signal its quality. In this way, our work relates to the study of social herding and information cascades, as presented in the seminal work of Banerjee [2] and Bikhchandani et al. [4]. The main idea in these papers is that when customers have imperfect information and make decisions sequentially, it may be rational for them to ignore their private information and instead mimic the actions of those who went before them. This can induce a "cascade" in which each later customer takes the same action, even if it is in fact a poor one. The rather pessimistic message from these papers is that the outcome that results from a sequence of individually rational decisions may be arbitrarily worse than what results from a socially optimal decision rule.

There are effectively two types of information cascade: those in which customers repeatedly patronize a firm that is producing items of disappointingly low quality, and those in which customers abandon a firm producing high quality items. In our model, the former cannot happen, but the latter is a possibility. Rather than lament this inefficiency, we take a much cheerier perspective. In particular, our results indicate that the threat of such a cascade may cause firms to set higher quality levels than they otherwise would. Viewed in this light, when the underlying state of the world (i.e. product quality in our model) is not exogenously determined but rather strategically selected, the possibility of herd behavior may actually *enhance* consumer welfare.

Even closer to our work is a set of papers that deal specifically with models in which customers must decide whether to purchase items whose quality they cannot observe. Examples include Smallwood and Conlisk [15], Shapiro [14], Rogerson [12], Allen [1], Wolinsky [18], Cooper and Ross [6], Hörner [9], Bar-Isaac [3], Bose et al. [5], Ifrach et al. [10]. These models differ on a number of dimensions, such as the information possessed by firms and customers, whether customers behave strategically, and the role of item price.

Both Wolinsky [18] and Allen [1] construct equilibria where price signals quality perfectly to consumers. While these may help us understand some markets, they preclude the study of learning effects. At the other end of the spectrum, Cooper and Ross [6], Hörner [9] and Bar-Isaac [3] construct equilibria where posted prices depend only on information already available to consumers, i.e. the history of product reviews. Although prices affect firm incentives in these models, customers do not use them for inference.

These examples highlight the difficulty of building models where price plays an interesting role in the customer learning process. If firms are informed, consumers are rational, and firms can set prices, equilibrium prices are typically either fully revealing or do not reveal any new information to customers. To see why this is true, consider pure strategy equilibria when there are only two firm types. If prices set by the two types of firms depend on information that is not available to customers, then so long as the prices are not equal, customers can use them to infer firm type. Because we wish to focus on *learning* in reputation games, we choose not to incorporate price directly into our model. In this sense, our work applies to markets in which prices do not notably differentiate products and customers must look elsewhere in order to infer product quality.

Bar-Isaac [3] and Bose et al. [5] avoid the problem of firm prices revealing quality by studying models in which firms have no more information than consumers do. Alternatively, Smallwood and Conlisk [15] and Ifrach et al. [10] do not model customers as strategic. Instead, they exogenously specify consumer behavior, and then study the market trajectory that results from various firm choices. Ifrach et al. [10] argue that assuming rationality "introduces a formidable analytical and computational onus on each agent that may be hard to justify as a model of actual choice behavior." While this is no doubt true in many cases, a strength of this paper is that customer best response dynamics in our model turn out to be quite simple to describe.

Another critical feature of our model is that, like Rogerson [12] and Shapiro [14], we study a scenario where firms choose product quality at the beginning of the game, and it remains fixed throughout. This model is appropriate for industries in which product quality is derived from high-cost training or long-term investment in capital. Other industries would be more appropriately modeled by allowing firms to choose their quality level dynamically. There are a number of interesting questions in this case, which we discuss briefly towards the end of the paper.

Although the papers discussed above differ in many respects, they agree on at least one point, expressed succinctly by Smallwood and Conlisk [15]: "consumers pay considerably for being ill-informed." Wolinsky [18] finds that as the signal to customers becomes more informative, equilibrium prices drop. In Rogerson [12], a more informative signal results in more high quality goods in the market. Allen [1] finds that making quality unobservable may not change equilibrium behavior, but if it does, it either results in a lower quantity being offered at a higher price than before, or collapses the market entirely. Cooper and Ross [6] observe that when some customers cannot determine product quality, the price offered is the same as under full information, but the average quality in the market is lower.

Shapiro [14] reports that any "self-fulfilling quality level" must lie below the quality of goods produced in the complete information case, and that as information about the firm's quality spreads more rapidly (i.e. approaching full information), the self-fulfilling quality level rises. Though there are many differences between Shapiro's model and the one we consider here, the most important is that in his model, reputation evolves deterministically, gradually shifting from the customer's initial expectations to the true underlying quality. Stochasticity is the key to our result that firms may produce higher quality goods when quality cannot be directly monitored: even firms producing items that exceed customer expectations have reason to fear a bad review.

3   Model Introduction

3.1   Game Rules

We consider the following general framework. Each firm has a type T, which can be either high, H, or low, L. Customers cannot observe a firm's type, but it is common knowledge that each firm is high-type with probability $\alpha \in (0,1)$ and low-type with probability $1-\alpha$. At the beginning of the game, each firm chooses Q, which represents the quality of the items that it will produce throughout the game. We assume that Q takes values in the range $[\underline{q}, \overline{q}] \subseteq [0,1]$. If a firm of type T chooses $Q = q$, they incur a one-time cost $C_T(q)$. We assume that cost functions are strictly increasing, and that high-type firms have strictly lower marginal costs of quality than low-type firms, i.e. $0 < C'_H(q) < C'_L(q)$.

After firms have chosen their quality, a signal S is drawn for each firm. Then, potential customers arrive sequentially and decide whether or not to purchase from the firm. If a customer buys from a firm with quality level q, he gets a reward of q and the firm gets a reward of 1. Otherwise, the customer gets a reward of $r \in (0,1]$, which we refer to as the customer's "reserve value" or "outside option." We assume that customers know their reserve values, and it is common knowledge that these values are drawn independently and identically from a distribution with cdf F.

If the customer patronizes the firm, they leave a review of their experience, which can be either positive or negative. The probability of a positive review is equal to the quality of the item purchased, and the review is independent from all other randomness in the system. We assume that these reviews have been summarized into the sufficient statistic $X = (n_-, n_+)$, where n_-, n_+ are the number of negative and positive reviews left so far, respectively.

In this paper, we consider three information structures:

- Full observability: Customers see a signal $S = Q$ which completely reveals the firm's quality level. No additional inference can be drawn from X.
- Partial observability: Customers see the initial signal S and the reviews X.
- No observability: Customers do not see S or X when making their decision.

3.2   Equilibrium Concept

We focus on symmetric pure strategy Bayesian equilibria of this game. These are characterized by a firm strategy σ and a customer strategy ψ. The firm strategy maps type T to the quality level Q selected. Because there are only two firm types, σ is fully characterized by the values $q_H := \sigma(H)$ and $q_L := \sigma(L)$. Customer strategies map the observed elements of the triple (r, S, X) to an action $A \in \{0,1\}$ where $A = 1$ represents a decision to purchase the item. We assume that firms and customers are risk-neutral and that customers are short-lived. Furthermore, we restrict our attention to equilibria in which the following holds:

Assumption 1. *Whenever a customer is indifferent about whether to patronize the firm, he does so. Whenever a firm is indifferent among a set of quality levels, it chooses Q to be the highest quality in this set.*

We make this assumption to simplify the notation and analysis; without it, firm best response sets may be empty. In what follows, we sometimes refer to the set of "equilibria" of a game. When doing so, we mean "symmetric pure strategy Bayesian equilibria in which Assumption 1 holds."

We define $E_\sigma[\cdot]$ (respectively, $E_\psi[\cdot]$) be the expectation of its argument when all firms play strategy σ (customers play strategy ψ). Analogously, define $P_\sigma(\mathcal{A})$ ($P_\psi(\mathcal{A})$) to be the probability of $\mathcal{A}$ if all firms play strategy σ (all customers play strategy ψ). We let $\mathcal{G}$ be the set of games described above.

3.3   Model Analysis

The majority of this paper considers separating equilibria (i.e. equilibria in which $q_H \neq q_L$) in partial information games. To set the stage for the discussion in sections 4 and 5, we establish here two results that hold for both the one-shot model and the infinite horizon model. The first states that high-type firms always choose a quality level that is at least as high as the one chosen by low-type firms. The second proposition addresses equilibria of "full observability" and "no observability" games (which are used as benchmarks in the remainder of the paper), as well as pooling equilibria of games with partial observability.

Proposition 1. *For any game $G \in \mathcal{G}$, if $\sigma = (q_L, q_H)$ is an equilibrium strategy, then $q_H \geq q_L$.*

This holds because if a low-type firm were weakly better off from playing $q_L > q_H$, then the assumption $C'_H(q) < C'_L(q)$ implies that a high-type firm strictly prefers q_L to q_H.

Proposition 2. *Let $G \in \mathcal{G}$.*
(i) If quality is unobservable, the only equilibrium is $(\underline{\sigma}, \underline{\psi})$ where $\underline{\sigma}$ satisfies $q_H = q_L = \underline{q}$ and $\underline{\psi}(S, X) = 1\{r \leq \underline{q}\}$.
(ii) If quality is fully observable, in equilibrium each customer purchases the item if and only if the observed signal exceeds their outside option ($r \leq S$). Therefore, firms either choose $Q = \underline{q}$ or $Q = r$.
(iii) If quality is partially observable, the pair $(\underline{\sigma}, \underline{\psi})$ described in part (i) is always an equilibrium. Furthermore, if $\overline{q} < 1$, this is the unique pooling equilibrium.

Claim (i) follows because when quality is unobservable, consumer choices are independent from the quality set by the firm. Therefore, for any fixed customer strategy, firms prefer to minimize costs by stetting the lowest quality level $\underline{q}$. The best response for customers in this case is to purchase if and only if $r \leq \underline{q}$. The second part of the proposition follows from the fact that if $\underline{q} < r$ and the firm produces at a level in the interval $(\underline{q}, r)$, it receives no customers. Similarly, there is no benefit to producing above r. Finally, (iii) follows since, when $q_H = q_L$, customers know the quality of the item they're considering. Because we have specified that indifferent customers purchase the item, this means that customer decisions don't depend on the observed review. Thus, there is no incentive for firms to choose any quality level higher than $\underline{q}$.

4 One-Period Model

4.1 Description of Equilibria

We now discuss a simple model in which at most a single customer patronizes each firm. For now, we focus on the case where costs are linear, with $C_T(q) = c_T q$, and where product quality is partially observable. This means that the customer sees S before making a decision, and that $S = 1$ with probability Q and $S = 0$ otherwise.

Some may ask where the first review S comes from. Although there are possible explanations (for example, it may be written by a professional reviewer whose job it is to try the products of new firms), this is not an essential feature of our model. Indeed, all of the intuition and techniques used below would apply to a model in which two customers consider the firm in sequence. In that model the first customer may choose not to buy from the firm, but if he does make a purchase, he leaves a review that the second customer sees. Our choice to make the first review "automatic" serves only to clarify the exposition.

We search for equilibria by fixing customer behavior ψ, computing the firm's best response, and then checking to see if this induces the specified behavior ψ. Note that ψ specifies for each r, what a customer with reserve r will do upon seeing a positive review, and what they will do upon seeing a negative review. We define $p_+(\psi)$ to be the probability that a customer playing ψ buys from the firm if $S = 1$ (i.e. the expectation of $\psi(r, S = 1)$ over possible reserves r), and $p_-(\psi)$ the corresponding probability when observing $S = 0$. Then the firm's best response to ψ is to select

$$q_T \in \arg\max_q \{q p_+(\psi) + (1 - q) p_-(\psi) - c_T q\}, T \in \{L, H\}$$

Note that the objective above is linear in q, so we have a simple characterization of firm best responses. If $p_+(\psi) - p_-(\psi) - c_T < 0$, choosing $Q = \underline{q}$ is uniquely optimal. Otherwise, $\overline{q}$ is in the firm's best response set. It follows that the only separating equilibria satisfying Assumption 1 have $q_L = \underline{q}, q_H = \overline{q}$.

For any firm strategy σ, the optimal customer response is to purchase if and only if the expected quality of the item given the observed signal exceeds the value of their outside option, i.e. $E_\sigma[Q|S = s] \geq r$. In particular, if σ satisfies $q_L = \underline{q}, q_H = \overline{q}$, then

$$E_\sigma[Q|S=1] = \frac{\alpha \overline{q}^2 + (1 - \alpha)\underline{q}^2}{\alpha \overline{q} + (1 - \alpha)\underline{q}} \text{ and } E_\sigma[Q|S = 0] = \frac{\alpha(1 - \overline{q})\overline{q} + (1 - \alpha)(1 - \underline{q})\underline{q}}{\alpha(1 - \overline{q}) + (1 - \alpha)(1 - \underline{q})}.$$

The leads to the following:

Proposition 3. *If $G \in \mathcal{G}$ is a one-period game with linear costs, the only possible firm strategy in a separating equilibrium of G is $q_H = \overline{q}$ and $q_L = \underline{q}$. This outcome is supported in equilibrium if and only if $c_H \leq c_0 < c_L$, where $c_0 = F\left(\frac{\alpha \overline{q}^2 + (1-\alpha)\underline{q}^2}{\alpha \overline{q} + (1-\alpha)\underline{q}}\right) - F\left(\frac{\alpha(1-\overline{q})\overline{q} + (1-\alpha)(1-\underline{q})\underline{q}}{\alpha(1-\overline{q}) + (1-\alpha)(1-\underline{q})}\right)$ is a constant that depends on model primitives.*

Most of this proposition has already been established above. To complete the proof, note that if ψ is the customer best response to $(q_L, q_H) = (\underline{q}, \overline{q})$ satisfying Assumption 1, then

$$p_+(\psi) = F\left(\frac{\alpha \overline{q}^2 + (1-\alpha)\underline{q}^2}{\alpha \overline{q} + (1-\alpha)\underline{q}}\right), p_-(\psi) = F\left(\frac{\alpha(1-\overline{q})\overline{q} + (1-\alpha)(1-\underline{q})\underline{q}}{\alpha(1-\overline{q}) + (1-\alpha)(1-\underline{q})}\right).$$

Note that when c_L and c_H are both higher or both lower than c_0, no separating equilibrium exists in this model. When both firm types have high costs, attracting new customers costs more than the benefit it provides. When both firm types have low costs, all firms would like to play $\overline{q}$. If $\overline{q} < 1$, however, this cannot be an equilibrium, as discussed in Proposition 2. The problem is that in this case, customers do not get any information from the signal S; firm costs do not sufficiently differentiate firms of opposite types.

Intuitively, the farther apart c_H and c_L are, the "more likely" a separating equilibrium is to occur. We formalize this by noting the following consequence of Proposition 3.

Remark 1. If $[c_H^1, c_L^1] \subseteq [c_H^2, c_L^2]$ and $(\alpha, \underline{q}, \overline{q}, F)$ are such that a separating equilibrium exists in the partial-information game when firm costs are c_H^1 and c_L^1, then the same $(\alpha, \underline{q}, \overline{q}, F)$ also admit a separating equilibrium when firm costs are c_H^2 and c_L^2.

4.2   Welfare Comparison

Here we compare producer and consumer surplus under the different information settings. The producer surplus is taken to be the equilibrium expected profit of the average firm, while consumer surplus is the equilibrium expected utility of an average consumer. We consider the case where customers have a common reserve r, so that F is a point mass. We will later discuss how the results and intuition extend to other suitably restricted distributions F. We summarize our findings as follows:

Proposition 4. *If all customers share a common outside option $r \in \mathbb{R}$:*
(i) Both producer and consumer surplus are minimized when quality is completely unobservable.
(ii) High type firms always prefer fully observable quality, while low-type firms may prefer that quality is only partially observable.
(iii) Consumer surplus under partial observability is always at least as large as in the full information model. In any separating equilibrium, consumers are strictly better than in full information except in the case where the value of the outside option r is exactly $\frac{\alpha \overline{q}^2 + (1-\alpha)\underline{q}^2}{\alpha \overline{q} + (1-\alpha)\underline{q}}$.
(iv) There are equilibria of games with partially observable quality in which both consumer and producer surplus exceed their levels in the equilibrium of a corresponding game with fully observable quality.

Claim (i) follows from Proposition 2 (i), which revealed that the unique equilibrium in the completely unobservable case is when $q_L = q_H = \underline{q}$. To see why (ii) is true, note first that since $q_L \geq q_H$ in equilibrium by by Proposition 1, it must be that $E_\sigma [Q|S = 0] \leq E_\sigma [Q|S = 1] \leq q_H$. That is, customers always expect to receive a lower quality good than what is truly being offered by the high-type firms. It follows by monotonicity of F that high-type firms earn no more than $F(q_H) - C_H(q_H)$ in this equilibrium. Furthermore, $F(q_H) - C_H(q_H) \leq \max_q \{F(q) - C_H(q)\}$, the amount that high-type firms earn in full information.

Now, consider claim (iii). In the full information setting, no firm will ever produce a quality level $q > r$, since by instead producing a quality strictly between r and q, it could reduce costs while still ensuring business. Therefore, customers earn exactly r in any equilibrium of the full information game.

In the partial information game, customers can still earn a certain payoff of r, and therefore expect to earn at least r in any equilibrium. In particular, given a fixed firm strategy σ, a customer with outside option r who responds optimally to σ earns $P_\sigma(S = 1) \max (E_\sigma[Q|S = 1], r) + P_\sigma(S = 0) \max (E_\sigma[Q|S = 0], r) \geq r$. Moreover, this surplus strictly exceeds r as long as $E_\sigma[Q|S = 1] > r$.

As discussed in Proposition 3, the only possible separating equilibrium satisfies $q_H = \bar{q}$, $q_L = \underline{q}$. When firms play this strategy, $E_\sigma[Q|S = 1] = \frac{\alpha \bar{q}^2 + (1-\alpha)\underline{q}^2}{\alpha \bar{q} + (1-\alpha)\underline{q}}$, which must be at least r in order for σ to be supported in equilibrium. Thus, unless this exactly equals r, customers are strictly better off than in the full information game.

Finally, note that although high-type firms prefer full information, low-type firms may prefer the partial information game. This is because a "lucky" signal may cause customers to mistakenly purchase from them. When $\bar{q}$ is large and c_H is sufficiently small, high-type firms are barely worse off in the partial information game, implying that firms are better off on average. For a specific numerical example take $r = 0.5$, $\underline{q} = 0.3$, $\bar{q} = 0.8$, $\alpha = 0.5$, $c_L = 6$, $c_H = 0.3$.

4.3   Discussion

To simplify the analysis we have focused on the case where firms have linear costs, and in the welfare analysis, on the case where consumers share a common outside option. The intuition and results extend beyond these restrictive cases, however.

When quality is fully observable, a firm's marginal benefit from increasing quality comes through the corresponding increase in $F(q)$. For many distributions, this benefit diminishes quickly beyond some threshold, giving firms little incentive to produce above that level. When quality is partially observable, the benefit to increasing Q comes from the rise in the probability of receiving a positive review (and the associated jump in the probability of attracting a customer). For many cost structures, and many choices of F, the differing incentives under these two information structures can lead to equilibria in which high-type firms set higher quality levels in the partial information game. We give one such example below.

Example 1. Suppose that customer reserve values are uniformly distributed on $[1/4, 3/4]$, and that $(\underline{q}, \overline{q}) = (1/4, 1)$. Suppose that firm costs are linear in quality with $c_L > 2$ and $c_H < 4/3$. Then the full information equilibrium is $q_L = 1/4, q_H = 3/4$, whereas by Proposition 3, the game with partial observability has an equilibrium in which $q_L = 1/4, q_H = 1$.

5   Infinite Horizon Model

We now consider a model in which an infinite sequence of homogeneous consumers visit each firm and firms seek to maximize expected discounted profit, with discount factor $\delta < 1$. We let X_t denote the review history seen by the tth customer to consider a firm. Furthermore, we take the signal S to be uninformative, so customer inference is based only on X_t. We look for an equilibrium by fixing the firm strategy $\sigma = (q_L, q_H)$, determining the optimal customer response, and then verifying that given the induced customer behavior, no firm has an incentive to deviate from σ.

5.1   Customer Best Response

Fix a choice of σ satisfying $q_H > q_L$. The best response of a customer with reserve r is to purchase if and only if they expect that Q is at least r, given history X. Define the rejection set $R_\sigma(r)$ to be the set of review histories (n_-, n_+) such that when firms play strategy σ, the optimal decision of a customer with reserve r is not to purchase. In other words,

$$R_\sigma(r) = \{(n_-, n_+) \in \mathbb{N}^2 : E_\sigma[Q|X = (n_-, n_+)] < r\}.$$

Then the best customer response to σ is defined by $\psi(r, S, X) = \mathbf{1}(X \notin R_\sigma(r))$. It turns out that we can precisely describe these rejection sets. Note that

$$E_\sigma[Q|X] = P_\sigma(T = H|X)q_H + P_\sigma(T = L|X)q_L.$$

Since $P_\sigma(T = L|X) = 1 - P_\sigma(T = H|X)$, rearranging terms gives us that $X \notin R_\sigma(r)$ if and only if $P_\sigma(T = H|X) \geq \frac{r-q_L}{q_H-q_L}$. More algebra reveals that this is equivalent to the condition

$$\frac{P_\sigma(T = H|X)}{P_\sigma(T = L|X)} \geq \frac{r - q_L}{q_H - r}.$$

Fortunately, the ratio of conditional probabilities on the left has a nice form. By Bayes' theorem,

$$\frac{P_\sigma(T = H|X)}{P_\sigma(T = L|X)} = \frac{P(T = H)}{P(T = L)} \frac{P_\sigma(X|T = H)}{P_\sigma(X|T = L)} = \frac{\alpha}{1 - \alpha} \left(\frac{q_H}{q_L}\right)^{n_+} \left(\frac{1 - q_H}{1 - q_L}\right)^{n_-}.$$

Taking logarithms, we see that a consumer purchases if and only if

$$n_+ \log\left(\frac{q_H}{q_L}\right) - n_- \log\left(\frac{1 - q_L}{1 - q_H}\right) \geq \log\left(\frac{1 - \alpha}{\alpha} \frac{r - q_L}{q_H - r}\right). \tag{1}$$

Thus, the optimal customer strategy takes the following appealingly simple form. Each firm starts with a "reputation score" (the left side of (1)) of zero.

Every positive review improves the firm's reputation by a fixed constant, while negative reviews decrease its reputation by a different constant. Customers map their reserve to a reputation cut-off (the right side of (1)) and purchase precisely when the firm's score is at least their cutoff. Note that the different coefficients on n_+ and n_- reflect the fact that positive and negative reviews may contain different amounts of information. If, for example, most customers have positive experiences ($q_H > q_L > 1/2$), someone reading reviews may (rationally) be said to learn more from a new negative review than a new positive one.

Put slightly differently, the firm reputation score follows a random walk. Every time the firm gets a customer, its score increases by $\log(q_H/q_L)$ with probability Q and decreases by $\log((1 - q_L)/(1 - q_H))$ with probability $1 - Q$. The rejection region $R_\sigma(r)$ is the set of points lying below a line whose slope depends only on σ. We now point out two particularly simple instances of this decision rule.

Remark 2.
(i) If, before reading reviews, a consumer is indifferent about whether to purchase the product (i.e. $r = \alpha q_H + (1 - \alpha)q_L$), the optimal decision rule is to go if and only if $\frac{n_+}{n_- + n_+} \geq \beta$, where β is defined to be

$$\frac{\log((1 - q_L)/(1 - q_H))}{\log(q_H/q_L) + \log((1 - q_L)/(1 - q_H))}.$$

(ii) If $q_H + q_L = 1$ and $r \in (q_L, q_H)$, the optimal customer strategy is to go if and only if $n_+ - n_- \geq d(r)$, where

$$d(r) = \log\left(\frac{\alpha}{1 - \alpha}\frac{r - q_L}{q_H - r}\right) / \log\left(\frac{q_H}{q_L}\right).$$

Both of these are established via basic algebraic manipulation of (1). Note that part (i) says that an initially indifferent customer should purchase whenever the "average review" is at least β. The decision rule in part (ii), meanwhile, compares the number of positive and negative reviews in absolute terms.

5.2   Firm Revenue Calculation

Because the reserve r is identical across customers, as soon as a firm is rejected by a customer, we know that it will be rejected by each subsequent customer. Let τ be the first period in which $X_\tau \in R_\sigma(r)$, i.e. the period at which the firm has effectively gone out of business. If this never occurs, define $\tau = \infty$. The following proposition addresses a case in which it is possible to derive a closed-form expression for the firm's expected discounted profit.

Proposition 5. *Suppose that all customers go unless there are at least $m \in \mathbb{N}_+$ more negative reviews than positive ones. The firm's expected discounted profit when choosing $Q = q$ is given by*

$$\frac{1 - E[\delta^\tau | Q = q]}{1 - \delta} - C_T(q), \quad T \in \{L, H\}, \tag{2}$$

where

$$E[\delta^\tau | Q = q] = \left(\frac{1 - \sqrt{1 - 4q(1-q)\delta^2}}{2\delta q} \right)^m. \tag{3}$$

The expression (2) arises because the firm's discounted revenue is given by $1 + \delta + \cdots + \delta^{\tau-1} = \frac{1-\delta^\tau}{1-\delta}$. A proof of (3) appears in the Appendix.

The expression in (3) is strictly decreasing in q, suggesting that expected discounted revenue is continuously increasing in q on the entire interval $[\underline{q}, \overline{q}]$. This occurs because customers only purchase from, and therefore only review, firms with sufficiently high reputation scores. As a result, even firms choosing quality levels above the customer threshold must fear that several poor reviews will leave them unable to signal their quality. This gives firms an incentive to set quality levels exceeding r, just as they had in the one-period model.

The following example illustrates that decision rules of the form assumed by Proposition 5 can be supported in equilibrium.

Example 2. When $\alpha \in (1/2, 3/5), C_T(q) = c_T/(1-q), c_H = \frac{8}{9} - \frac{28}{9\sqrt{19}} \approx 0.175, c_L = \frac{9}{2} - \frac{18}{\sqrt{19}} \approx 0.371, \delta = 1/2, r = 1/2$, there is an equilibrium in which $q_H = 0.6, q_L = 0.4$, and customers purchase if and only if there are at least as many positive reviews as negative ones.

To verify this, note that given customer behavior, (2) and (3) imply that the firm best response σ is given by $q_H = 0.6, q_L = 0.4$. When firms choose these values, by Remark 2 (ii), customer behavior is optimal.

5.3   Welfare Analysis

Here, we discuss two welfare properties of equilibria in this infinite-horizon model. Proposition 6 says that customers who arrive later are better off. In Proposition 7 we show that customers are weakly better off in the partial information game than when quality is visible, and that some customers are strictly better of so long as high-type firms play above $\underline{q}$.

Proposition 6. *For any σ, if customers play a best response to σ, the expected surplus of the tth customer to consider the firm is non-decreasing in t. Furthermore, the tth customer is strictly better off than the first if the probability that the tth customer buys lies in $(0, 1)$.*

Proof. Let ψ be a best response to σ. The payout to a customer of type r who sees history X and plays ψ is $\max(E_\sigma[Q|X], r)$. Regardless of customer strategies, the sequence $\{E_\sigma[Q|X_t]\}_{t\geq 1}$ is a martingale. Because max is a convex function, Jensen's inequality implies that a customer of type r is weakly better off arriving in period t than in any earlier period. Averaging over r proves the first part of the proposition. Additionally, for any fixed r, if $0 < P_\psi(r \leq E_\sigma[Q|X_t]) < 1$, then Jensen's inequality is strict, i.e.

$$E_\psi[\max(E_\sigma[Q|X_t], r)] > \max(E_\psi[E_\sigma[Q|X_t]], r) = \max(E_\sigma[Q], r).$$

Note that the left side represents the expected profits of customer t, and the right side equals the expected profit of the first customer.

Proposition 7. *In an infinite-horizon model where customers have identical reserves $r \in (\underline{q}, \overline{q})$, the equilibrium expected surplus of the tth customer in the incomplete information game is at least as great as the equilibrium surplus of a customer in the full-information game for all t. Furthermore, unless the partial information equilibrium satisfies $q_H = q_L \in \{\underline{q}, 1\}$, this inequality is strict for sufficiently large t.*

Proof. As with Proposition 4, in the full information game firms have no incentive to set a quality above r, so all customers receive r and thus they are at least as well off in the partial information game. Since, by Proposition 2 (iii) the only possible pooling equilibria occur at $\underline{q}$ and 1, we must have $q_H > q_L$. If this is a best response, it must be that the first customer buys, which in turn implies that $E_\sigma[Q] = \alpha q_H + (1 - \alpha)q_L \geq r$. In any separating equilibrium, customers must use reviews to make decisions, i.e. there must exist review histories X that occur with positive probability such that $E_\sigma[Q|X] < r$. Thus, for sufficiently large t, the probability that the tth customer buys lies strictly between zero and one. By Proposition 6, this customer is strictly better off in the partial information game than in the game of full information.

6   Conclusion

Here we have presented a simple model of reputation and product quality in markets where consumers publicly share reviews of their experience. We emphasize a setting in which customers' experiences are intrinsically random, but are positively correlated with the quality of the product. We arrive at the insight that a noisy review process of this form may yield equilibria in which firms produce higher quality goods than they would if quality were directly observable. This occurs because even high quality firms are at risk of losing future customers due to bad initial reviews. The effect is compounded by a cascade-like phenomenon: when customers are unlikely to patronize the firm, it can be difficult or even impossible for firms to improve their reputation. Above, we illustrate these ideas through stylized models. Due to the tractability of our models, we consider this paper a promising foundation for future work.

Perhaps the most natural extension of our work is to consider models in which each firm's quality level is allowed to vary over time. Models of this form can pose significant technical challenges, as strategies and the corresponding inferential procedures become much more complicated. Successful analysis of a model incorporating these elements could shed light on the question of how a firm's long run quality level depends on its review history.

Another interesting avenue for future work would be to consider a model in which firms are competing more directly with one another. In our work, firm decisions only indirectly affect other firms, by changing customer inferences. An alternative model might allow customers search, at some cost, for a firm with better reviews. Such a model has a similar flavor to the one we study, except that customer reserves are not exogenously specified, but rather determined endogenously by competition.

References

[1] Allen, F.: Reputation and product quality. RAND Journal of Economics 15(3), 311–327 (1984), http://EconPapers.repec.org/
RePEc:rje:randje:v:15:y:1984:i:autumn:p:311-327

[2] Banerjee, A.V.: A simple model of herd behavior. The Quarterly Journal of Economics 107(3), 797–817 (1992),
http://qje.oxfordjournals.org/content/107/3/797.abstract,
doi:10.2307/2118364

[3] Bar-Isaac, H.: Reputation and survival: Learning in a dynamic signalling model. The Review of Economic Studies 70(2), 231–251 (2003),
http://restud.oxfordjournals.org/content/70/2/231.abstract,
doi:10.1111/1467-937X.00243

[4] Bikhchandani, S., Hirshleifer, D., Welch, I.: Learning from the behavior of others: Conformity, fads, and informational cascades. Journal of Economic Perspectives 12(3), 151–170 (1998),
http://ideas.repec.org/a/aea/jecper/v12y1998i3p151-70.html

[5] Bose, S., Orosel, G., Ottaviani, M., Vesterlund, L.: Monopoly pricing in the binary herding model. Economic Theory 37(2), 203–241 (2008),
http://www.jstor.org/stable/40282924 ISSN 09382259

[6] Cooper, R., Ross, T.W.: Prices, product qualities and asymmetric information: The competitive case. Review of Economic Studies 51(2), 197–207 (1984),
http://ideas.repec.org/a/bla/restud/v51y1984i2p197-207.html

[7] Fudenberg, D., Maskin, E.: The folk theorem in repeated games with discounting or with incomplete information. Econometrica 54(3), 533–554 (1986),
http://www.jstor.org/stable/1911307 ISSN 00129682

[8] Fudenberg, D., Levine, D., Maskin, E.: The folk theorem with imperfect public information. Econometrica 62(5), 997–1039 (1994),
http://www.jstor.org/stable/2951505 ISSN 00129682

[9] Hörner, J.: Reputation and competition. American Economic Review 92(3), 644–663 (2002), http://ideas.repec.org/a/aea/aecrev/v92y2002i3p644-663.html

[10] Ifrach, B., Maglaras, C., Scarsini, M.: Monopoly pricing in the presence of social learning. Working Papers 12-01, NET Institute (August 2012),
http://ideas.repec.org/p/net/wpaper/1201.html

[11] Luca, M.: Reviews, reputation, and revenue: The case of yelp. com. Com (September 16, 2011). Harvard Business School Working Paper (12-016) (2011),
http://hbswk.hbs.edu/item/6833.html

[12] Rogerson, W.P.: Reputation and product quality. The Bell Journal of Economics 14(2), 508–516 (1983), http://www.jstor.org/stable/3003651 ISSN 0361915X

[13] Rubinstein, A.: Equilibrium in supergames with the overtaking criterion. Journal of Economic Theory 21(1), 1–9 (1979),
http://ideas.repec.org/a/eee/jetheo/v21y1979i1p1-9.html

[14] Shapiro, C.: Consumer information, product quality, and seller reputation. The Bell Journal of Economics 13(1), 20–35 (1982),
http://www.jstor.org/stable/3003427 ISSN 0361915X

[15] Smallwood, D.E., Conlisk, J.: Product quality in markets where consumers are imperfectly informed. The Quarterly Journal of Economics 93(1), 1–23 (1979),
http://www.jstor.org/stable/1882595 ISSN 00335533

[16] Spence, M.: Job market signaling. The Quarterly Journal of Economics 87(3), 355–374 (1973), http://www.jstor.org/stable/1882010 ISSN 00335533

[17] Wehrum, K.: How businesses can respond to criticism on yelp (June 2009), http://www.inc.com/magazine/20090601/how-businesses-can-respond-to-criticism-on-yelp.html

[18] Wolinsky, A.: Prices as signals of product quality. The Review of Economic Studies 50(4), 647–658 (1983), http://www.jstor.org/stable/2297767 ISSN 00346527

A   Computing Expected Firm Revenue

In this section, we prove Proposition 3, i.e. that when customers all choose to buy from the firm unless $n_- - n_+ \geq m \in \mathbb{N}$, and τ is defined as in Section 5,

$$E[\delta^\tau] = \left(\frac{1 - \sqrt{1 - 4q(1-q)\delta^2}}{2q\delta} \right)^m \quad \text{for } \delta \in (0, 1). \tag{4}$$

We start by noting that the firm's reputation score can be re-normalized to be a simple random walk starting at 0, where τ is the first time that the walk reaches $-m$ (or ∞, if it never does). The firm chooses Q, i.e. the probability that the walk moves up.

Let $X_i \in \{-1, +1\}$ be i.i.d. and take the value $+1$ with probability q, $Y_n = \sum_{i=1}^n X_i$, and $\tau = \min\{n : Y_n = -m\}$. Define $\phi(\theta) = E[e^{\theta X_1}] = qe^\theta + (1-q)e^{-\theta}$, and let $\psi(\theta) = \log \phi(\theta)$. It follows that for any θ, $e^{\theta Y_n - n\psi(\theta)}$ is a martingale. Then for all n,

$$E[e^{\theta Y_{\tau \wedge n} - (\tau \wedge n)\psi(\theta)}] = 1. \tag{5}$$

Let $\theta_0 = \min\{\theta : \phi(\theta) = 1\}$. We can compute directly that if $q \leq 1/2, \theta_0 = 0$, and if $q > 1/2$ $\theta_0 = \log(\frac{1-q}{q}) < 0$. We will let $n \to \infty$ in Equation (5) and show that for $\theta < \theta_0$,

$$e^{m\theta} = E[\phi(\theta)^{-\tau}]. \tag{6}$$

Once this has been established, we use the fact that $\phi(\theta) = qe^\theta + (1-q)e^{-\theta}$ is decreasing on $(-\infty, \theta_0)$ and onto $(1, \infty)$. In particular, given $\delta \in (0, 1)$ we can find a unique $\theta \in (-\infty, \theta_0)$ such that $1/\delta = \phi(\theta)$, i.e. $qe^{2\theta} - e^\theta/\delta + (1-q) = 0$. Apply the quadratic formula to solve explicitly for θ; it is defined by $e^\theta = \frac{1 - \sqrt{1 - 4q(1-q)\delta^2}}{2q\delta}$. Substituting $1/\delta$ for $\phi(\theta)$ in (6), this implies that $E[\delta^\tau] = \left(\frac{1 - \sqrt{1 - 4q(1-q)\delta^2}}{2q\delta} \right)^m$, as claimed.

We now justify (6). Note that for $\theta < \theta_0$, $\psi(\theta) > 0$, so $e^{-(\tau \wedge n)\psi(\theta)} \leq 1$. Additionally, $S_{\tau \wedge n} \geq -m$. Combining these shows that $0 \leq e^{\theta Y_{\tau \wedge n} - (\tau \wedge n)\psi(\theta)} \leq e^{-m\theta}$. By the dominated convergence theorem, we may exchange limit and expectation to obtain

$$E[\lim_{n \to \infty} e^{\theta Y_{\tau \wedge n} - (\tau \wedge n)\psi(\theta)}] = 1. \tag{7}$$

If $q \leq 1/2, \tau < \infty$ almost surely, and thus the left side of (7) is $e^{-m\theta}E[\phi(\theta)^{-\tau}]$, which can be rearranged to yield (6).

If instead $q > 1/2$, $P(\tau = \infty) > 0$ so

$$E[\lim_{n\to\infty} e^{\theta Y_{\tau \wedge n} - (\tau \wedge n)\psi(\theta)}] = E[\lim_{n\to\infty} e^{\theta Y_n - n\psi(\theta)}; \tau = \infty] + E[e^{\theta Y_\tau - \tau\psi(\theta)}; \tau < \infty].$$

On the event $\{\tau = \infty\}$, however, $Y_n < 0$. Since $\theta < 0$, we have that $e^{\theta Y_n - n\psi(\theta)} \leq e^{-\theta m}\phi(\theta)^{-n}$. Note that since $\phi(\theta) > 1$ for $\theta < \theta_0$, $e^{-\theta m}\phi(\theta)^{-n} \to 0$ as $n \to \infty$. It follows that the left side of (7) equals $e^{-m\theta}E[\phi(\theta)^{-\tau}; \tau < \infty]$, which in turn equals $e^{-m\theta}E[\phi(\theta)^{-\tau}]$, completing the proof that (6) holds whenever $\theta < \theta_0$.

The Complexity of Computing the Random Priority Allocation Matrix

Daniela Saban[1] and Jay Sethuraman[2,*]

[1] Graduate School of Business, Columbia University, New York, NY
dhs2131@columbia.edu
[2] IEOR Department, Columbia University, New York, NY
jay@ieor.columbia.edu

Abstract. Consider the problem of allocating n objects to n agents who have strict ordinal preferences over the objects. We study the Random Priority (RP) mechanism, in which an ordering over the agents is selected uniformly at random; the first agent is then allocated his most-preferred object, the second agent is allocated his most-preferred object among the remaining ones, and so on. The output of this mechanism is a bi-stochastic allocation matrix, in which entry (i, a) indicates the probability that agent i obtains object a (whenever objects are indivisible), or the fraction of object a allocated to agent i (when objects are divisible). Our main result is that the allocation matrix associated with the RP mechanism is hard to compute, in a sense that can be made precise using the theory of computational complexity. An important consequence is that an efficient algorithm to compute the allocation matrix exactly is unlikely. In addition, we examine two decision problems associated with the RP allocation: deciding whether an agent gets an object with probability 1, and deciding whether an agent gets an object with positive probability. We provide a polynomial-time algorithm to solve the former and show that the latter is hard to decide. This hardness result has two strong implications. First, it is not possible to design an efficient algorithm to get a good (multiplicative) approximation to the RP allocation matrix (under suitable complexity-theoretic assumptions). Second, for an assignment problem with inadmissible objects, it is hard to decide whether or not a given subset of objects is matched in *some* Pareto efficient matching.

A full version of this paper is available at:
http://www.columbia.edu/~js1353/pubs/rpcomplexity.pdf

* Research supported by NSF grants CMMI-0916453 and CMMI-1201045.

Y. Chen and N. Immorlica (Eds.): WINE 2013, LNCS 8289, p. 421, 2013.

Vickrey Auctions for Irregular Distributions

Balasubramanian Sivan[1] and Vasilis Syrgkanis[2]

[1] Microsoft Research
bsivan@microsoft.com
[2] Computer Science Dept., Cornell University
vasilis@cs.cornell.edu

Abstract. The classic result of Bulow and Klemperer [1] says that in a single-item auction recruiting one more bidder and running the Vickrey auction achieves a higher revenue than the optimal auction's revenue on the original set of bidders, when values are drawn i.i.d. from a regular distribution. We give a version of Bulow and Klemperer's result in settings where bidders' values are drawn from non-i.i.d. irregular distributions. We do this by modeling irregular distributions as some convex combination of regular distributions. The regular distributions that constitute the irregular distribution correspond to different population groups in the bidder population. Drawing a bidder from this collection of population groups is equivalent to drawing from some convex combination of these regular distributions. We show that recruiting one extra bidder from each underlying population group and running the Vickrey auction gives at least half of the optimal auction's revenue on the original set of bidders.

Keywords: Bulow-Klemperer, irregular distributions, prior-independent, Vickrey auction.

1 Introduction

Simplicity and detail-freeness are two much sought-after themes in auction design. The celebrated classic result of Bulow and Klemperer [1] says that in a standard single-item auction with n bidders, when the valuations of bidders are drawn i.i.d from a distribution that satisfies a regularity condition, running a Vickrey auction (second-price auction) with one extra bidder drawn from the same distribution yields at least as much revenue as the optimal auction for the original n bidders. The Vickrey auction is both simple and detail-free since it doesn't require any knowledge of bidder distributions. Given this success story for i.i.d. regular distributions, we ask in this paper, what is the analogous result when we go beyond i.i.d regular settings? Our main result is a version of Bulow and Klemperer's result to non-i.i.d irregular settings. Our work gives the first positive results in designing simple mechanisms for irregular distributions, by parameterizing irregular distributions, i.e., quantifying the amount of irregularity in a distribution. Our parameterization is motivated by real world market structures and in turn indicates that most realistic markets will not be highly irregular with respect to this metric. Our results enable the first positive approximation bounds on the revenue of the second-price auction with an anonymous reserve in both i.i.d. and non-i.i.d. irregular settings.

Before explaining our results, we briefly describe our setting. We consider a single-item auction setting with bidders having quasi-linear utilities. That is the utility of a

Y. Chen and N. Immorlica (Eds.): WINE 2013, LNCS 8289, pp. 422–435, 2013.

bidder is his value for the item if he wins, less the price he is charged by the auction. We study auctions in the Bayesian setting, i.e. the valuations of bidders are drawn from known distributions[1]. We make the standard assumption that bidder valuations are drawn from independent distributions.

Irregular distributions are common. The technical regularity condition in Bulow and Klemperer's result is quite restrictive, and indeed irregular distributions are quite common in markets. For instance, any distribution with more than a single mode violates the regularity condition. The most prevalent reason for a bidder's valuation distribution failing to satisfy the regularity condition is that a bidder in an auction is randomly drawn from a heterogeneous population. The population typically is composed of several groups, and each group has its characteristic preferences. For instance the population might consist of students and seniors, with each group typically having very different preferences from the other. While the distribution of preferences within any one group might be relatively well-aligned and the value distribution might have a single mode and satisfy the regularity condition, the distribution of a bidder drawn from the general population, which is a mixture of such groups, is some convex combination of these individual distributions. Such a convex combination violates regularity even in the simplest cases.

For a variety of reasons, including legal reasons and absence of good data, a seller might be unable to discriminate between the buyers from different population groups and thus has to deal with the market as if each buyer was arriving from an irregular distribution. However, to the least, most sellers do know that their market consists of distinct segments with their characteristic preferences.

Measure of Irregularity. The above description suggests that a concrete measure of irregularity of a distribution is the number of regular distributions required to describe it. We believe that such a measure could be of interest in both designing mechanisms and developing good provable revenue guarantees for irregular distributions in many settings. It is a rigorous measure of irregularity for any distribution since any distribution can be well-approximated almost everywhere by a sufficient number of regular ones and if we allow the number of regular distributions to grow to infinity then any distribution can be exactly described[2] Irregular distributions that typically arise in practice are combinations of a small number of regular distributions and this number can be considered almost a constant with respect to the market size. In fact there exist evidence in recent [8, 6] and classical [12] microeconomic literature that irregularity of the value distribution predominantly arises due to market segmentation in a small number of parts (e.g. loyal customers vs. bargain-hunters [8], luxury vs. low income buyers [6] etc). Only highly pathological distributions require a large number of regular distributions to be described — such a setting in a market implies that the population is heavily segmented and each segment has significantly different preferences from the rest.

---

[1] One of the goals of this work is to design detail-free mechanisms, i.e., minimize the dependence on knowledge of distributions. Thus most of our results make little or no use of knowledge of distributions. We state our dependence precisely while stating our results.

[2] This follows from the fact that a uniform distribution over an interval is a regular distribution and every distribution can be approximated in the limit using just uniform distributions.

Motivated by this, we consider the following setting: the market/population consists of k underlying population groups, and the valuation distribution of each group satisfies the regularity condition. Each bidder is drawn according to some probability distribution over these groups. That is bidder i arrives from group t with probability $p_{i,t}$. Thus if F_t is the cumulative distribution function (cdf) of group t, the cdf of bidder i is $F_i = \sum_t p_{i,t} F_t$. For example, consider a market for a product that consists of two different population groups, say students and seniors. Now suppose that two bidders come from two cities with different student to senior ratios. This would lead to the probability $p_{i,t}$'s to be different for different i's. This places us in a non-i.i.d. irregular setting. All our results also extend to the case where these probabilities $p_{i,t}$ are arbitrarily correlated.

Example 1 (An illustrative example). Consider an eBay seller of an ipad. One could think of the market as segmented mainly in two groups of buyers: young and elder audience. These two market segments have completely different value distributions. Suppose for instance, that the value distribution of young people is distributed as a normal distribution $N(\mu_y, \sigma)$ while the elder's is distributed as a normal distribution $N(\mu_e, \sigma)$ with $\mu_y > \mu_e$. In addition, suppose that the eBay buyer population is composed of a fraction p_y young people and $p_e < p_y$ of elders. Thus the eBay seller is facing an irregular valuation distribution that is a mixture of two Gaussian distribution with mixture probabilities p_y and p_e (see Figure 1). Even more generally, this mixture could be dependent on the city of the buyer and hence be different for different buyers.

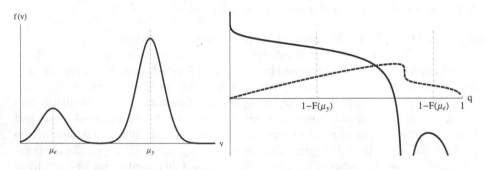

Fig. 1. Left figure depicts pdf of the bimodal distribution of valuations of Example 1, while the right figure depicts the revenue (dashed) $R(q) = q \cdot F^{-1}(1 - q)$ where q is the probability of sale, and the marginal revenue curve $\frac{dR(q)}{dq}$ for this distribution

The eBay seller has two ways of increasing the revenue that he receives: a) increasing the market competition by bringing extra bidders through advertising (possibly even targeted advertising), and b) setting appropriately his reserve price in the second price auction that he runs. Observe that he has no means of price discriminating. Even running Myerson's auction which is non-discriminatory for i.i.d. settings leads to randomization. In particular, randomization leads to the undesirable feature of sometimes serving an agent with smaller value. This raises the main questions that we address in this paper: how should he run his advertising campaign? How many people more (either through targeted or non-targeted advertising) should he bring to the auction to get a good approximation to the optimal revenue? What approximation of the optimal revenue is he guaranteed by running a Vickrey auction with a single anonymous reserve?

Giving a sneak preview of our main results, our paper gives positive results to all the above questions: 1) bringing just one extra young bidder in the auction (targeted advertising) and running a Vickrey auction with no reserve would yield revenue at least $1/2$ of the optimal revenue (Theorem 2), 2) bringing 2 extra bidders drawn from the distribution of the combined population (non-targeted advertising) would yield at least $\frac{1}{2}\left(1 - \frac{1}{e}\right)$ of the optimal revenue (Theorem 4), 3) running a Vickrey auction among the original n bidders with an anonymous reserve price can yield an 8-approximation of the optimal revenue (Theorem 5).

Our Results

First result (Section 3): Targeted Advertising for non-i.i.d. irregular settings. We show that by recruiting an extra bidder from each underlying group and running the Vickrey auction, we get a revenue that is at least half of the optimal auction's revenue in the original setting. While the optimal auction is manifestly impractical in a non-i.i.d. irregular setting due to its complicated rules, delicate dependence on knowledge of distribution and its discriminatory nature[3], the Vickrey auction with extra bidders is simple and detail-free: it makes no use of the distributions of bidders. The auctioneer must just be able to identify that his market is composed of different groups and must conduct a targeted advertising campaign to recruit one extra bidder from each group. This result can be interpreted as follows: while advertising was the solution proposed by Bulow and Klemperer [1] for i.i.d. regular distributions, *targeted advertising* is the right approach for non-i.i.d. irregular distributions.

Tightness. While we do not know if the the factor 2 approximation we get is tight, Hartline and Roughgarden [7] show that even in a non-i.i.d. regular setting with just two bidders it is impossible to get better than a $4/3$-approximation by duplicating the bidders, i.e., recruiting n more bidders distributed identically to the original n bidders. This lower bound clearly carries over to our setting also: there are instances where recruiting only one bidder from each different population group cannot give anything better than a $4/3$-approximation.

Second result (Main result, Section 4): Just one extra bidder for hazard rate dominant distributions. If the k underlying distributions are such that one of them stochastically dominates, hazard-rate wise, the rest, then we show that recruiting just one extra bidder from the hazard rate dominant distribution and running the Vickrey auction gets at

---

[3] The optimal auction in a non-i.i.d. setting will award the item to the bidder with the highest virtual value and this is not necessarily the bidder with the highest value. In addition, typically a different reserve price will be set to different bidders. This kind of discrimination is often illegal or impractical. Also, the exact form of the irregular distribution will determine which region's of bidder valuations will need to be "ironed", i.e. treated equally.

least half of the optimal revenue for the original setting. A distribution F hazard rate dominates a distribution G iff for every x in the intersection of the support of the two distributions the hazard rate $h_F(x)(= \frac{f(x)}{1-F(x)}$, where $f(\cdot)$ and $F(\cdot)$ are the pdf and cdf respectively) is at most $h_G(x)(= \frac{g(x)}{1-G(x)}$, where $g(\cdot)$ and $G(\cdot)$ are the pdf and cdf respectively). We denote such a domination by $F \succeq_{hr} G$.

Further, hazard rate dominance requirement is not uncommon: for instance, if all the k underlying distributions were from the same family of distributions like the uniform, exponential, Gaussian or even power law, then one of them is guaranteed to hazard rate dominate the rest. Though several common distributions satisfy this hazard rate dominance property, it has never been previously exploited in the context of approximately optimal auctions.

Third result (Section 5): Non-targeted advertising for i.i.d. irregular distributions. When the bidders are identically distributed, i.e., the probability $p_{i,t}$ of distribution t getting picked for bidder i is the same for all i (say p_t), we show that if each $p_t \geq \delta$, then bringing $\Theta\left(\frac{\log(k)}{\delta}\right)$ extra bidders drawn from the original distribution (and not from one of the k underlying distributions) yields a constant approximation to the optimal revenue. Further in the special case where one of the underlying regular distributions hazard rate dominates the rest and its mixture probability is δ then $\Theta\left(\frac{1}{\delta}\right)$ bidders drawn from the original distribution are enough to yield a constant approximation. This shows that when each of the underlying population groups is sufficiently thick, then recruiting a few extra bidders from the original distribution is all that is necessary.

Remark 1. For the latter result it is not necessary that the decomposition of the irregular distribution that we use should resemble the actual underlying population groups. Even if the market is highly fragmented with several population groups, as long as there is mathematically some way to decompose the irregular distribution into the convex combination of a few regular distributions our third result still holds.

Fourth result (Section 6): Vickrey with a Single (Anonymous) Reserve. Suppose we are unable to recruit extra bidders. What is the next simplest non-discriminatory auction we could hope for? The Vickrey auction with a single reserve price. We show that when the non-i.i.d irregular distributions all arise as different convex combinations of the k underlying regular distributions, there exists a reserve such that the Vickrey auction with this single reserve obtains a $4k$ approximation to the optimal revenue. Though the factor of approximation is not small, it is the first non-trivial approximation known for non-i.i.d irregular distributions via Vickrey with anonymous reserve. In addition, as we already explained, in typical market applications we expect the number of different population groups k to be some small constant.

What is the best one can hope for a non-i.i.d irregular setting? Chawla, Hartline and Kleinberg [2] show that for general non-i.i.d irregular distributions it is impossible to get a $o(\log n)$ approximation using Vickrey auction with a single reserve price, and it is unknown if this lower bound is tight, i.e., we do not yet know of a $\Theta(\log n)$ approximation. However the $o(\log n)$ impossibility exists only for arbitrary non-i.i.d irregular settings and doesn't apply when you assume some natural structure on the irregularity of the distributions, which is what we do.

To put our results in context: Single reserve price Vickrey auctions were also analyzed by Hartline and Roughgarden [7] for non-i.i.d *regular* settings, that showed that there exists a single reserve price that obtains a 4-approximation. Chawla et al. [3] show that when bidders are drawn from non-i.i.d irregular distributions, a Vickrey auction with a distribution-specific reserve price obtains a 2-approximation. Thus if there are k different distributions, k different reserve prices are used in this result. This means that if we insist on placing a single (anonymous) reserve price, this result guarantees a $O(1/k)$ approximation. In particular, when all distributions are different, i.e. $k = n$, this boils down to a $O(1/n)$ approximation.

In contrast, our result shows that even when all the distributions are different, as long as every irregular distribution can be described as some convex combination of k regular distributions, Vickrey with a single reserve price gives a factor $4k$ approximation. Further the factor does not grow with the number of players n.

Remark 2. We also show that if the bidders are distributed with identical mixtures and the mixture probability is at least δ then Vickrey auction with a single reserve achieves a $\Theta\left(1 + \frac{\log(k)}{n\delta}\right)$ approximation. If one of the regular distribution hazard rate dominates the rest and has mixture probability δ, then Vickrey with a single reserve achieves a $\Theta\left(1 + \frac{1}{n\delta}\right)$ approximation.

Observe that if all k regular distributions in the mixture have equal probability of arriving, then our results shows that a Vickrey auction with a single reserve achieves at least a $\Theta\left(1 + \frac{k\log(k)}{n}\right)$ of the optimal revenue. This approximation ratio becomes better as the number of bidders increases, as long as the number of underlying regular distributions remains fixed. If the number of underlying distributions increases linearly with the number of bidders, then the result implies a $\Theta(\log(n))$ approximation, matching the lower bound of [3].

Related Work. Studying the trade-off between simple and optimal auctions has been a topic of interest for long in auction design. The most famous result is the already discussed result of Bulow and Klemperer [1] for single-item auctions in i.i.d regular settings. Hartline and Roughgarden [7] generalize [1]'s result for settings beyond single-item auctions: they consider auctions where the set of buyers who can be simultaneously served form the independent set of a matroid; further they also relax the i.i.d constraint and deal with non-i.i.d settings. Dhangwatnotai, Roughgarden and Yan [5] study revenue approximations via VCG mechanisms with multiple reserve prices, where the reserve prices are obtained by using the valuations of bidders as a sample from the distributions. Their results apply for matroidal settings when the distributions are regular, and for general downward closed settings when the distributions satisfy the more restrictive monotone hazard rate condition. As previously discussed, Chawla et al. [3] show that for i.i.d irregular distributions, Vickrey auction with a single reserve price gives a 2-approximation to the optimal revenue and for non-i.i.d distributions Vickrey auction with a distribution-specific reserve price guarantees a 2-approximation; Chawla et al. [2] show that it is impossible to achieve a $o(\log n)$ approximation via Vickrey auction with a single reserve price for non-i.i.d irregular distributions. Single-item Vickrey auctions with bidder specific monopoly reserve prices were also studied in Neeman [11] and Ronen [13]. Approximate revenue maximization via VCG mechanisms with supply limitations were studied in Devanur et al. [4] and Roughgarden et al. [14].

2   Preliminaries

Basic model. We study single item auctions among n bidders. Bidder i has a value v_i for a good, and the valuation profile for all the n players together is denoted by $\mathbf{v} = (v_1, v_2, \ldots, v_n)$. In a sealed bid auction each player submits a bid, and the bid profile is denoted by $\mathbf{b} = (b_1, b_2, \ldots, b_n)$. An auction is a pair of functions $(\mathbf{x}, \mathbf{p})$, where x maps a bid vector to outcomes $\{0, 1\}^n$, and p maps a bid vector to $\mathbf{R}_+^n$, i.e., a non-negative payment for each player. The players have quasi-linear utility functions, i.e., their utilities have a separable and linear dependence on money, given by $u_i(v_i, \mathbf{v}_{-i}) = v_i x_i(\mathbf{v}) - p_i(\mathbf{v})$. An auction is said to be dominant strategy truthful if submitting a bid equal to your value yields no smaller utility than any other bid in every situation, i.e., for all $\mathbf{v}_{-i}$, $v_i x_i(\mathbf{v}) - p_i(\mathbf{v}) \geq v_i x_i(b_i, \mathbf{v}_{-i}) - p_i(b_i, \mathbf{v}_{-i})$. Since we focus on truthful auctions in this paper $\mathbf{b} = \mathbf{v}$ from now on.

Distributions. We study auctions in a Bayesian setting, i.e., the valuations of bidders are drawn from a distribution. In particular, we assume that valuation of bidder i is drawn from distribution F_i, which is independent from but not necessarily identical to F_j for $j \neq i$. For ease of presentation, we assume that these distributions are continuous, i.e., they have density function f_i. We assume that the support of these distributions are intervals on the non-negative real line, with non-zero density everywhere in the interval.

Regularity and irregularity. The hazard rate function of a distribution is defined as $h(x) = \frac{f(x)}{1 - F(x)}$. A distribution is said to have a Monotone Hazard Rate(MHR) if $h(x)$ is monotonically non-decreasing. A weaker requirement on distributions is called regularity: the function $\phi(x) = x - \frac{1}{h(x)}$ is monotonically non-decreasing. We do not assume either of these technical conditions for our distributions. Instead we assume that the market of bidders consists of k groups and each group has a regular distribution G_i over valuations. Each bidder is drawn according to some (potentially different) convex combination of these k regular distributions, i.e., $F_i(x) = \sum_{t=1}^{k} p_{i,t} G_t(x)$. Such a distribution $F_i(\cdot)$ in most cases significantly violates the regularity condition.

In fact, mathematically, any irregular distribution can be approximated by a convex combination of sufficiently many regular distributions and as we take the number of regular distributions to infinity then it can be described exactly. Thus the number of regular distributions needed to describe an irregular distribution is a valid measure of irregularity that is well-defined for any distribution.

Revenue Objective. The objective in this paper to design auctions to maximize expected revenue, i.e., the expectation of the sum of the payments of all agents. Formally, the objective is to maximize $\mathbb{E}_\mathbf{v}[\sum_i p_i(\mathbf{v})]$. Myerson [10] characterized the expected revenue from any auction as its expected virtual surplus, i.e. the expected sum of virtual values of the agents who receive the item, where the virtual value of an agent is $\phi(v) = v - \frac{1}{h(v)}$. Formally, for all bidders i, $\mathbb{E}_\mathbf{v}[p_i(\mathbf{v})] = \mathbb{E}_\mathbf{v}[\phi_i(v_i) x_i(\mathbf{v})]$. The equality holds even if we condition on a fixed v_{-i}, i.e., $\mathbb{E}_{v_i}[p_i(v_i, \mathbf{v}_{-i})] = \mathbb{E}_{v_i}[\phi(v_i) x_i(v_i, \mathbf{v}_{-i})]$.

3   Targeted Advertising and the Non-i.i.d. Irregular Setting

In this section we give our version of Bulow and Klemperer's result [1] for non-i.i.d irregular distributions.

Theorem 1. *Consider an auction among n non-i.i.d irregular bidders where each bidder's distribution F_i is some mixture of k regular distributions $\{G_1, \ldots, G_k\}$ (the set of regular distributions is the same for all bidders but the mixture probabilities could be different). The revenue of the optimal auction in this setting is at most twice the revenue of a Vickrey auction with k extra bidders, where each bidder is drawn from a distinct distribution from $\{G_1, \ldots, G_k\}$.*

Proof. Bidder i's distribution $F_i(x) = \sum_{t=1}^{k} p_{i,t} G_t(x)$ can be thought of as being drawn based on the following process: first a biased k-valued coin is flipped that decides from which distribution G_t player i's value will come from (according to the probabilities $p_{i,t}$), and then a sample from G_t is drawn. Likewise, the entire valuation profile can be thought of as being drawn in a similar way: first n independent, and possibly non-identically biased, k-valued coin tosses, decide the regular distribution from each bidder's value is going to be drawn from. Subsequently a sample is drawn from each distribution.

Let the random variable q_i be the index of the regular distribution that bidder i's value is going to be drawn, i.e., q_i is the result of the coin toss for bidder i. Let q denote the index profile of all players. Let $p(q) = \prod_{i=1}^{n} p_{i,q_i}$ be the probability that the index profile q results after the n coin tosses. Let $G(q) = \times_i G_{q_i}$ be the joint product distribution of players' values conditioned on the profile being q.

Let M_q be the optimal mechanism when bidders' distribution profile is q. Let $\mathcal{R}_M^q$ be the expected revenue of mechanism M_q. Let $R_M^q(\mathbf{v})$ denote the revenue of the mechanism when bidders have value $\mathbf{v}$. The revenue of the optimal mechanism M which cannot learn and exploit the actual distribution profile q is upper bounded by the revenue of the optimal mechanism that can first learn q. Therefore we have,

$$\mathcal{R}_M \leq \sum_{q \in [1..k]^n} p(q) \mathbb{E}_{\mathbf{v} \sim G(q)}[R_M^q(\mathbf{v})] \tag{1}$$

Now, $\mathbb{E}_{\mathbf{v} \sim G(q)}[R_M^q(\mathbf{v})]$ corresponds to the optimal expected revenue when bidder i's distribution is the regular distribution G_{q_i}. Let $k(q)$ denote the number of distinct regular distributions contained in the profile q. Note that $k(q) \leq k$ for all q. Thus the above expectation corresponds to the revenue of a single-item auction where players can be categorized in $k(q)$ groups and bidders within each group t are distributed i.i.d. according to a regular distribution G_t. Theorem 6.3 of [14] applies to such a setting and shows that the optimal revenue for each of these non-i.i.d regular settings will be at most twice the revenue of Vickrey auction with one extra bidder for each distinct distribution in the profile q. Hence,

$$\mathcal{R}_M \leq \sum_{q \in [1..k]^n} p(q) \mathbb{E}_{\mathbf{v} \sim G(q)}[R_M^q(\mathbf{v})] \leq 2 \sum_{q \in [1..k]^n} p(q) \mathbb{E}_{\mathbf{v} \sim G(q)}[R_{SP_{n+k(q)}}(\mathbf{v})] \tag{2}$$

$$\leq 2 \sum_{q \in [1..k]^n} p(q) \mathbb{E}_{\mathbf{v} \sim G(q)}[R_{SP_{n+k}}(\mathbf{v})] \tag{3}$$

Since, the Vickrey auction with k extra bidders doesn't depend on the index profile q the RHS of (3) corresponds to the expected revenue of SP_{n+k} when bidders come from the initial i.i.d irregular distributions. $\square$

The above proof actually proves an even stronger claim: the revenue from running the Vickrey auction with k extra bidders is at least half approximate even if the auctioneer could distinguish bidders by learning the bidder distribution profile q and run the corresponding optimal auction R_M^q.

Lower bound. A corner case of our theorem is when each bidder comes from a different regular distribution. From Hartline and Roughgarden [7] we know that a lower bound of $4/3$ exists for such a case. In other words there exists two regular distributions such that if the initial bidders came each from a different distribution among these, then adding two extra bidders from those distributions will not give the optimal revenue but rather a $4/3$ approximation to it. The same lower bound proves that if bidders came from the same mixture of these two regular distributions (i.e. are i.i.d), then the expected revenue of the auction that first distinguishes from which regular distribution each bidder comes from and then applies the optimal auction, yields higher revenue than adding two extra bidders from the two distributions and running a Vickrey auction.

4 Just one Extra Bidder for Hazard Rate Dominant Distributions

In this section we examine the setting where among the k underlying regular distributions there exists one distribution that stochastically dominates the rest in the sense of hazard rate dominance. Hazard rate dominance is a standard dominance concept used while establishing revenue guarantees for auctions (see for example [9]) and states the following: A distribution F hazard rate dominates a distribution G iff for every x in the intersection of the support of the two distributions: $h_F(x) \le h_G(x)$. We denote such a domination by $F \succeq_{hr} G$.

In such a setting it is natural to ask whether adding just a single player from the dominant distribution is enough to produce good revenue guarantees. We actually show that adding only one extra person coming from the dominant distribution achieves exactly the same worst-case guarantee as adding k extra bidders one from each underlying distribution.

Theorem 2. *Consider an auction among n non-i.i.d irregular bidders where each bidder's distribution F_i is some mixture of k regular distributions $\{G_1, \ldots, G_k\}$ such that $G_1 \succeq_{hr} G_t$ for all t. The revenue of the optimal auction in this setting is at most twice the revenue of a Vickrey auction with one extra bidder drawn from G_1.*

The proof is based on a new lemma for the regular distribution setting: bidders are drawn from a family of k regular distributions such that one of them hazard-rate dominates the rest. This lemma can be extended to prove Theorem 2 in a manner identical to how Theorem 6.3 of Roughgarden et al. [14] was extended to prove Theorem 1 in our paper. We don't repeat that extension here, and instead just prove the lemma. The lemma uses the notion of commensurate auctions defined by Hartline and Roughgarden [7].

Lemma 1. *Consider a non-i.i.d. regular setting where each player's value comes from some set of distributions* $\{F_1, \ldots, F_k\}$ *such that* $F_1 \succeq_{hr} F_t$ *for all t. The optimal revenue of this setting is at most twice the revenue of Vickrey auction with one extra bidder drawn from* F_1.

Proof. Let **v** denote the valuation profile of the initial n bidders and let v^* the valuation of the extra bidder from the dominant distribution. Let $R(\mathbf{v}, v^*)$ and $S(\mathbf{v}, v^*)$ denote the winners of the optimal auction (M) and of the second price auction with the extra bidder (SP_{n+1}) respectively. We will show that the two auctions are commensurate (see [7]) which is sufficient for proving the lemma. Establishing commensurateness boils down to showing that:

$$\mathbb{E}_{\mathbf{v}, v^*}[\phi_{S(\mathbf{v}, v^*)}(v_{S(\mathbf{v}, v^*)})|S(\mathbf{v}, v^*) \neq R(\mathbf{v}, v^*)] \geq 0 \tag{4}$$

$$\mathbb{E}_{\mathbf{v}, v^*}[\phi_{R(\mathbf{v}, v^*)}(v_{R(\mathbf{v}, v^*)})|S(\mathbf{v}, v^*) \neq R(\mathbf{v}, v^*)] \leq \mathbb{E}_{\mathbf{v}, v^*}[p_{S(\mathbf{v}, v^*)}|S(\mathbf{v}, v^*) \neq R(\mathbf{v}, v^*)] \tag{5}$$

where p_S is the price paid by the winner of the second price auction. The proof of equation (5) is easy and very closely follows the proof in [7] above.

We now prove equation (4). Since $F_1 \succeq_{hr} F_t$ we have that for all x in the intersection of the support of F_1 and F_t: $h_1(x) \leq h_t(x)$, which in turn implies that $\phi_1(x) \leq \phi_t(x)$, since $\phi_t(x) = x - \frac{1}{h_t(x)}$. By the definition of winner in Vickrey auction we have $\forall i : v_{S(\mathbf{v}, v^*)} \geq v_i$. In particular, $v_{S(\mathbf{v}, v^*)} \geq v^*$. If v^* is in the support of $F_{S(\mathbf{v}, v^*)}$, then the latter, by regularity of distributions, implies that $\phi_{S(\mathbf{v}, v^*)}(v_{S(\mathbf{v}, v^*)}) \geq \phi_{S(\mathbf{v}, v^*)}(v^*)$. Now $F_1 \succeq_{hr} F_t$ implies that $\phi_{S(\mathbf{v}, v^*)}(v^*) \geq \phi_1(v^*)$ (since by definition v^* must also be in the support of F_1). If v^* is not in the support of $F_{S(\mathbf{v}, v^*)}$, then since $v^* < v_{S(\mathbf{v}, v^*)}$ and all the supports are intervals, it must be that v^* is below the lower bound L of the support of $F_{S(\mathbf{v}, v^*)}$. Wlog we can assume that the support of F_1 intersects the support of every other distribution. Hence, since v^* is below L and the support of F_1 is an interval, L will also be in the support of F_1. Thus L is in the intersection of the two supports. By regularity of $F_{S(\mathbf{v}, v^*)}$, F_1 and by the hazard rate dominance assumption, we have $\phi_{S(\mathbf{v}, v^*)}(v_{S(\mathbf{v}, v^*)}) \geq \phi_{S(\mathbf{v}, v^*)}(L) \geq \phi_1(L) \geq \phi_1(v^*)$. Thus in any case $\phi_{S(\mathbf{v}, v^*)}(v_{S(\mathbf{v}, v^*)}) \geq \phi_1(v^*)$. Hence, we immediately get that:

$$\mathbb{E}_{\mathbf{v}, v^*}[\phi_{S(\mathbf{v}, v^*)}(v_{S(\mathbf{v}, v^*)})|S(\mathbf{v}, v^*) \neq R(\mathbf{v}, v^*)] \geq \mathbb{E}_{\mathbf{v}, v^*}[\phi_1(v^*)|S(\mathbf{v}, v^*) \neq R(\mathbf{v}, v^*)]$$

Conditioned on **v** the latter expectation becomes:

$$\mathbb{E}_{v^*}[\phi_1(v^*)|S(\mathbf{v}, v^*) \neq R(\mathbf{v}, v^*), \mathbf{v}]$$

But conditioned on **v**, $R(\mathbf{v}, v^*)$ is some fixed bidder i. Hence, the latter expectation is equivalent to: $\mathbb{E}_{v^*}[\phi_1(v^*)|S(\mathbf{v}, v^*) \neq i]$ for some i. We claim that for all i the latter expectation must be positive. Conditioned on **v**, the event $S(\mathbf{v}, v^*) \neq i$ happens only if v^* is sufficiently high, i.e., there is a threshold $\theta(\mathbf{v})$ such that $S(\mathbf{v}, v^*) \neq i$ happens only if $v^* \geq \theta(\mathbf{v})$ (if i was the maximum valued bidder in the profile **v** then $\theta(\mathbf{v}) = v_i$, else $\theta(\mathbf{v}) = 0$.) By regularity of distributions, $v^* \geq \theta(\mathbf{v})$ translates to $\phi_1(v^*) \geq \phi_1(\theta)$. So we now have to show that: $\mathbb{E}_{v^*}[\phi_1(v^*)|\phi_1(v^*) \geq \phi_1(\theta)] \geq 0$. Since the unconditional expectation of virtual value is already non-negative, the expectation conditioned on a lower bound on virtual values is clearly non-negative. $\qquad\square$

Examples and Applications. There are many situations where a hazard-rate dominant distribution aactually exists in the market. We provide some examples below.

Uniform, Exponential, Power-law distributions. Suppose the k underlying distributions were all uniform distributions of the form $U[a_i, b_i]$. The hazard rate $h_i(x) = \frac{1}{b_i - x}$. Clearly, the distribution with a larger b_i hazard-rate dominates the distribution with a smaller b_i. If the k underlying distributions were all exponential distributions, i.e., $G_i(x) = 1 - e^{-\lambda_i x}$, then the hazard rate $h_i(x) = \lambda_i$. Thus the distribution with the smallest λ_i hazard rate dominates the rest. If the k underlying distributions were all power-law distributions, namely, $G_i(x) = 1 - \frac{1}{x^{\alpha_i}}$, then the hazard rate $h_i(x) = \frac{\alpha_i}{x}$. Thus the distribution with the smallest α_i hazard-rate dominates the rest.

A general condition. If all the k underlying regular distributions were such that for any pair i, j they satisfy $1 - G_i(x) = (1 - G_j(x))^{\theta_{ij}}$, then it is easy to verify that there always exists one distribution that hazard-rate dominates the rest of the distributions. For instance, the family of exponential distributions, and the family of power-law distributions are special cases of this general condition.

5   Non-targeted Advertising and the i.i.d. Irregular Setting

In this section we consider the setting where all the bidders are drawn from the same distribution F. We assume that F can be written as a convex combination of k regular distributions $F_1, \dots, F_k$, i.e. $F = \sum_{t=1}^{k} p_t F_t$ and such that the mixture probability p_t for every distribution is at least some constant δ: $\forall t \in [1, \dots, k] : p_t \geq \delta$. A natural question to ask in an i.i.d. setting is how many extra bidders should be recruited from the original distribution to achieve a constant fraction of the optimal revenue (i.e., by running a non-targeted advertising campaign)?

In this section answer the above question as a function of the number of underlying distributions k and the minimum mixture probability δ. We remark that our results in this section don't require the decomposition of F into the F_t's resemble the distribution of the underlying population groups. Even if the number of underlying population groups is very large, as long as there is some mathematical way of decomposing F into k regular distributions with a minimum mixture probability of δ, our results go through. Hence, one can optimize our result for each F by finding the decomposition that minimizes our approximation ratio.

Theorem 3. *Consider an auction among n i.i.d. irregular bidders where the bidders' distribution F can be decomposed into a mixture of k regular distributions $\{G_1, \dots, G_k\}$ with minimum mixture probability δ. The revenue of the optimal auction in this setting is at most $2\frac{k+1}{k}$ the revenue of a Vickrey auction with $\Theta\left(\frac{\log(k)}{\delta}\right)$ extra bidders drawn from distribution F.*

Proof. Suppose that we bring n^* extra bidders in the auction. Even if the decomposition of the distribution F doesn't correspond to an actual market decomposition, we can always think of the value of each of the bidders drawn as follows: first we draw a number t from 1 to k according to the mixture probabilities p_t and then we draw a value from distribution G_t.

Let $\mathcal{E}$ be the event that all numbers 1 to k are represented by the n^* random numbers drawn to produce the value of the n^* extra bidders. The problem is a generalization of the coupon collector problem where there are k coupons and each coupon arrives with probability $p_t \geq \delta$. The relevant question is, what is the probability that all the coupons are collected after n^* coupon draws? The probability that a coupon t is not collected after n^* draws is: $(1 - p_t)^{n^*} \leq (1 - \delta)^{n^*} \leq e^{-n^*\delta}$. Hence, by the union bound, the probability that some coupon is not collected after n^* draws is at most $ke^{-n^*\delta}$. Thus the probability of event $\mathcal{E}$ is at least $1 - ke^{-n^*\delta}$. Thus if $n^* = \frac{\log(k) + \log(k+1)}{\delta}$ then the probability of $\mathcal{E}$ is at least $1 - \frac{1}{k+1}$.

Conditional on event $\mathcal{E}$ happening we know that the revenue of the auction is the revenue of the initial auction with at least one player extra drawn from each of the underlying k regular distributions. Thus we can apply our main theorem 1 to get that the expected revenue conditional on $\mathcal{E}$ is at least $\frac{1}{2}$ of the optimal revenue with only the initial n bidders. Thus:

$$\mathcal{R}_{SP_{n+n^*}} \geq \left(1 - \frac{1}{k+1}\right) \mathbb{E}_{v,\tilde{v} \sim F^{n+n^*}}[\mathcal{R}_{SP_{n+n^*}}(v,\tilde{v}) | \mathcal{E}] \geq \left(1 - \frac{1}{k+1}\right) \frac{1}{2} \mathcal{R}_M$$

$\square$

Theorem 4. *Consider an auction among n i.i.d. irregular bidders where the bidders' distribution F can be decomposed into a mixture of k regular distributions $\{G_1, \ldots, G_k\}$ such that G_1 hazard rate dominates G_t for all $t > 1$. The revenue of the optimal auction in this setting is at most $2\frac{e}{e-1}$ the revenue of a Vickrey auction with $\frac{1}{p_1}$ extra bidders drawn from distribution F.*

Proof. Similar to theorem 3 conditional on the even that an extra player is drawn from the hazard rate distribution, we can apply Lemma 1 to get that this conditional expected revenue is at least half the optimal revenue with the initial set of players. If we bring n^* extra players then the probability of the above event happening is $1 - (1 - p_1)^{n^*} \geq 1 - e^{-n^*p_1}$. Setting $n^* = \frac{1}{p_1}$ we get the theorem. $\square$

Prior-Independent Mechanisms. The two theorems above imply prior-independent mechanisms for the i.i.d. irregular setting based on a reasoning similar to the one used by [5] in converting Bulow-Klemperer results to prior-independent mechanims in the i.i.d. regular setting. Specifically, instead of bringing k extra i.i.d. bidders we could use the maximum value of a random subset of k existing bidders as a reserve on the remaining $n - k$ bidders. The theorems above then imply that this prior-independent mechanism yields a constant approximation with respect to the optimal mechanism among the $n - k$ bidders. Further, since the bidders are all i.i.d., and the k bidders were chosen before their valuation are drawn, the expected optimal revenue among the $n - k$ bidders is at least $1 - \frac{k}{n}$ of the optimal revenue among the n bidders. Thus as long as the number of bidders k required by Theorems 3 and 4 is smaller than n, this approach yields a prior-independent mechanism with a meaningful revenue approximation guarantee. Hence, Theorem 3 implies that if $n\delta \geq c\log(k)$ (i.e. the expected number of players from each population is at least $c\log(k)$) the random sampling mechanism described above is $2\frac{k+1}{k}\frac{c}{c-1}$ approximate. Similarly, Theorem 4 implies that it is $2\frac{e}{e-1}\frac{c}{c-1}$-approximate, if $n \cdot p_1 \geq c$, i.e. the expected number of players from the hazard-rate dominant distribution at least c.

6   Vickrey with Single Reserve for Irregular Settings

In this section we prove revenue guarantees for Vickrey auction with a single reserve in the general irregular setting.

Theorem 5. *Consider an auction among n non-i.i.d irregular bidders where each bidder's distribution F_i is some mixture of k regular distribution $\{G_1, \ldots, G_k\}$ (the set of regular distributions is the same for all bidders but the mixture probabilities could be different). The revenue of the optimal auction in the above setting is at most $4k$ times the revenue of a second price auction with a single reserve price which corresponds to the monopoly reserve price of one of the k distributions G_i.*

Proof. We use the same notation as in Section 3. In particular, we let q denote the index profile of distributions for all players and $p(q) = \prod_{i=1}^{n} p_{i,q_i}$ be the probability that an index profile arises. Let $G(q) = \times_i G_{q_i}$ be the product distribution that corresponds to how players values are distributed conditional on the coin tosses having value q.

Let M_q be the optimal mechanism when bidders' distribution profile is q. Let $\mathcal{R}_M^q$ be the expected revenue of mechanism M_q. By equation (2) in Section 3 we have,

$$\mathcal{R}_M \leq \sum_{q \in [1..k]^n} p(q) \mathbb{E}_{\mathbf{v} \sim G(q)}[R_M^q(\mathbf{v})] \leq 2 \sum_{q \in [1..k]^n} p(q) \mathbb{E}_{\mathbf{v} \sim G(q)}[R_{SP_{n+k(q)}}(\mathbf{v})] \quad (6)$$

Consider the auction $SP_{n+k(q)}$. If instead of adding the $k(q)$ extra bidders, we place a random reserve drawn from the distribution of the maximum value among the $k(q)$ extra bidders, and ran the Vickrey auction. Call the later $SP_n(R(q))$. If the winner of the auction $SP_{n+k(q)}$ is one among the original n bidders, clearly $SP_{n+k(q)}$ and $SP_n(R(q))$ will have the same expected revenue. Further, the expected revenue of $SP_{n+k(q)}$ conditioned on the winner being one among the original n bidders is no smaller than the expected revenue of $SP_{n+k(q)}$ conditioned on the winner being one among the newly added $k(q)$ bidders. Also, the probability that the winner comes from the newly added $k(q)$ bidders is at most $1/2$. Thus $SP_n(R(q)) \geq \frac{1}{2} SP_{n+k(q)}$. Combining this with Equation (6), we have

$$\mathcal{R}_M \leq 2 \sum_{q \in [1..k]^n} p(q) \mathbb{E}_{\mathbf{v} \sim G(q)}[R_{SP_{n+k(q)}}(\mathbf{v})] \leq 4 \sum_{q \in [1..k]^n} p(q) \mathbb{E}_{\mathbf{v} \sim G(q)}[R_{SP_n(R(q))}(\mathbf{v})]$$

$$= \sum_{t=1}^{k} 4 \sum_{q \in [1..k]^n} p(q) \mathbb{E}_{\mathbf{v} \sim G(q)}[R_{SP_n(R(q),t)}(\mathbf{v})] \quad (7)$$

$$\leq 4k \sum_{q \in [1..k]^n} p(q) \mathbb{E}_{\mathbf{v} \sim G(q)}[R_{SP_n(R(q),t^*)}(\mathbf{v})] \quad (8)$$

In equation (7), the revenue $R_{SP_n(R(q))}(\mathbf{v})$ is written as $\sum_{t=1}^{k} R_{SP_n(R(q),t)}(\mathbf{v})$, i.e., as the sum of contributions from each population group. Given this split, there exists a polulation group t^* that gets at least $\frac{1}{k}$ fraction of all groups together, and thus at least $\frac{1}{4k}$ fraction of the optimal mechanism, which is what is expressed through inequality (8).

Now the auction $SP_n(R(q))$ from the perspective of the group t^* is just the Vickrey auction run for group t^* alone with a single random reserve of $\max\{R(q),$ Maximum value from groups other than $t^*\}$. However within the group t^* since we are in a i.i.d regular setting it is optimal to run Vickrey auction for the group t^* alone with the monopoly reserve price of that group. That is if we replace the single reserve of $\max\{R(q),$ Maximum value from groups other than $t^*\}$ with the optimal (monopoly) reserve price for t^*, Vickrey auction for group t^* with such a reserve gives no lesser revenue, and this holds for every q! Finally, when we add in the agents from other groups, single-item Vickrey auction's revenue for the entire population with monopoly reserve price of group t^* is no smaller than the revenue of single-item Vickrey auction for group t^* alone with the monopoly reserve price of group t^*. Chaining the last two statements proves the theorem. $\square$

References

[1] Bulow, J., Klemperer, P.: Auctions vs negotiations. American Economic Review 86(1), 180–194 (1996)

[2] Chawla, S., Hartline, J., Kleinberg, R.D.: Algorithmic pricing via virtual valuations. In: ACM Conference on Electronic Commerce, pp. 243–251 (2007)

[3] Chawla, S., Hartline, J., Malec, D.L., Sivan, B.: Multi-parameter mechanism design and sequential posted pricing. In: STOC, pp. 311–320 (2010)

[4] Devanur, N., Hartline, J., Karlin, A., Nguyen, T.: Prior-independent multi-parameter mechanism design. In: Chen, N., Elkind, E., Koutsoupias, E. (eds.) WINE 2011. LNCS, vol. 7090, pp. 122–133. Springer, Heidelberg (2011)

[5] Dhangwatnotai, P., Roughgarden, T., Yan, Q.: Revenue maximization with a single sample. In: Proceedings of the 11th ACM Conference on Electronic Commerce, EC 2010, pp. 129–138 (2010)

[6] Guimaraes, B., Sheedy, K.D.: Sales and monetary policy. American Economic Review 101(2), 844–876 (2011)

[7] Hartline, J., Roughgarden, T.: Simple versus optimal mechanisms. In: ACM Conference on Electronic Commerce, pp. 225–234 (2009)

[8] Johnson, J.P., Myatt, D.P.: Multiproduct quality competition: Fighting brands and product line pruning. The American Economic Review 93(3), 748–774 (2003)

[9] Kirkegaard, R.: Ranking asymmetric auctions using the dispersive order (2011) (manuscript)

[10] Myerson, R.: Optimal auction design. Mathematics of Operations Research 6, 58–73 (1981)

[11] Neeman, Z.: The effectiveness of english auctions. Games and Economic Behavior 43(2), 214–238 (2003)

[12] Robinson, J.: The Economics of Imperfect Competition. McMillan and Co. (1933)

[13] Ronen, A.: On approximating optimal auctions. In: ACM Conference on Electronic Commerce, pp. 11–17 (2001)

[14] Roughgarden, T., Talgam-Cohen, I., Yan, Q.: Supply-limiting mechanisms. In: ACM Conference on Electronic Commerce, pp. 844–861 (2012)

Strategy-Proof and Efficient
Offline Interval Scheduling and Cake Cutting

Yuan Tian*

Department of Economics
University of Chicago

Abstract. I construct a class of strategy-proof and Pareto efficient mechanisms of cake cutting in the context of offline interval scheduling. Motivating applications include the circulation of the single copy of a book owned by a public library among all self-interested and rational potential players with piecewise uniform value densities over continuous time intervals. This class of mechanisms accommodate both anonymous and non-anonymous configurations and can serve as a flexible platform to implement allocations of heterogeneous goods according to distributional objectives such as arbitrary guaranteed shares of reported demand.

The cake-cutting literature has recently made significant progress toward a better understanding of strategy-proofness. Two-person incentive compatible cake-cutting problems were studied by Maya and Nisan in [1], while Mossel and Tamuz in [2] and Chen et al. in [3] proposed solutions to such problems with any arbitrary number of players, with the former focusing on stochastic mechanisms. A more recent paper [4] by Aziz and Ye discussed various cake-cutting algorithms for piecewise constant and piecewise uniform value densities and examined the compatibility of Pareto efficiency, strategy-proofness, and fairness. Procaccia [5] contains a great introduction to the cake-cutting literature including the issue of strategy-proofness. A wonderful review of classical cake-cutting algorithms focusing on fairness issues can be found in Procaccia [6].

This paper proposes a class of mechanisms generalizing the strategy-proof, Pareto efficient, and envy-free mechanism developed in [3] by temporarily compromising on fairness. Representing the mechanism design problem as a constrained optimization problem, I manage to characterize all feasible cuts of the cake with a collection of linear constraints regarding the maximum total payoffs of any subset of players by slightly generalizing the network-flow proof in [3]. More importantly, I demonstrate that strategy-proofness can be obtained from the additivity and concavity of the social planner's (in this case, the library's) utility in terms of the payoffs of the players (the readers). The essential result in constructing strategy-proofness is a non-inferiority condition that roughly states the following: when a player increases her demand to a superset of her original demand, a specific group—depending on the identity of

* I thank Balázs Szentes, Alex Frankel, and Varun Gupta for continuous encouragement and advice. I also thank Eric Budish, Ian Fillmore, Sam Hwang, and Hugo Sonnenschein for helpful discussions. tiany@uchicago.edu

Y. Chen and N. Immorlica (Eds.): WINE 2013, LNCS 8289, pp. 436–437, 2013.

the player changing her report—of players will all be allocated (weakly) larger pieces[1].

Partially relaxing the fairness requirement suggests some interesting and potentially positive extensions of the current literature. In the context of scheduling, maintaining piecewise uniform valuations, the allocation of a certain resource (the book, in this case) at an early instant should not depend on the report of players who arrive at a later time, suggesting a type of dynamic consistency as a necessity in the design of online cake-cutting mechanisms[2]. Such dynamic consistencies may be incompatible with fairness[3]. In addition, under piecewise constant valuations, the incompatibility of strategy-proofness and even weaker notions of fairness than envy-freeness, such as equal-treatment-of-equals, has been widely documented in both computer science (see [4]) and economics (see Zhou [9]) literature. Earlier works including [10] by Bogomolnaia and Moulin focus on maintaining fairness by mildly yielding on strategy-proofness. The mechanisms developed in this paper provide an adaptable potential platform upon which strategy-proof and Pareto efficient mechanisms, while unable to achieve envy-freeness, can be developed to perform better on fairness than simple mechanisms such as random serial dictatorship.

References

1. Maya, A., Nisan, N.: Incentive compatible two player cake cutting. In: Goldberg, P.W. (ed.) WINE 2012. LNCS, vol. 7695, pp. 170–183. Springer, Heidelberg (2012)
2. Mossel, E., Tamuz, O.: Truthful fair division. In: Kontogiannis, S., Koutsoupias, E., Spirakis, P.G. (eds.) SAGT 2010. LNCS, vol. 6386, pp. 288–299. Springer, Heidelberg (2010)
3. Chen, Y., Lai, J.K., Parkes, D.C., Procaccia, A.D.: Truth, justice, and cake cutting. Games and Economic Behavior 77(1), 284–297 (2013)
4. Aziz, H., Ye, C.: Cake cutting algorithms for piecewise constant and piecewise uniform valuations. ArXiv E-Prints (September 2013)
5. Procaccia, A.D.: Cake cutting: Not just child's play. Communications of the ACM 56(7), 78–87 (2013)
6. Procaccia, A.D.: 13. Cake Cutting Algorithms. In: Handbook of Computational Social Choice. Cambridge University Press (2014)
7. Quah, J.K.H.: The comparative statics of constrained optimization problems. Econometrica 75(2), 401–431 (2007)
8. Walsh, T.: Online cake cutting. In: Algorithmic Decision Theory, pp. 292–305. Springer (2011)
9. Zhou, L.: On a conjecture by gale about one-sided matching problems. Journal of Economic Theory 52(1), 123–135 (1990)
10. Bogomolnaia, A., Moulin, H.: A new solution to the random assignment problem. Journal of Economic Theory 100(2), 295–328 (2001)

---

[1] The establishment of the non-inferiority condition draws inspirations from the monotone comparative statics literature such as [7].

[2] See [8] for an early exploration in this direction.

[3] Counterexamples can be constructed showing that Mechanism 1 in [3] cannot accommodate such dynamic consistencies. However, such counterexamples do not conclusively show the incompatibility between dynamic consistency and fairness.

Author Index